Key Concepts in Geography

Key Concepts in Geography

Second Edition

Edited by
**Nicholas J. Clifford, Sarah L. Holloway,
Stephen P. Rice and Gill Valentine**

Los Angeles | London | New Delhi
Singapore | Washington DC

Editorial arrangement © Nicholas J. Clifford, Sarah L. Holloway, Stephen P. Rice, Gill Valentine, 2009

Chapter 1 © Mike Heffernan 2009
Chapter 2 © Keith Richards 2009
Chapter 3© Ron Johnston 2009
Chapter 4 © Alison Blunt 2009
Chapter 5 © Nigel Thrift 2009
Chapter 6 © Martin Kent 2009
Chapter 7 © John B. Thornes 2009
Chapter 8 © Peter J. Taylor 2009
Chapter 9 © Noel Castree 2009
Chapter 10 © Ken Gregory 2009
Chapter 11 © Tim Burt 2009
Chapter 12 © Andrew Herod 2009
Chapter 13 © Cindi Katz 2009
Chapter 14 © Stephan Harrison 2009
Chapter 15 © Murray Gray 2009

Chapter 16 © Karen M. Morin 2009
Chapter 17 © Franklin Ginn and David Demeritt 2009
Chapter 18 © Roy Haines-Young 2009
Chapter 19 © James R. Faulconbridge & Jonathan V. Beaverstock 2009
Chapter 20 © Nicholas J. Clifford 2009
Chapter 21 © Katie D. Willis 2009
Chapter 22 © Robert Inkpen 2009
Chapter 23 © Shaun French 2009
Chapter 24 © Graham A. Tobin and Burrell E. Montz 2009
Chapter 25 © David Bell 2009
Chapter 26 © Michael Church 2009

First published 2003
This second edition first published 2009
Reprinted 2009, 2011, 2012

SAGE Publications Ltd
1 Oliver's Yard
55 City Road
London EC1Y 1SP

SAGE Publications Inc.
2455 Teller Road
Thousand Oaks, California 91320

SAGE Publications India Pvt Ltd
B 1/I 1 Mohan Cooperative Industrial Area
Mathura Road, Post Bag 7
New Delhi 110 044

SAGE Publications Asia-Pacific Pte Ltd
3 Church Street
#10-04 Samsung Hub
Singapore 049483

Library of Congress Control Number

British Library Cataloguing in Publication data

A catalogue record for this book is available from the British Library

ISBN 978-1-4129-3021-5
ISBN 978-1-4129-3022-2 (pbk)

Typeset by C&M Digitals (P) Ltd, Chennai, India
Printed in Great Britain by the MPG Books Group

Contents

Notes on Contributors

Jonathan V. Beaverstock is Professor of Economic Geography at Nottingham University with interests that examine globalization and world cities, international financial centres and the organizational strategies of transnational advanced producer service firms in the world economy.

David Bell is Senior Lecturer in Critical Human Geography and leader of the Urban Cultures and Consumption research cluster in the School of Geography at the University of Leeds, UK. His recent books include *The Cybercultures Reader* (edited with Barbara M. Kennedy, second edition, Routledge, 2008) and *The Cultures of Space Travel* (edited with Martin Parker, Blackwell, 2008).

Alison Blunt is Professor of Geography at Queen Mary College, University of London, and Adjunct Professor in the Department of Geography, Norwegian University of Science and Technology, Trondheim. Her research interests include feminist and postcolonial geographies; geographies of home, identity, migration and diaspora; and imperial travel and domesticity. Her current research includes projects on 'Diaspora cities: imagining Calcutta in London, Toronto and Jerusalem' (funded by The Leverhulme Trust) and 'Gender and the built environment' (funded by Urban Buzz). Recent books include *Domicile and Diaspora: Anglo-Indian Women and the Spatial Politics of Home* (Blackwell, 2005) and, co-authored with Robyn Dowling, *Home* (Routledge, 2006). Alison is incoming editor of *Transactions* (from January 2008).

Tim Burt is Professor of Geography and Master of Hatfield College at the University of Durham. He has degrees from Cambridge, Carleton (Ottawa) and Bristol, and previously worked at Huddersfield Polytechnic and Oxford University before moving to Durham in 1996. His main research interests are in hydrology, geomorphology and climatic records.

Noel Castree is Professor of Geography at the University of Manchester. His interests are in the political economy of environmental change, wage workers and the geography of class struggle, and the dynamics of the academic labour process. He is author of *Nature* (Routledge, 2005), co-author (with N. Coe, K. Ward and M. Samers) of *Spaces of Work: Global Capitalism and the Geographies of Labour* (Sage, 2003), and co-editor (with Bruce Braun) of *Remaking Reality: Nature at the Millennium* (Routledge, 1998) and *Social Nature* (Blackwell, 2001).

Michael Church is Professor Emeritus at the University of British Columbia. He is a Fellow of the Royal Society of Canada and has received many other honours, including the G.K. Warren Prize from the US National Academy of Sciences and the David Linton Award from the British Society for Geomorphology. His research interests focus on fluvial sediment transport and the interpretation of river channel changes, and on theoretical geomorphology. He has, throughout his career, also participated in practical work on water management in forestry, mining and fisheries, and has worked extensively on management issues on large rivers. He is a registered Professional Geoscientist in British Columbia, Canada.

Nicholas J. Clifford is Professor of River Science at the University of Nottingham. He received his undergraduate and postgraduate degrees from the University of Cambridge, specializing in physical geography. He is author of *Incredible Earth* (Dorling Kindersley, 1997); co-editor (with Gill Valentine) of *Key Methods in Geography* (Sage, 2003) and co-editor (with Jon French and Jack Hardisty) of *Turbulence: Perspectives on Flow and Sediment Transport* (John Wiley, 1993). His principal interests are river geomorphology and river restoration, environmental dynamics and monitoring, and the history and philosophy of Geography

David Demeritt is Professor of Geography at King's College London specializing in social theory and the environment. His research focuses, in particular, on the articulation of environmental knowledges, especially scientific and technical ones, with power and the policy process.

James R. Faulconbridge is a Lecturer in Economic Geography at Lancaster University whose research focuses on the new economic geographies of professional services in the world economy, specifically investigating the relational geographies of knowledge flows in the legal and advertising industries.

Shaun French is Lecturer in Geography at the University of Nottingham. His research interests are in the economic and social geographies of financial services and of e-commerce. His current research explores the relationship between processes of financialization, the politics of financial exclusion and life assurance.

Franklin Ginn is a PhD student at King's College London. His research builds on recent efforts to rethink nature within geography by investigating the human/non-human orders that are produced and performed in suburban gardens in south London.

Murray Gray is Reader Emeritus in the Department of Geography, Queen Mary College, University of London. A glacial geomorphologist by training, his current work is focused on developing the paradigm of geodiversity. He is author of over 80 articles and books, including *Geodiversity: Valuing and Conserving Abiotic Nature* (John Wiley, 2004). He is Visiting Professor at the University of Minho, Portugal, lecturing on geodiversity and geoconservation, and has given keynote lectures on geodiversity in USA, Cananda and throughout Europe. He is a member of the Broads Authority and Chairman of its Planning Committee.

Ken Gregory was Warden of Goldsmiths College, University of London, from 1992 to 1998 and is now Emeritus Professor of the University of London and Visiting Professor of the University of Southampton. He was made CBE in 2007 for services to Geography and Higher Education, is a Fellow of University College London, and a former Secretary and Chairman of the British Geomorphological Research Group. His research interests are in river channel change and management, palaeohydrology, and the development of physical geography. Recent publications include *The Changing Nature of Physical Geography* (Edward Arnold, 2000), *River Channel Management* with Peter Downs (Edward Arnold, 2004), and he edited *Palaeohydrology: Understanding Global Change* with Gerardo Benito (John Wiley, 2003), *Physical Geography* (Sage, 2005) and was a lead editor for *Environmental Sciences: A Companion Primer* (Sage, 2008).

Roy Haines-Young is Professor of Environmental Management and Director of the Centre for Environmental Management at the University of Nottingham. The focus of his research is on sustainability at the landscape scale and, through the Centre for Environmental Management, he is currently involved in research projects with a number of public sector organizations (including DEFRA, Natural England and the European Environment Agency). Although trained as a natural scientist, Roy has sought to develop a stronger social content in his recent work. In part, this has arisen from his experiences of working extensively with policy-makers in central government and with agencies, and, from this, he has developed an awareness of the need to develop a science that is responsive to the wider public debates about environmental issues.

Stephan Harrison is Associate Professor of Quaternary Science at the University of Exeter. His main research interests lie in mountain geomorphology, particularly glacier fluctuation histories and the geomorphological impact of recent glacier retreat on valley-side slopes. His work in contemporary glacial environments allows him to test models of Quaternary landform evolution. He is involved in research on the upland geomorphology of the British Isles, southern Chile and Kazakhstan. Dr Harrison also has research interests in the philosophy of physical geography. He has written on the ontology of quantum theory as an argument against realist philosophy in geography, and argued for the identification of emergent properties in landscapes as an alternative to the reductionist model-building paradigm. He is also editor of *Patterned Ground*.

Mike Heffernan is Professor of Historical Geography at the University of Nottingham. He is interested in the historical, cultural and political geographies of Europe and North America from the eighteenth to the twentieth centuries and the history of geographical thought in the same period. His most recent publication is *The European Geographical Imagination* (Franz Steiner, 2007).

Andrew Herod is Professor of Geography and Adjunct Professor of International Affairs and of Anthropology at the University of Georgia, Athens, USA. He is author of *Labor Geographies: Workers and the Landscapes of Capitalism* (Guilford Press, 2001), editor of *Organizing the Landscape: Geographical Perspectives on Labor Unionism* (University of Minnesota Press, 1998), and co-editor of *The Dirty Work of Neoliberalism: Cleaners in the Global Economy* (Blackwell, 2006), *Geographies of Power: Placing Scale* (Blackwell, 2002), and *An Unruly World?: Globalization, Governance and Geography* (Routledge, 1998). He writes frequently on issues of labour and globalization. In his non-academic life, he is an elected member of the government of Athens-Clarke County, Georgia.

Robert Inkpen is a Principal Lecturer in Geography at the University of Portsmouth and has previously worked for the Building Research Establishment. His main research interests are the decay and conservation of heritage building materials, the role of networks in the development of geography and the history and philosophy of physical geography. His recent work explores the relationships between individuals and their context and how these complex relationships aid our understanding of both physical and human environments. He is the author of *Science, Philosophy and Physical Geography* (Routledge, 2005).

Ron Johnston is a Professor in the School of Geographical Sciences at the University of Bristol, having previously worked at Monash University and the Universities of Canterbury, Sheffield and Essex. His main research interests are in electoral and urban social geography, and in the history of geography. He is co-author of *Geography and Geographers: Anglo-American Human Geography since 1945* (Hodder, 2004) and co-editor of *The Dictionary of Human Geography* (4th edition, Wiley Blackwell, 2000). In 1999 he was elected a Fellow of the British Academy and awarded the Prix Vautrin Lud.

Cindi Katz teaches at the Graduate School of the City University of New York. Her book, *Growing Up Global: Economic Restructuring and Children's Everyday Lives* (University of Minnesota Press 2004), won the Meridian book award from the Association of American Geographers. She was editor (with Sallie Marston and Katharyne Mitchell) of *Life's Work: Geographies of Social Reproduction* (Wiley, 2004), and (with Janice Monk) of *Full Circles: Geographies of Women over the Lifecourse* (Routledge, 1993)

Martin Kent is Professor of Biogeography at the University of Plymouth, and has published widely in the fields of plant ecology and biogeography. His book *Vegetation Description and Analysis: A Practical Approach* (John Wiley, 1992), with Dr Paddy Coker, reached a wide audience and is being revised at the present time. Current research interests include spatial analysis of plant community boundaries, machair vegetation in the Outer Hebrides, vegetation burial in sand dunes and the ecology and conservation of evergreen broadleaved forest in Eastern China. Examples of recent publications may be found in *Journal of Vegetation Science, Forest Ecology and Management, Journal of Biogeography, Progress in Physical Geography* and *Annals of Botany*.

Burrell E. Montz is Professor of Geography and Environmental Studies and Associate Director of the Center for Integrated Watershed Studies at Binghamton University, New York State. With more than 25 years of experience with research in natural hazards, Dr Montz has published more than 50 articles, proceedings papers and book chapters. With Graham Tobin, she co-authored *Natural Hazards: Explanation and Integration* (Guilford Press, 1997). Her current research centres on various hazard topics, including the effectiveness of structural and non-structural mitigation measures, the flow and use of warning system information, and applications of GIS to understanding vulnerability to multiple hazards.

Karen M. Morin is Professor of Geography at Bucknell University in Pennsylvania. Much of her work focuses on postcolonialism, gender relations and North American historical geography. She is co-editor of *Women, Religion and Space: Global Perspectives on Gender and Faith* (Syracuse University Press, 2007), and author of *Frontiers of Femininity: A New Historical Geography of the 19th century American West* (Syracuse University Press, 2008).

Keith Richards is a graduate of the Department of Geography at the University of Cambridge, where he is now Professor of Geography, and has been Head of Department and Director of the Scott Polar Research Institute. He is a fluvial geomorphologist with interests in river channel forms and processes in a wide range of environments; river management, river and floodplain restoration, and inter-relationships of hydrological and ecological processes in floodplain environments; hydrological and sediment production and transfer processes in drainage basins; glacial hydrology; and the philosophy and methodology of geography and the environmental sciences. He is a former Secretary and Chairman of the British Geomorphological Research Group, and Vice-Chairman (Research) of the Royal Geographical Society–Institute of British Geographers.

Peter J. Taylor is Professor of Geography at Loughborough University. Among his recent books are: *World City Network: A Global Urban Analysis* (Routledge, 2004), *Political Geography: World-Economy, Nation-State, Locality* (Prentice Hall, 2006, fifth edition with Colin Flint), and *Cities in Globalization: Practices, Policies and Theories* (Routledge, 2006, edited with Ben Derudder, Pieter Saey and Frank Witlox). His current researches focus upon relations between world cities and he is founder of the Globalization and World Cities (GaWC) Study Group and Network which operates out of Loughborough University and Virginia Tech. He was elected a Fellow of the British Academy in 2004 and awarded an Honorary Doctorate by Oulu University in 2006.

John B. Thornes was Research Chair in Physical Geography at King's College London, and previously Head of the Geography Departments at King's College and Royal Holloway College, University of London, and at Bristol University. He was Honorary Fellow of Queen Mary College, London, former President of the Institute of British Geographers, Vice President of the Royal Geographical Society and Chair of the British Geomorphological Research Group. His research interests were mainly in semi-arid geomorphology, especially the consequences of the invasion of grasslands by woody vegetation in South Africa, and the impact of grazing on soil erosion.

Nigel Thrift is a geographer known for his work on international finance, on time, and on non-representational theory. During his academic career he has been the recipient of a number of distinguished academic awards, including the Royal Geographical Society Victoria Medal for contributions to geographic research. Professor Thrift is an Academician of the Academy of Learned Societies for Social Sciences, and was made a Fellow of the British Academy in 2003. Formerly Pro Vice-Chancellor for Research at the University of Oxford, he is now Vice-Chancellor of the University of Warwick.

Graham A. Tobin is Professor of Geography at the University of South Florida. Dr Tobin has published books, chapters, articles, and technical reports on natural hazards, water resources management and policy, and environmental contamination. He has received research awards from the Southeastern Division of the Association of American Geographers, and the Askounes-Ashford Distinguished Scholar Award from the University of South Florida. His current research activities are concerned with evacuation strategies and health problems associated with volcanic eruptions, socio-economic and environmental impacts of flooding and hurricanes, and pollution reduction strategies in urban areas.

Katie D. Willis is Reader in Development Geography at Royal Holloway, University of London. Her main research interests are in the fields of gender, migration and health, focusing mainly on the urban areas of the Global South. She has undertaken significant periods of fieldwork in Mexico, China and Singapore. She is the author of *Theories and Practices of Development* (Routledge, 2005) and is editor of *Geoforum* and *International Development Planning Review*.

Preface

Defining the core of geography is harder than one might expect. Sociologists have society, biologists living things, economists the economy, and physicists matter and energy. But what is at the very core of geography? What are its key concepts? For the general public, the answer is very often maps and perhaps encyclopaedic knowledge of other people and places, from the world's longest rivers to the names of capitals in far-away countries. While such knowledge will serve you well in a quiz, it is less likely to provide the resources that you need to answer a university examination question well. Geography at university is about much more than maps, facts and figures, although these are also necessary of course.

When studying geography at university, then, many students are obliged to explain to their family and friends that they are learning about the Earth, about humans' relationships with the Earth, and about peoples' relationships with one another – all of which we know vary across time and space. Such a definition of geography does not lead to one clearly identifiable concept which lies at the heart of the discipline, in the same the way that the past becomes the focus of history, or elementary and compound substances the core of chemistry. Rather than having one central organizing concept, geography has many. As students of geography – and in this we include those of us paid to work in the discipline as well as the undergraduates we teach – we can take a step back from this kind of definition of geography and identify a number of concepts which lie at the centre of our discipline. Key among these are space, place, time, scale, landscape, nature, systems, globalization, development and risk.

The aim of this book is to help you to understand the use (and abuse) of these concepts within the discipline of geography. Naming them as core to our discipline might lead the unsuspecting reader to assume that they are all well defined and theorized concepts. On some occasions this is indeed the case. Castree (Chapter 9), for example, provides an insightful review of the multiple ways in which place has been conceptualized in human geography. In doing so, he reveals place to be a much theorized concept, although, importantly, it is also a contested concept. On other occasions, however, these concepts are implicitly assumed but not defined, or are only sporadically considered (see, for example, Kent, Chapter 6, on space in physical geography). When we asked potential authors to write these chapters some ran a metaphorical mile, but others replied that they would enjoy the challenge of thinking about something that is ever present but implicit in their work, essential but unsaid.

In naming this book *Key Concepts in Geography*, and giving a list of key concepts as we did above, you should not let us dupe you into thinking that our choice of concepts is either self-evident or unproblematic. The fact that this second edition of our text covers a much wider range of concepts than the first edition is but one example of the changing interpretation of what count as key concepts in geography. Thus while most geographers would agree that the concepts we have included are important to geography – even if individuals might choose to question the relevance of some to their own work – the fact remains that our choice of concepts here is inevitably partial, reflecting our diverse positions as editors and the contested nature of the discipline in which we all work and study (see Chapters 1 – 4).

If further evidence were needed of the temporary and unstable nature of these concepts, the degree of overlap (and in some cases contradiction) between chapters provides this. Thinking about this in more detail, we can see that the end of one concept and the beginning of another is not neatly defined. In human geography, for example, Taylor (Chapter 8) argues cogently that you cannot think about time without thinking about space, while in defining space, Thrift (Chapter 5) also talks about place and images, the respective topics of the chapters by Castree (Chapter 9) and Morin (Chapter 16). Moreover, these authors do not always have the same take on the concepts in question, illustrating that the concepts are not stable and bounded, but like geography itself, open, temporary and mobile. Rather than being problematic, this is what makes geography a dynamic and fascinating discipline.

Lying as they do at the core of geography, these concepts are an important aspect of all undergraduate degree programmes. Here we have chosen to bring both sides of the discipline together, and discuss human and physical geographers' understandings of key concepts in one volume. At one level, this is a practical decision, as it combines all the approaches an undergraduate studying human and physical geography would need together in one volume. However, it is also more than this. One of the claims geographers can often be heard to make about our discipline is that it, uniquely, bridges the divide between the physical and social sciences (and to a lesser extent the humanities), thus bringing us unrivalled insights into human–environment interactions. In turn, we often impose on our undergraduates modules in both human and physical geography with the aim of providing a rounded geographical education. However, relatively few academic geographers combine an interest in physical and human geography in their own research. Thus, there is scope for both students and teachers to engage with the 'other' side of the discipline.

This volume attempts to bring the two sides together in thinking about our key concepts and in doing so to produce a text which will be of use to all students of geography, whether first-year undergraduates or established academics. The keen-eyed reader will observe, however, that within the volume human and physical geography remain separated as we have devoted two chapters to each concept, one for either side of the discipline. This lack of integration within the book reflects the different treatments the concepts have been given in human and physical geography. While some concepts are explicitly theorized in both sides of the discipline (e.g. time and scale), others are much more explicit in one and implicit in the other (e.g. space and systems) as concepts go in and out of academic fashion.

What is striking in the chapters is the degree to which we draw upon and reinforce other disciplines. In order to put our treatment of these concepts in context, the first section of the book 'Geographical Traditions: Emergence and Divergence', traces the origins of geography and the ways in which it has interacted with the physical science, social science and humanities traditions at different points in time. These chapters illustrate the degree of two-way traffic in ideas between geography and an exceedingly wide range of disciplines, from chemistry to literary criticism. Some of the chapters in the subsequent section of the book then show how these interdisciplinary links have influenced geographers' understandings of their own key concepts, for example as developments in ecology are reawakening interest in space in physical geography, as ideas about risk developed in the social sciences are invigorating thinking on risk in human geography, and ideas in cultural studies are helping us to rethink the concept of landscape in cultural geography.

The apparent vitality of interdisciplinary linkages, alongside the evident intradisciplinary tensions, might lead some to declare the end of geography. Such a declaration, however, would be premature. A more thorough reading of this volume will show that there are indeed common threads in, and cross-linkages between, human and physical geography, as the two sides of the discipline exchange and rework ideas. For example, see Richards (Chapter 2) on the debunking of science as positivism and the true nature of scientific endeavour as a multifaceted process that is relevant in investigations of nature, environment *and* society. That these similarities are sometimes less evident than our differences merely reinforces the importance of the discipline. Geography does indeed occupy a unique position at the intersection of a variety of different traditions, and this often uncomfortable position brings us unique insights because of, rather than despite, our differences.

This passion for geography inspires this volume. In it, we aim to provide an accessible introduction to the ways geography has been shaped by (and shapes) the physical science, social science and humanities traditions, to the concepts which lie at the heart of the discipline, and to the ways in which geographical knowledges can be relevant in our changing world. The chapters are written by experts in their field: the different subject-matter they tackle, as well as their own particular style of communicating, are evident in the ways they have chosen to write the chapters. Some authors have chosen to write an historical review of the way a diverse range of geographers have thought about a particular concept or tradition, ending with the diversity of thought evident today. Others have written pieces with a more contemporary focus, analysing the competing strands evident in current writing, and identifying new ways forward. Some pieces appear neutral, others are more overtly polemical.

You should see these chapters as a way into geography's key concepts. Do not under any circumstance expect to begin reading a chapter over your cornflakes and to become an expert on that particular concept by lunchtime. While we do aim to provide you with an accessible introduction to these issues, the material is intellectually demanding and will require effort on your part. These chapters are the beginning of a process and not the end. For this reason, we have included a section on further reading at the end of each chapter. These have been annotated to guide your travels through the literature, allowing you to follow up themes and counter-themes which have sparked your interest. It is in this way, and only in this way, that you will truly read for your degree in geography.

On a final, and very sad note, we have to record the loss of Professor John Thornes, who died during the production of this book. John will be missed by friends and colleagues. He was, perhaps, the finest geomorphologist of his generation: a consummate theoretician whose ideas were always grounded by an enthusiastic engagement with fieldwork. As a researcher and as a teacher, he recognized the imporance of the contributions that geographers can make across the sciences and the humanities, and his contribution here (Chapter 7) is testimony to the breadth and ambition of his thinking.

Nicholas J. Clifford, Sarah L. Holloway,
Stephen P. Rice and Gill Valentine

Acknowledgements

We would like to thank Robert Rojek at Sage for commissioning this book and Vanessa Harwood for her support during production. We are grateful to Mark Szegner for redrawing many of the figure.

The book would not have been completed without the support of our respective families and partners SR wishes to thank Georgina and NJC thanks Kate and the boys.

The following gave their permission to reproduce material: Figures 7.3 and 7.4, Island Press; Figure 7.5, Oxford University Press; Figure 14.1, The American Meteorological Society (AMS); Figure 15.2 (Sugden, 1970), Blackwell Publishing; Figure 17.2, Ministry of Defence.

Every attempt has been made to obtain permission to reproduce copyright material. If any proper acknowledgement has not been made, we would invite copyright holders to inform us of the oversight.

GEOGRAPHICAL TRADITIONS:
EMERGENCE AND DIVERGENCE

1

Histories of Geography

Mike Heffernan

ⓓefinition

There is no single history of 'geography', only a bewildering variety of different, often competing versions of the past. One such interpretation charts the transition from early-modern navigation to Enlightenment exploration to the 'new' geography of the late nineteenth century and the regional geography of the interwar period. This contextualist account – like all other histories of geography – reflects the partialities of its author.

INTRODUCTION

The deceptively simple word 'geography' embraces a deeply contested intellectual project of great antiquity and extraordinary complexity. There is no single, unified discipline of geography today and it is difficult to discern such a thing in the past. Accordingly, there is no single history of 'geography', only a bewildering variety of different, often competing versions of the past. Physical geographers understandably perceive themselves to be working in a very different historical tradition from human geographers, while the many perspectives employed on either side of this crude binary division also have their own peculiar historical trajectories (see, as examples, Chorley et al., 1964, 1973; Glacken, 1967; Beckinsale and Chorley, 1991; Livingstone, 1992).

Until recently, the history of geography was written in narrow, uncritical terms and was usually invoked to legitimize the activities and perspectives of different geographical constituencies in the present. The discipline's past was presented in an intellectual vacuum, sealed off from external economic, social, political or cultural forces. More recently, however, the history of geography has been presented in a less introspective, self-serving and teleological fashion. Drawing on skills, techniques and ideas from the history of science, a number of scholars (some based in geography departments, others in departments of history or the history of science) have revealed a great deal about various kinds of geography in different historical and national contexts. We now have a substantial body of historical research on the development of geography in universities and learned societies, in primary and secondary schools, and within the wider cultural and political arenas. This research has focused mainly on Europe and North America and extends the longer and richer vein of scholarship on the history of cartography (on the latter, see Harley, 2001). Summarizing this research is no easy task and the following represents only a crude, chronologically simplified outline account of geography's history from the sixteenth to the mid-twentieth centuries.

FROM NAVIGATION TO EXPLORATION: THE ORIGINS OF MODERN GEOGRAPHY

The classical civilizations of the Mediterranean, Arabia, China and India provided many of the geographical and cartographical practices that European geographers would subsequently deploy (Harley and Woodward, 1987, 1992–4). That said, the origins of modern geography can be dated back to western Europe in the century after Columbus. The sixteenth century witnessed far-reaching economic, social and political upheavals, linked directly to the expansion of European power beyond the continent's previously vulnerable limits. By c. 1600, a new, mercantilist Atlantic trading system was firmly established, linking the emerging, capitalist nation-states of western Europe with the seemingly unlimited resources of the American 'New World'. Whether this expansion proceeded from internal changes associated with the transition from feudalism to capitalism (as most historians of early-modern Europe have argued) or whether it preceded and facilitated these larger transformations (as revisionist historians insist) is a 'chicken-or-egg' question that was extensively debated at the end of the twentieth century (Blaut, 1993; Diamond, 1997). All we can say for certain is that early-modern innovations in shipbuilding, naval technology and navigation progressively increased the range of European travel and trade, particularly around the new Atlantic rim, and in so doing transformed European perceptions of the wider world as well as the European self-image (Livingstone, 1992: 32–62).

Firmly rooted in the practical business of long-distance trade, early-modern geography – 'the haven-finding art', as Eva Taylor (1956) memorably called it – encompassed both the technical, mathematical skills of navigation and map-making as well as the literary and descriptive skills of those who wrote the numerous accounts of the flora, fauna, landscapes, resources and peoples of distant regions (see, for early accounts, Taylor, 1930, 1934). As Figure 1.1 suggests, based on evidence from France, geographical descriptions of the non-European

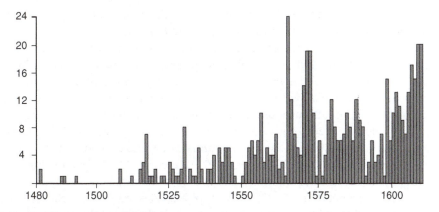

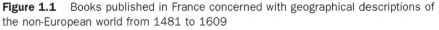

Figure 1.1 Books published in France concerned with geographical descriptions of the non-European world from 1481 to 1609

Source: Atkinson (1927, 1936)

world became steadily more popular through the sixteenth century – the staple fare of the expanding European libraries which were also the principal repositories for the politically important archive of maps produced by Europe's growing army of cartographers (see, for example, Konvitz, 1987; Buisseret, 1992; Brotton, 1997). Equipped with this developing body of geographical fact (liberally sprinkled though it was with speculative fiction), the larger European universities began to offer specialized courses in geography and related pursuits, including chorography, navigation and cartography (see, for example, Bowen, 1981; Cormack, 1997).

The epistemological foundations of modern science were established during the seventeenth century, the era of the so-called 'Scientific Revolution'. This 'revolution' coincided with, and was partly explained by, widespread religious and political upheaval in Europe and had its own geographies that have recently been explored (Livingstone and Withers, 2005: 23–132). The inchoate science of geography, although generally viewed as a practical, navigational skill that merely facilitated scientific discovery (Livingstone, 1988, 1990, 1992: 63–101, 2003), was gradually implicated in, and ultimately transformed by, wider moral, philosophical and political debates about the possibilities of human development within and beyond Europe, the relative merits of the different societies, cultures and civilizations around the world, and the geographical limits on supposedly universal human rights and attributes (see, more generally, Broc, 1981; Livingstone, 1992: 102–38; Livingstone and Withers, 1999; Mayhew, 2000; Withers and Mayhew, 2002). By the eighteenth century – the era of the European Enlightenment – the teaching of geography had become critically important in the creation of new, distinctively modern forms of popular national and imperial identities (Withers, 2001, 2007). At the same time, an interest in travel as an educational activity, beneficial in and of itself, spread from the European aristocracy into the ranks of the newly enriched urban bourgeoisie. From this emerged the 'Grand Tour' of the Mediterranean heartlands of the ancient world so beloved of wealthier and lettered European men and women (Chard and Langdon, 1996; Chard, 1999).

Partly as a result of these developments, the simple idea of geography as navigation gave way to a new formulation: geography as exploration. This was, to be sure, a shift of emphasis rather than a fundamental transformation but it reflected and engendered an entirely new geographical language and rationale. While scientific discoveries might emerge as more or less fortunate by-products of navigation, such discoveries were seen as the planned and considered objectives of the kind of purposeful, self-consciously scientific exploration that developed during the eighteenth century, backed up by new cartographic and navigational techniques and by the substantial resources of modern nation-states (see, for example, Sobel, 1996; Edney, 1997; Burnett, 2000): '[W]hat distinguishes geography as an intellectual activity from ... other branches of knowledge', claims David Stoddart (1986: 29), 'is a set of attitudes, methods, techniques and questions, all of them developed in Europe towards the end of the eighteenth century.' Elsewhere in the same text, Stoddart (1986: 33) is even more specific about the point of departure for the new geography of exploration. The year 1769, when James Cook first sailed into the Pacific, was a genuine turning point in the development of modern geography, claims Stoddart, and not simply because Cook's journeys opened up the Australian landmass with its unique flora and fauna to the inquisitive European gaze. Unlike earlier generations of navigators, claims Stoddart, Cook's explorations were specifically intended to achieve scientific objectives, to be carried out by the illustrious international savants who accompanied him.

The idea that geography developed from navigation to exploration through the early-modern period should not be seen as evidence of a progressive or virtuous evolution from a speculative commercial practice to an objective scientific pursuit. Columbus and Cook were both sponsored by European nation-states eager to exploit the resources that might be uncovered by their voyages. Despite the rhetoric of scientific internationalism, Cook's explorations reflected, a fortiori, the same imperial objectives that had motivated earlier sea-faring navigators. Neither can it be claimed that the wilder speculations of early-modern navigators and their chroniclers were more extravagant than the fantasies of later generations of explorers and their ghostwriters (see Heffernan, 2001). Geography as a practical navigational and cartographic activity was not supplanted by geography as an organized scientific pursuit based on detailed assessments of the human and environmental characteristics of different regions; rather, the same activity acquired new layers of meaning and a new scientific language through which its findings could be expressed.

The new Enlightenment geography was probably best exemplified by Alexander von Humboldt, the Prussian polymath who was born as Cook and his fellow explorers were charting their way across the Pacific. An inveterate explorer and a prolific author, von Humboldt was a complex figure: the archetypal modern, rational and international scientist, his ideas were also shaped by the late eighteenth-century flowering of European romanticism and German classicism. His travels, notably in South America, were inspired by an insatiable desire to uncover and categorize the inner workings of the natural world, and his many published works, especially the multi-volume *Kosmos*, which appeared in the mid-nineteenth century, sought to establish a systematic science of geography that could analyse the natural and the human worlds together and aspire to describe and explain all regions of the globe (Godlewska, 1999b; Buttimer, 2001). His only rival in this ambitious discipline-building project was his German near contemporary, Carl Ritter, a more sedentary writer of relatively humble origins whose unfinished 19-volume *Erdkunde*, also published in the mid-nineteenth century, reflected its author's Christian worldview but was inspired by the same objective of creating a generalized world geography, even though the analysis was to advance no further than Africa and Asia.

INSTITUTIONALIZING EXPLORATION: THE GEOGRAPHICAL SOCIETIES

At this juncture, the European exploratory impetus was still largely dependent on the personal resources of the individuals involved. By the end of the eighteenth century, however, new institutional structures began to emerge within and beyond the agencies of the state dedicated to sponsoring exploration and geographical discovery. In 1782, Jean-Nicolas Buache was appointed geographer to the court of Louis XVI in France and attempted unsuccessfully to launch a geographical society to co-ordinate French exploration (Lejeune, 1993: 21–2). Stung into action by this failed initiative, a group of London scientists and businessmen, led by Sir Joseph Banks (President of the Royal Society) and Major James Rennell (Chief Surveyor of the East India Company) launched the Association for Promoting the Discovery of the Interior Parts of Africa in 1788. Over the next decades, the African Association sponsored several pioneering expeditions, including those of Mungo Park, Hugh Clapperton and Alexander Gordon Laing (Heffernan, 2001; Withers 2004).

The French Revolution and the Napoleonic wars brought a halt to the best forms of Enlightenment geographical inquiry (Godlewska, 1999a) but gave a fresh impetus to the strategically important sciences of cartography and land survey. By 1815, affluent, educated and well-travelled former soldiers were to be found in virtually every major European city, and these men were the natural clientele for the first geographical societies, the building blocks of the modern discipline. The earliest such society was the Société de Géographie de Paris (SGP), which held its inaugural *séance* in July 1821. A fifth of the 217 founder members were born outside France, including von Humboldt and Conrad Malte-Brun, the Danish refugee who became the society's first Secretary-General (Fierro, 1983; Lejeune, 1993). A second, smaller geographical society was subsequently established in Berlin, the Gesellschaft für Erdkunde zu Berlin (GEB), at the instigation of the cartographer Heinrich Berghaus in April 1828, with a foundation membership of just 53, including von Humboldt and Carl Ritter, who became the society's inaugural president (Lenz, 1978).

The establishment in 1830 of the Royal Geographical Society (RGS) of London, under the patronage of William IV, marked a significant new departure. Several London societies committed to fieldwork and overseas travel already existed, including the Linnean Society for natural history (established in 1788), the Palestine Association (1804), the Geological Society (1807), the Zoological

Society (1826) and the Raleigh Club (1826), the last named being a dining club whose members claimed collectively to have visited every part of the known world. The RGS was to provide a clearer London focus for those with an interest in travel and exploration. Even at its foundation, it was far larger than its existing rivals in Paris and Berlin. The 460 original fellows included John Barrow, the explorer and essayist, and Robert Brown, the pioneer student of Australian flora. Within a year, the RGS had taken over the Raleigh Club, the African Association and the Palestine Association to gain a virtual monopoly on British exploration (Brown, 1980).

The pre-eminence of the RGS as the focal point of world exploration increased over subsequent decades. By 1850, there were nearly 800 fellows (twice the number in Berlin and eight times more than Paris, where the SGP membership had slumped) and, by 1870, the fellowship stood at 2,400. Most fellows were amateur scholars but a number of prominent scientists also joined the society's ranks, including the young Charles Darwin, who was elected after his return from the voyage of the *Beagle* in 1838. The dominant figure in the RGS during the middle years of the nineteenth century was Sir Roderick Murchison, who was president on three separate occasions: 1843-5, 1851-3 and 1862-71. A talented publicist and entrepreneur, Murchison advocated geographical exploration as a precursor to British commercial and military expansion (Stafford, 1989). While other societies offered only *post hoc* awards and medals for successfully completed voyages, the RGS used its substantial resources to sponsor exploration in advance and on a large scale by providing money, setting precise objectives, lending equipment and arbitrating on the ensuing disputes. It also published general advice through its *Hints to Travellers*, which began in 1854 (Driver, 2001: 49-67), and developed what was probably the largest private map collection in the world.

The success of the RGS reflected the strength of British amateur natural science, the wealth of the country's upper middle class (which provided the bulk of the fellowship) and the confidence that a large navy and overseas empire gave to prospective British explorers (Stoddart, 1986: 59-76). By concentrating on exploration and discovery, the RGS exploited a vicarious national passion for muscular 'heroism' in exotic places that was enthusiastically promoted by the British press. The explorer was the ideal masculine hero of Victorian society (the notion of a female geographer seemed almost a contradiction in terms), selflessly pitting himself against the elements and hostile 'natives' in remote regions for the greater glory of science and nation. Africa loomed especially large in the public imagination and the exploration of the 'Dark Continent', particularly the quest for the source of the Nile, provided an exciting and popular focus for the society's activities. All the major African explorers of the day – Burton, Speke, Livingstone, Stanley – were influenced in some degree by the RGS, although their relationships with the society were not always cordial (Driver, 2001: 117-45). As the blank spaces on the African map were filled in, the RGS more than any other organization was able to bask in the reflected glory, while always shifting its focus to new regions. By the late nineteenth and early twentieth centuries, under the powerful influence of Sir Clements Markham and Lord Curzon, attention was directed mainly towards central Asia, the polar ice caps and the vertical

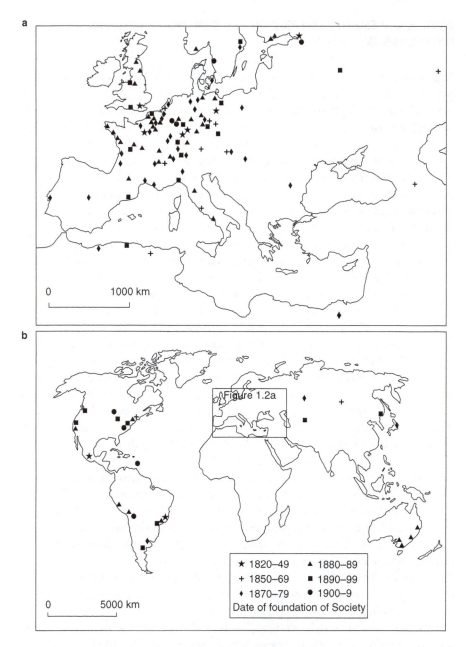

Figure 1.2 European and non-European geographical societies, by date of foundation

Source: Kolm (1909)

challenges of high mountain ascents in the Himalayas. The geographical societies in Paris and Berlin also expanded rapidly after 1850, under the direction of the Marquis de Chasseloup-Laubat (Napoleon III's former Naval and Colonial Minister) and Heinrich Barth (the leading German African explorer) and new societies sprang up elsewhere in Europe and in the burgeoning cities of Northand South America (Figure 1.2).

SCIENCE AND EMPIRE: EXPLORATION, THE 'NEW' GEOGRAPHY AND THE MODERN IMPERIAL NATION

The vision of the discipline promoted by these mid-nineteenth century geographical societies exemplified the soaring ambition of the European imperial mind (Driver, 1992, 2001; Bell et al., 1994; Godlewska and Smith, 1994). The navigational and cartographic skills of the geographer during the 'heroic' age of exploration and discovery paved the way for European military and commercial colonization of the Americas, Asia and Africa. The principal geographical 'tool' was, of course, the map. By representing the huge complexity of a physical and human landscape in a single image, geographers and cartographers provided the European imperial project with arguably its most potent device. European exploration and mapping of the coastlines of the Americas, Africa, Asia and the Pacific, and the subsequent terrestrial topographic surveying of these vast continents, were self-evidently exercises in imperial authority. To map hitherto 'unknown' regions (unknown, that is, to the European), using modern techniques in triangulation and geodesy, was both a scientific activity dependent on trained personnel and state-of-the art equipment and also a political act of appropriation which had obvious strategic utility (Edney, 1997; Burnett, 2000; Harley, 2001).

The shift in the European balance of power following the Franco-Prussian war of 1870 gave an unexpected boost to geography. Aggressive colonial expansion outside Europe was identified as one way to reassert a threatened or vulnerable national power within Europe, and the later decades of the nineteenth century were characterized by a surge of colonial expansion (particularly the so-called 'Scramble for Africa') as each imperial power sought comparative advantage over its enemies, both real and imagined. This frenzied land grab emphasized the practical utility of geography and cartography. By the end of the nineteenth century, the 'high-water mark' of European imperial expansion, geography had become 'unquestionably the queen of all imperial sciences ... inseparable from the domain of official and unofficial state knowledge' (Richards, 1993: 13; see also Said, 1978, 1993). In Germany, 19 new geographical societies had been established, including associations in the former French towns of Metz (1878) and Strasbourg (1897). In France, there were 27 societies, one in virtually every French city, and no fewer than four in French Algeria. A number of the French provincial societies were devoted to commercial geography and sought to encourage trading links with the French empire (Schneider, 1990). At this point, one-third of the world's geographers were based in France (Figure 1.3). The British were by no means immune to this late-century geographical fever and the RGS remained the largest and wealthiest geographical society in the world. A handful of provincial societies were established in the UK during the late nineteenth century, notably in Edinburgh (the Royal Scottish Geographical Society) and Manchester (both 1884) but, unlike the countries of continental Europe, the RGS retained its dominance of the British geographical movement (MacKenzie, 1994).

Backed by a new generation of civic educational reformers, a 'new' geography began to emerge in schools and universities, with Germany and France leading the way. The German university system had been significantly reformed during the nineteenth century (based in part on the ideas of Wilhelm von

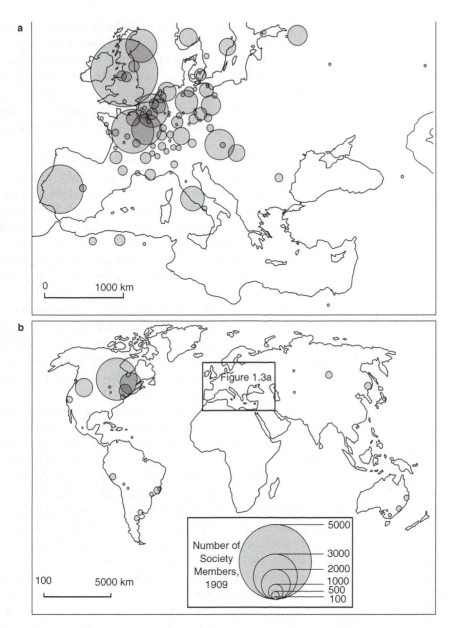

Figure 1.3 Size of European and non-European geographical societies, late nineteenth century

Source: Kolm (1909)

Humboldt, Alexander's brother) and geography already had a powerful presence in the tertiary and secondary educational programmes. The same republican politicians in Paris who championed colonial expansion as a route to national rejuvenation were also convinced that France needed to learn from Germany by completely revising its school and university system to inculcate the patriotic values that had seemed shockingly absent from the French armies of 1870. A carefully constructed geography curriculum was identified as the key to such a system. This would introduce the next generation to the beauty, richness and variety of France's *pays* while informing them of their nation's role and responsibility in the wider world. The French universities would need to train the next generation of geography teachers, and a dozen new chairs were established during the 1880s and 1890s for this purpose (Broc, 1974). Germany, eager to sustain its reputation as the leading intellectual centre of the discipline, responded with a similar educational drive.

The fiercely independent British universities initially resisted this trend, to the dismay of the geographers in the RGS. A chair of geography had been established at University College London as early as 1833 (filled by Captain Alexander Maconochie, the RGS's first Secretary), but this lapsed after three years and a full-time British university post in geography was not created until 1887 when an Oxford University readership was awarded to Halford Mackinder, partly financed by the RGS's (Stoddart, 1986: 41–127). The RGS, along with the Geographical Association (established in 1893 to promote geography in schools) and other educational organizations such as the British Association for the Advancement of Science (Withers et al., 2006), worked hard to change attitudes. Sir Harry H. Johnston, the explorer, colonial administrator and prominent RGS fellow, argued that geography should become a compulsory school subject, for it was only through detailed geographical description, complete with authoritative and regularly updated topographical and thematic maps, that a region could be known, understood and therefore fully possessed by those in authority (Heffernan, 1996: 520). By dividing the world into regions and ordering the burgeoning factual information about the globe into regional segments, he insisted, geography offered one solution to the yearned-for objective of classifying and understanding the human and environmental characteristics of the entire globe. Through geography the world could, at last, be visualized and conceptualized as a whole, a process facilitated by the use of new techniques of geographical representation, particularly photography (Ryan 1997; Schwartz and Ryan, 2003).

The 'new' school and university geography was no less an imperial science than its exploratory predecessor, as a cursory glance at the textbooks of the period makes clear (Hudson, 1977). The principal representatives of academic geography – notably Mackinder at Oxford and Friedrich Ratzel in Leipzig – sought not only to explain the human and natural features of the world, but also to justify the existence of European empires (Heffernan, 2000a). Ratzel, in particular, was profoundly influenced by the writings of Charles Darwin and insisted (as did many so-called 'social Darwinists' within and beyond geography) that the principles of 'natural selection' applied equally to the natural, social and political realms (Stoddart, 1986: 158–79; Bassin, 1987). Nation-states, like species, struggled for space and resources, and the 'fittest' were able to impose their will on less fortunate 'races'. For many geographers of this period, including those who were fashioning a distinctively American geographical tradition, such as Ellsworth Huntington and Ellen Churchill Semple, the dominance that certain peoples exerted over others was either divinely preordained or the product of environmentally conditioned racial characteristics (Keighren, 2006). Building on Enlightenment ideas about the environmentally determined nature of different peoples, a new brand of 'scientific racism' infused geographical theory in the later nineteenth and early twentieth centuries. The expansive, imperial 'races' of Europe and the European settler communities in the Americas benefited from unique climatic and environmental circumstances, it was claimed, and these advantages had created energetic, expansive civilizations. The very different climates and environments of the colonial periphery had created inferior societies and weaker civilizations in need of an ordering and benign European presence (Peet, 1985).

Such distasteful ideas reflected a prevailing orthodoxy but they also pro-voked spirited debate. While environmental determinism and scientific racism were often mutually reinforcing ideas, they could also contradict one another. Some racial theorists assumed that different 'races' were fixed in an unchanging 'natural' hierarchy, the contemporary manifestation of quite distinct evolution-ary sequences from different points of departure (polygenesis). 'External' environmental factors could have only limited impact on this preordained racial system. This argument presupposed the need for a permanent imperial presence of intellectually and racially superior rulers in order to manage the irredeemably inferior peoples and environments of the colonial world (Livingstone, 1992: 216–60). By emphasizing the overriding significance of climatic and physical geo-graphical factors on the process of social and economic progress and the essen-tial unity of humanity (monogenesis), many environmental determinists tended to focus on the possibilities of human development through the judicious inter-vention in the natural world. If scientifically advanced European societies could overcome the worst aspects of the challenging environments of the colonial periphery by draining the pestilential marsh or irrigating the barren desert, this would not only improve colonial economic productivity but would also, in time, improve the nature of the local societies and cultures. Environmental and moral 'improvement' were thus intimately interlinked and both were dependent on the 'benign' intervention of a 'superior' external force. In time, and if coupled with appropriate cultural and educational policies, colonized peoples would be allowed to take control of their own resources and manage their own affairs.

In dramatic contrast to these views were the radical geographical theories of Petr Kropotkin and Elisée Reclus, leading figures in the Russian and French anarchist circles, respectively. For Kropotkin and Reclus, the new science of geography suggested ways of developing a new harmony of human societies with the natural world, freed from the pernicious influence of class-based, nationalist politics (Blunt and Wills, 2000).

THE NEW WORLD: GEOGRAPHY AND THE CRISIS OF THE EARLY TWENTIETH CENTURY

The onset of the twentieth century provoked a rather anxious debate about the future of the 'great powers'. Many believed that 1900 would mark a turning point in world history, the end of a 400-year period of continuous European expansion. The unexplored and unclaimed 'blank' spaces on the world map were rapidly diminishing, or so it seemed, and the sense of a 'global closure' was palpable.

Different versions of this *fin-de-siècle* lament were rehearsed in several contexts. The German geomorphologist, Albrecht Penck, used the idea of global closure in the early 1890s to justify his inspirational but sadly inconclusive scheme for a new, international 1:1 million map of the world (Heffernan, 2002). At the same time, the American historian, Frederick Jackson Turner, delivered a famous lecture at the Columbian Exposition in Chicago in 1893 (an event designed to commemorate the quatercentenary of Colombus's voyage to the Americas), which suggested the need for the newly established transcontinental USA to seek out new imperial frontiers beyond the traditional limits of the national homeland, particularly in the Pacific. And in 1904, Halford Mackinder addressed the RGS on the likely end of the 'Columbian era' of maritime, trading empires and the emergence of a twentieth-century world order dominated by cohesive land-based empires (such as the USA), bound together by railways. Mackinder dubbed the great Eurasian landmass – the largest expanse of territory on the planet – as the 'geographical pivot of history' and argued that whichever power could control the limitless resources of this huge region would dominate world affairs in the coming century. The 'closed' system Mackinder described would be extremely dangerous, he implied, because the frontiers on the new empires would straddle the globe (for a comparison of Turner and Mackinder, see Kearns, 1984).

The outbreak of the First World War, the first truly global conflict, confirmed many of these fears. Although it reached its peak of savage intensity on the Western Front, Mackinder later insisted that the war had erupted from precisely the territorial struggle he had foreseen in 1904. Germany's pitch for global hegemony had been based on the idea of winning what Ratzel had famously called 'living space' (*Lebensraum*) in the east, at the expense of Russia, the region Mackinder now called 'the heartland' of the 'world island' (Mackinder, 1919). Mackinder was not asked to advise the British delegation which negotiated the peace treaties in Paris in 1919 (to his considerable frustration), but leading geographers from other countries were prominently involved in the redrawing of the postwar political map. The larger geographical societies in all belligerent countries had been fully mobilized by the intelligence services of each state (not least because of their extensive map collections) and had generated a mass of new geographical information and cartography for their paymasters. In the USA, the President of the American Geographical Society, Isaiah Bowman, was an important adviser to President Woodrow Wilson during the peace negotiations and had previously recruited many of America's leading geographers (including William Morris Davis and Ellen Churchill Semple) on to the so-called House Inquiry to help formulate US policy on postwar Europe and the wider world. Bowman also wrote the main geographical text on the postwar order, *The New World* (1921). Several French geographers, led by Paul Vidal de la Blache, fulfilled a similar role as members of the Comité d'Études that advised the French government during the war and at the peace conferences. The RGS, for its part, was also prominently involved as a metropolitan 'centre of calculation' for both the Naval and War Office intelligence services (Heffernan, 2000b).

In these countries, geography emerged from the carnage of the First World War with its reputation significantly enhanced. New geography appointments quickly followed in the leading schools and universities, notably in

Britain, where the teaching of geography still lagged behind the continent and where university courses had previously been taught by a single lecturer. The first British honours schools of geography were established during the war itself (in Liverpool in 1917 and at the LSE and Aberystwyth in 1918) or immediately afterwards (at University College London and Cambridge in 1918, Manchester in 1923 and Sheffield in 1924) (Stoddart, 1986: 45–6). Although the RGS had overseen the initial appointments to geography positions in British universities, the subsequent expansion of the discipline eroded the society's control of the British geographical agenda. Anxious to develop a more rigorous, scientific geography to match the developments taking place in other countries, British university geographers established their own independent organization in 1933, the Institute of British Geographers (IBG), which only recently re-merged with the older society.

By the interwar years, the 'new' geography that had arisen before 1914 had evolved into a sophisticated and popular discipline, prominent at all levels in the educational system. In the universities, a host of new subdisciplines arose, most of which continue to the present, but two wider interwar trends are worthy of special mention. The first was the conviction that geography should be an integrative, regional science. Physical and human geography should always be brought together in the analysis of specific regions, it was repeatedly argued, and the otherwise vague and undeveloped idea of the region emerged as the single most important intellectual contribution of interwar geography, particularly in Britain and France. The importance of the region can easily be explained. For the geographers who rose to prominence after 1918, the traditional nation-state was a suspect entity, the focus and the engine of the discredited nineteenth-century nationalism that had culminated with the disasters of 1914–18. The region, whether subnational or supranational, offered the prospect of radically alternative forms of government in the future. The French school of geography (dominated until his death in 1918 by Paul Vidal de la Blache and continued after the war by his many students) saw the region as the discipline's fundamental building block. Alongside the numerous regional monographs that were produced with assembly-line efficiency by French geographers, the Vidalians also proffered various schemes for devolved regional government from below (based in part on Vidal's own recommendations from 1910) and for integrated, European government from above (the most prophetic coming from Albert Demangeon). Similar ideals inspired regional geographers in Britain, including A.J. Herbertson, C.B. Fawcett, L. Dudley Stamp and H.J. Fleure, most of whom were influenced by the radical idealism of the Scottish natural scientist, planner and general polymath, Patrick Geddes (Livingstone, 1992: 260–303; Heffernan, 1998: 98–106, 128–31). The same agenda shaped the development of geography in other national contexts; in Germany, to be sure, but also the USA where the school of cultural geography established by Carl Sauer at Berkeley celebrated the idea of historical and geographical particularism and the unique qualities of diverse regions (see Chapter 16 on the importance of Sauer's work). Each of these national schools had distinguishing traits, but all shared a common conviction in the civic utility of geography as an educational, field-based and interpretative discipline (on geography and fieldwork, see Ploszajska, 1998; Lorimer, 2003; Withers and Finnegan, 2003; Lorimer and Spedding, 2005).

The second, very different, interwar trend was associated with fascist Italy and Nazi Germany, where a new generation of academic geographers sought to relaunch their discipline as an overtly political science dedicated to questioning the geopolitical order established in Paris after 1918. The Italian and German geopolitical movements (developed by Giorgio Roletto and Karl Haushofer and associated with the journals *Geopolitica* and *Zeitschrift für Geopolitik*) had much in common, including a penchant for bold, black-and-white propaganda cartography and hard-hitting, journalistic articles. Despite their overtly nationalist stance, both movements imagined a future integrated Europe though of a very different kind than was proposed by French and British regional geographers. The influence of Italian geopolitical theorists on government policy was minimal, and the impact of their German equivalents on Nazi programmes was even smaller, despite the close relationship between Haushofer and Rudolf Hess, one of Hitler's chief acolytes. Haushofer and his fellow academics had remarkably little to say on the central question of race and this, more than anything else, limited their appeal to Hitler and his Nazi ideologues (Heffernan, 1998: 131–49; Dodds and Atkinson, 2000).

The Second World War spelt the end of the geopolitical movements of Italy and Germany (and also brought about the temporary collapse of political geography *tout court*, and not only in these two countries). While the interwar regional geographical tradition continued into the post-1945 era, this too came under increasing pressure from new developments, particularly the quantitative geographical inquiry pioneered in the USA and in Britain during the 1960s and 1970s (see Chapter 3). Although it had arisen from a practical concern with the region as an alternative level of government and administration, the particularism of interwar regionalism, with its focus on the uniqueness of place, sat uncomfortably beside the new idea of geography as a law-seeking, 'spatial science'. Instead of the old, more historical form of regionalism, a new and more rigorously scientific regional science developed strongly during the postwar years to play its part alongside the many other branches of geographical research and teaching (Johnston, 1997).

CONCLUSION

The preceding survey is a personal account and should certainly not be read as a story of radical departures or revolutionary changes. The rough sequence of events charted here – the transition from early-modern navigation to Enlightenment exploration to the 'new' geography of the late nineteenth century and the regional geography of the interwar period – represents a process of accretion rather than displacement; an evolution in which traditions merged, overlapped and persisted alongside later developments to create an ever more complex picture. It is impossible to distil from these stories an essential core theme that has always animated geographical inquiry, but one thing is clear: geography, whether defined as a university discipline, a school subject or a forum for wider debate, has always existed in a state of uncertainty and flux. While some have lamented this as a sign of disciplinary weakness, it might equally be argued that the absence of conceptual conformity has been one of the discipline's great strengths. If the developments of the last few decades can be taken as a guide, it would seem that this is one 'geographical tradition' that is destined to continue.

SUMMARY

- The deceptively simple word 'geography' embraces a deeply contested intellectual project of great antiquity and extraordinary complexity. There is no single, unified discipline of geography today and it is difficult to discern such a thing in the past.

- A rough sequence of events can be charted from the early-modern navigation, to Enlightenment exploration, to the 'new' geography of the late nineteenth century and the regional geography of the interwar period.

- It is impossible to distil from these stories an essential core theme that has always animated geographical inquiry. This could be seen either as a sign of disciplinary weakness or as a strength.

Further Reading

The literature on the history of geography is large and varied. The best starting point is David Livingstone's (1992) *The Geographical Tradition: Episodes in the History of a Contested Enterprise*, which is excellent on wider intellectual and philosophical contexts. David Stoddart's (1986) *On Geography and its History* is a spirited defence of geography's place within the natural sciences. On the Enlightenment, Robert Mayhew's (2000) *Enlightenment Geography: The Political Languages of British Geography 1650–1850* and Anne Godlewska's (1999) *Geography Unbound: French Geographic Science from Cassini to Humboldt* have different perspectives but survey the British and French experiences very effectively, while Charles Withers' (2007) *Placing the Enlightenment* provides a comprehensive and perceptive general survey. The collections edited by David Livingstone and Charles Withers on *Geography and Enlightenment* (1999) and *Geography and Revolution* (2005) contain useful introductory essays and strong chapters on specific topics. Charles Withers' (2001) *Geography, Science and National Identity* provides an outstanding illustration of geography's civic educational role. Anne Godlewska and Neil Smith's (1994) *Geography and Empire* is good on the imperial theme in general and can be supplemented, for the nineteenth and early twentieth centuries, by Morag Bell et al. (1994) *Geography and Imperialism, 1820–1940*. Felix Driver's (2001) *Geography Militant: Cultures of Exploration and Empire* is a sparkling and highly imaginative study on the nineteenth century.

Note: Full details of the above can be found in the references list below.

References

Atkinson, G. (1927) *La Littérature géographique française de la Renaissance: Répertoire biblio-graphique*. Paris: Auguste Picard.

Atkinson, G. (1936) *Supplément au Répertoire bibliographique se rapportant à la Littérature géo-graphique française de la Renaissance*. Paris: Auguste Picard.

Bassin, M. (1987) 'Imperialism and the nation-state in Friedrich Ratzel's political geography', *Progress in Human Geography*, 11: 473–95.

Beckinsale, R. and Chorley, R. (1991) *The History of the Study of Landforms, or The Development of Geomorphology. Vol. III. Historical and Regional Geomorphology, 1890–1950*. London: Routledge.

Bell, M., Butlin, R. and Heffernan, M. (eds) (1994) *Geography and Imperialism, 1820–1940*. Manchester: Manchester University Press.

Blaut, J. (1993) *The Colonizer's Model of the World: Geographical Diffusionism and Eurocentric History*. London: Guilford Press.

Blunt, A. and Wills, J. (2000) *Dissident Geographies: An Introduction to Radical Ideas and Practice*. London: Prentice-Hall.

Bowman, I. (1921) *The New World*. New York: World Book Co.

Bowen, M. (1981) *Empiricism and Geographical Thought: From Francis Bacon to Alexander von Humboldt*. Cambridge: Cambridge University Press.

Broc, N. (1974) 'L'établissement de la géographie en France: diffusion, institutions, projets (1870–1890)', *Annales de Géographie*, 83: 545–68.

Broc, N. (1981) *La Géographie des Philosophes: Géographes et Voyageurs français au XVIIIe siècle*. Paris: Éditions Ophrys.

Brotton, J. (1997) *Trading Territories: Mapping the Early-Modern World*. London: Verso.

Brown, E. (ed.) (1980) *Geography Yesterday and Tomorrow*. Oxford: Oxford University Press.

Buisseret, D. (ed.) (1992) *Monarchs, Ministers and Maps: The Emergence of Cartography as a Tool of Government in Early-Modern Europe*. Chicago, IL: University of Chicago Press.

Burnett, D. (2000) *Masters of All They Surveyed: Exploration, Geography, and a British El Dorado*. Chicago, IL: University of Chicago Press.

Buttimer, A. (2001) 'Beyond Humboldtian science and Goethe's way of science: challenges of Alexander von Humboldt's geography', *Erdkunde*, 55: 105–20.

Chard, C. (1999) *Pleasure and Guilt on the Grand Tour: Travel Writing and Imaginative Geography, 1600–1830*. Manchester: Manchester University Press.

Chard, C. and Langdon, H. (eds) (1996) *Transports: Travel, Pleasure and Imaginative Geography, 1600–1830*. New Haven, CT: Yale University Press.

Chorley, R., Beckinsale, R. and Dunn, A. (1964) *The History of the Study of Landforms, or The Development of Geomorphology. Vol. I. Geomorphology before Davis*. London: Methuen.

Chorley, R., Beckinsale, R. and Dunn, A. (1973) *The History of the Study of Landforms, or The Development of Geomorphology. Vol. II. The Life and Works of William Morris Davis*. London: Methuen.

Cormack, L. (1997) *Charting an Empire: Geography and the English Universities 1580–1620*. Chicago, IL: University of Chicago Press.

Diamond, J. (1997) *Guns, Germs and Steel: The Fates of Human Societies*. London: Jonathan Cape.

Dodds, K. and Atkinson, A. (eds) (2000) *Geopolitical Traditions: A Century of Geopolitical Thought*. London: Routledge.

Driver, F. (1992) 'Geography's empire: histories of geographical knowledge', *Environment and Planning D: Society and Space*, 10: 23–40.

Driver, F. (2001) *Geography Militant: Cultures of Exploration and Empire*. Oxford: Blackwell.

Edney, M. (1997) *Mapping an Empire: The Geographical Construction of British India, 1765–1843*. Chicago, IL: University of Chicago Press.

Fierro, A. (1983) *La Société de Géographie de Paris (1826–1946)*. Geneva and Paris: Librairie Groz and Librairie H. Champion.

Glacken, C. (1967) *Traces on the Rhodian Shore: Nature and Culture in Western Thought from Ancient Times to the End of the Eighteenth Century*. Berkeley and Los Angeles, CA: University of California Press.

Godlewska, A. (1999a) *Geography Unbound: French Geographic Science from Cassini to Humboldt*. Chicago, IL: University of Chicago Press.

Godlewska, A. (1999b) 'From Enlightenment vision to modern science: Humboldt's visual thinking', in D. Livingstone and C. Withers (eds) *Geography and Enlightenment*. Chicago, IL: University of Chicago Press, pp. 236–75.

Godlewska, A. and Smith, N. (eds) (1994) *Geography and Empire*. Oxford: Blackwell.

Harley, J. (ed. P. Laxton) (2001) *The New Nature of Maps: Essays in the History of Cartography*. Baltimore, MD: Johns Hopkins University Press.

Harley, J. and Woodward, D. (eds) (1987) *The History of Cartography. Vol. I. Cartography in Prehistoric, Ancient, and Medieval Europe and the Mediterranean*. Chicago, IL: University of Chicago Press.

Harley, J. and Woodward, D. (eds) (1992–4) *The History of Cartography. Vol. II. Book 1. Cartography in the Traditional Islamic and South Asian Societies. Book 2. Cartography in the Traditional East and Southeast Asian Societies*. Chicago, IL: University of Chicago Press.

Heffernan, M. (1996) 'Geography, cartography and military intelligence: the Royal Geographical Society and the First World War', *Transactions, Institute of British Geographers*, 21: 504–33.

Heffernan, M. (1998) *The Meaning of Europe: Geography and Geopolitics*. London: Arnold.

Heffernan, M. (2000a) 'Fin de siècle, fin du monde: on the origins of European geopolitics, 1890–1920', in K. Dodds and D. Atkinson (eds) *Geopolitical Traditions: A Century of Geopolitical Thought*. London: Routledge, pp. 27–51.

Heffernan, M. (2000b) 'Mars and Minerva: centres of geographical calculation in an age of total war', *Erdkunde*, 54: 320–33.

Heffernan, M. (2001)' "A dream as frail as those of ancient Time": the in-credible geographies of Timbuctoo', *Environment and Planning D: Society and Space*, 19: 203–25.

Heffernan, M. (2002) 'The politics of the map in the early 20th century', *Cartography and Geographical Information Systems*, 29: 207–26.

Hudson, B. (1977) 'The new geography and the new imperialism', *Antipode*, 9: 12–19.

Johnston, R. (1997) *Geography and Geographers: Anglo-American Human Geography since 1945*. London: Arnold.

Kearns, G. (1984) 'Closed space and political practice: Frederick Jackson Turner and Halford Mackinder', *Environment and Planning D: Society and Space*, 22: 23–34.

Keighren, I. (2006) `Bringing geography to the book: charting the reception of Influences of Geographic Environment', *Transactions, Institute of British Geographers*, NS 31: 525–40.

Kolm, G. (1909) 'Geographische Gessellschaften, Zeitschriften, Kongresse und Austellungen', *Geographisches Jahrbuch*, 19: 403–13.

Konvitz, J. (1987) *Cartography in France, 1660–1848: Science, Engineering and Statecraft*. Chicago, IL: University of Chicago Press.

Lejeune, D. (1993) *Les Sociétés de Géographie en France et l'Expansion coloniale au XIXe siècle*. Paris: Albin Michel.

Lenz, K. (1978) 'The Berlin Geographical Society 1828–1978', *Geographical Journal*, 144: 218–22.

Livingstone, D. (1988) 'Science, magic and religion: a contextual reassessment of geography in the sixteenth and seventeenth centuries', *History of Science*, 26: 269–94.

Livingstone, D. (1990) 'Geography, tradition and the scientific revolution: an interpretative essay', *Transactions, Institute of British Geographers*, 15: 359–73.

Livingstone, D. (1992) *The Geographical Tradition: Episodes in the History of a Contested Enterprise*. Oxford: Blackwell.

Livingstone, D. (2003) *Putting Science in its Place: Geographies of Scientific Knowledge*. Chicago, IL: University of Chicago Press.

Livingstone, D. and Withers, C. (eds) (1999) *Geography and Enlightenment*. Chicago, IL: University of Chicago Press.

Livingstone, D. and Withers, C. (eds) (2005) *Geography and Revolution*. Chicago, IL: University of Chicago Press.

Lorimer, H. (2003) 'Telling small stories: spaces of knowledge and the practice of geography', *Transactions, Institute of British Geographers*, 28: 197–217.

Lorimer, H. and Spedding, N. (2005) 'Locating field science: a geographical family expedition to Glenn Roy, Scotland', *British Journal for the History of Science* 38: 13–34.

MacKenzie, J. (1994) 'The provincial geographical societies in Britain, 1884–1894', in M. Bell et al. (eds) *Geography and Imperialism, 1820–1940*. Manchester: Manchester University Press, pp. 31–43.

Mackinder, H. (1919) *Democratic Ideals and Reality: A Study in the Politics of Reconstruction*. London: Constable.

Mayhew, R. (2000) *Enlightenment Geography: The Political Languages of British Geography 1650–1850*. Basingstoke: Macmillan.

Peet, R. (1985) 'The social origins of environmental determinism', *Annals of the Association of American Geographers*, 75: 309–33.

Ploszajska, T. (1998) 'Down to earth: geography and fieldwork in English schools, 1870–1914', *Environment and Planning D: Society and Space*, 16: 757–74.

Richards, T. (1993) *The Imperial Archive: Knowledge and the Fantasy of Empire*. London: Verso.

Ryan, J. (1997) *Picturing Empire: Photography and the Visualization of the British Empire*. London: Reaktion.

Said, E. (1978) *Orientalism*. London: Routledge.

Said, E. (1993) *Culture and Imperialism*. London: Jonathan Cape.

Schneider, W. (1990) 'Geographical reform and municipal imperialism in France, 1870–80', in J. MacKenzie (ed.) *Imperialism and the Natural World*. Manchester: Manchester University Press, pp. 90–117.

Schwartz, J. and Ryan, J. (eds) (2003) *Picturing Place: Photography and the Geographical Imagination*. London: I.B. Tauris.

Sobel, D. (1996) *Longitude: The True Story of a Lone Genius who Solved the Greatest Scientific Problem of his Time*. London: Fourth Estate.

Stafford, R. (1989) *Scientist of Empire: Sir Roderick Murchison, Scientific Exploration and Victorian Imperialism*. Cambridge: Cambridge University Press.

Stoddart, D. (1986) *On Geography and its History*. Oxford: Blackwell.

Taylor, E. (1930) *Tudor Geography 1485–1583*. London: Methuen.

Taylor, E. (1934) *Late Tudor and Early Stuart Geography 1583–1650*. London: Methuen.

Taylor, E. (1956) *The Haven-Finding Art: A History of Navigation from Odysseus to Captain Cook*. London: Hollis & Carter.

Withers, C. (2001) *Geography, Science and National Identity: Scotland since 1520*. Cambridge: Cambridge University Press.

Withers, C. (2004) 'Mapping the Niger, 1798–1832: truth, testimony and "ocular demonstration" in the late Enlightenment', *Imago Mundi*, 56: 170–93.

Withers, C. (2007) *Placing the Enlightenment: Thinking Geographically about the Age of Reason*. Chicago, IL: University of Chicago Press.

Withers, C. and Finnegan, D. (2003) 'Natural history societies, fieldwork and local knowledge in nineteenth-century Scotland: towards an historical geography of civic science', *Cultural Geographies*, 10: 334–53.

Withers, C., Finnegan, D. and Higgitt, R. (2006) 'Geography's other histories? Geography and science in the British Association for the Advancement of Science, 1831–c.1933', *Transactions, Institute of British Geographers*, NS 31: 433–51.

Withers, C. and Mayhew, R. (2002) 'Rethinking "disciplinary" history: geography in British universities, c. 1580–1887', *Transactions, Institute of British Geographers*, 27: 11–29.

2

Geography and the Physical Sciences Tradition

Keith Richards

Definition

The physical sciences provide a role model for many disciplines, but the model is a contested one. Some of the successes of the physical sciences have been the product of lengthy gestation over hundreds of years, and the methods employed and the philosophical frameworks underpinning them have changed in response to emerging understandings. Thus there is no single tradition, except that of pluralism. Within *this* 'tradition', there is a rich source for geography of methods drawing on observation, measurement, various forms of experimentation, theory development, and testing. These diverse but closely related practices are the true legacy of the sciences, and one to which geography can contribute as well as from which it may draw.

INTRODUCTION

This chapter examines the relationship between the philosophy and method of the physical sciences and those of geography, with a view to exploring some similarities and differences. The physical sciences have been very successful in providing us with understanding of many aspects of the world, as a result of powerful procedures for revealing the nature of that world. The basic argument of the chapter is that the so-called scientific method on which the physical sciences have been based is really only a model itself, and that the balance of procedures in scientific method has changed as knowledge and understanding have developed, and may also change as a particular investigation proceeds. This is illustrated with many examples, but the Gas Laws and a research problem in eco-hydrology are emphasized as particular cases. There are many different dimensions to the scientific method, and no single 'tradition'. This has implications for the conduct of debate in geography, where particular assumed traditions may sometimes be misrepresented, especially in support of an anti-quantitative stance. It also suggests that there are clear possibilities for naturalism – a common methodology applicable to a wide range of kinds of scientific inquiry, including environmental and social – because the pluralist nature of the procedures of the physical scientific method have many counterparts in these other sciences.

This chapter begins by stressing the dangers of assuming the existence of a single tradition. The importance of identifying researchable questions as an early component of scientific problem-solving is then reviewed. In turn, aspects of scientific practice are considered in detail. It is emphasized that practical research procedures, adapted to present conditions of understanding, constitute a more important tradition, or rather, legacy, than the oversimplified labels attached to particular methodological contexts (e.g. positivism). The procedures of science which are reviewed include: observation and measurement; experimentation; the theoretical issues of hypothesis generation and testing; identification of mechanisms behind regularities; and establishing criteria for warranting the claims of competing theories. In the conclusion, the importance of the pluralism demonstrated by these procedures is emphasized, and an argument is made for naturalism, and a 'rational criticism' model of the scientific method.

CONTESTED HISTORIES

In this review of some relationships between geography and the physical sciences, it is useful to expose two challenges at the outset. The first is that any discussion of methodology is fraught with difficulty, because the terms employed are so unstable. This instability can result in both confusion and artifice; the former when protagonists unwittingly adopt different meanings, the latter when a particular meaning is chosen deliberately, even politically, to support a specific position or cause. The second challenge arises because traditions are often not what they seem. They frequently turn out to be more modern (a term used deliberately) than the use of the word 'tradition' implies, to have a history in a politics of representation, also to be unstable in content, and to take different forms for different actors. For these reasons, discussing any 'tradition' in geography is liable to seem like walking across quicksand, while at the same time it is likely to expose some of the discipline's internal tensions and divisions.

One might begin by asking what 'science' actually is – and even this turns out to be a less than straightforward question to answer. In the English language, the word is commonly assumed to refer to the physical sciences (leading a former British minister of education to try and deny a national research funding agency the title 'The Social *Sciences* Research Council'). But in France *la science* and in Germany *wissenschaft* have much broader implications. Even if we consider one linguistic context, and examine the history of something we have called 'science' in English, it appears to have changed markedly in its nature over time. Woolgar (1988) traces three main phases, although these could easily be subdivided and qualified. A period of 'amateur' science, in which doctors and clergymen played significant roles as natural historians, characterizes the seventeenth and eighteenth centuries, after which science increasingly became 'academic' in nature as it moved into the developing universities. Then, in the mid-twentieth century it became 'professional' as government, industry and the military dominated scientific agendas. The purpose of science has changed, its sociology has altered, and the extraction of a single 'tradition' is accordingly not only difficult, but even quite inappropriate. An additional implication is that any important scientific discoveries, especially those that have unfolded over a significant period of time (such as the Gas Laws, discussed below), are likely to have been affected by this changing context in which science is undertaken and practised. Finally, there is a historiography to scientific traditions which we should recognize; in geomorphology, we label whole approaches as 'Davisian' or 'Gilbertian', using historical figures as convenient symbolic representations of whole 'traditions', parts of which they might not recognize themselves were they to reappear (e.g. Sack, 1992). Such symbolic labelling also gives the impression of step changes in a history which may often be rather more gradual, and with multiple origins (consider Wallace, Darwin and evolutionary theory). Thus, for all these reasons, we should begin with the sceptical view that any methodological 'tradition' to which we might refer is itself a selection, not a given; a device, not a reality.

SCIENCE AND PROBLEM-SOLVING

Turning more formally to the question of definition, science may be defined as 'systematic and formulated knowledge; the pursuit of this; the principles regulating such pursuit; any branch of such knowledge' (*Oxford English Dictionary*). On this basis, whether there is a physical science tradition hardly matters; geography is itself a science (at least, it is if we consider our discipline to pursue and codify knowledge in such a systematic manner). Perhaps the interesting point about the formal definition of science is that it almost immediately introduces *method* – science as an activity or entity seems to be defined less by *what* it is, than by *how* it is done.

However, before dealing with the details of scientific method, there is one additional, critical characteristic to review, and this is science's sense of 'problem' or 'question'. This is itself an aspect of the scientific method which we activate early in any investigation, although it is not often treated explicitly as part of that 'method'. But as Bird (1989: 2) notes, science is a problem-solving activity, and 'the scientific method starts with some kind of problem: we might go as far as to say that problem orientation is the *raison d'être* of scientific enquiry. Problem recognition is a very important part of scientific endeavour, and often a very difficult intellectual exercise.' Anyone who has tried to develop a research project will know the significance of the final phrase in this quotation. It is often easy to find a general area in which research would be interesting, even valuable, but less straightforward to turn that interest into a researchable set of questions. This is the essence of problem formulation, and a scientific tradition which has made science such a successful enterprise.

An example will illustrate this difficulty of turning a general issue into a researchable problem: the decline of floodplain woodland in many river valleys in Europe and, indeed, globally. Channelization and river-flow management have separated rivers from their floodplains, leading to reductions of flood frequency, of rates of channel migration, and of recharge of floodplain sedimentary aquifers. In turn, this has caused reduced opportunities for the regeneration of woody riparian species, a loss of habitat, and a reduction in biodiversity. One initial question we may need to ask is: 'Does society value the natural functions of floodplains sufficiently to wish to reverse these trends?' In one sense this question has already been answered because society has established institutions capable of monitoring the decline, and the existence of knowledge of the trend is a partial answer. However, social research may be needed to establish the sense in which this is regarded as a general 'problem' that needs to be placed high on the agenda of the environmental sciences. At this point, the problem must be converted into more specific researchable questions, the answers to which help to explain the changed relationship between rivers and floodplains and the observed loss of ecosystem health. Examples might be: 'Does the timing of high flows under a controlled flow regime no longer match the timing of seed production?'; 'Does reduced flooding and recharge prevent the establishment and growth of seedlings?'; or 'Is seedling growth affected by the rate of draw-down of the floodplain water table?' Then, we can develop various strategies to attempt to answer the questions we have posed; we may, for example, deploy the range of scientific methods, which include observation, measurement, experimentation, and theorization. During the discussion of scientific method below, we will return to this example to assess how these approaches can be helpful, using the last of these three questions as a case study.

One way in which geography frequently parts with a modern physical science tradition is in the manner in which it poses questions, given the nature of its general subject-matter. Geography is concerned with phenomena that unfold in an unconstrained social and environmental space, across a wide range of scales. In the physical sciences, the subject-matter can often be controlled in the laboratory, and accordingly there is a tradition of posing questions which are answerable, given the methods known to be both available in such a context, and capable of providing answers. Part of the very success of science therefore lies in its characteristic tradition of asking 'well-posed' (as opposed to 'ill-posed') questions. An example of this can be drawn from an application of probability theory. We may ask: 'How many people are required in order to have a better than evens chance that two or more of them have the same birthday?' This is an 'ill-posed question', which we cannot answer until we are provided with a number of assumptions and conditions. If we assume that each individual's birthday is equally likely to be on any of the 365 days of the year, and that one person's birthday has no effect on anybody else's, then the answer is that just 23 people suffice. This is rather a small number, which seems surprising. Indeed, a TV personality once denied the answer on the grounds that he had asked an audience of 100 people whether anyone had the same birthday as him, and the answer was 'no'. In fact, however, he was then asking a different question. The answer to his new question: 'How many people are required in order that there is a better than evens chance that someone has my birthday?', is actually 253 (Grimmett, G.R., personal communication).

It is also often true of the physical sciences that research 'programmes' exist in which series of such problems or questions are posed in a planned, systematic, and even evolutionary manner. It might be argued that geography again parts from the physical sciences in this respect, because it seems to have been much less able to develop this systematic approach to the identification and planning of specific, tractable research questions. Concern about an apparently dilettante approach to the identification of problems underlies Stoddart's (1987) plea for geography to rediscover larger questions. However, perhaps this is partly a characteristic of geography as an 'open-system' science, which operates in the uncontrollable natural and social world where programmatic and answerable questions are more difficult to define; and partly it may reflect another geographical tradition, that of exploration – which may need planning, but must also respect serendipity. Notwithstanding such reasons, practitioners of the discipline need to confront the possibility that this may also in part be a failing, in so far as it may reflect the lack of a coherent sense of systematic problem-formulation and problem-solving.

ASPECTS OF SCIENTIFIC PRACTICE

From the opening sections of this chapter, it should come as no surprise that the scientific method is in some senses itself unstable, since the nature of science as an activity is historically contingent, and indeed also geographically contingent (Livingstone, 2005). There are several reasons for this instability, both external and internal to the specific practice of science. The former relate to the social context of scientific activity; scientific agendas are driven by changing political, industrial and business agendas, with science, technology and society forming a single domain (Latour, 1987), and with changes emerging from the feedbacks among them. The internal sources of instability arise because the appropriate methodology to employ may need to change as partial answers are found to the questions raised by the scientific problem in hand. Thus, what we have come to know about a problem may change our view of what more is knowable, and of the means of acquiring this additional knowledge (Richards, 1996). This implies an interdependence between ontology, which is concerned with the nature of phenomena and existence, and epistemology, which is concerned with the means of gaining knowledge about them. There is no linear connection between these, but rather, a spiral of change over time. Since the acquisition of knowledge may alter our ontological view of the world, the method we employ subsequently may have to change. That this occurs implies that the scientific method cannot be considered as a normative set of rules, but only as a more or less temporary model of how things may be done. Thus, 'the' scientific method is no more than the best, or simply the most appropriate, available description of the way in which

---

Box 2.1   Generalized relationships between philosophies and aspects of scientific method

Positivism	*Representing*: observation, measurement (Auguste Comte, 1798–1857)
Logical empiricism	*Intervening*: experimentation; laws of constant conjunction (David Hume, 1711–76)
Critical rationalism	*Theorizing*: hypothesize, experiment, test, falsify, refine (Karl Popper, 1902–94)
Realism	Uncovering hidden structures and mechanisms (Roy Bhaskar, 1944– ; Andrew Sayer)
(Post-)modern science	Perception reflects the perspective of the observer; the uncertainty principle (Albert Einstein, 1879–1955; Werner Heisenberg, 1901–76); at the limits of observation and intervention; chaotic behaviour of non-linear dynamical systems; sensitivity to initial conditions
Rational criticism?	

---

reliable scientific knowledge can be acquired at a particular time, or at a particular stage in the investigation of a problem. As a model, therefore, it is subject to criticism, test, and reformulation just like any other model.

Given this view of the scientific method, debates about positivism, critical rationalism and realism may seem less important than consideration of the methodological contributions that these have made to the range of practices available to scientists, and therefore also to geographers. These practices are the abiding legacies of science, if not its tradition, and it is appropriate to consider their role in geographical enquiry. Box 2.1 suggests, very broadly, the kinds of contributions to scientific practice associated with particular ontological and epistemological traditions. At the beginning of any scientific investigation, it is likely that we will draw on one of the central contributions of positivism to the range of scientific practices, its focus on *observables*. This contribution reflected the ontological project of positivism, which was to separate science from metaphysics, by arguing that the former should be restricted to the study of those phenomena capable of being 'represented' (Hacking, 1983) through either observation or measurement, while it was the latter that was concerned with matters of faith. This, however, does not mean that positivism is exclusively associated with quantification, an erroneous assumption that often appears in geographical discussions. When this is perpetrated in error it is, perhaps, a reflection of the degree to which the practice of 'measurement', necessarily constrained by precise operational definitions, has almost effaced the more subtle and open-ended process of 'observation'. But it may also be perpetrated as a form of disciplinary politics, when positivism in its considerable diversity is rendered as a straw man to support an anti-quantitative stance which both oversimplifies 'positivism' by reducing it to

Figure 2.1 An example of the association of recruitment of seedlings (of *Populus nigra*) with patches of fine sediment in a braided reach of the River Drôme, southeast France, which has a dominantly gravel-sand sediment load and substrate

Photograph by Francine Hughes

numerical data, and ignores the potential value, at some stage of many investigations, of such data and the tools for their analysis.

Representing: observation and measurement

Thus, if we return to our question about the growth of seedlings, we may begin by using our (field) observations of riparian and floodplain environments. We may have noticed that in an Alpine river in which we are interested, the hydropower dams release water in ways that result today in the timing, magnitude and duration of the spring high-water period all differing from the historic situation when flows were natural; and we may also notice that riparian trees that seeded to coincide with the *natural* annual high-water period, now do so too soon. We may also observe that where there are patches of damp, fine sediment on bar surfaces, germination and initial establishment of seedlings tend to occur preferentially in these locations (Figure 2.1), indicating the relationship between the physical nature of the floodplain substrate, hydrology, and regeneration of the population of certain species. Then (and only then, after we have observed and considered), we might undertake some measurements, for example by installing instruments to monitor groundwater levels in the floodplain, to confirm that its moisture status closely tracks the now-controlled fluctuations of river flow (Figure 2.2). These observations and measurements do not always answer our question. Indeed, they often only suggest new ones. However, they provide us with information on patterns that begin to trigger ideas about underlying structures,

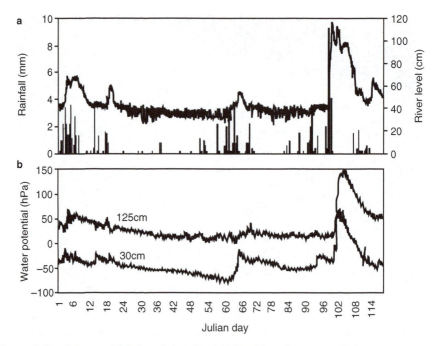

Figure 2.2   Data on rainfall and river level in the River Ouse near St Ives, Cambridgeshire, UK (a), and on water potential at two depths beneath the surface (30 and 125 cm) in the floodplain adjacent to the water-level recorder (b). At 30 cm, the potential is generally negative except during and after rainfall, especially from Julian day 100. At 125 cm, the potential is positive, indicating that the water table is generally above this tensiometer; at around Julian day 104, the potential is equivalent to a water depth of greater than 125 cm, reflecting the occurrence of flooding

mechanisms and processes about which we can theorize before undertaking new observations and measurements. Sometimes, of course, it is difficult to extract a pattern from what we see in the world, because there are many pieces of conflicting evidence in a world that is unconstrained, and in which some effects are cancelled by others. It is therefore difficult to theorize; but we can help ourselves to solve this problem by not only 'representing', but also by 'intervening', with experimental manipulation, as discussed in the next section.

Several further issues arise when considering measurement; those of resolution, instrumentation, precision and accuracy, and dimensionality all provide interrelated grounds for comparison of the physical sciences and geography. The reductionist tendency in science has driven it to examine phenomena of ever-smaller scale, demanding increasing resolution of techniques for observation and measurement. Linked with this is the primacy of the visual sense in observation, and the development of instruments to assist in 'seeing' these phenomena. Hacking (1983) provides an excellent account of the implications of this process of 'representing' the world for our appreciation of its nature. We cannot 'see' the human cell, but we can have a theoretical model of its form and structure; it begins as a theoretical entity, therefore, not an immediately observable one.

What can convince us that it is 'real' is that correspondence occurs between the form and structure predicted by the biological theory of the cell, and the form and the structure in the images of the cell revealed by either the optical or physical theory embodied in the design of various forms of microscope. Similar issues arise in the study of velocity variations in rivers. Here, the traditional propeller- or cup-based electro-mechanical current meter has been supplanted by electromagnetic and acoustic Doppler current meters, as the subject of concern has shifted from time-averaged velocities to the structure of turbulence, requiring measurement of velocity variations with frequencies of 10–20 Hz. A problem, however, is that there are no theoretical descriptions of the structure of turbulence in particular circumstances – it has no generally definable character – against which to evaluate the evidence revealed by these instruments. Furthermore, two different instruments measuring at the same location can generate data with different temporal structures (Richards et al., 1997), resulting in a lack of confidence about the reality of those structures.

An area in which the physical sciences have a tradition which geography lacks is in their close alliance with the technology required to further scientific enquiry. This has been manifest in different ways at different times. For example, those working in the Cavendish Laboratory in Cambridge in the 1930s and early 1940s were adept at cobbling together apparatus for their experiments on subatomic particles and radio-active elements, using bits and pieces left over from other experiments, and deploying practical skills of technical construction that were allied to their theoretical understanding of the scientific problems at hand. Today, the technology required to build scanning electron or scanning tunnelling microscopes is beyond a single research group or laboratory, and such equipment is provided by commercial scientific instrument makers, albeit with detailed design input from the scientists intending to use the apparatus. This use of 'off-the-shelf' instrumentation is prevalent in measurement practice in geographical research, and characterizes the study of turbulence discussed above. Here, the manufacturers' decisions are less commonly influenced by the scientific users, and the result can be data collected in specific research circumstances which are contaminated with design decisions made for general construction. The most obvious of these are the physical properties of the sensor head (its shape and size), and the properties of the signal conditioning and filtering built into the electronic circuits. The best, and most 'scientific', research into turbulence has been that which seeks a detailed understanding of the effects of these factors on the data collected, and seeks redesign when necessary (Roy et al., 1997).

In science, measurements are normally quoted with due consideration for their precision and accuracy (Figure 2.3); until recently this has been more honoured in the breach in the geographical literature. This is partly because there is rarely independent evidence of the 'true' value of a variable, and therefore difficulty in determining precision and accuracy. Consider the measurement of rainfall using (point) raingauge data and (spatially-distributed) rainfall radar; calibrating the latter with measurements from the former requires an extensive sampling exercise before an approximation to a 'true' value can be defined. Nevertheless, the use of photogrammetric methods to derive digital terrain models (DTMs) has led to increasing emphasis on precision and accuracy. The DTM

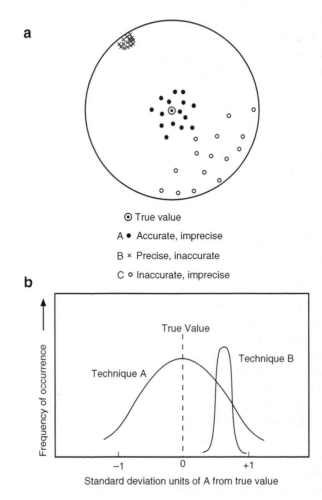

Figure 2.3   The concepts of accuracy and precision are illustrated in (a) with the example of an archery target. The bull's-eye is the true value; attempts labelled A are accurate (they cluster around the true value) but imprecise (the variability is high). In (b) the concepts are illustrated using hypothetical frequency distributions; here, measurement method A gives an accurate but imprecise distribution of values, while method B gives a distribution which is precise but inaccurate. After Davidson (1978)

is also interesting because it provides three-dimensional information on the morphology of the earth's surface. Early quantitative geomorphology used one-dimensional indexes to represent the topographic form, such as the relative relief and drainage density. These provided a much-needed alternative to imprecise and generalized qualitative descriptions of, for example, 'rolling landscapes'. However, the collapse of a three-dimensional entity (the landscape) into one-dimensional measurements loses a great deal of information, and these variables proved equally imprecise, and not very accurate. The example of measuring flow velocity in rivers also draws attention to the change in dimensionality of much measurement in geography. Where the electro-mechanical current meter measured a time-averaged velocity in one direction (the main flow direction), the modern instruments that measure at high frequencies also measure in two or three orthogonal directions. This means that attention can be given to the vertical and cross-stream components of velocity, and to questions of secondary circulation. It also means that the accuracy and precision of the measurements is not contaminated by folding into a single velocity component the velocities in the other two orthogonal directions. This is therefore an example of the rigour of geographical method adapting to new technology that permits a closer approximation to the traditional norms of the physical science.

Intervening: experimentation

By 'intervening' (Hacking, 1983) we imply 'experimenting' – controlling, simplifying and manipulating the world so that we can see the patterns in a part of it more clearly. A typical illustration of a physical science experiment is that which reveals the pattern that was eventually codified as the Gas Law, which states:

$$pV/T = nR$$

where p is pressure, V is volume, T is temperature, n is the number of moles (the number of atoms or molecules) and R is the gas constant. If a known quantity of a given gas is heated in a container of fixed volume, the pressure increases, at a rate peculiar to that gas represented by its gas constant. In such an experiment, measurements of the temperatures and pressures would define a Humean 'law of constant conjunction', which is an empirical generalization deriving from the experimental measurements. This is a classic example of empirical chemistry, although its history spans two centuries, from Torricelli's invention of the mercury barometer in 1643, through the experiments of Robert Boyle and Jacques Charles, to the definition of Amadeo Avogadro's Number (which was based on a principle not accepted until 1860). The law is a very powerful one, and although it was derived under controlled laboratory conditions, it allows us to account for changes in an uncontrolled world; without it, it would be difficult to explain adiabatic air movements in the atmosphere, and meteorology would be impoverished.

If we return to the example of floodplain ecology, where we have observed and measured some possible effects of altered hydrological regime on the regeneration of woody riparian plant species, we could conceive of an 'intervention' to assess our intuition here more rigorously. This might involve a field experiment; we could enclose some patches of floodplain (to exclude animal interference), and seed them under different but carefully controlled conditions. Because every site will experience slight variations in character – for example, in soils and sediments, or in competitive species already existing in the seedbank in the soil – an experiment of this kind requires replication, to employ statistical control of these possible extraneous influences. Such replication requires very careful experimental design, and agricultural science and ecology have done much to create the framework for this design following the pioneering work of Ronald Fisher (1890–1962) (Underwood, 1997). Although there are many variants on this theme, a key point is that experimental units (plots) must be so located

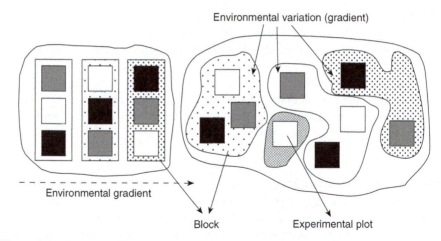

Figure 2.4   Examples of randomized plot experiments. In the left-hand case, there is a relatively smooth environmental gradient and, at intervals along the gradient, blocks are defined within which random plots are chosen. In the right-hand case, the three 'blocks' are actually patches with a typical character, within which plots are randomly located

along the gradient of an hypothesized control variable as to avoid bias in the event of co-variance of a further unknown independent variable with that control. This requires randomization of replicate plots within 'blocks' defined at different points on the gradient of the control variable (Figure 2.4), and subsequent analysis of variance (ANOVA), a very powerful statistical tool. An alternative approach, however, is to seek a more physical than statistical form of experimental control, by extracting a piece of the world into a laboratory (in this case, a greenhouse), and using apparatus like that shown in Figure 2.5 to grow seedlings under conditions in which soils are as standard as possible, and the rates of water-table lowering are controlled. This enables the growth of seedlings to be monitored and related to the varying behaviour of the water-table in different soil types, and allows us to reinforce our conclusions about the effects of changes in hydrology and soil moisture on the regeneration performance of different species. Figure 2.6 illustrates some results of an experiment of this kind, with *Alnus incana* seedlings showing differential growth between two different substrates (a silty fine sand and a gravelly coarse sand), and with different rates of water-table lowering.

Returning to the scepticism about the stability of traditions, it is worth noting that the concept of the 'experiment' is itself highly variable (as the very different statistical and physical ecological experiments described above indicate). There are many kinds and purposes of experiment, as is clearly demonstrated in Box 2.2. Harré (1981) uses Boyle's Gas Law experiments to make a 'measure of the Force of the Spring of the Air compressed and dilated' to exemplify the practice of 'finding the form of a law inductively', and identifies this as a kind of experiment which constitutes a formal aspect of scientific method. However, some experiments are more for (im)proving a technique than discovering a property of the world, and even more striking, others are for 'finding the

Figure 2.5   Rhizopods – soil-filled tubes connected at the base to central water reservoirs which allow control of the water table. Seedlings growing in the tubes show the effect of different water-table lowering treatments

Photograph by Francine Hughes

hidden mechanism of a known effect'. This latter kind of experiment, since it appears not to relate to observable entitities, cannot really be a positivist method; thus, it is hardly acceptable to regard a desire to conduct experiments as a sign of a positivist ontology. Experiments can include approaches as diverse as questionnaires and interviews, as long as these are rigorously designed to investigate a particular problem, and the issues of sampling and bias are given appropriate consideration (see Pawson, 1989). The tradition of experimental activity is therefore as diverse as science itself. However, experiments will not always *explain* why the regularities in behaviour that they reveal arise, which is why scientific enquiry into a particular problem has to adapt both its ontology and its epistemology as it uncovers information, and we gain a deeper understanding of the world as a result. This is where theory (Box 2.1) enters the range of practices which constitute scientific activity, and where the critical rationalism of Popper and the realism of Harré and Bhaskar part company from positivism; science is clearly not simply about observables.

An experimental approach common in geography is one which combines extensive and intensive research (Sayer, 2000), often at different stages in an investigation but not necessarily in a sequential manner – there is often a switching back and forth between these modes. Extensive research is empirical and concrete, and often requires large samples to be gathered so as to demonstrate the variability of the phenomena, and the degree to which specific cases studied in depth are representative. Intensive research, by contrast, is abstract and theoretical, involves small samples, and seeks to uncover the underlying causal mechanisms that generate the patterns revealed by extensive investigation. One simple

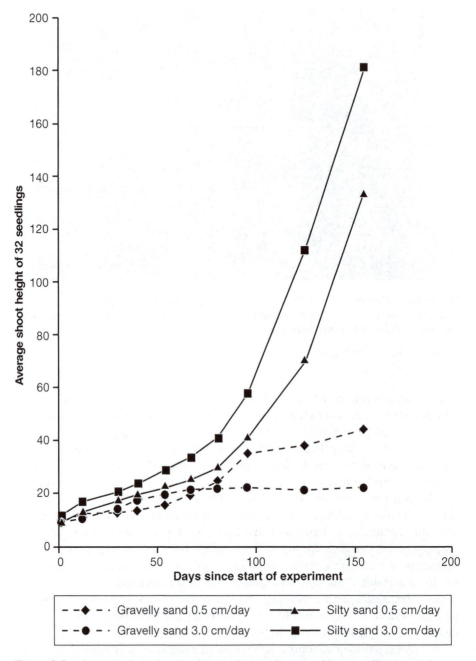

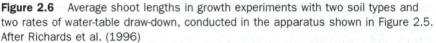

Figure 2.6   Average shoot lengths in growth experiments with two soil types and two rates of water-table draw-down, conducted in the apparatus shown in Figure 2.5. After Richards et al. (1996)

illustration of these two elements is the randomly-sampled household question-naire and the semi-structured interview with key respondents. Another is the systematic monitoring of solute concentrations in the discharge of water from a small catchment, combined with analysis of the mineral stability fields for

---

Box 2.2   The uses of experiment as identified by Harré (1981)

A. As formal aspects of method

1   To explore the characteristics of a naturally-occurring process
2   To decide between rival hypotheses
3   To find the form of a law inductively
4   As models to simulate an otherwise unresearchable process
5   To exploit an accidental occurrence
6   To provide null or negative results

B. In the development of the content of a theory

7   Through finding the hidden mechanism of a known effect
8   By providing existence proofs
9   Through the decomposition of an apparently simple phenomenon
10   Through demonstration of underlying unity within apparent variety

C. In the development of technique

11   By developing accuracy and care in manipulation
12   Demonstrating the power and versatility of apparatus

---

interstitial water in the hillslope soils in order to explain variation in the slope drainage water chemistry. However, this is similar to the combination of empirical evidence and theoretical analysis that characterizes a classic physical science problem such as the behaviour of gases subjected to variable pressure and temperature (A1, B10 and B7 in Box 2.2); geography and the physical science tradition are thus not far apart.

The intensive phase of geographical enquiry often leads to case-study research. Such small sample investigation can nevertheless lead to generalization because the aim is to theorize about and understand the underlying mechanisms, and observe how they cause observable events in one location. A case-study experiment requires recontextualization – particular mechanisms interact with the specific contextual character of the location under study, to cause observable events which give us clues about the way nature works. An example is the observed event of a change in channel pattern on the South Platte River after 1850, from a braided to a meandering form, following reservoir impoundment in its headwaters. This control of flow reduced the stream power at the mean annual flood. The threshold between meandering and braided patterns occurs at a stream power of about 50 W m^{-2}, but after impoundment, the stream power in the South Platte declined to about 2.5 W m^{-2} at the mean annual flood. Thus, the post-1850 conditions were conducive to a change from dynamic braided patterns to more stable meandering. The contextualization of a general statement about channel pattern in this specific case is similar to the covering law model of the physical sciences; events are accounted for by the combination of a general statement (the covering law) and a specific statement (the initial conditions). Again, therefore, there are echoes of a physical science tradition in the methods of geography. In fact, a great deal of scientific activity requires defining the circumstances in which general statements (laws) are applicable, and can be defined. This is even true of the experimental apparatus used in investigating the behaviour of gases. Place-based case-study research is to geography what the laboratory is to science. The only question is whether geographers can describe the open-system character of their field areas with both sufficient and necessary detail to account for the events generated by particular mechanisms or structures in this context, and to enable other geographers to replicate the findings of their case study.

Theory: hypothesis generation, testing and falsification

One way in which theorizing enters the scientific process is through the control we impose on observation, measurement and experiment. Popper's critical rationalism (Box 2.1) denies inductivism, and concludes that observation and its counterparts are theory-laden. We do not gather observations randomly, and *then* structure them according to some *post hoc* logic; the logic is mobilized at the outset because our theoretical preconceptions must determine what we choose to observe, what and how we measure, and the design of the experiments by which we collect data from which we may then generalize. Furthermore, experiments are created with a (theorized) purpose, one of which may often be to test a hypothesis, which will itself have been constructed as an outcome of theoretical consideration. However, such an experiment cannot logically be intended to *prove* or verify the hypothesis, because there is always a possibility that further observations will be made which are contrary to the theory or model thus far supported by the empirical evidence. But once a statement has been disproved, it cannot be 'un-disproved', and so experiments are designed to *falsify* a hypothesis. In the randomized plot field experiments noted above (Figure 2.4), the results of which are investigated using analysis of variance, this is achieved by stating a null hypothesis, such as 'there is no effect of moisture availability on the growth rate of seedlings of *Alnus incana*', and then seeking to disprove this in order to accept the logical alternative that an effect does exist.

If a null hypothesis is accepted, this suggests that the model that generated its logical alternative is wrong, which forces renewed theorization. If it is rejected, this is not the end either, because new hypotheses and experiments must be devised to provide even more stringent tests of the model. The limits of the Gas Laws were identified, for example, by experiments under temperature and pressure conditions at which the gases liquefied. As Underwood (1997: 19) notes, the cycle illustrated in Figure 2.7 is never ending:

> there is no way out of the procedure once you have started it, until you die or change research fields. ... Re-examination, novel testing, and more rigorous experimental analysis are part of the framework of investigation. No single study – whether it rejects or retains some null hypothesis – is sufficient to declare a problem solved.

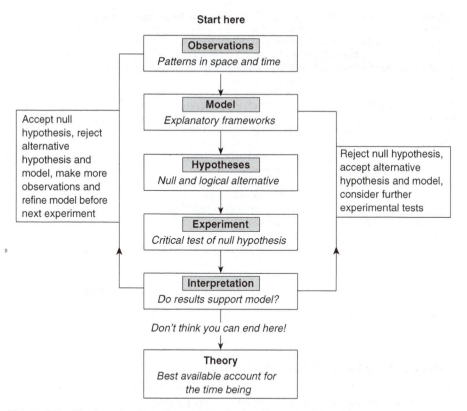

Start here

Observations
Patterns in space and time

Model
Explanatory frameworks

Accept null hypothesis, reject alternative hypothesis and model, make more observations and refine model before next experiment

Hypotheses
Null and logical alternative

Reject null hypothesis, accept alternative hypothesis and model, consider further experimental tests

Experiment
Critical test of null hypothesis

Interpretation
Do results support model?

Don't think you can end here!

Theory
Best available account for the time being

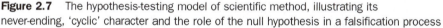

Figure 2.7   The hypothesis-testing model of scientific method, illustrating its never-ending, 'cyclic' character and the role of the null hypothesis in a falsification process

The reference to rigour here is a reminder that whole books (including that by Underwood) have been written on the subtleties required of experimental design to ensure logical hypothesis testing, and the pitfalls are numerous; unstructured predictions, lack of replication, pseudo-replication, Type I and Type II errors (Table 2.1), lack of experimental power, conflation of interaction effects with error variance, autocorrelation in temporal data, are just a selection. Consider Figure 2.6 again; these experimental data appear reasonably convincing, but later differences in growth are determined in part by those that arise early in the time series, possibly for abnormal reasons. There is also an interaction between the effects of substrate and rate of water-table lowering. What all of these problems imply is that experimental tests must be based on a clear understanding of under-lying theory, and the logical implications that allow the generation of testable hypotheses. Experimentation is never pure empiricism; it relies on theoretical depth and careful logic. But the converse is also true; theory requires evaluation against empirical evidence, to demonstrate its value as a mirror held up to the world. The lessons learnt from books on experimental design in the field sciences are that such empirical evaluation is an extremely arduous process, and in some areas of geography, the tradition of applying the necessary rigour has been abandoned.

Table 2.1   Type I and II errors, and illustrations of their implications

A Type I error in a test on pollution may lead to continued sampling and control, but this means
there is no harm to the environment. A Type II error in a drug-testing process could result in a
harmful drug being marketed. Type II errors are often more critical, but more attention is paid to
Type I errors.

Outcome of statistical test is to:	Null hypothesis (H_0) is True	(unknown to us) False
Reject H_0	Type I error: rejection of true null hypothesis (Risk: falsely accept that an industrial discharge does not pollute)	Correct conclusion: false H_0 is rejected
Accept H_0	Correct conclusion: true H_0 is accepted	Type II error: accept a false null hypothesis (Risk: falsely accept that a drug has no side-effects, and market it)

Theory: mechanisms behind regularities

A second role for theory arises because, having shown experimentally that there is
a simple relationship between temperature and pressure in a fixed volume of gas
(at least, for a defined range of these controls), the question then arises as to 'why?',
or 'how?', nature engineers this trend. This is a request for an explanatory account,
not just a description. To answer these kinds of question, a theoretical approach is
required which addresses unobservable mechanisms (see Box 2.1 for 'realism'). In
the case of the Gas Laws, this theoretical explanation of the regularity is provided
by the kinetic theory of gases. The pressure exerted by a gas is caused by collision
of its molecules on the walls of its container; heat the gas, and the speed of the molecules increases, the frequency of collision increases, and the pressure rises; this is
the essence of Daniel Bernoulli's suggestion in 1743 that was eventually quantified
through the work of James Clerk Maxwell and Ludwig Boltzmann between 1860
and 1880. Typing 'gas law demonstration' into an internet search engine will find
several 'virtual laboratory' experiments that show this mechanism quite well, in
addition to indicating the empirical regularity it generates. The regularity can be
distorted very easily, however, by varying both the volume and temperature simultaneously, showing that the revelation of the regularity, and the theorizing that it
triggers, both depend on the skill deployed in designing the experiment.

How, then, does this relate to the question of floodplain ecology. Considering
Figure 2.6, a little thought reveals that seedling growth rates are probably not
really *determined* by the rate of water-table lowering; this is a surrogate variable
for something else. Especially when the water-table lowers very rapidly, a
seedling's growth depends on its capacity to extract water from the soil pore spaces
in the unsaturated zone above the water-table. Clearly this is likely to be more
difficult when the water-table is receding rapidly, so there is at least a *correlation*
between the depth of the water-table and the moisture status of the unsaturated

zone above it, if not a direct causal connection. What determines the capacity of the plant to extract moisture from the soil is the interrelationship of the rate of transpiration, the osmotic pressure at the roots, and the soil moisture tension. When there is little moisture in the soil pores, the surface tension of the films of water held against the soil particles becomes very high, and even if the transpiration pull is also large, the diffusion of moisture into the roots is inhibited as cell walls fail and the plant wilts. Thus, the critical factors determining growth rates are the limiting osmotic pressure exerted through the roots, relative to the soil moisture tension. In an unsaturated soil, soil moisture tension prevents plants from accessing water when the moisture content is below the wilting point, and this depends on the soil moisture characteristic represented in the tension–moisture relationship for the soil (Figure 2.8a), and on the height above the water-table. The 'real' control variable is the soil moisture tension, for which the water-table depth is a surrogate. The tension can of course be measured, using a tensiometer (Figure 2.8b), but it can also be theoretically predicted using numerical models of the saturated–unsaturated soil moisture dynamics above and below the water-table. In their simplest forms, such models are one-dimensional finite-difference approximations of a column of soil (Figure 2.9), which numerically solve the Darcian equation for the soil water flow velocity (v) at depths h:

$$v = -K \, (\partial \varphi / \partial h)$$

where K is the hydraulic conductivity, which is related to the soil water tension and therefore the moisture content (as in Figure 2.8a), and φ is the soil water potential (positive beneath the water-table where the water is under pressure, negative above the water-table, where there is tension). The continuity equation is also required to account for the changing flux and storage of water in each cell of the model. Such theoretically-based models are now very sophisticated, handling three-dimensional flow and multiple soil types. However, their use regularly reveals shortcomings in knowledge, such as that relating to the effect of large macro-pores in causing drainage at flow velocities faster than Darcy's law predicts, or that relating to the effects of plant roots in extracting water from storage at different depths. Thus theoretically-based models also have an important role to play in probing the limits of existing understanding.

Theory: demarcation criteria and the warranting of claims to truth

An important issue in scientific enquiry is that of determining which theory is closest to being a 'true' account of the aspect of the world being investigated (note: not 'the true account'). This is why empirical investigation, albeit closely linked with theory, forms such an important part of science. In some areas of geography, there is now almost a view that theory can be an end in itself, and a related view that all accounts are equally valid. However, this is fraught with danger; there must be some criteria for determining the trustworthiness of an account, since as Phillips (1992: 114) notes, 'a swindler's story is coherent and convincing'. More tellingly, Phillips adds that to apply to a theoretical argument about the world the criteria of value that we might apply to literature, poetry, art or music is to make a category mistake; these may be things of beauty, they may

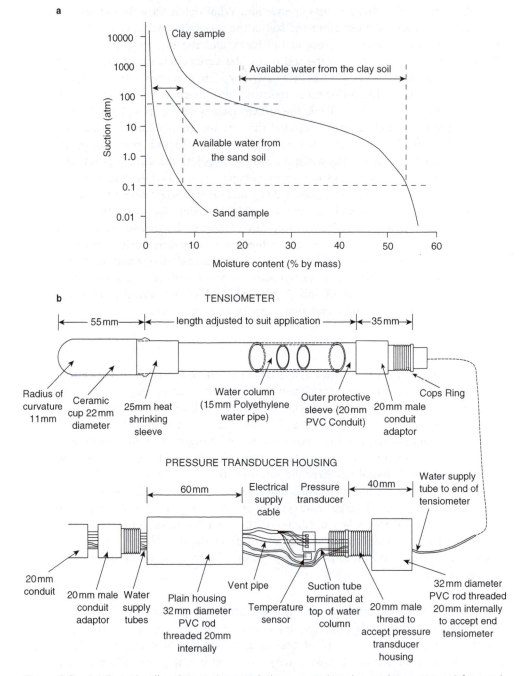

Figure 2.8 (a) Typical soil moisture characteristics curves (tension–moisture curves) for sand and clay soils, illustrating the different amounts of water available to plants, defined by the suction range across which plants can extract water from the soil-pore spaces; (b) a typical design of a tensiometer based on a pressure transducer, which measures suction when the water in the instrument is extracted through the porous tip to equilibrate with the soil moisture potential. This instrument is designed to be topped-up with water from the surface and to be connected to a continuously monitoring data logger

Design by Adrian Hayes; diagram by Owen Tucker

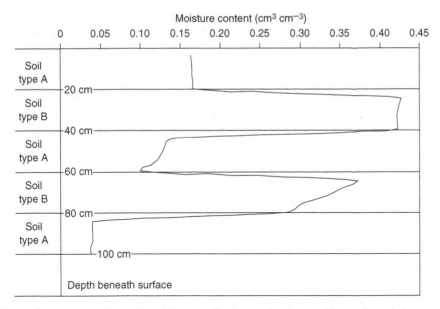

Figure 2.9   A modelled soil-moisture profile in a soil column with sandy and clayey horizons, with the finite-difference computational structure (25 cells each 4 cm deep) shown at the left. After Richards et al. (1996)

be evocative, and they may illuminate the human condition, but they cannot be said to be 'true'. But equally, to apply the standards of literary criticism to geographical theorizing, or to deny the legitimacy of any claim to truthfulness, is also to deny the application of theories in intervention or policy, and to undermine any political project the research may have. This does not have to imply the risk of imposing rigid forms of policy, bolstered by claims to truthfulness made by science, for as we have seen above, in science there is always room for refinement and for the recontextualization of existing knowledge.

   To reverse the argument, however, we may also note that there is in some geographical writing a narrow interpretation of the criteria employed by science for warranting claims to truth; in the simplified model of the scientific method commonly presented in geographical literature, the criterion for selection of a preferred theory is that of predictive success. This relates to a misreading of Figure 2.7, in which hypotheses are seen to be tested by experiments, leading to a comparison of the prediction from the hypothesis with the data generated by the experiment. The misreading is in the assumption that measurement of 'success' is provided by a strong correlation between prediction and observation. As explained above, such an outcome of experimentation merely leads to another experiment, until a falsification can bring about modification of the underlying theory; it is therefore empirical failure that brings successful theoretical development. This Popperian revision of the meaning of experimental success implies that warranting the truth claims of a given theory is a much more subtle and complex process in the sciences, just as it is in other branches of knowledge.

Phillips (1992) provides some valuable insights into the procedures employed to test the validity, or the proximity to truth, of hypotheses, models or theories, in a discussion of the warranting of knowledge claims in qualitative social science research. This is part of a sustained argument against 'fabled threats' posed to naturalistic social science (that is, social science in which the ontology and epistemology is similar to that in the physical or natural sciences; see also Chapter 3, this volume). There are several practices that aid us in demarcating one theory as better than another, in addition to the conventional one of predictive success (which, as argued above, gives only an illusion of validity). The first is a judgement about the *explanatory power* of a theory. This introduces the asymmetry between explanation and prediction; the capacity to predict does not imply an ability to explain, as any statistical forecasting model in economics or hydrology demonstrates. A predictive test of a theory that purports to provide explanation is inherently inconsistent, therefore, and something is required in addition. Of course, measuring the quality of an explanation is difficult, but two further criteria assist us in this; these are *structural corroboration* and *referential adequacy*. The first requires that the various parts of a theory give each other mutual support, and the second that the theory suggests new ideas about other, related phenomena. A glaciological example of this described by Richards (1996) involved experiments on surface melting, water balance, subglacial hydrology hydrochemistry and basal water pressure, which all combined to provide support for a theory of the seasonal evolution of the glacier and its drainage system. The research itself evolved as new theoretical connections were identified and suggested additional experiments. However, these explanatory qualities alone cannot provide a reliable warrant for theory and they must be supplemented by more familiar strategies: *multiplicative replication*, and the *search for negative evidence*. The first involves repeated testing which generates comparable results, often with others undertaking the additional experiments. This is consistent with the practice in the physical sciences when new and unexpected results emerge and the scientific community seeks to replicate them; pathological events are discovered by this procedure, such as the spurious claims to have achieved atomic fusion at room temperature (Close, 1992). The search for negative evidence is, of course, precisely the Popperian ideal. Thus, although this procedure for warranting the knowledge claims of social science seems to lack rigour, and to be messy and imprecise, this is balanced by a spurious emphasis on predictive success in oversimplified accounts of demarcation in the physical sciences. The multiple methods of assessment required by all of the sciences in their struggles to explain aspects of a world that hides its complex secrets from us have many elements in common, and the 'traditions' (or legacies) of one are also those of the many.

CONCLUSION

Thus, we have moved in this discussion, and in the associated interdisciplinary example involving hydrology and ecology, through a chain of practices from observation, through measurement, hypothesis generation and experimentation, to theoretical and physical analysis, just as physicists have done over the years in elucidating the behaviour of gases. Science requires this wide range of practices, and geographical enquiry is equally reliant upon it, and has in some cases developed its own strategies that have close parallels in the physical sciences (e.g. field-based case studies in place of laboratory investigation). The history of the Gas Laws shows that the only scientific tradition that now has meaning and value is that of plurality of method; this is what marks the practice of the physical sciences, and the flexibility, innovation and power that it bestows has led to their greatest successes. As Box 2.1 mischievously suggests, a (post-)modern science now exists, for which a simplified, positivistic tradition is almost completely irrelevant. Particle physics and astronomy theorize about elementary particles and gravitational waves well beyond the limits of observation (but try to solve that problem); science widely grapples with non-linear dynamic processes with apparently stochastic outputs but deterministic internal structures; and physical science has had to deal with the problematic relationship between observer and observed since the quantum mechanical revolution and Schrödinger's cat. Thus, the negative interpretation of a 'straw' positivistic science as part of an anti-quantitative purpose in some geographical writing is both anachronistic and inaccurate.

Geography, and geographical research, are all the better when they espouse a pluralist ontological and epistemological tradition of the kind that the physical sciences have evolved over centuries of uncovering the nature of the world. And, just as the practices and procedures have evolved as science has generated new knowledge and understanding, so, within a given research project, there is a continual movement between the various available practices as additional exploratory investigation vies with more developed theorizing about mechanisms in the established areas of research. All of this implies that 'naturalism' need not be feared as an imperialist imposition of method by the physical sciences, but rather the normal set of potential procedures and practices for undertaking a logical, rigorous investigation that can be regarded as 'scientific', *sensu lato*. And, as Box 2.1 concludes, what we now require is a new tradition of rational criticism to cope with the diversity of modern scientific enquiry, whether physical, natural, environmental or social.

SUMMARY

- Traditions are rarely quite what they are claimed to be.

- Science has varied in its nature over time, so it is difficult to select a 'tradition'.

- Science is concerned with problems and problem-solving; an important early stage of the procedure of investigation is thus identifying researchable questions.

- 'The' scientific method is itself a model, and the procedures of scientific enquiry have evolved as knowledge and understanding has developed.

- Initially, procedures are required to 'represent' the world, through observation and measurement.

- It may then be necessary to 'intervene' by controlling, simplifying and manipulating part of the world in some form of experiment, in order to indentify regularities of behaviour. This may require carefully designed field experiments or laboratory manipulation. The Gas Laws, and greenhouse experiments on variations in seedling growth rate in response to hydrological control, are examples of the kinds of experiment that generate information on regularities.

- Experiments should be used to falsify theories, and this requires theorizing, hypothesis testing, and the use of a null hypothesis as a falsifiable alternative to the logical hypothesis derived from theory.

- Once regularities have been revealed, they are explained through a process of theorizing about underlying mechanisms.

- A complex array of theoretical and practical procedures is required (in addition to the conventional approach based on prediction) in order to assess the relative merits of competing theories.

- All of these procedures are employed in geography, either directly or in a suitably adapted form.

- Thus, geography has had at its disposal the same pluralist methodology that characterizes the physical sciences but which is, in fact, common to many areas of science (physical, natural, environmental and social).

Further Reading

Bird (1989), in **The Changing Worlds of Geography**, provides a balanced general introduction to ideas and methods in geography. A readable account of relationships between physical geography and science is presented by Inkpen (2005), while Castree (2005) and Harrison (2005) respectively question in successive chapters whether geography is a science, and what kind of science physical geography is! An intriguing demonstration of the diversity of the sciences and their practice can be found in Harré (1981) **Great Scientific Experiments**. Phillips' (1992) volume **The Social Scientist's Bestiary** is an excellent critical analysis (and debunking) of the arguments against naturalistic social sciences. Richards (1996), in a useful anthology edited by Rhoads and Thorn, discusses the role of the case study and its methodological import. An interesting example of the historiography of physcial geography can be found in Sack (1992), where she discusses the debate about Davisian and Gilbertian geomorphology. A readable account of the nature of science and scientific activity, and the relationship of science with society can be found in Woolgar (1988) *Science: The Very Idea*.

Note: Full details of the above can be found in the reference list below.

References

Bird, J. (1989) *The Changing Worlds of Geography: A Critical Guide to Concepts and Methods.* Oxford: Oxford University Press.

Castree, N. (2005) 'Is geography a science?', in N. Castree, A. Rodgers and D. Sherman (eds) *Questioning Geography.* Oxford: Blackwell, pp. 57–79.

Close, F. (1992) *Too Hot to Handle: The Story of the Race for Cold Fusion.* London: Penguin.

Davidson, D.A. (1978) *Science for Physical Geographers.* London: Edward Arnold.

Hacking, I. (1983) *Representing and Intervening.* Cambridge: Cambridge University Press.

Harré, R. (1981) *Great Scientific Experiments.* Oxford: Oxford University Press.

Harrison, S. (2005) 'What kind of science is physical geography?', in N. Castree, A. Rodgers and D. Sherman (eds) *Questioning Geography.* Oxford: Blackwell, pp. 80–95.

Inkpen, R. (2005) *Science, Philosophy and Physical Geography.* London: Routledge.

Latour, B. (1987) *Science in Action: How to Follow Scientists and Engineers through Society.* Cambridge, MA: Harvard University Press.

Livingstone, D.N. (2005) 'Science, text and space: thoughts on the geography of reading', *Transactions, Institute of British Geographers*, 30 (4): 391–401.

Pawson, R. (1989) *A Measure for Measures: A Manifesto for Empirical Sociology.* London: Routledge.

Phillips, D.C. (1992) *The Social Scientist's Bestiary: A Guide to Fabled Threats to, and Defences of, Naturalistic Social Sciences.* Oxford: Pergamon Press.

Richards, K.S. (1996) 'Samples and cases: generalisation and explanation in geomorphology', in B.L. Rhoads and C.E. Thorn (eds) *The Scientific Nature of Geomorphology.* Chichester: John Wiley & Sons, pp. 171–90.

Richards, K.S., Brooks, S.M., Clifford, N.J., Harris, T.R.J. and Lane, S.N. (1997) 'Theory, measurement and testing in "real" geomorphology and physical geography', in D.R. Stoddart (ed.) *Process and Form in Geomorphology.* London: Routledge, pp. 265–92.

Richards, K.S., Hughes, F.M.R., El-hames, A.S., Harris, T., Pautou, G., Peiry, J.-L. and Girel, J. (1996) 'Integrated field, laboratory and numerical investigations of hydrological influences on the establishment of riparian tree species', in M.G. Anderson, D.E. Walling and P.D. Bates (eds). *Floodplain Processes.* Chichester: John Wiley & Sons, pp. 611–35.

Roy, A.G., Biron, P.M. and Lapointe, M.F. (1997) 'Implications of low-pass filtering on power spectra and autocorrelation functions of turbulent velocity signals', *Mathematical Geology*, 29 (5): 653–68.

Sack, D. (1992) 'New wine in old bottles: the historiography of a paradigm change', *Geomorphology*, 5: 251–63.

Sayer, R.A. (2000) *Realism and Social Science.* London: Sage.

Stoddart, D.R. (1987) 'To claim the high ground: geography for the end of the century', *Transactions, Institute of British Geographers*, 12: 327–36.

Underwood, A.J. (1997) *Experiments in Ecology: Their Logical Design and Interpretation Using Analysis of Variance.* Cambridge: Cambridge University Press.

Woolgar, S. (1988) *Science: The Very Idea.* Chichester: Ellis Horwood.

3

Geography and the Social Science Tradition

Ron Johnston

D **efinition**

Social science is the study of human society and activity; its member disciplines include economics, political science and sociology. These social sciences expanded rapidly after 1945, using scientific methods to analyse problems and suggest how they may be solved. Before the 1970s few human geographers identified their discipline as a social science, but many now do. This shift was initially linked to the adoption of a positivist ontology and its associated 'scientific method', but many contemporary human geographers who identify as social scientists have challenged this orthodoxy, drawing on a diverse range of theories and approaches, including Marxism, feminism, postmodernism and post-structuralism, to create a very broad and diverse contemporary discipline.

INTRODUCTION

Geography in general, and human geography in particular, has moved among the major divisions of academic life within universities over the last century. Before the 1970s, very few human geographers identified their discipline as a social science: two decades later, most did. That shift was neither 'natural' nor necessarily obvious: it resulted from conflicts over the discipline's identity and over the willingness of 'the social sciences' to accept geographers within their orbit. This chapter traces some of those conflicts and the changes in geography that they involved, with particular reference to the situation in the UK and North America.

GEOGRAPHY'S ORIGINS

Geography as both an intellectual and a practical activity has a long history (Livingstone, 1992); geographical material was being taught in the ancient British universities by the late sixteenth century (Cormack, 1997; Withers, 2002; Withers and Mayhew, 2002) and in several American colleges in the early nineteenth (Koelsch, 2002). But in both countries it became a recognized segment of the academic discipline of labour – only with separate university departments of, and degrees in, geography – only in the early twentieth century. In the UK, by 1950 virtually every university had a geography department (and a professor, indicative of the discipline's status), but most of these were small, with few graduates each year and no more than five staff members (Johnston, 2003a). In the USA, there was no time when there was a geography department in the majority of universities.

Much of the early pressure for the discipline's establishment in British universities came from the Royal Geographical Society (RGS) (founded in 1830), whose major concerns were with the promotion of British imperialism and associated notions of citizenship (Ploszajska, 1999; Driver, 2001; see also Chapter 1 in this volume and Schulten, 2001 on early geography in North America): it focused attention on Cambridge and Oxford. Elsewhere, the demand for geographical teaching came from a variety of sources. In some universities its introduction was linked to a major donor's wishes. In others, economists made the case for courses in commercial geography (Chisholm, 1886; Wise, 1975; Barnes, 2001a).[1] Indeed, the first professorship in geography was held by L.W. Lyde (a classical scholar and author of numerous school texts with sales of over 4 million), who taught courses for the Department of Economics at University College London.[2] In others, a separate geography presence emerged from the geology departments to cover the study of contemporary landscapes. Whatever the origin, in most UK universities the main rationale for full geography degrees was to train students who would then teach the discipline in the country's public and grammar schools. In these ways, geography as an academic discipline was established by individuals from a variety of backgrounds (Johnston, 2005b).[3] The USA had no central body pressurizing universities to introduce geography courses and departments, and those established reflected local demands; most departments originated (such as the oldest, at the University of California, Berkeley) with the appointments of geographers to teach courses for either geology or economics/commerce students – as was the case also in the universities of the then British Dominions (Australia, Canada, New Zealand, South Africa).

There was a very strong symbiosis between secondary schools and universities in promoting geography in the UK – as in Germany (Schelhas and Hönsch, 2002): the schools provided the university students, many of whom returned to be schoolteachers after graduation. This symbiosis was enhanced by the Geographical Association, founded in 1893 to promote geographical education at all levels, but which focused on schools. It remains an extremely important pressure group (Balchin, 1993): without it (spearheaded for much of its existence by a few senior academics), it is very unlikely that geography would be as large a discipline in the country's universities as it is now (Rawling, 2001;

Walford, 2001). In this, the UK situation contrasts with the American where, although there was early pressure for training geography teachers in the 'normal schools' in some states, the discipline was not in the high-school curricula and thus very few proceeding to university had much knowledge of it.[4] Student interest there had to be captured by professors offering attractive and interesting introductory courses within much broader curricula than was the case with the highly specialized UK honours degrees; indeed, very few American undergraduates today go to university with the specific intention of majoring in geography.

Geography's origins were reflected in how it was practised for the first half of the twentieth century. The roots in geology were the basis for the development of physical geography – especially geomorphology, as with the influence of the Harvard geologist, William Morris Davis (Chorley et al., 1973). Those in economics stimulated interest in patterns of economic activity – of agriculture, industry and trade[5] – whereas links with anthropology (very strong at Aberystwyth and Belfast in the UK, for example) generated work on less developed societies. This was enhanced by the creation of geography departments in the universities in the then British Empire, which were at least partly staffed by expatriates who did research on the local area; a number of British geographers also developed regional interests based on their experiences during the Second World War.

These divergent scholars shared concerns with the interrelations between the physical environment and human activity. For some, the environment was a determining influence on human activity; to others, increasingly the majority, it was a strong constraint, but the ultimate determinant was human free will. Whichever position was taken, however, the outcome was the same: a mosaic of areas with particular environmental characteristics and human activities (see Chapter 9 on place and human geography). Such areas were regions, separate areas with distinct landscapes (both natural and human) that distinguished them from their neighbours. Geographers saw the main rationale of their discipline as identifying, describing and accounting for the characteristics of these areas (at a variety of scales and on a range of criteria; see also Chapter 1 and Johnston, 2005a). The region was the core geographical concept; defining regions – largely through map comparison techniques – occupied the heart of the discipline's methodology; and studying regions was the ultimate purpose of a training in geography. (Many honours degree courses in the UK, and especially their final years, were dominated by regional courses until the 1960s.[6])

SOCIAL SCIENCE ORIGINS

This orientation of the discipline meant that there was little contact with the social sciences. In the first half of the twentieth century only economics from that group of disciplines was established in most UK universities, but there were few links between its theoretical approaches and geographers' empirical concerns. After 1945, neoclassical economists sought accounts for market operations through deductive model-building while geographers mapped patterns of economic activity and related them to the physical environment. With few exceptions, geographers made no use of professional economists' tools in their research and teaching (as illustrated by Rawstron, 2002: see also Barnes, 2001a, on the history of economic geography in both the UK and the USA).

The other social sciences were but minor presences in UK universities before the Second World War; only anthropology was firmly established in some institutions (on which see Kuper, 1996: geography's links with anthropology are discussed in Taylor, 1993). Geography was institutionalized into UK academic life long before either sociology or political science in the twentieth century. The Royal Geographical Society was founded, as we have seen, in 1830; the Geographical Association (GA) in 1893; and the Institute of British Geographers (IBG) in 1933.[7] By contrast, the Political Studies Association was founded in 1950, and the British Sociological Association in 1951. There were just two sociology departments in British universities in the mid-1950s and, of the 54 academic sociologists across 16 universities a decade later, fully 16 of them were at the LSE (Platt, 2003; Halsey, 2004). In the USA, sociology and political science departments were founded much earlier in many universities and their presence made it difficult for geography departments to be established there, a situation exacerbated when separate teaching of geography in high schools was squeezed out by the establishment of social studies programmes (Schulten, 2001).

In the 1950s, therefore, there were some links between geographers and anthropologists, and a few with economists, but geography largely existed outside the social sciences, instead occupying a claimed bridging-point between the arts and the sciences, combining the study of human activity within its environmental context through a focus on regions. Its nearest academic neighbour, according to one much-referenced North American scholar at the time (Hartshorne, 1939), was history. Both disciplines employed 'exceptionalist' approaches: historians studied particular time periods whereas geographers studied particular places. Both provided explanatory accounts of their periods/places through a synthesis of available material; both eschewed generalization; and they came together – especially in the UK where a strong tradition developed around one scholar, H.C. Darby (Prince, 2000) – in the study of historical geography.

The mid-1950s saw the onset of a major change in human geography, its identity and links with the social sciences. Those other disciplines were already growing rapidly, reflecting their perceived relevance and applied worth. Economics became increasingly important as states became larger actors in and regulators of economies and as individual businesses became more professional in their operations, with ownership and management shifting from individuals and families to company shareholders. Economists played major roles in wartime governments, for example, and remained important thereafter, as the expanded state took on wider peacetime roles in economic management.

Economists' roles also increased within the growing state apparatus because of the growth of the welfare state, which provided economic and social protection for the vulnerable, invested in the future through universal schooling and widening university education, and redistributed wealth to produce a more equal society – a dominant ideological force of the times in the UK. Sociologists played important roles too, providing intellectual foundations for the more applied disciplines of 'social administration' and 'social policy' as well as through the importance of their core concept of class to those promoting a redistributive state. The applied relevance of political science, which emerged as a separate discipline from roots in history and philosophy, came through desires to understand the working of the state apparatus and ensure the efficient operation of state bureaucracies – public administration (both national and local). And as globalization increased, with all the associated political tensions and conflicts, and with the Cold War stimulated by the ideological gulf between east and west, so the study of international relations increased in importance.

These three disciplines at the core of the social sciences – economics, sociology and political science – became major components of the academic world from the 1950s on. Anthropology failed to expand at the same pace, however, as interest shifted from 'primitive' to 'modern' societies and the stimuli to studying the former were reduced with decolonization and 'modernization' (Peel, 2006). Other disciplines which overlapped the social sciences similarly increased in academic importance – notably psychology, which assumed increased importance in understanding and managing human behaviour in a range of contexts.

This demand for the social sciences, from users and potential students, stimulated growth at the universities – though less so in England's ancient establishments than elsewhere. (Sociology and political science have only recently achieved departmental status at Oxford, for example, although a major centre for postgraduate research in those disciplines was established at Nuffield College in the 1940s.) The LSE became a major UK centre for social science teaching and research, having been a pioneer in those areas for more than half a century.[8] Furthermore, almost all the new UK universities established in the early 1960s invested heavily in the core social science disciplines, whose postgraduate training and research activities were funded when a Social Science Research Council (SSRC) was established in the mid-1960s.

BELATED MEETING: HUMAN GEOGRAPHY AND SOCIAL SCIENCE

Where was geography when all this was going on? What was its contribution to the war effort and to the burgeoning demand for social science expertise thereafter? With regard to the former, geographers were involved in a range of intelligence-gathering and provision activities – much of it in the UK associated with mapping, air-photo interpretation, and the production of handbooks on countries where military operations were likely (Balchin, 1987; Clout, 2003). In the USA, a large number of geographers was assembled in Washington to work in the Office of Strategic Services, alongside other social scientists (see Kollmorgen, 1979; Barnes, 2006; Barnes and Farrish, 2006). Some of those involved concluded that their contributions were not of high quality, stimulating campaigns for changes in the nature of the discipline (Ackerman, 1945, 1958): one of those convinced of this need – Edward Ullman – was among the early promoters of a 'new geography' less than a decade later (see below).

One potential area for geographers to apply their knowledge and expertise was identified as the growing activity of town and country planning (called city and regional planning in the USA). There were increasing concerns for the most efficient use of land, for example, to ensure an adequate food supply during wartime. Subsequently, attention shifted to the need to distribute economic activities efficiently rather than allowing an overconcentration in certain areas, which would make them vulnerable to air attack: the US interstate freeway system was a response to this. The need to protect high-quality agricultural land, to prevent urban sprawl, to reduce concentration on certain regions and to distribute land uses within cities efficiently (notably though their transport systems) stimulated planning legislation. This was enacted in the late 1940s in the UK, with both national planning (extending the regional policies of the 1930s enacted to deal with the problems of industrial 'depressed areas') and a requirement for all local authorities to produce local plans within a national framework.

Geographers' knowledge and expertise about regions could provide information for the production of national and local land-use plans – and a major Land Utilization Survey mounted by Dudley Stamp at the LSE in the 1930s (by far the largest geographical 'research project' in the first half of the century) provided both valuable information and a template for such data-gathering exercises (Stamp, 1946). But could geographers be more than just data-gatherers and displayers? From the early 1930s, one of the first urban geographers, Robert Dickinson, argued for geographers focusing attention not on 'formal regions' (the separate parts of the landscape mosaic defined largely by their physical characteristics), but on 'functional regions', the tributary areas of towns and cities that formed the basic spatial framework within which society was organized. He argued that local government should be restructured to fit this pattern of functional organization, and that intra-urban planning should recognize the 'natural areas' of cities. In 1947 he published a major book based on US and European as well as British sources – *City Region and Regionalism* – which promoted these goals. City regions, according to his Preface, are 'aspects of the inherent spatial or geographical structure of society upon which planning must be based' and he presented the book as '... not about planning. It is concerned with certain aspects of the inherent spatial or geographical structure of society on which planning must be based' (Dickinson, 1947: xiii). By the end of the 1950s, many geography graduates were entering the planning profession but, although some occupied leadership roles (see Willetts, 1987), most were at the level of data-gatherers and displayers: the leadership in spatial planning of the 'brave new world' was provided by architects, surveyors and engineers (see Hall, 2003). Had geographers missed the boat? A new set of disciplines had come to the fore, from which they were largely excluded, although a few, seeing the potential, allied themselves closely to the new disciplines and transferred their allegiance accordingly.

In the 1950s and 1960s a new generation of geographers sought to reorientate their discipline towards the social sciences (Johnston and Sidaway, 2004). Much of the early impetus occurred in the USA, where a number of British geographers went for postgraduate training and other experiences, bringing the new ideas back to their country in the early 1960s (Johnston, 1997a; Johnston and Sidaway, 2004). A major centre for this 'revolution' was the Department of Geography at the University of Washington, Seattle, where a group of postgraduates converged to study with Ullman, but who switched to work with William Garrison. They rapidly spread their ideas through circulating discussion papers and making conference presentations (Martin and James (1993: 372) call it spontaneous, although networks linking the various groups were soon established). Other groups were established elsewhere (at the University of Iowa, for example, at Northwestern University and the University of Chicago – both in Chicago – and at Ohio State University). Very soon 'revolutionary' success was being claimed (Burton, 1963), and a new suite of disciplinary practices was being spread – not only through the USA but also in the UK, where it was led and diffused from Cambridge (and later Bristol) by two relatively junior lecturers, Dick Chorley (a physical geographer) and Peter Haggett (a human geographer).

The 'exceptionalism' of regional geography was rejected by these 'revolutionaries' – as providing 'mere description' – and the newly emerging social sciences were lauded, in part at least because their approaches and methods were closer to those of the natural sciences than to the arts. Three aspects of the new work were especially attractive to postwar generations of scholars:

1   *Its concern for scientific rigour* which involved geographers interrogating literatures in the philosophy of science and knowledge, which they had previously largely ignored. Much current geographical practice was portrayed as theoretically weak and lacking the objectively neutral approach associated with the 'natural sciences'. Schaefer's (1953) damning critique of 'Hartshornian orthodoxy' argued that geographers should focus on identifying the laws underpinning spatial arrangements. This involved adopting the hypothetico-deductive 'scientific method', fully explored by Harvey's pioneering examination of the methodologies associated with this philosophy of science, *Explanation in Geography*, which concluded with the statement 'by our theories you will know us' (Harvey, 1969: 486): explanation and prediction were to be human geography's research goals.

2   *An argument that quantitative methods formed a necessary component of this more rigorous approach* to the portrayal and analysis of information, including geographical information, although not all the early proponents of this cause necessarily tied it directly to the philosophical claims regarding 'scientific method'. The adoption of standard statistical procedures was seen by some simply as the correct way to use data (as in Gregory's *Statistical Methods and the Geographer* (1963)). To be rigorous, geographers had to be quantitative.

3   *A realization that rigorously obtained research results could be applied to a wide range of problems.* Many geographers were concerned that their discipline lacked status among decision-makers (see Coppock, 1974; Steel, 1974). The social science disciplines were much more influential because they took a more rigorous approach to problem-solving associated with 'scientific method' and 'quantification'. Geographers should promote their expertise in the creation of spatial order – increasingly needed with the growth in spatial planning – but should do this as scientists (which increasingly physical geographers were becoming too).

Those attracted to this cause explored the literature (past and present) for inspiration. They found it in the general concept of spatial organization, the spatially ordered arrangement of human activities. Exceptional among those stimuli was the work of a German geographer, Walter Christaller (1966), who developed central place theory in the early 1930s to understand settlement patterns. Individuals who journey to shops and offices for goods and services want to minimize the time and cost involved: the needed facilities should be as close to their homes as possible and clustered together so that they can make as many of their purchases as possible in the same place. And the owners of businesses want to maximize their turnover – with people spending as much as possible in the shops and offices and minimizing their transport costs. An efficient distribution of services was in the interests of both suppliers and customers. Christaller showed that this would result in a distribution of service centres across a uniform plane (i.e. with no topographical barriers) in an hexagonal arrangement, with the smaller centres (providing fewer services) nested within the market areas of the larger, although the details of that arrangement would depend on whether the goal was to minimize the number of settlements or the total length of roads. (On central place theory and its early influence, see Berry (1967) and Barnes (2001b).) A little later, thanks largely to a visit he made to Seattle in 1959, the work of a Swedish geographer, Torsten Hägerstrand, on spatial diffusion (Hägerstrand, 1968) was similarly a major stimulus for further research (as exemplified in Morrill, 2005).

Other works – all by non-geographers – also provided stimuli. Economists such as Hoover, Palander, Lösch and Weber, for example, suggested that manufacturing industries would be located so as to minimize their input costs (among which a major variable element was the costs of transporting them to the plant from a range of sources) as well as their distribution costs (getting the final goods to the market): least-cost location was the goal, which could be modelled as a form of spatial economics. (On these theories, see Garrison, 1959–60.) And a nineteenth-century German landowner-economist (von Thünen) derived a similar model for the location of agricultural production, suggesting a zonal patterning of different activities consistent with the costs of transporting the output to markets. Economists built on this to suggest similar zonal organization of land-uses within cities, which would be correlated with the pattern of land values. Such sources stimulated not only hypothesis-testing research at Seattle and elsewhere (Haggett first encountered much of the literature when teaching economic geography at University College London in the mid-1950s; some location theory, such as Hoover's, was already being taught there and at the LSE) but also applied work, such as that done by Garrison and his students on the impact of transportation improvements (Garrison et al., 1959).

Work on spatial patterns was complemented by studies of flows of goods, people and information. Their modelling was also based on principles of least effort, assuming that people wish to minimize travel costs. The Newtonian gravity model was adapted, using the analogy that the larger the places of origin and destination the greater the movement between them, but that this would decrease, the greater the distance separating them. The various models of patterns and flows were brought together and a new discipline, regional science, was launched, though it failed to gain the status its founder (Walter Isard) sought (see Isard, 2003; Barnes, 2003, 2004). These location–allocation models integrated locations and flows, suggesting both optimum locations for facilities and efficient flows between them.

Geographers – especially those trained after the Second World War – were attracted to these models, as foundations for hypotheses that could be tested, using rigorous, quantitatively-based procedures to show both that locational decision-making was economically rational and that planning for new facilities and routes could be based on such models. In addition, they 'rediscovered' models of the internal spatial organization of cities into 'natural areas' developed by sociologists and others at the University of Chicago (Dickinson was the first to notice them, in the 1930s: Johnston, 2002). These various sources were brought together in innovative, influential textbooks which discussed both the patterns and the methods for analysing them, such as Haggett's *Locational Analysis in Human Geography* (1965), Chorley and Haggett's *Models in Geography* (1967), Morrill's *The Spatial Organization of Society* (1970) and Abler, Adams and Gould's *Spatial Organization* (1971). In different ways these emphasized the theme earlier pronounced by Watson (1955) that 'geography is a discipline in distance'. Cox (1976) argued that this new orientation brought geographical interests into line with contemporary society: in the pre-industrialized world, 'vertical' relationships between society and nature predominated as influences on regional patterns; in the industrial world, the horizontal relationships between and within societies were salient – and their study involved geography joining the social sciences.

Over the next couple of decades, the volume of work in this mould expanded greatly, applying and modifying the 'classic models', developing statistical and mathematical procedures for analysing spatial organization, exploring the underlying philosophy of the 'scientific method' (positivism: Harvey, 1969), and arguing that their models could be used as planning tools for cities and regions (Wilson, 1974). Substantive interests expanded, too, and a subfield of 'behavioural geography' evolved to embrace the 'scientific' study of human spatial behaviour and decision-making through the quantitative analysis of data obtained from questionnaires and similar instruments (Johnston, 2003b; Golledge, 2006; Golledge and Stimson, 1996).

GAINING RECOGNITION

Human geography was very substantially remade during the 1950s–1970s, therefore, though not without considerable conflict with those who sought to defend the *status quo* in, especially, regional and historical geography (Johnston, 1997a; Johnston and Sidaway, 2004). As such, the remodelled discipline presented itself as a social science, claiming a clear niche within that area of activity with its focus of location and space (identifying itself as spatial science or locational analysis). But it was too late to gain entry to most of the UK's new universities of the 1960s: of them, only Sussex had a (relatively small) geographical presence virtually from the outset, and one was added at Lancaster in the early 1970s (because geography departments could attract students). An attempt by the RGS to promote geography with the founding bodies for the new institutions was unsuccessful; its claims for the discipline failed to match the scientific mood of the times (Johnston, 2003a; Johnston, 2004a). A few of the others (East Anglia, Lancaster, Stirling and Ulster) included geographers within multidisciplinary environmental science schools, but human geographers were in the minority there relative to their physical geography colleagues (who had also been seduced by the three characteristics of the 'scientific method' listed above, and were remodelling their part of the discipline too (see Chapter 2 on geography and the physical science tradition)).

Geography was also excluded from the Social Science Research Council (SSRC) when it was established. A group challenged this, presenting a case based on the 'new' geography (which was contested by some heads of geography departments and others, who wanted to maintain the *status quo* and did not identify with the social sciences). This was accepted, but geography, unlike the original disciplines, was not accorded separate committee status within the SSRC; instead it was linked with planning (Chisholm, 2001; Johnston, 2004b). Having achieved that status, the chief author of the case published a number of books promoting the new view of the discipline (Chisholm, 1971, 1975; Chisholm and Manners, 1973; Chisholm and Rodgers, 1973). Similar attempts were made in the USA, and two *ad hoc* committees made the case for recognition of geography both within the country's main research academy (NAS–NRC, 1965) and its social science community (Taaffe, 1970; Gauthier (2002: 577) records that the before the report was written the committee chaired by Taaffe 'faced a serious challenge ... geography had initially not been selected to participate in the survey, because the other panels in economics, sociology, psychology, anthropology, and political science did not view the field as a viable social science'). These were bolstered by a further attempt to sustain and enhance their position three decades later (NRC, 1997; see Johnston, 1997b, 2000). Even so, several US geography departments with graduate schools were closed in the last third of the twentieth century (including prestigious institutions such as the Universities of Chicago, Michigan and Pennsylvania, plus Columbia, Northwestern and Yale universities; Harvard's department was closed in 1948 (Smith, 1987)[9]), by the end of which only one of the country's Ivy League universities – Dartmouth – had a geography department. As Koelsch (2002: 270) expressed it: 'the closing of geography in the major private universities sent a powerful signal that geography is no longer valued by academic administrators in institutions that traditionally have turned out the country's economic decision-makers and its cultural and political élite'.

Although social scientific recognition has been achieved, nevertheless geography is still considered peripheral to some aspects of academic life. In almost every country there is one or more national academy, an elected body of the country's main scholars. In the UK, the two main bodies are the Royal Society (for the sciences) and the British Academy for the humanities and social sciences. Only five geographers have ever been elected to the Royal Society (Fleure, Wooldridge, Rhind, Battarbee and Wilson: the last three were elected in the last five years; Wooldridge was elected in the 1950s and Fleure the 1940s) and no geographer became a Fellow of the British Academy until 1967, when the historical geographer, Clifford Darby, was elected. Today there are some 30 fellows – a further five are now deceased – and four (Darby, Coppock, Haggett and Kain) have served as Academy Vice-Presidents. In the USA, there are two major comparable institutions. The National Academy of Sciences currently has some 1,900 active members, of whom just 11 are geographers in a 'Human environmental sciences' section (with five previous members now deceased); only one geographer – Brian Berry – has served on its council. (There is also one foreign member who is a geographer, a Nigerian – Akin Mabogunje.) The other body is the American Academy of Arts and Sciences, which currently has some 4,000 fellows, of whom only 12 are geographers in its Archaeology, Anthropology, Sociology, Geography and Demography section (with one overseas member – Peter Haggett).[10]

OPENING OUT

But things did not stand still. The social sciences were changing fast during the last three decades of the century, and geographers were changing with them. They discovered stimuli in aspects of the core disciplines that they had previously largely ignored. In economics, for example, there were both welfare (Chisholm, 1966) and Marxian (Harvey, 1973, 1982; the first of these books – *Social Justice and the City* – was extremely influential in stimulating a new focus to much Anglo-American human geography) approaches to be explored. Sociologists, including the Chicago School, had studied a much broader range of subjects, with a wider range of methods, than those initially identified and adopted by geographers (as Jackson and Smith (1984) cogently argued). And a range of multidisciplinary approaches, such as world-systems analysis, offered new arenas within which a spatial perspective could be crafted (Taylor, 1982).

At the same time the quantitative/positivist 'revolution', which many welcomed for its 'conceptual rigour' (Davies, 1972), itself came under attack. By reducing most decision-making to economic criteria, subject to immutable 'laws' regarding least-costs, profit-maximization and distance-minimizing, geographers, it was claimed, were ignoring (even denigrating) the role of culture and individuality in human conditioning and behaviour. By suggesting the use of those 'laws' as the bases for spatial planning, they were simply seeking to reproduce the *status quo* – of capitalist domination. And by assuming universal patterns of behaviour they were patronizing those who chose to operate differently.

Out of these arguments grew three main strands of work, developments of which involved geographers in much wider-ranging discussions than heretofore about the philosophy of science and knowledge-production: issues of epistemology and ontology (to which many geographers were introduced in Gregory's pioneering book (1978)), as well as methodology, became central to debates over the discipline's rationale. One was *Marxist-inspired* (often termed radical), which explored not only the workings of the economy from that perspective, and added a spatial dimension to it (notably in Harvey, 1982 and Smith, 1984), but also the class conflict which underpins Marxian analyses of the economy and is central to a major area of sociological and political science literature (see Sayer and Walker, 1992). For such work, the positivist 'scientific method' was irrelevant since it assumes constant conditions within which economic decisions are taken whereas, for Marxist scholars, continuous change is the norm. Among alternative approaches within this broad 'radical strain', the most popular (either explicitly or implicitly) was critical realism (Sayer, 1984). This accepts that there are general (or immanent) tendencies operating within capitalism (or any other societal context), but that they are only latent until implemented by individual human agents making decisions in context – as illustrated by Massey's (1984) classic study of the changing geography of economic activity in Great Britain. Since those decisions change the context – in Massey's analogy, a new round of decision-making imposes a new layer on the map of locational activity – then the contingent circumstances within which future decisions are made must change too. Furthermore, the decision-makers themselves change as they learn from the making and consequences of previous decisions. There is a continuous interplay between structure and agency, or context and decision-maker, which Giddens (1984) termed structuration in a major contribution to sociological theory that was also influential among geographers. Thus, for realists, it is possible to explain why an event occurred – why a factory was located at a particular site, for example – but not as an example of a general law of location: explanation refers to specific events in context when decision-makers react to circumstances in order to meet certain imperatives (such as making a profit) within the constraints of their particular situations (what they know; what they believe their competitors will do; how they manipulate that knowledge, etc. An early attempt to incorporate such factors to spatial decision-making was Pred (1967–9).

The second strand drew particularly on work in sociology, especially though not exclusively work on gender and the growth of *feminist scholarship*. The core of the argument was that individuals occupy multiple positions within society, not just the class position which is at the core of Marxian analyses. Feminist geographers argued that not only was geography a male-dominated discipline, but also that its concerns reflected masculine positions (Rose, 1993). Women were subordinated and ignored, and their goal was both to remove that ignorance and demonstrate that gender divisions in society could not be reduced to class position. From this emerged a wider concern with 'positionality', which embraced not only gender divisions within society but also ethnic, racial and national, plus age, disability, sexual orientation and other criteria on which individuals' identities were based – such as the position of those living in postcolonial situations. Thus even gender had to be subdivided recognizing, for example, the different positions (and politics) of white and black women, of women in developed and developing world contexts, in various religions and so forth (McDowell, 1993, 2003, 2006).

Appreciating those divisions, and people's positionality within them – and the many hybrids that emerge through, for example, mixing in multi-ethnic cities – cannot be achieved by the abstract theorizing of either spatial science or Marxian analysis. It calls for interpretative methodologies aimed at understanding through empathy, gained through a variety of methods developed in other social sciences – such as participant observation, focus groups, in-depth interviewing, the examination of archived resources (novels, diaries, biographies, works of art, maps, landscapes and homescapes, etc.) – which allow access to how people interpret their place(s) in the world, and how they act accordingly. This was the case with the burgeoning subdiscipline of critical geopolitics in the 1990s, for example, which, through links to parallel developments in international relations, sought to appreciate how influential political thinkers and politicians develop and propagate mental maps of the world as structures for action (Dodds and Atkinson, 2000; Gregory, 2004).

Much of this work came to be associated, more explicitly in some cases than others, with what become known as postmodernism, again a major development in the social sciences (outside economics). This argues that there are no absolute truths and therefore no grand theories that can provide both explanations and guides to action (political or otherwise). Truths are the beliefs on which people act, so there are multiple truths – none of which can claim primacy over others, although the 'value' of competing truths can be assessed ethically (Smith, 2000). People learn their truths from others, either directly or through indirect sources (such as books). Such learning is context-dependent and, since most live relatively spatially constrained lives, the spaces within which they learn are their homes, their neighbourhoods, their workplaces, the formal organizations they participate in and so on. Appreciation of the role of context has brought places back to centre-stage in much human geographical research (and has been introduced to other social science and humanities disciplines), not in the former regional tradition with contexts defined by environmental features but, rather, in a much more plastic way: places are made, remade and dissolved; they may overlap, or they may be bounded and defended (see Chapter 9 on place and human geography).

This revived interest in places, and a shift of focus away from space within the discipline, is a feature also of the third strand. Geographers are playing significant roles within a burgeoning field of *cultural studies*, which brings together scholars from the humanities and social sciences in new ways of approaching the study of human behaviour in context (see Chapter 4 for a discussion of geography and the humanities). This work ranges over many aspects of behaviour, including the micro-scale of the individual body, seeking to understand the meanings that underpin actions – many of which are never recorded during the processes of everyday life. The relationships between people and nature are also being reconsidered, breaking down the perceived artificial boundaries between these long-considered binary opposites (Whatmore, 2002). Here again, new approaches are being explored for the interrogation of actions, including places as their arena. Indeed, such is the geographical contribution to cultural studies that some identify a 'spatial turn' within the humanities (Anderson et al., 2002); other geographers continue to explore the interactions between humans and their environment in more 'traditional' ways (Turner, 2002).

CONCLUSIONS: HUMAN GEOGRAPHY – SOCIAL SCIENCE AT LAST

Geography came late to the social sciences, therefore, and by the time that human geographers sought to ally with them they found they were excluded. In response, while remaking their own discipline they also had to make strong claims that it was now clearly a social science. To do this, they initially emphasized a particular aspect of the social sciences, privileging economic over other forces as determinants of human behaviour, and emphasizing models of spatial behaviour – of organization and flows – in which those forces dominated. They achieved some success in this strategy. A stream of work was introduced which remains strong, although it has changed over the last four decades. Rigorous analysis of quantitative data remains at the core of what is known as the spatial analysis tradition (Johnston, 2003b; Fotheringham, 2006). Formal models based on idealized spatial patterns derived from oversimplified principles have largely been jettisoned, however, though interestingly they were taken up by a school of economists in the 1990s, in a 'new economic geography' which geographers (with some exceptions) claim they disowned 20 years ago (Clark et al., 2000); the two 'strands' are being brought together through a journal – *Journal of Economic Geography* – which incorporates both.

Alongside the spatial analysts, with their increased technical sophistication and reliance on advanced technology (including GIS), other geographers discovered a wide range of approaches to explanation and understanding within the social sciences. Some have adopted approaches to explanation which differ from the positivism on which the original spatial analysts relied: others have argued that explanation is not feasible and only understanding is possible. They interact with very different areas of social science from the spatial analysts, and they too have won recognition and regard among their interdisciplinary peers. Nevertheless, it remains the case that in general (as can be seen from perusal of the literature referenced in articles in the leading journals of various disciplines – on which see Johnston, 2003c) human geographers are net importers, drawing more on other disciplines than vice versa.

'Positionality' is as central to academic life as to all other areas of society. Individual academics are schooled in particular approaches to the overall goal of understanding and changing society, within their own context – their own 'place'. Human geographers have their collective 'place' – a perspective based around the key concepts of place, space, environment and scale (Massey et al., 1999) – which they promote, and within the discipline different groups of geographers emphasize different concepts. From those bases, some located in 'real places' (particular graduate schools, for example), they interact with other social scientists, bringing separate perspectives to bear on shared subject-matter. Interactions among the practitioners create wholes that are greater than the sums of the parts, communities with new hybrid perspectives on worlds and how they should be studied. For the last three decades at least, human geographers have been party to these negotiations, largely abandoning their origins as a discipline built on firm foundations in the physical sciences, having come late to the conference table.

SUMMARY

- Over the last half-century human geography has moved from its position on the boundary between the arts and the physical sciences to become firmly established as a social science.

- The core social sciences (economics, sociology and politics) grew rapidly after the Second World War because of their relevance in understanding and managing the emerging global economy and changing social and political relations.

- In the 1950s and 1960s a new generation of human geographers sought to reorientate the discipline towards those social sciences.

• Initially, that reorientation involved a concern with scientific rigour and the adoption/adaptation of quantitative methods to analyse spatial patterns and develop models of spatial organization.

• This 'scientific orthodoxy' was subsequently challenged and contemporary human geographers who identify as social scientists draw on a more diverse range of theories and approaches, including Marxism, feminism, postmodernism and post-structuralism.

Further Reading

For overviews of the history of geography, see Martin's (2005) *All Possible Worlds*, Livingstone's (1992) *The Geographical Tradition*, Johnston and Sidaway's (2004) *Geography and Geographers: Anglo-American Human Geography since 1945* and the essays in Dunbar's (2002) *Geography: Discipline, Profession and Subject since 1870*. Much of the discipline's nature and development is charted, and its terminology outlined, in the many essays and entries in Johnston et al.'s (2000) *The Dictionary of Human Geography*. A useful anthology of relevant materials is Agnew et al.'s (1996) *Human Geography: An Essential Anthology*.

Note: Full details of the above can be found in the references list below.

NOTES

1 Chisholm's book appeared in 20 separate editions, the last (rewritten by Sir Dudley Stamp) in 1980.
2 University College briefly had a chair in geography in the 1830s, occupied by Alexander Maconochie, who was also influential in establishment of the RGS (Ward, 1960).
3 One, Kenneth Mason – the first professor of geography at Oxford – had no academic degrees, having been a military surveyor and explorer.
4 One of the leading American geographers of the early twentieth century – Mark Jefferson – taught at a 'normal school' (Martin, 1968).
5 The University of Melbourne had two geography departments until the late 1960s: the oldest was a Department of Economic Geography in the Faculty of Commerce; the other, established in the early 1960s, was in the Faculty of Arts.
6 As an undergraduate between 1959 and 1962, the courses I took in the first year were all compulsory; none were regional in orientation. The second year included compulsory courses on Great Britain and on Ireland, and there was one optional course – I did the regional geography of India. In the final year, in addition to a dissertation, there was one compulsory course (on the geography of France and Germany), one major option (I did applied geography) and one minor option (I did the regional geography of southwest Asia). There were also two

papers in the final exams (on map interpretation – using French and German maps – and a general essay) for which there were no courses. There were some systematic courses in physical geography, but none in human geography (e.g. nothing on urban geography or industrial geography, etc.).

7 The RGS's goal has always been to promote geography and its study in all walks of life, whereas the GA focused on geographical education and the IBG was a learned society for researchers – most of its members were either academic geographers or postgraduate researchers: the IBG and the RGS merged in 1995. The comparable organizations in the USA are the Geographical Society of America, the National Council for Geographical Education and the Association of American Geographers (which merged with the Association of Professional Geographers in the 1940s); there is also the National Geographical Society, renowned for its popular magazine *National Geographic*.

8 Sir Halford Mackinder, who founded the School of Geography at Oxford in 1887, was Director of the LSE in the early twentieth century before developing a career as an MP and diplomat.

9 Harvard reintroduced geography in 2006 with the establishment of a Center for Geographic Analysis, which stresses 'modern' spatial analysis and the deployment of GIS as 'one of the technology platforms in Harvard's Institute for Quantitative Social Science': http://www.gis.harvard.edu/icb/icb.do.

10 There are currently 19 geographers who have been elected as Fellows of the Academy of the Social Sciences in Australia, out of a total of some 400, and yet geography is rapidly losing identity in the universities there (Holmes, 2002; Johnston, 2006).

References

Abler, R.F., Adams, J.S. and Gould, P.R. (1971) *Spatial Organization: The Geographer's View of the World*. Englewood Cliffs, NJ: Prentice-Hall.

Ackerman, E.A. (1945) 'Geographic training, wartime research, and immediate professional objectives', *Annals of the Association of American Geographers*, 35: 121–43.

Ackerman, E.A. (1958) *Geography as a Fundamental Research Discipline*. Research Paper 53. Chicago, IL: Department of Geography, University of Chicago.

Agnew, J., Livingstone, D.N. and Rogers, A. (eds) (1996) *Human Geography: An Essential Anthology*. Oxford: Blackwell.

Anderson, K., Domosh, M., Pile, S. and Thrift, N.J. (eds) (2002) *Handbook of Cultural Geography*. London: Sage.

Balchin, W.G.V. (1987) 'United Kingdom geographers in the Second World War', *The Geographical Journal*, 153: 159–80.

Balchin, W.G.V. (1993) *The Geographical Association: The First Hundred Years, 1893–1993*. Sheffield: The Geographical Association.

Barnes, T.J. (2001a) 'In the beginning was economic geography: a science studies approach to disciplinary history', *Progress in Human Geography*, 25: 521–44.

Barnes, T.J. (2001b) 'Lives lived and lives told: biographies of geography's quantitative revolution', *Environment and Planning D: Society and Space*, 19: 409–29.

Barnes, T.J. (2003) 'What's wrong with regional science? A view from science studies', *Canadian Journal of Regional Science*, 26: 3-26.

Barnes, T.J. (2004) 'The rise (and decline) of American regional science: lessons for the new economic geography?', *Journal of Economic Geography*, 4: 107-29.

Barnes, T.J. (2006) 'Geographical intelligence: American geographers and research and analysis in the Office of Strategic Services, 1941–1945', *Journal of Historical Geography*, 32: 149–68.

Barnes, T.J. and Farish, M. (2006) 'Between regions: science, militarism, and American geography from World War to Cold War', *Annals of the Association of American Geographer*, 97: 807–26.

Berry, B.J.L. (1967) *The Geography of Market Centers and Retail Distribution*. Englewood Cliffs, NJ: Prentice-Hall.

Burton, I. (1963) 'The quantitative revolution and theoretical geography', *The Canadian Geographer*, 7: 151–62.

Chisholm, G.G. (1886) *Handbook of Commercial Geography*. London: Longman.

Chisholm, M. (1966) *Geography and Economics*. London: Bell.

Chisholm, M. (1971) *Research in Human Geography*. London: Heinemann.

Chisholm, M. (1975) *Human Geography: Evolution or Revolution*. London: Penguin.

Chisholm, M. (2001) 'Human geography joins the Social Science Research Council: personal recollections', *Area*, 33: 428–30.

Chisholm, M. and Manners, G. (eds) (1973) *Spatial Policy Problems of the British Economy*. Cambridge: Cambridge University Press.

Chisholm, M. and Rodgers, B. (eds) (1973) *Studies in Human Geography*. London: Heinemann.

Chorley, R.J., Beckinsale, R.P. and Dunn, A.J. (1973) *The History of the Study of Landforms. Vol. II. The Life and Work of William Morris Davis*. London: Methuen.

Chorley, R.J. and Haggett, P. (eds) (1967) *Models in Geography*. London: Methuen.

Christaller, W. (1966) *Central Places in Southern Germany* (trans. C.W. Baskin from 1933 original in German). Englewood Cliffs, NJ: Prentice-Hall.

Clark, G.L., Feldman, M.P. and Gertler, M.S. (eds) (2000) *The Oxford Handbook of Economic Geography*. Oxford: Oxford University Press.

Clout, H. (2003) 'Place description, regional geography and area studies: the chorological inheritance', in R.J. Johnston and M. Williams (eds) *A Century of British Geography*. Oxford: Oxford University Press.

Coppock, J.T. (1974) 'Geography and public policy: challenges, opportunities and implications', *Transactions, Institute of British Geographers*, 63: 1–16.

Cormack, L. (1997) *Charting an Empire: Geography at the English Universities, 1580–1620*. Chicago, IL: University of Chicago Press.

Cox, K.R. (1976) 'American geography: social science emergent', *Social Science Quarterly*, 57: 182–207.

Davies, W.K.D. (1972) 'The conceptual revolution in geography', in W.K.D. Davies (ed.) *The Conceptual Revolution in Geography*. London: University of London Press, pp. 9–17.

Dickinson, R.E. (1947) *City Region and Regionalism*. London: Routledge & Kegan Paul.

Dodds, K.J. and Atkinson, D. (eds) (2000) *Geopolitical Traditions: A Century of Geopolitical Thought*. London: Routledge.

Driver, F. (2001) *Geography Militant: Cultures of Exploration in an Age of Empire*. Oxford: Blackwell.

Dunbar, G.S. (ed.) (2002) *Geography: Discipline, Profession and Subject since 1870*. Amsterdam: Kluwer.

Fotheringham, A.S. (2006) 'Quantification, evidence and positivism', in S. Aitken and G. Valentine (eds) *Approaches to Human Geography*. London: Sage, pp. 237–50.

Garrison, W.L. (1959–60) 'Spatial structure of the economy I, II and III', *Annals of the Association of American Geographers*, 49–50: 238–9, 357–73 and 471–82.

Garrison, W.L., Berry, B.J.L., Marble, D.F., Nystuen, J.D. and Morrill, R.L. (1959) *Studies of Highway Development and Geographic Change*. Seattle, WA: University of Washington Press.

Gauthier, H.L. (2002) 'Edward "Ned" Taaffe (1921–2001)', *Annals of the Association of American Geographers*, 92: 573–83.

Giddens, A. (1984) *The Constitution of Society*. Cambridge: Polity Press.

Golledge, R.G. (2006) 'Philosophical bases of behavioural research in geography', in S. Aitken and G. Valentine (eds) *Approaches to Human Geography*. London: Sage, pp. 75–85.

Golledge, R.G. and Stimson, R.J. (1996) *Spatial Behavior: A Geographic Perspective*. New York: Guilford Press.

Gregory, D. (1978) *Ideology, Science and Human Geography*. London: Hutchinson.

Gregory, D. (2004) *The Colonial Present: Afghanistan, Palestine, Iraq*. Oxford: Blackwell.

Gregory, S. (1963) *Statistical Methods and the Geographer*. London: Longman.

Hägerstrand, T. (1968) *Innovation Diffusion as a Spatial Process*. Chicago, IL: University of Chicago Press.

Haggett, P. (1965) *Locational Analysis in Human Geography*. London: Edward Arnold.

Hall, P. (2003) 'Geographers and the urban century', in R.J. Johnston and M. Williams (eds) A *Century of British Geography*. Oxford: Oxford University Press.

Halsey, A.H. (2004) *A History of Sociology in Britain: Science, Literature and Society*. Oxford: Oxford University Press.

Hartshorne, R. (1939) *The Nature of Geography*. Lancaster, PA: Association of American Geographers.

Harvey, D. (1969) *Explanation in Geography*. London: Edward Arnold.

Harvey, D. (1973) *Social Justice and the City*. London: Edward Arnold.

Harvey, D. (1982) *The Limits to Capital*. Oxford: Blackwell.

Holmes, J.H. (2002) 'Geography's emerging cross-disciplinary links: processes, causes, outcomes and challenges', *Australian Geographical Studies*, 42: 299–306.

Isard, W. (2003) *History of Regional Science and the Regional Science Association International: The Beginnings and Early History*. Berlin: Springer.

Jackson, P. and Smith, S.J. (1984) *Exploring Social Geography*. London: Allen & Unwin.

Johnston, R.J. (1997a) *Geography and Geographers: Anglo-American Human Geography since 1945*. London: Edward Arnold.

Johnston, R.J. (1997b) 'Where's my bit gone? Reflections on rediscovering geography', *Urban Geography*, 18: 353–9.

Johnston, R.J. (2000) 'Intellectual respectability and disciplinary transformation? Radical geography and the institutionalisation of geography in the USA since 1945', *Environment and Planning A*, 32: 971–90.

Johnston, R.J. (2002) 'Robert E. Dickinson and the growth of urban geography: an evaluation', *Urban Geography*, 22: 702–36.

Johnston, R.J. (2003a) 'The institutionalisation of geography as an academic discipline', in R.J. Johnston and M. Williams (eds) *A Century of British Geography*. Oxford: Oxford University Press, pp. 45–92.

Johnston, R.J. (2003b) 'Order in space: geography as a discipline in distance', in R.J. Johnston and M. Williams (eds) *A Century of British Geography*. Oxford: Oxford University Press, pp. 303–46.

Johnston, R.J. (2003c) 'Geography: a different sort of discipline?', *Transactions, Institute of British Geographers*, NS 29: 133–41.

Johnston, R.J. (2004a) 'Institutions and disciplinary fortunes: two moments in the history of UK geography in the 1960s – I: geography in the 'plateglass universities'', *Progress in Human Geography*, 28: 57–78.

Johnston, R.J. (2004b) 'Institutions and disciplinary fortunes: two moments in the history of UK geography in the 1960s – II: human geography and the Social Science Research Council', *Progress in Human Geography*, 28: 204–26.

Johnston, R.J. (2005a) 'Geography – coming apart at the seams', in N. Castree, A. Rogers and D. Sherman (eds) *Questioning Geography: Fundamental Debates*. Oxford: Blackwell, pp. 9–25.

Johnston, R.J. (2005b) 'Learning our history from our pioneers: UK academic geographers in the *Oxford Dictionary of National Biography*', *Progress in Human Geography*, 29: 651–67.

Johnston, R.J. (2006) 'Research quality assessment and geography in Australia: can anything be learned from the UK experience?', *Geographical Research*, 44: 1–11.

Johnston, R.J., Gregory, D., Pratt, G. and Watts, M. (eds) (2000) *The Dictionary of Human Geography* (4th edn). Oxford: Blackwell.

Johnston, R.J. and Sidaway, J.D. (2004) *Geography and Geographers: Anglo-American Human Geography since 1945*. London: Edward Arnold.

Koelsch, W. (2002) 'Academic geography, American style: an institutional perspective', in G.S. Dunbar (ed.) *Geography: Discipline, Profession and Subject since 1870: An International Survey*. Dordrecht: Kluwer, pp. 281–316.

Kollmorgen, W.N. (1979) 'Kollmorgen as a bureaucrat', *Annals of the Association of American Geographers*, 69: 77–89.

Kuper, A. (1996) *Anthropology and Anthropologists*. London: Routledge.

Livingstone, D.N. (1992) *The Geographical Tradition: Episodes in the History of a Contested Enterprise*. Oxford: Blackwell.

Martin, G.J. (1968) *Mark Jefferson: Geographer*. Ypsilanti, MI: Eastern Michigan University Press.

Martin, G.J. (2005) *All Possible Worlds: A History of Geographical Ideas* (4th edn). New York: Wiley.

Martin, G.J. and James, P.E. (1993) *All Possible Worlds: A History of Geographical Ideas* (3rd edn). New York, Wiley.

Massey, D. (1984) *Spatial Divisions of Labour: Social Structures and the Geography of Production*. London: Macmillan.

Massey, D., Allen J. and Sarre, P. (eds) (1999) *Human Geography Today*. Cambridge: Polity Press.

McDowell, L. (1993) 'Space, place and gender relations (two parts)', *Progress in Human Geography*, 17: 157–79 and 305–18.

McDowell, L. (2003) 'Geographers and sexual difference: feminist contributions', in R.J. Johnston and M. Williams (eds) *A Century of British Geography*. Oxford: Oxford University Press, pp. 603–24.

McDowell, L. (2006) 'Difference and place', in S. Aitken and G. Valentine (eds) *Approaches to Human Geography*. London: Sage, pp. 205–10.

Morrill, R.L. (1970) *The Spatial Organization of Society*. Belmont, CA: Wadsworth.

Morrill, R.L. (2005) 'Hägerstrand and the "quantitative revolution": a personal interpretation', *Progress in Human Geography*, 29: 333–6.

NAS–NRC (1965) *The Science of Geography*. Washington, DC: National Academy of Science–National Research Council.

NRC (1997) *Rediscovering Geography*. Washington, DC: National Research Council.

Peel, J.D.Y. (2006) 'Not really a view from without: the relations between sociology and anthropology', in A.H. Halsey and W.G. Runciman (eds) *British Sociology Seen from Within and Without*. Oxford: Oxford University Press, pp. 70–93.

Platt, J. (2003) *The British Sociological Association: A Sociological History*. Durham: Sociology Press.

Ploszajska, T. (1999) *Geographical Education, Empire and Citizenship: Geographical Teaching and Learning in English Schools*. London: Historical Geography Research Group of the Royal Geographical Society.

Pred, A.R. (1967–9) *Behavior and Location: Foundations for a Geographic and Dynamic Location Theory, Parts I and II*. Lund: C.W.K. Gleerup.

Prince, H.C. (2000) *Geographers Engaged in Historical Geography in British Higher Education 1931–1991*. Historical Geography Research Series 36. London: Historical Geography Research Group of the Royal Geographical Society.

Rawling, E.M. (2001) *Changing the Subject: The Impact of National Policy on School Geography, 1980–2000*. Sheffield: The Geographical Association.

Rawstron, E.M. (2002) 'Textbooks that moved a generation', *Progress in Human Geography*, 26: 831–6.

Rose, G. (1993) *Feminism and Geography*. Cambridge: Polity Press.

Sayer, A. (1984) *Method in Social Science: A Realist Approach*. London: Hutchinson.

Sayer, A. and Walker, R.A. (1992) *The New Social Economy: Reworking the Division of Labour*. Oxford: Blackwell.

Schaefer, F.K. (1953) 'Exceptionalism in geography: a methodological examination', *Annals of the Association of American Geographers*, 43: 226–49.

Schelhas, B. and Hönsch, I. (2002) 'History of German geography: worldwide reputation, and strategies of nationalisation and institutionalisation', in G.S. Dunbar (ed.) *Geography: Discipline, Profession and Subject since 1870: An International Survey*. Dordrecht: Kluwer, pp. 9–44.

Schulten, S. (2001) *The Geographical Imagination in America, 1880–1950*. Chicago, IL: University of Chicago Press.

Smith, D.M. (2000) *Moral Geographies: Ethics in a World of Difference*. Edinburgh: Edinburgh University Press.

Smith, N. (1984) *Uneven Development: Nature, Capital and the Production of Space*. Oxford: Blackwell.

Smith, N. (1987) '"Academic wars over the field of geography": the elimination of geography at Harvard, 1947–1951', *Annals of the Association of American Geographers*, 77: 157–72.

Stamp, L.D. (1946) *The Land of Britain: Its Use and Misuse*. London: Longman.

Steel, R.W. (1974) 'The Third World: geography in practice', *Geography*, 59: 189–97.

Taaffe, E.J. (1970) *Geography*. Englewood Cliffs, NJ: Prentice-Hall.

Taylor, P.J. (1982) 'A materialist framework for human geography', *Transactions, Institute of British Geographers*, NS 7: 15–34.

Taylor, P.J. (1993) 'Full circle, or a new meaning for the global?', in R.J. Johnston (ed.) *The Challenge for Geography: A Changing Word, a Changing Discipline*. Oxford: Blackwell, pp. 181–97.

Turner, B.L. (2002) 'Contested identities: human–environment geography and disciplinary implications in a restructuring academy', *Annals of the Association of American Geographers*, 92: 52–74.

Walford, R. (2001) *Geography in British Schools 1885–2000*. London: Woburn Press.

Ward, R.G. (1960) 'Captain Alexander Maconochie, R.N., 1787–1860', *The Geographical Journal*, 126: 459–68.

Watson, J.W. (1955) 'Geography: a discipline in distance', *Scottish Geographical Magazine*, 71: 1–13.

Whatmore, S. (2002) *Hybrid Geographies*. London: Sage.

Willetts, E.C. (1987) 'Geographers and their involvement in planning', in R.W. Steel (ed.) *British Geography 1918–1945*. Cambridge: Cambridge University Press, pp. 100–16.

Wilson, A.G. (1974) *Urban and Regional Models in Geography and Planning*. Chichester: Wiley.

Wise, M.J. (1975) 'A university teacher of geography', *Transactions, Institute of British Geographers*, 66: 1–16.

Withers, C.W.J. (2002) 'A partial biography: the formalization and institutionalization of geography in Britain since 1887', in G.S. Dunbar (ed.) *Geography: Discipline, Profession and Subject since 1870*. Amsterdam: Kluwer, pp. 79–119.

Withers, C.W.J. and Mayhew, R.J. (2002) 'Rethinking "disciplinary" history: geography in British universities, c. 1580–1887', *Transactions, Institute of British Geographers*, NS 27: 11–29.

4

Geography and the Humanities Tradition

Alison Blunt

D efinition

The humanities encompass the study of human creativity, knowledge, beliefs, ideas, imagination and experience. Such work has inspired a wide range of humanistic, cultural and historical geographical research. At the same time, ideas about space, place and imaginative geographies have inspired work across the humanities. This chapter explores the creative interfaces between geography and the humanities.

INTRODUCTION

Geography and the humanities are intimately connected, although it is only relatively recently that such connections have been explicitly explored in theory and in practice, and both by geographers and by those working in other disciplines. The term 'humanities' refers to the study of human life and humanity more widely and is usually associated with particular subjects, approaches and methods (see Williams, 1976, for more on 'humanity' and related terms). Subjects usually (but not exclusively) located within the humanities include literary studies, languages, history, art, philosophy, archaeology and cultural studies. Research within the humanities is often concerned with human creativity, interpretation, meanings, values and experience, in both historical and contemporary contexts and over both imaginative and material terrains. Rather than view the humanities and social sciences as clearly distinct from each other, it is more helpful to think of a series of interfaces between them. Indeed, the diverse traditions and approaches within human geography have shaped a wide range of research and teaching located at these interfaces, making productive connections that span sources, methods and ideas both within and between the humanities and social sciences. Much has changed since Denis Cosgrove wrote that 'the idea of human geography as a *humanity* is scarcely a mature or fully developed one' (1989: 121).

In the years since Cosgrove made this argument – and in large part inspired by his own research on visual images, iconography and landscape – the idea of human geography as a humanity has taken root to an unprecedented extent. Early attempts to draw creative connections between geography and the humanities have long antecedents (particularly in the work of Carl Sauer on pre-modern cultural landscapes), but are usually attributed to the work of humanistic geographers in the 1970s. As D.W. Meinig (1983: 315) explains:

> while we have long had a 'human geography', we have never before had an explicitly 'humanistic geography' with such a self-conscious drive to connect with that special body of knowledge, reflection, and substance and human experience and human expression, about what it means to be a human being on this earth. (For an overview of humanistic geography, see Cloke et al., 1991.)

Writing to counter positivist spatial science and structural Marxism (see Chapter 3), humanistic geographers sought to replace 'rational economic man' with a fully human subject, whose thoughts, experiences, values, emotions, agency and creativity made him a unique individual within a wider humanity (and it *was* usually 'him' in humanistic work; see below). David Ley and Marwyn Samuels, for example, wanted 'man put back together with all the pieces in place, including a heart and even a soul, with feelings as well as thoughts, with some semblance of secular and perhaps transcendental meaning' (1978: 2–3). At the same time, humanistic geographers claimed that part of what made people human was an intense, sensual, and often passionate, attachment to place. As Yi-Fu Tuan (1976: 269) wrote:

> How a mere space becomes an intensely human place is a task for the humanistic geographer; it appeals to such distinctively humanist interests as the nature of experience, the quality of the emotional bond to physical objects, and the role of concepts and symbols in the creation of place identity.

The work of humanistic geographers was very influential in foregrounding human experience and a sense of place to counter the abstractions of spatial science. Alongside Marxist geographical work and the emerging discipline of cultural studies, humanistic geography laid important foundations for a revitalized cultural geography, particularly from the 1980s onwards (Jackson, 1989; Crang, 1998; Cook et al., 2000). But many critiques of humanistic geography centred on its limited engagement not only with the humanities but also with the very idea of humanity itself. In relation to the humanities, Stephen Daniels (1985: 150) writes that '[t]he naïve approach of most humanistic geographers to the arts, especially their neglect of the artistic form, reveals their poor working knowledge of the humanities, especially literary and art criticism and history'. Not only did humanistic geography largely lack a theoretical and methodological engagement with other research in the humanities – particularly in humanistic studies of geography and literature and the surprising disregard of the visual arts – but humanistic geography also offered little to other scholars in the humanities seeking to understand literary texts and visual images in material as well as imaginative contexts. In relation to humanity more widely, Gillian Rose (1993) critiques the masculinism of humanistic geography that privileges but does not interrogate the primacy of the male subject. In particular, Rose shows how humanistic geographers have celebrated the home as the site of authentic meaning and value but failed to analyse the home as a gendered space shaped by different and unequal relations of power. Alongside such feminist critiques that show the importance of understanding the gendered dynamics of human subjectivity, postcolonial geographical work critiques ethnocentric – and often transparently white – visions of the 'human' in human and humanistic geography (for more on challenging the assumed transparency of whiteness, see Bonnett, 1999; for more on postcolonial geographies, see Blunt and McEwan, 2002).

Emerging, in part, from such critiques of humanistic geography, cultural and historical geographers today engage with the humanities in more critical and diverse ways. Cultural and historical geographers (and, as will become clear, the work of many geographers is *both* cultural and historical) have developed a more critical and theoretical engagement with other research in the humanities, often in the light of post-structuralism, feminism and postcolonialism. There has also been a greater recognition and interrogation of the complexities of humanity itself, both in terms of the politics of identity and also in terms of the moral and ethical considerations that are a crucial part of understanding the human world (Proctor and Smith, 1999; Smith, 2000; Lee and Smith, 2004). At the same time, rather than view space merely as a container for human life and experience, or merely to celebrate a sense of place and individual perceptions of place, there has been an increasingly critical understanding of the production of space and place mediated by different relations of power (see Chapters 5 and 9). More than ever before, scholars working in other disciplines in the humanities are thinking and writing in explicitly spatial terms, most notably in terms of imaginative geographies and the multiple and contested spaces of identity, which are often articulated through spatial images such as mobility, location, borderlands, exile and home. The rest of this chapter addresses these points more fully in relation to three main interfaces between geography and the humanities: geography, text and writing; geography, landscape and the visual arts; and geography, embodiment and performativity.

GEOGRAPHY, TEXT AND WRITING

If 'geography' means 'writing the world', geographers write the world and think about the world *and* writing in different ways. Geographers have increasingly turned their attention not only to 'writing' and the 'world' being written about, but also to the wider politics and poetics of representation (Barnes and Duncan,

1992; Duncan and Ley, 1993; Barnes and Gregory, 1997). For many humanistic geographers, the study of 'geography and literature' offered one way of reading human attachments to, and perceptions of, place (Tuan, 1978; Pocock, 1981; Mallory and Simpson-Housley, 1987). For Tuan (1978: 205):

> Literary art serves the geographer in three principal ways. As though experiment on possible modes of human experience and relationship, it provides hints as to what a geographer might look for when he studies, for instance, social space. As artifact it reveals the environmental perceptions and values of a culture. ... Finally, as an ambitious attempt to balance the subjective and the objective it is a model for geographical synthesis.

Seeking to foster a 'meeting of the disciplines' between geography and literature, Mallory and Simpson-Housely (1987: xii) hoped to 'bridge the gap between the geographer's factual descriptions and the writer's flights of imagination, hence giving the world – both in geographical and literary terms – a more unified shape'. But in both these examples, geography and literary studies – sometimes seeming to parallel 'reality' and 'imagination' – remained largely separate from each other. While literature (particularly novels) provided a new source for geographical analysis, the analysis itself often remained social scientific and paid little attention to literary theory and criticism. Rather than study literary forms, conventions and language, humanistic geographers often regarded 'literature simply [as] perception' (Pocock, 1981: 15). As Daniels (1985: 149) writes, 'For a humanist to abstract the perceptive insight of an artist from the language in which it is embodied is paradoxically to diminish an artist's humanity'.

Such early encounters between geography and literature also led to an increased awareness of geographical writing itself. Many geographers stressed the importance of vivid, compelling descriptions of the world, infusing an understanding of place with human meanings, values and experiences. So, for example, Pierce Lewis described his love of the Michigan sand dunes in sensual and embodied prose:

> My love affair with [the] Michigan dunes ... had everything to do with violent immediate sensations: the smell of October wind sweeping in from Lake Michigan, of sand blown hard against bare legs, the pale blur of sand pluming off the dune crest against a porcelain-blue sky, Lake Michigan a muffled roar beyond the distant beach, a hazy froth of jade and white. As I try to shape words to evoke my feelings, I know why the Impressionists painted landscapes as they did – not literally, but as fragments of color, splashes of pigment, bits of shattered prismatic light. One is meant to feel those landscapes, not to analyze them. I loved those great dunes in my bones and flesh. It was only much later that I learned to love them in my mind as well. (1985: 468; also see Meinig, 1983)

As well as such 'thick' description, a closer attention to creative geographical writing also took a more experimental form, best shown in the work of Gunnar Olsson (1980, 1981), who focused attention on the slippages between language and meaning. While such experimental writings revealed the instabilities and ambiguities of meaning, they have also been critiqued for their opacity. As Daniels (1985: 148) writes: 'You don't have to be a linguistic puritan to conclude that the literary styles of humanistic geographers are no less gratuitous than the quantitative or algebraic flourishes of some positivist geographers'.

Since the late 1980s, a number of geographers have engaged more directly and fully with literary theory and have developed a more critical understanding of text, language, reading and writing (Brosseau, 1994; Sharp, 2000; Kneale, 2003). Often inspired by post-structuralism – particularly in terms of discourse analysis and deconstruction – cultural geographers have turned their attention more fully to the politics of representation. Rather than analyse texts as unproblematic representations of the world or merely in terms of perception, cultural geographers ask more critical questions about texts and the contexts in which they are written and read. A number of geographers have also developed a more metaphorical understanding of text, as shown by the work of James and Nancy Duncan on reading landscapes as texts rather than reading texts merely as 'literary artifacts' (Duncan and Duncan, 1988; Duncan, 1990). Geographers have also begun to analyse a much wider range of writing, including travel writing (Blunt, 1994; Duncan and Gregory, 1999), tour guides (Gilbert, 1999; Howell, 2001), geography textbooks (Ploszajska, 2000), children's fiction (Phillips, 1997, 2001), letters, memoirs and diaries (Blunt, 2000; Gowans, 2001), science fiction (Kitchin and Kneale, 2002) and other novels (Sharp, 1994; Brosseau, 1995). One key theme has been an attempt to explore geographies of writing in both imaginative and material contexts and in the very form of writing itself. So, for example, diaries can be read as 'spatial stories', detailing events and feelings not only at the time but also in the place where they occur, while travel writing is inherently geographical in its depiction of mobility within and between the spaces of home and away. Much of this more recent geographical work on writing has been influenced by postcolonialism and feminism, interrogating identity in terms of gender and race and, to a lesser extent, class and sexuality. Most importantly, geographers increasingly recognize that writing – and representation more broadly – is located within a nexus of power relations.

While geographers are increasingly inspired by work across the humanities in the study of writing, a number of literary theorists have explored the imaginative geographies produced in part through writing. Perhaps most famously in this context, the literary critic and postcolonial theorist, Edward Said, has written about the imaginative geographies of 'self' and 'other' in his study of Orientalism (Said, 1995; for more on imaginative geographies and Said's work, see Gregory, 1995; Rose, 1995; Driver, 1999). Said shows the interplay of power, knowledge and representation through written and visual depictions of an exotic East. Focusing on texts such as travel writings, scholarly accounts and novels, Said argues that Orientalism produced knowledges about colonized people and places as inferior, irrational and 'other' to a powerful, rational, Western 'self'. As he shows, such writings, and the imaginative geographies they helped to shape, had material consequences in the exercise and legitimation of imperial rule. More recently, Said has written a memoir of his early life in Jerusalem, Cairo, Lebanon and the USA. Aptly titled *Out of Place* (Said, 1999), geography lies at the heart of Said's memoir. As he explains: 'Along with language, it is geography – especially in the displaced form of departures, arrivals, farewells, exile, nostalgia, homesickness, belonging and travel itself – that is at the core of my memories' (Said, 1999: xvi). Imaginative and material geographies are clearly central to interpreting text and writing not only within but also far beyond the discipline of geography.

GEOGRAPHY, LANDSCAPE AND THE VISUAL ARTS

While humanistic geographers saw literature as an important source for describing human emotion, experience and a sense of place, they paid far less attention to the visual arts (Daniels, 1985). More recently, cultural geographers have increasingly recognized that vision, visual representation and 'the gaze' – in other words, practices of looking and observing – are inherently spatial. Interpreting the spatiality of the gaze means paying attention to *where* observation takes place as well as what is being observed; thinking critically about the distance between the observer and the subject of observation; seeking to locate and embody the gaze and to challenge claims for objective detachment; and analysing various framing strategies that put boundaries around what is being observed (and for more on how to do this in practice, see Rose, 2001). Geographers have always been interested in visual representations of the world, particularly through maps and mapping. Rather than view maps in positivist terms as objective reflections of reality, many geographers now argue that all maps are socially constructed forms of knowledge that are partial, infused with different meanings and shaped by different relations of power (Harley, 1988; Woods, 1994; Pinder, 2003). Geographers have, for example, studied the ways in which maps have been produced and used not only as objects of imperial power but also of postcolonial resistance (see Jacobs, 1996, for an example of Aboriginal 'counter-cartography' in Australia). In his work on avant-garde visions of the city, David Pinder (1996; also see Pinder, 2005) explores Utopian remappings of urban space as a space of human freedom and creativity, in part by overturning conventions not only of cartography but also of urban planning and design. Such critical, interpretative and often deconstructive readings of maps are closely related to the visual and textual analysis associated with work in the humanities.

Unlike the textual descriptions of landscape in humanistic geography, many cultural geographers have studied landscape as a 'way of seeing' and have analysed visual depictions of landscape in painting and photography and, to a lesser extent, in film (Cosgrove and Daniels, 1988; Aiken and Zonn 1994; Ryan, 1997; Nash, 1999a; Schwartz and Ryan, 2003). As Cosgrove and Jackson (1987: 96) explain, 'the landscape concept is itself a sophisticated cultural construction', and one that is closely tied to power relations. Denis Cosgrove (1985) traces the 'idea of landscape' to landscape painting and property ownership in the Renaissance, arguing that visual depictions of landscape – revolutionized by the geometric representation of space – were bound up with the interests of the powerful. Geographers study the iconography of landscape, which involves studying the symbolic significance of the landscape itself and analysing symbolic markers within a landscape, such as monuments and place-names, in their social and political context (Cosgrove and Daniels, 1988; Johnson, 1995; Nash, 1999b).

Geographers also study the ways in which ideas of landscape are closely bound up with imaginative and material geographies of nation and empire. David Matless (1998; see also Daniels, 1993), for example, analyses a diverse range of visual and textual sources in his study of landscape and Englishness from the early twentieth century to the present day. Focusing on debates about nation, citizenship and heritage, he shows how ideas of Englishness were intimately bound up with ideas of landscape. Other work shows how such ideas of Englishness and landscape have had far-reaching effects over imperial space. In her discussion of the picturesque style of painting and viewing landscape, Catherine Nash explains that:

> [a] sense of the superiority of English landscape aesthetics was linked to a broader certainty that English ways were the best ways of doing things and of their natural superiority and authority over other people and places. Ideas of landscape were involved in the multiple ways in which European expansion within imperial trade and colonization was naturalized and legitimated. (1999a: 220; also see Seymour, 2000)

A number of feminist geographers have argued that ideas of landscape, and the geographical analysis of visual images more widely, are profoundly gendered both in theory and practice. Gillian Rose (1993: 61) critiques both 'social-scientific' and 'aesthetic' masculinism, with the latter claiming 'complete sensitivity to a mysterious yet crucial world'. Focusing on the work of 'new' cultural geographers, Rose argues that there are uneasy tensions between pleasure in, and knowledge about, landscape that reflect and repeat profoundly gendered distinctions between a feminine 'nature' and a masculine 'culture'. As she writes: 'Pleasure in the landscape is often seen as a threat to the scientific gaze, and it is argued that the geographer should not allow himself to be seduced by what he sees' (Rose, 1993: 72). Feminist geographers argue that not only are vision, visual representation, 'the gaze' and ideas about landscape inherently spatial, but that they are also gendered in important, but often unacknowledged, ways. As Rose (1993: 109) puts it, 'cultural geography's erotics of knowledge' are masculinist and heterosexist, gazing on a feminized landscape in voyeuristic, distanced and disembodied ways that render the specificity of its own gaze unmarked and invisible. Countering such a masculinist gaze, Catherine Nash has studied the gendered and sexualized interplay of landscape, body and nation in modern Ireland (Nash, 1994, 1996), while Rose has more recently written about photographs of, and by, women in the nineteenth and early twentieth centuries (1997, 2000) and about contemporary family photographs (2003, 2004).

I want to contrast two images to illustrate the gendered geographies of the gaze in more detail. Both images date from 1857 and portray gendered spaces of home, nation and empire through English eyes. The first is a painting called 'The sinews of old England' by George Elgar Hicks (Figure 4.1). It shows a family positioned on the threshold between home and world. The title and the painting represent English national identity in clearly embodied terms that centre on the strength and work of the man. But in this image such an embodied, masculine and working-class national identity also depends on feminized domesticity. The

Figure 4.1 'The sinews of old England', by George Elgar Hicks, 1857

Source: Unknown (thought to be in a Private collection)

man is pictured with his wife and child, and the viewer can look into the tidy and well-ordered home beyond the threshold. While the man gazes into the distance away from home and beyond the gaze of both his wife and the viewer, his wife looks up at him lovingly. While the man does not return this gaze, we, as viewers of the image, see the family as a whole within its pastoral and domestic context. This mid-Victorian painting is a classic representation of public and private space and the white, heterosexist gender order on which such spatial and social divisions relied.

Compare Hicks' painting with another image from 1857 (Figure 4.2). This second image is a cartoon entitled 'How the Mutiny came to English homes' and is located in India at the start of the uprising that threatened to overthrow British rule (for more on home, nation and empire, see Blunt and Dowling, 2006). While Hicks' painting offered a glimpse over the threshold into a tidy and ordered home of peace and calm, this image represents the home as a place of danger and violation. With a young child playing next to her and a baby at her

Figure 4.2   'How the Mutiny came to English homes', 1857
Source: Unknown

breast, a British woman is located at the centre of domestic and familial calm that
has just been shattered by the invasion of two Indian insurgents. The violent
presence of these men is the only indication that the home is in India rather than
Britain. The rebels appear set to destroy the defenceless woman, her children
and the home itself. Moreover, the box labelled 'England' on the *chaise-longue*
suggests the vulnerability of national and imperial power. The absent British hus-
band and father – present only in a portrait on the wall – is unable to protect his
wife and family. This image stands in stark contrast to Hicks' painting of national
identity embodied by a family on the threshold between home and world. It represents the threat to domestic, national and imperial life during the uprising, with
the severity of the threat shown by the vulnerability of a British wife, mother and
children. Both images show the political importance of the home in terms of
national and imperial identity, and both images also show the gendered subjects
and spaces shaping national and imperial identity. If, as Hicks' painting suggests,
the 'sinews of old England' were embodied by masculine strength and feminine
domesticity, the cartoon shows such sinews – and the security and identity of 'old
England' itself – being torn apart.

    While geographers now study visual images in more critical ways, a
number of art historians and film theorists have begun to make the spatiality of
visual images central to their work. Feminist work has been particularly influential in seeking to embody and contextualize the gendered spatiality of spectator-
ship and the gaze. As one example of such work, Griselda Pollock (1988) analyses
the paintings of two female Impressionists in explicitly spatial ways. Focusing on
the work of Berthe Morisot and Mary Cassatt in late nineteenth-century Paris,

Pollock explores the spaces of their art in three connected ways. First, she writes about the spaces represented in their paintings, which were often interior, domestic spaces. Then she explores the spatial order within their work, showing how spaces are often separated by a balcony or a balustrade, marking spaces in gendered ways and positioning both the painter and the viewer in close proximity to the subject of the painting. Finally, she examines the social spaces of representation in terms of where and what a woman was able to paint. She quotes from the diary of Marie Baskirteff, another artist in Paris in the nineteenth century, who wrote that her lack of independence and mobility constrained her work:

> What I long for is the freedom of going about alone, of coming and going ... of stopping and looking at the artistic shops, of entering churches and museums, of walking about old streets at night; that's what I long for; and that's the freedom without which one cannot become a real artist. Do you imagine that I get much good from what I see, chaperoned as I am, and when, in order to go to the Louvre, I must wait for my carriage, my lady companion, my family? (quoted in Pollock, 1988: 70)

Imaginative and material geographies are also clearly central to interpreting landscape and the visual arts not only within, but also far beyond, the discipline of geography.

GEOGRAPHY, EMBODIMENT AND PERFORMATIVITY

Unlike humanistic geographical work that celebrated human creativity, agency and individuality, other geographical work has engaged more critically with questions of identity by interrogating both the politics of identity and the political intersections of place and identity (Rose, 1995). An important theme within such work has been to understand and interpret identities in embodied terms (Butler, 1999; McDowell, 1999; Longhurst, 2000; Valentine, 2001). Ideas about embodiment have important implications not only for studying identity but also for the production of knowledge more widely. Resisting disembodied claims to objectivity, many scholars argue that knowledge itself should be embodied. In the context of geographical knowledge more specifically, James Duncan and Derek Gregory (1999) argue that studies of travel writing should encompass the embodied practices of travel itself, while Felix Driver (2001) argues that fieldwork and exploration also need to be understood as embodied practices rather than read solely in textual terms. Retelling a famous story of exploration, John Wylie (2002a) studies the Antarctic expeditions of Scott and Amundsen in terms of their embodied understanding of the landscape and, more specifically, of ice (also see Wylie, 2002b and 2005 for other studies of embodied landscapes and mobilities, this time walking in England). Contrasting English and Norwegian encounters with Antarctic ice, Wylie shows how the landscape itself came to shape different mobilities across it. Elsewhere, Wylie (2002c) discusses the different modes of dwelling and moving embodied by Scott and Amundsen in their polar expeditions. By paying attention to their 'sensualities and sensibilities' (2002c: 263), Wylie (2002c: 259) vivdly describes exploration as an embodied practice, as shown by the following passage:

> Each night in the tents, eyes blinded by the luminous intimacy of the landscape are treated with zinc sulphate and cocaine, then swaddled in rags and tea leaves. Frozen feet and hands are placed upon the warm chests and stomachs of con-senting companions in a series of awkward embraces, unlikely arrangements of bodily parts. Antarctica demands, above all, that the frontiers of one's body be rigorously established and maintained.

Ideas about embodied practices are often closely tied to ideas about performance and performativity, as shown by geographical studies of city workplaces (McDowell, 1997), gay and lesbian identities (Bell and Valentine, 1995) and more theoretical accounts about the spatialities of performance (Dewsbury et al., 2002; Latham and Conradson, 2003). Often closely bound up with research on embod-iment and performativity, a growing field of work explores and articulates sen-sual, emotional and/or affective geographies (Thrift, 2004; Davidson et al., 2005; Thien, 2005). Geographical studies of embodiment, performativity and emotion offer important new ways of thinking about geography in relation to the human-ities. At the same time, such studies also represent one of the most productive interfaces between social scientific and humanities-based geographical approaches, methods and analysis, whereby interviews, ethnographic research and focus group discussions can be combined with textual, visual and aural analysis of different sources and practices (see Blunt et al., 2003 for a range of examples of cultural geography in practice). Geographers have begun to engage with cultural forms and practices beyond literature and the visual arts that include creative fields such as dance, theatre and music (Thrift, 1997, 1999; Leyshon et al., 1998; Malbon, 1999; Cresswell, 2006). Geographers and many others working in the arts and humanities have also paid more critical attention to the spaces of display and performance, both within and beyond spaces that include art galleries, museums, music venues and nightclubs. Closely tied to this critical engagement with space, embodiment and performativity is a growing interest in cultural *practices* as well as cultural forms, artefacts, texts and images. The diversity of such interest is reflected in a section in each issue of the journal *Cultural Geographies* (formerly *Ecumene*) dedicated to 'Cultural geographies in practice', which 'offers a space for critical reflection on how practices within the artistic, civic and policy fields inform and relate to the journal's cultural geo-graphic concerns'. I want to discuss two examples that illustrate this interest before turning to some theoretical and methodological challenges posed by studying embodiment and performativity.

     The first example concerns an artwork that takes place beyond the space of the gallery, and one that is concerned with sound, vision, movement and embodiment. In his study of an audio walk through east London by artist Janet Cardiff, David Pinder (2001: 2) shows that '[t]he artwork literally takes place in the streets, finding its meaning through its embodied enaction. In effect, it is per-formed or co-created by participants'. While the artwork is 'a highly specific experience [that] is different according to mood, circumstances, events as well as the identity of the individual walker' (2001: 15), it also raises wider questions about the city itself. As Pinder argues, the audio walk 'raises critical questions about reading and representing urban ambiences, about the interweaving of memories and urban space, and about the construction of senses of self through urban space-times' (2001: 15). Addressing similar themes about embodied engagement and performance, but in a different context and both within and far beyond the space of the gallery, Ian Cook discusses the art installation by Shelley Sacks called 'Exchange values'. As Cook (2000: 338) begins:

> They're in your face. Banana skins. Dried. Cured. Blackened. Flattened. Sewn together in a panel. Stretched. Taut. On a frame. Right in your face. And it smells. It's rich. Gorgeous. You can't move too far away. Get too distanced. If you want to keep the headphones on. The ones that are attached to the little metal box below the panel and the frame. The one with the number on. E490347.

As Cook explains, the number E490347 identifies Vitalis Emmanuel, a banana farmer from St Lucia. He grows bananas for export, and this number appears on the boxes that protect them as they are transported around the world. Some of his bananas were handed out to people in Nottingham on condition that they ate them on the spot and gave back the skins. The banana skins form the basis of the installation, and the metal box below contains a tape that plays the voices of 20 farmers including Vitalis Emmanuel. Cook (2000: 342) describes the installation in terms of connective and collective creativity between, for example:

> Shelley Sacks, the people who helped her to give out those bananas in Nottingham and to collect the skins, the people who helped to dry, cure and stitch those skins onto panels, the 20 St Lucian banana growers, and all those people – politicians, business people, extension officers and others – who helped her to find them. Add to that the representative groups campaigning for changing relationships between producers and consumers.

Such connections not only stretch over space but also create 'a reflective space. A space of possibility. Where connections can be seen. Felt. Thought through' (2000: 342).

As these two examples show, ideas about embodiment and performativity, cultural practices and sensual geographies are inspiring a diverse and exciting body of geographical work. These ideas also pose some important theoretical and practical challenges. A central task for geographers working within a humanities tradition is to examine their own research in practice by engaging more fully and directly with methodological debates. Nigel Thrift's ideas about non-representational theory – and the challenges of putting such ideas into practice – are particularly important in this context. Thrift (1999: 318) argues that cultural geography 'often seems to have taken representation as its central focus to its detriment' and proposes a 'non-representational theory' of dance and other embodied practices. As Catherine Nash (2000: 656) writes: 'Thrift is advocating a new and demanding direction for cultural geography, away from the analysis of texts, images and discourses, and towards understanding the micro-geographies of habitual practices, departing from deconstructing representations to explore the non-representational'. Nash ends her review of work on 'performativity in practice' with some wariness about 'a retreat from feminism and the politics of the body in favour of the individualistic and universalizing sovereign subject' (2000: 662). Echoes of a humanistic geography clearly remain important and contested in humanities-based geographical work today.

CONCLUSION

Much has changed since Cosgrove (1989) lamented that the idea of geography as a humanity was still largely undeveloped and immature. From the work of humanistic geographers in the 1970s, through the revitalization of cultural geography since the 1980s, the idea of geography as a humanity has taken much stronger root. The three parts of this chapter show the different ideas, sources and approaches involved in geographical studies of text and literature, landscape and the visual arts, and embodiment and performativity. In each case, several key themes emerge as geographers today engage with the humanities in more critical and diverse ways. First, cultural and historical geographers have developed a closer theoretical engagement with other research in the humanities, particularly in relation to studies of literature and the visual arts and, more recently, in relation to studies of dance, theatre and music. Second, there has also been a greater recognition and interrogation of the complexities of humanity itself, both in terms of the politics of identity and in envisaging a humane geography. Finally, there has been an increasingly critical understanding of the power-laden production of space and place. Geographical ideas about the spatiality of identity and the politics of place have become increasingly important in other disciplines across the humanities. While the humanities continue to inspire a vibrant range of geographical research, geographical concerns are also increasingly important in stimulating and articulating other research across the humanities today.

SUMMARY

- Humanities-based geographical research has become increasingly important since the work of humanistic geographers in the 1970s and cultural geographers since the 1980s.

- More than ever before, scholars working in other disciplines in the humanities are engaging with geographical ideas.

- Examples of humanities-based geographical work include historical and cultural research and studies of literature, the visual arts, and other cultural forms and practices.

- Geographers working within a humanities tradition address critical questions about the production of space, the politics of representation and embodied knowledges and identities.

Further Reading

A good place to start is with some early humanistic geographical writings: Ley and Samuels' (1978) edited collection, **Humanistic Geography: Prospects and Problems**, Daniels' (1985) critique of humanistic geography and Cosgrove's (1989) argument about the importance of thinking about geography as a humanity. Two edited collections – Barnes and Duncan (1992) and Duncan and Ley (1993) – and pieces by Kneale (2003) and Sharp (2000) are good starting points for more on geography, text and writing. Read Cosgrove (1985) in **Transactions**, Matless's (1998) **Landscape and Englishness** and Nash (1996) in **Gender, Place and Culture** for more on geography, landscape and the visual arts. Then read Catherine Nash's (2000) review article about geography, embodiment and performativity, and see David Pinder's (2001) study of embodied geographies in the city and Ian Cook's (2000) article on the exhibition 'Exchange values'. Also see other papers in the regular section 'Cultural geographies in practice' in the journal **Cultural Geographies**.

Note: Full details of the above can be found in the references list below.

References

Aiken, S.C. and Zonn, L.E. (eds) (1994) *Place, Power, Situation and Spectacle: A Geography of Film*. Lanham, MD: Rowman & Littlefield.

Barnes, T. and Duncan, J. (eds) (1992) *Writing Worlds: Discourse, Text and Metaphor in the Representation of Landscape*. London: Routledge.

Barnes, T. and Gregory, D. (eds) (1997) *Reading Human Geography: The Poetics and Politics of Inquiry*. London: Arnold.

Bell, D. and Valentine, G. (eds) (1995) *Mapping Desire: Geographies of Sexualities*. London: Routledge.

Blunt, A. (1994) *Travel, Gender and Imperialism: Mary Kingsley and West Africa*. New York: Guilford Press.

Blunt, A. (2000) 'Spatial stories under siege: British women writing from Lucknow in 1857', *Gender, Place and Culture*, 7: 229–46.

Blunt, A. and Dowling, R. (2006) *Home*. London: Routledge.

Blunt, A., Gruffudd, P., May, J., Ogborn, M. and Pinder, D (eds) (2003) *Cultural Geography in Practice*. London: Arnold.

Blunt, A. and McEwan, C. (eds) (2002) *Postcolonial Geographies*. London: Continuum.

Bonnett, A. (1999) *White Identities*. Harlow: Prentice-Hall.

Brosseau, M. (1994) 'Geography's literature', *Progress in Human Geography*, 18: 333–53.

Brosseau, M. (1995) 'The city in textual form: Manhattan Transfer's New York', *Ecumene*, 2: 89–114.

Butler, R. (1999) 'The body', in P. Cloke et al. (eds) *Introducing Human Geographies*. London: Edward Arnold, pp. 238–45.

Cloke, P., Philo, C. and Sadler, D. (1991) *Approaching Human Geography*. London: Paul Chapman.

Cook, I. (2000) 'Social sculpture and connective aesthetics: Shelley Sacks's "Exchange values" ', *Ecumene*, 7: 337–43.

Cook, I., Crouch, D., Naylor, S. and Ryan, J. (eds) (2000) *Cultural Turns/Geographical Turns*. London: Prentice-Hall.

Cosgrove, D. (1985) 'Prospect, perspective and the evolution of the landscape idea', *Transactions, Institute of British Geographers*, 10: 45–62.

Cosgrove, D. (1989) 'Geography is everywhere: culture and symbolism in human landscapes', in D. Gregory and R. Walford (eds) *Horizons in Human Geography*. Basingstoke: Macmillan, pp. 118–35.

Cosgrove, D. and Daniels, S. (eds) (1988) *The Iconography of Landscape*. Cambridge: Cambridge University Press.

Cosgrove, D. and Jackson, P. (1987) 'New directions in cultural geography', *Area*, 19: 95–101.

Crang, M. (1998) *Cultural Geography*. London: Routledge.

Cresswell, T. (2006) *On the Move: Mobility in the Modern Western World*. London: Routledge.

Daniels, S. (1985) 'Arguments for a humanistic geography', in R.J. Johnston (ed.) *The Future of Geography*. London: Methuen, pp. 143–58.

Daniels, S. (1993) *Fields of Vision: Landscape Imagery and National Identity in England and the United States*. Cambridge: Polity Press.

Davidson, J., Bondi, L. and Smith, M. (eds) (2005) *Emotional Geographies*. Aldershot: Ashgate.

Dewsbury, J.D., Wylie, J., Harrison, P. and Rose, M. (2002) 'Enacting geographies', *Geoforum*, 32: 437–41.

Driver, F. (1999) 'Imaginative geographies', in P. Cloke et al. (eds) *Introducing Human Geographies*. London: Edward Arnold, pp. 209–16.

Driver, F. (2001) *Geography Militant: Cultures of Exploration and Empire*. Oxford: Blackwell.

Duncan, J. (1990) *The City as Text: The Politics of Landscape Interpretation in the Kandyan Kingdom*. Cambridge: Cambridge University Press.

Duncan, J. and Duncan, N. (1988) '(Re)reading the landscape?', *Environment and Planning D: Society and Space*, 6: 117–26.

Duncan, J. and Gregory, D. (eds) (1999) *Writes of Passage: Reading Travel Writing*. London: Routledge.

Duncan, J. and Ley, D. (eds) (1993) *Place/Culture/Representation*. London: Routledge.

Gilbert, D. (1999) 'London in all its glory – or how to enjoy London: guidebook representations of imperial London', *Journal of Historical Geography*, 25: 279–97.

Gowans, G. (2001) 'Gender, imperialism and domesticity: British women repatriated from India, 1940–1947', *Gender, Place and Culture*, 8: 255–69.

Gregory, D. (1995) 'Imaginative geographies', *Progress in Human Geography*, 19: 447–85.

Harley, B. (1988) 'Maps, knowledge and power', in D. Cosgrove and S. Daniels (eds) *The Iconography of Landscape*. Cambridge: Cambridge University Press, pp. 277–312.

Howell, P. (2001) 'Sex and the city of bachelors: popular masculinity and public space in nineteenth-century England and America', *Ecumene*, 8: 20–50.

Jackson, P. (1989) *Maps of Meaning: An Introduction to Cultural Geography*. London: Unwin Hyman.

Jacobs, J. (1996) *Edge of Empire: Postcolonialism and the City*. London: Routledge.

Johnson, N. (1995) 'Cast in stone: monuments, geography and nationalism', *Environment and Planning D: Society and Space*, 13: 51–65.

Kitchin, R. and Kneale, J. (eds) (2002) *Lost in Space: Geographies of Science Fiction*. London: Continuum.

Kneale, J. (2003) 'Secondary worlds: reading novels as geographical research', in A. Blunt, P. Gruffudd, J. May, M. Ogborn and D. Pinder (eds) *Cultural Geography in Practice*. London: Edward Arnold, pp. 39–51.

Latham, A. and Conradson, D. (2003) 'The possibilities of performance', *Environment and Planning A*, 35: 1901–6.

Lee, R. and Smith, D.M. (eds). (2004) *Geographies and Moralities: International Perspectives on Development, Justice and Place*. Oxford: Blackwell.

Lewis, P. (1985) 'Beyond description', *Annals of the Association of American Geographers*, 75: 465–77.

Ley, D. and Samuels, M. (eds) (1978) *Humanistic Geography: Prospects and Problems*. London: Croom Helm.

Leyshon, A., Matless, D. and Revill, G. (eds) (1998) *The Place of Music*. New York: Guilford Press.

Longhurst, R. (2000) *Bodies: Exploring Fluid Boundaries*. London: Routledge.

Malbon, B. (1999) *Clubbing: Dancing, Ecstasy, Vitality*. London: Routledge.

Mallory, W.E. and Simpson-Housley, P. (eds) (1987) *Geography and Literature: A Meeting of the Disciplines*. Syracuse, NY: Syracuse University Press.

Matless, D. (1998) *Landscape and Englishness*. London: Reaktion.

McDowell, L. (1997) *Capital Culture: Gender at Work in the City*. Oxford: Blackwell.

McDowell, L. (1999) *Gender, Identity and Place: Understanding Feminist Geographies*. Cambridge: Polity Press.

Meinig, D.W. (1983) 'Geography as an art', *Annals of the Association of American Geographers*, 8: 314–28.

Nash, C. (1994) 'Remapping the body/land: new cartographies of identity, gender, and landscape in Ireland', in A. Blunt and G. Rose (eds) *Writing Women and Space: Colonial and Postcolonial Geographies*. New York: Guilford Press, pp. 227–50.

Nash, C. (1996) 'Reclaiming vision: looking at landscape and the body', *Gender, Place and Culture*, 3: 149–69.

Nash, C. (1999a) 'Landscapes', in P. Cloke et al. (eds) *Introducing Human Geographies*. London: Edward Arnold, pp. 217–25.

Nash, C. (1999b) 'Irish placenames: post-colonial locations', *Transactions, Institute of British Geographers*, 24: 457–80.

Nash, C. (2000) 'Performativity in practice: some recent work in cultural geography', *Progress in Human Geography*, 24: 653–64.

Olsson, G. (1980) *Birds in Egg/Eggs in Bird*. London: Pion.

Olsson, G. (1981) 'On yearning for home: an epistemological view of ontological transformations', in D. Pocock (ed.) *Humanistic Geography and Literature*. London: Croom Helm, pp. 121–9.

Phillips, R. (1997) *Mapping Men and Empire: A Geography of Adventure*. London: Routledge.

Phillips, R. (2001) 'Politics of reading: decolonizing children's geographies', *Ecumene*, 8: 125–50.

Pinder, D. (1996) 'Subverting cartography: the situationists and maps of the city', *Environment and Planning* A, 28: 405–27.

Pinder, D. (2001) 'Ghostly footsteps: voices, memories and walks in the city', *Ecumene*, 8: 1–19.

Pinder, D. (2003) 'Mapping worlds: cartography and the politics of representation', in A. Blunt, P. Gruffudd, J. May, M. Ogborn and D. Pinder (eds) *Cultural Geography in Practice*. London: Edward Arnold, pp. 172–90.

Pinder, D. (2005) *Visions of the City*. Edinburgh: Edinburgh University Press.

Ploszajska, T. (2000) 'Historiographies of geography and empire', in B. Graham and C. Nash (eds) *Modern Historical Geographies*. Harlow: Longman, pp. 121–45.

Pocock, D. (ed.) (1981) *Humanistic Geography and Literature*. London: Croom Helm.

Pollock, G. (1988) *Vision and Difference: Femininity, Feminism and the Histories of Art*. London: Routledge.

Proctor, J. and Smith, D. (eds) (1999) *Geography and Ethics: Journeys in a Moral Terrain*. London: Routledge.

Rose, G. (1993) *Feminism and Geography*. Cambridge: Polity Press.

Rose, G. (1995) 'Place and identity: a sense of place', in D. Massey and P. Jess (eds) *A Place in the World?* Oxford: Oxford University Press, pp. 87–132.

Rose, G. (1997) 'Engendering the slum: photography in East London in the 1930s', *Gender, Place and Culture*, 4: 277–300.

Rose, G. (2000) 'Practising photography: an archive, a study, some photographs and a researcher', *Journal of Historical Geography*, 26: 555–71.

Rose, G. (2001) *Visual Methodologies*. London: Sage.

Rose, G. (2003) 'Family photographs and domestic spacings: a case study', *Transactions, Institute of British Geographers*, 28: 5–18.

Rose, G. (2004) "Everyone's cuddled up and it just looks really nice": an emotional geography of some mums and their family photos', *Social and Cultural Geography*, 5: 549–64.

Ryan, J. (1997) *Picturing Empire*. London: Reaktion.

Said, E. (1995) *Orientalism*. London: Penguin. (First published 1978.)

Said, E. (1999) *Out of Place: A Memoir*. London: Granta.

Schwartz, J. and Ryan, J. (eds) (2003) *Picturing Place: Photography and the Geographical Imagination*. London: I.B. Tauris.

Seymour, S. (2000) 'Historical geographies of landscape', in B. Graham and C. Nash (eds) *Modern Historical Geographies*. Harlow: Longman, pp. 193–217.

Sharp, J. (1994) 'A topology of "post" nationality: (re)mapping identity in *The Satanic Verses*', *Ecumene*, 1: 65–76.

Sharp, J. (2000) 'Towards a critical analysis of fictive geographies', *Area*, 32: 327–34.

Smith, D. (2000) *Moral Geographies: Ethics in a World of Difference*. Edinburgh: Edinburgh University Press.

Thien, D. (2005) 'After or beyond feeling? A consideration of affect and emotion in geography', *Area*, 37: 450–6.

Thrift, N. (1997) 'The still point: resistance, expressive embodiment and dance', in S. Pile and M. Keith (eds) *Geographies of Resistance*. London: Routledge, pp. 124–51.

Thrift, N. (1999) 'Steps to an ecology of place', in D. Massey et al. (eds) *Human Geography Today*. Cambridge: Polity Press, pp. 295–322.

Thrift, N. (2004) 'Intensities of feeling: the spatial politics of affect', *Geografiska Annaler Series B*, 86: 57–78.

Tuan, Y.-F. (1976) 'Humanistic geography', *Annals of the Association of American Geographers*, 66: 266–76.

Tuan, Y.-F. (1978) 'Literature and geography: implications for geographical research', in D. Ley and M. Samuels (eds) *Humanistic Geography: Prospects and Problems*. London: Croom Helm, pp. 194–206.

Valentine, G. (2001) *Social Geographies: Space and Society*. Harlow: Pearson.

Williams, R. (1976) *Keywords: A Vocabulary of Culture and Society*. London: Fontana.

Woods, D. (1994) *The Power of Maps*. New York: Guilford Press.

Wylie, J. (2002a) 'Earthly poles: the Antarctic voyages of Scott and Amundsen', in A. Blunt and C. McEwan (eds) *Postcolonial Geographies*. London: Continuum, pp. 169–83.

Wylie, J. (2002b) 'An essay on ascending Glastonbury Tor', *Geoforum*, 32: 441–54.

Wylie, J. (2002c) 'Becoming-icy: Scott and Amundsen's South Polar voyages, 1910–1913', *Cultural Geographies*, 9: 249–65.

Wylie, J. (2005) 'A single day's walking: narrating self and landscape on the South West Coast Path', *Transactions, Institute of British Geographers*, 30: 234–47.

KEY CONCEPTS

5

Space: The Fundamental Stuff of Geography

Nigel Thrift

D efinition

As with terms like 'society' and 'nature', space is not a common-sense external background to human action. Rather, it is the outcome of a series of highly problematic temporary settlements that divide and connect things up in to different kinds of collectives which are slowly provided with the means which render them durable and sustainable.

INTRODUCTION

'Space' is often regarded as the fundamental stuff of geography. Indeed, so fundamental that the well-known anthropologist Edward Hall once compared it to sex: 'It is there but we don't talk about it. And if we do, we certainly are not expected to get technical or serious about it' (cited in Barcan and Buchanan, 1999: 7). Indeed, it would be fairly easy to argue that most of the time most geographers do tend to get rather embarrassed when challenged to come out with ideas about what the supposed core of their subject is, and yet they continue to assert its importance. Rather like sex, they argue, without space we would not be here. So is all this just mass disciplinary hypocrisy? Not really. It is more about the extreme difficulty of describing certain aspects of the medium which is the discipline's message.

This brief introduction to the topic of space aims to tell you what space is and why we need to study it. It will do this as straightforwardly as possible, but it is important to point out that one of the problems that geographers have with space is that something that appears as though it really ought to be quite straightforward very often isn't – after all, we all have trouble at times in getting from A to B!

Even nowadays, of course, some geographers still persist in believing that it ought to be possible to explain space in such simple terms that you should be able to understand what is going on straight off. But increasingly, this kind of simple-minded approach has come to be understood as more likely to be part of a desperate attempt to try to render down the wonderful complexity and sheer richness of the world in ways which mimic the predictable worlds of those privileged few who have the ability to make things show up in the way they want them to (Latour, 1997). In this piece, in contrast, while I will certainly attempt to write about space clearly, you should not think that this will be the end of the matter. You will need to read more and think more to really start to get a grip on the grip that space exerts on all our lives – and, as we shall see, the ways that we can alter that grip in order to make new kinds of spaces.

Space has been written about in lots of ways. There are, for example, books upon books which document the different kinds of conceptions of space that can be found in disciplines like philosophy or physics (e.g. Crang and Thrift, 2000). But, I want to keep well away from most of these accounts for now, though they will figure indirectly in quite a lot of what I have to say. Rather, I want to write about how modern geography thinks about space. That could cover pages and pages and so I will have to condense these thoughts into a manageable form. I will therefore make what some will regard as the outrageously simple claim that currently human geographers are chiefly writing about four different kinds of space.

However different the writings about these different kind of spaces may appear to be, they all share a common ambition: that is they abandon the idea of any pre-existing space in which things are passively embedded, like flies trapped in a web of co-ordinates – the so-called *absolute* view of space – for an idea of space as undergoing continual construction as a result of the agency of things encountering each other in more or less organized circulations. This is a *relational* view of space in which space is no longer viewed as a fixed and absolute container within which the world proceeds. Rather, space is seen as a co-production of those proceedings, as a process in process. To begin with, I will artificially separate these four spaces out but, as I will point out in the conclusion, the exciting thing about geography today is that we are learning how to put them together in combinations that are beginning to produce unexpected insights.

FIRST SPACE: EMPIRICAL CONSTRUCTIONS

Talking of putting things together, let's start with the empirical construction of space. It takes only a few minutes of reflection to start listing down all the things that we rely on to keep our spaces going – houses, cars, mobiles, knives and forks, offices, bicycles, computers, clothes and dryers, cinemas, trains, televisions, garden paths – but because these things are usually so mundane we tend to overlook them. So we often forget just what an extraordinary achievement the fabric of our daily lives actually is. Indeed, it is only recently that geographers have started to think systematically about the humble texts, instruments and devices that make up so much of what we are.[1] Let's take just one example of the kind of space that we make every day: the space of measurement. We are so used to looking at road signs measured out in terms of metres and kilometres or consulting a map or looking up an address or working out how long a journey will take that we forget what an extraordinary historical achievement these very ordinary practices are. They didn't suddenly come into existence over night but were the subject of progressive standardizations and co-ordinations that have taken centuries to put in to place. And they required extraordinary investments too. They required the invention of specialized devices that could measure the same things at the same places, culminating in today's satellite-based global positioning system (GPS). They required a whole knowledge of measurement that itself had to be able to be transported around the world in devices, books and journals. They required, latterly, endless boring committees that were able to agree that the same measures would be measured in the same way in different places and then integrated with each other. And they demanded a good deal of brute force. After all, many of the ways space is measured out around the world were imposed by imperial conquest, not prettily negotiated. Nevertheless, it is important to realize the sheer load of human effort that has gone in to making measured space and the often near to insane enterprises that have made this space possible. Let us remember, with a certain amount of awe, the attempts to give birth to a new unit of measure, the metre, under the first French Republic (Guedj, 2001). Between 1792 and 1799 the astronomers Pierre Mechain and Jean-Baptiste Delambre travelled from one end of France to the other measuring the length of the Paris meridian in order to determine the exact length of the standard metre, which the National Assembly had decreed would be one ten millionth of the quarter meridian. The enterprise was an extraordinary one, involving the dragging of large pieces of equipment up hill and down dale, but it laid the basis for the whole decimal metric system which is now so familiar.[2]

What is remarkable about the present time is the way in which this empirical construction of space is currently taking another leap forward. In the late nineteenth century, there was a widespread standardization of time. Driven by the increasing speed of transport and communications and more exact timekeeping instruments, states agreed on a common standard of time (based on the Greenwich meridian) and on a set of time zones spanning the globe in each of which time would be agreed to be uniform. Now, in the twenty-first century, something very similar is taking place in space. Driven by the demands of modern logistics and new, more exact ways of registering space (most especially the combination of GPS, geographical information systems (GIS) and radio frequency identifier tags (RFID)) it will soon be possible to locate everything – yes, everything – using standards of measurement, some of which (as we have seen) were already being laid down in the eighteenth century. Through the standardization of space made possible by these technologies (and the large bureaucracies that employ them), each object and activity taking place on the globe will, at least in principle, be able to be exactly located. The result will be that we will live in a world of perpetual contact, in which it will be possible to track and trace most objects and activities on a continuous basis, constantly adjusting time and space in real time, so producing what is now called micro- or hyperco-ordination (Katz and Aakhus, 2002). Numerous examples of hyperco-ordination already exist in the logistics industry, where it is necessary to continually adjust delivery schedules, but they are also becoming common in our daily lives, for example in the way in which we use mobile phone text messaging to continually adjust meetings with friends or satellite navigation systems to continually recalculate the route as we change our minds about where to go next.

SECOND SPACE: FLOW SPACE

The second way we need to think of space is as a series of carefully worked-up connections through which what we know as the world interacts. These connections consist of pathways which bind often quite unalike things together, usually on a routine, circulating basis. They can range all the way from the movements of office workers around offices to the movements that these office workers themselves order – of trade, of travel, even of arms. They can range all the way from the movements of a few already slightly drunk teenagers around the bars of Benidorm to the global flows of tourists of which they are a part. They can range all the way from the very restricted movements of prisoners let out of their cells only one hour in every 24 to the vast disciplinary apparatus of states dispensing laws and correction on an increasingly international scale. And so on. Trying to think about a world based on these flows of goods and people and information and money has occupied the attention of geographers to an increasing extent because their presence has become increasingly evident as the world has become increasingly knitted together by them, a tendency that sometimes goes by the name of globalization (see Chapter 19 on globalization and human geography).

The problem is that these pathways are difficult to represent conceptually. We can map them, we can list them, we can write about them, all key means by which the pathways themselves are able to achieve order. But how can we go a little further and create representational spaces which are still attached to these mundane means of achieving order but also pack an added analytical bite? For a long time in geography, the accepted way was to mimic a standard means by which the world is organized and draw boundaries around areas which were assumed to contain most of a particular kind of action and between which there was interaction. Once geographers had drawn lines round and labelled these large blocks, they then held them responsible for producing characteristic forces or powers. So, for example, we might say that this block of interaction was a capitalist space or an imperialist space, a neoliberal space or a dependent space, a city space or a community space, and that it had particular inherent qualities. Such a strategy of regionalization is obviously useful. It captures and holds still a particular aspect of the world and it is doubtful that we could ever do without it. But it is always an approximation and it has some serious disadvantages, most notably the tendency to assume that boundary equals cause, but also the tendency to freeze what is often a highly dynamic situation. So, geographers began to become more and more impatient with these kinds of representation, not so much because they were wrong but because they seemed to leave so much out of contention.

Nowadays, therefore, geographers tend to look for representations that can take more of the world in. One way of doing this has simply been to disaggregate these bounded spaces into smaller subordinate ones called 'scales', usually with some of the same qualities, but also with other qualities that operate only (or operate more strongly) at that scale (see Chapter 12 on scale and human geography). But it is questionable whether such a mode of proceeding does anything more than continuing the same method of drawing lines round and labelling blocks of interaction, though in slightly different form by allowing the possibility of the creation of new blocks, or the migration of powers from one block to another. So many geographers are now trying a different tack. Instead of trying to draw boundaries around flows, they are asking 'what if we regarded the world as made up of flows and tried to change our style of thought to accommodate that depiction' (Urry, 2000: 23)? It is no easy task to represent these 'spaces of flows' (Castells, 2000) but we can now see a whole series of approaches that are trying to start with movement and flow as origin rather than endpoint and which stress mutable, travelling identities over fixed notions of belonging (see Cresswell, 2006; Sheller and Urry, 2006). For example, there is so-called actor-network theory which tries to trace out circulations in which the actor is the 'network' itself; things moving together through networks have powers (including the power to make stable spaces) that they could never have when separated. There is the voluminous work on commodity chains which tries to map out the way that commodities are assembled along pathways that cross the world. There is work by feminist and postcolonial theorists which is searching for spatial figures which can convey the ambition to build different, more fluid kinds of space which can simultaneously engage periphery and centre, continually suggesting multiple routes of entry and exit. And a new more expansive vocabulary is coming in to being that can match these several ambitions: events as well as structures, lines of flight as well as lines, transformation and becoming as well as system and being – all means of freeing thought from the straitjacket of the container thinking of absolute space and replacing it with the process thought of relational space.

In turn, all kinds of new spaces of differentiation are being constructed, sometimes fleeting and sometimes concerted experiments in living different kinds of life which, rather than providing definitive answers, are a set of questions about what kinds of space can be in a world of flows. And the questions are, as Elizabeth Grosz (2001: 130) puts it, 'How then can space function differently from the ways in which it has always functioned? What are the possibilities of inhabiting otherwise? Of being extended otherwise? Of living relations of nearness and farness differently?' All around the world geographers are now both studying and taking part in the spatial *experiments* which can begin to answer these questions. These experiments range far and wide; all the way from the kinds of experiments that are associated with reworking what we mean by 'wild' or 'natural' to the kinds of experiments that are meant to perform everyday life differently to the kinds of experiments which are trying to map new meanings and practices of 'global' (Blunt and Wills, 2000; Abrams and Hall, 2006).

No one quite knows what they are doing. But that is the point of good experiments: they are risky because they leave room for the world to speak back.

THIRD SPACE: IMAGE SPACE

The third kind of space consists of what we might call pictures or, perhaps better, given all of the associations that the word conjures up, images (see Chapter 16 on landscape and human geography). In the past, mention of the image might well have conjured up the notion of a formal painting. But nowadays, images come in all shapes and sizes – from paintings to photographs, from portraits to postcards, from religious icons to pastoral landscapes, from collages to pastiches, from the simplest graphs to the most complex animations. What is certain is that images are a key element of space because it is so often through them that we register the spaces around us and imagine how they might turn up in the future. The point is even more important because increasingly we live in a world in which pictures of things like news events can be as or more important than the things themselves, or can be a large part of how a thing is constituted (as in the case of a brand or a media celebrity). Part of the reason for the pervasiveness of images is that we now live in a world populated by all kinds of screen which produce a continuous feed of images. These screens are now so pervasive that we hardly notice their existence (McCarthy, 2001). So we find television screens populating not just homes but also bars, airports, shops, malls, and waiting rooms, while computer screens can be found in dealing rooms, offices, studies and bedrooms and, increasingly, as public access screens, in airports and stations as well as in internet cafés. This extraordinary proliferation of screens over the last 50 years has produced an image realm of extraordinary richness which is changing how we do space.

This change can be linked to another way in which our thinking about image space is changing. In the past, particular kinds of image very often created spaces in their likeness. So, for example, a particular notion of spatial symmetry helped to produce the Palladian landscape while a particular kind of modernist sensibility helped to produce the kinds of strictly laid out and ascetically ordered landscapes still to be found lingering on in many urban housing estates. But the proliferation of images has made it increasingly difficult to read off images straightforwardly on to space like this. And it has also pointed to an aspect of images which has often been heretofore neglected; that they are the result of complex processes of mediation which themselves bear meaning. For example, Bruno Latour (1998) shows how a finished piece of work like a religious painting can involve all kinds of intermediaries, each of which can be bearers of spatial meaning – varnishes, dealers, patrons, assistants, maps, measuring devices, graphs, charts, angels, saints, monks, worshippers – and each of which has its own complicated intervening geographies. Such an example also shows that there is no direct reference to the world contained in an image, but rather a never-ending set of transformations – or what Latour calls 'cooking steps' – each of which can involve quite different ways of seeing and working on the image.

If there are now so many image spaces swirling through us in so many different ways, it is clear that they must compete for our attention. And it is this aspect of image space that I want to point to in concluding this section. For what is clear is that the issue of attention is probably the most pressing one now facing the geography of images (see also Chapter 4 on geography and the humanities tradition). Caught in a snowstorm of images, why do we attend to some images rather than others? In the nineteenth century, the matter of attention was a key element of debates on space. It was subsequently taken up by writers like Walter Benjamin and Georg Simmel, who argued that the constant barrage of images was causing people to grow a kind of mental carapace which would protect them from this continuous bombardment, a carapace which was showing up in cities as new and studied social styles (like cynicism and a blasé attitude) which constructed certain routinized kinds of attention. However, the growth of the mass media, such as the cinema, had also provided the opportunity for new kinds of moving images to come into existence, which to some extent undercut these social styles and produced new apprehensions of space.

In the twenty-first century, we can see this debate being replayed as geographers consider the ways in which new image forms are again providing new social and cultural pathologies, but also new opportunities, as we have seen in the case of the sheer pervasiveness of the screen and the images supplied by it. We can wrap all these new image forms up in one big package called 'postmodernism' (Harvey, 1989), making all the images add up to one vast capitalist spectacle, but better by far to do what geographers are doing now and try to look at all the cooking steps of different kinds of image and their geographies, testing each step for its various potentials to tell us something new about how we see the world. This is a much harder slog, of course, one which requires a lot of methodological expertise (Rose, 2001). It also makes it much more difficult to write in terms of one stable big picture like postmodernism, rather than multiple, shifting arrangements. But then perhaps that is not such a bad thing. After all, one of the continuing dangers of work on images is to read too much significance into them, rather than considering them as just another set of mundane tools and practices of seeing which allow us to see some things and not others and so construct some spaces and not others (Anderson, 2003).

FOURTH SPACE: PLACE SPACE

The final kind of space is space understood as place: I say 'understood' loosely since the nature of place is anything but fully understood (see Chapter 9 on place and human geography). One reason for this is precisely that place so often seems to be caught up with the idea of a natural register. Whether it be the quiet glories of Thoreau's Walden Pond or the noisy cultural authenticity of an urban enclave, somehow place is more 'real' than space, a stance born out of the intellectual certainties of humanism and the idea that certain spaces are somehow more 'human' than others; these are the places where bodies can more easily live out (or at least approximate) a particular Western idea of what human being should be being. But, other geographers are moving away from this kind of certainty both about what 'human' and 'being' might be. They are more interested in testing the limits of 'human' and 'being' through experiment and, in the process, are starting to point to new kinds of space (see Chapter 4 on geography and the humanities tradition).

Whatever the case, all of those working on place seem to agree that place consists of particular rhythms of being that confirm and naturalize the existence of certain spaces. Often, they will use phrases like 'everyday life' to indicate the way that people, through following daily rhythms of being, just continue to expect the world to keep on turning up and, in doing so, help precisely to achieve that effect (Lefebvre, 2004). The problem is that rhythms of being can vary so enormously that such phrases often provide only the most tenuous hold on what happens. This problem of variation does not just exist because there are so many different rhythms of being, but also because when the minutiae of everyday interaction is closely looked at what we see is not just routines but also all kinds of creative improvisations which are not routine at all (though they may have the effect of allowing that routine to continue). So, in everyday life, what is striking is how people are able to use events over which they often have very little control to open up little spaces in which they can assert themselves, however faintly. Using talk, gesture, and more general bodily movement, they can open up pockets of interaction over which they can have control and which give them a feel for place (Laurier and Philo, 2006). Clearly, an important part of this process is that spatial awareness that we call place. For places not only offer resources of many different kinds (for example, spatial layouts which may allow certain kinds of interaction rather than others), but they also provide cues to memory and behaviour. In a very real sense, places are a part of the interaction.

One thing that does seem to be widely agreed is that place is involved with embodiment. It is difficult to think of places outside the body. Think, for example, of a country walk and place consists not just of eye surveying prospect but also the push and pull of walking up hill and down dale, the sound of birds and the wind in the trees, the touch of wall and branch, the smell of trampled grass and manure. Or think of a walk in the city and place consists not just of eye making contact with other people or advertising signs or buildings, but also the sound of traffic noise and conversation, the touch of ticket machine and hand rail, the smell of exhaust fumes and cooking food. Once we start to think of place in this kind of way, we also start to take notice of all kinds of things which previously were hidden from us. So, for example, there is now a thriving study of how sound (and especially music) conjures up place associations (Leyshon et al., 2000). And other senses, such as touch and smell, are also beginning to receive their due too.[3]

But there is a big problem here. What do we mean by 'body'? And this is where we get to the most intriguing prospect of all. For though it is possible to think of *the* body as flesh surrounded by an envelope of skin, all the current thinking suggests that this container thinking is too simple. It probably makes much more sense to think of the individual body as a part of something much more complex, as a link in a larger spatial dance with other 'dividual' parts of bodies and things and places which is constantly reacting to encounters and evolving out of them, not individual awareness but dividual 'a-where-ness'. And this larger dance is held together in particular by the play of 'affects' like love and hate, sympathy and antipathy, jealousy and despair, hope and disappointment, and so on.

Affect is often thought of as just a posh word for emotion but it is meant to point to something which is non-individual, an impersonal force resulting from the encounter, an ordering of the relations between bodies which results in an increase or decrease in the potential to act. Place (understood as a part of this complex process of embodiment) is a crucial actor in producing affects because, in particular, it can change the composition of an encounter by changing the affective connections that are made (Thrift, 2005). Thus, as we all know, certain places can and do bring us to life in certain ways, whereas others do the opposite. It is this expressive quality of place which has recently lead to the emphasis on performance in geography. For, through experiments with particular kinds of performance (from art to dance to drama), it may be possible to show some of this affective play and use the understandings (or should it be stancings) of place thereby gained to make places which can help to produce the same sense of empowerment and general creative potential that we currently most often identify with situations like standing on the top of a hill on a windy day drinking in the atmosphere or being moved by a great new piece of music. In other words, geographers working on place have started to join in a kind of politics which is intent on freeing up more of the potentials of place – and installing some quite new ones.

CONCLUSIONS: JOINED-UP SPACING

What is fascinating about the present time is that geographers are now attempting to fit these four kinds of space together, partly because models for doing this are erupting in the social sciences and humanities in a way that they never did before. In the past, many theoretical models of space that had an ambition to connect spaces of various kinds simply simulated the command and control models of the dominant systems around them. So, for example, many early Marxist models of capitalist space produced spatial connection by nesting some kinds of 'little' space in other 'big' spaces, declaring the 'little' spaces to be 'unique' and the big spaces to be 'general'. But nowadays this simple 'size' distinction does not hold. We are no longer sure what is big or little or what is general or unique. Instead, as we have seen in the case of each of the four spaces that we have examined, the hunt is on to think about space in quite different ways, ways which can prompt new 'a-where-nesses' (Massey, 2005, Thrift, 2006).

And this relates to the most important point that I want to make. This is that all of these ways of thinking space are attempts to rethink what constitutes *power* if we can no longer think of power as simply command and control (Allen, 2002). So new thinking about the empirical construction of space involves considering the prolonged hard grind of actually putting viable pathways together, especially when, as nowadays, they can stretch around the world and back. New thinking about unblocking space involves the difficult task of redescribing the world as flow and continuous transformation. New thinking about image space involves reconsidering how images are circulated and kept stable when that circulation involves large numbers of intermediaries. And new thinking about place space involves trying to understand the gaps in the rhythms of everyday life through which new performances are able to pass. What we are seeing are new spaces being imagined into being by reworking the spatial technologies that we hold dear and what is clear is that these acts of imagination are all profoundly political acts: what we often think of as 'abstract' conceptions of space are a part of the fabric of our being and transforming how we think about those conceptions means transforming 'ourselves'.

SUMMARY

- Space arises out of the hard and continuous work of building up and maintaining collectives by bringing different things (bodies, animals and plants, manufactured objects, landscapes) into alignment. All kinds of different spaces can and therefore do exist which may or may not relate to each other.

- For the purpose of simplification, it is possible to identify four different kinds of these constructed spaces: empirical, flow, image and place.

- Empirical space refers to the process whereby the mundane fabric of daily life is constructed.

- Flow space refers to the process whereby routine pathways of interaction are set up around which boundaries are often drawn.

- Image space refers to the process whereby the proliferation of images has produced new apprehensions of space.

- Place refers to the process whereby spaces are ordered in ways that open up affective and other embodied potentials.

Further Reading

Space has been written about in lots of ways. One book which documents some of the different conceptions of space that are drawn upon by different disciplines is Crang and Thrift's (2000) *Thinking Space*. Different takes on the nature of space within the discipline are evident in Anderson et al.'s (2003) *The Cultural Geography Handbook*, Gregory's (1994) *Geographical Imaginations*, Harvey's (1989) *The Condition of Postmodernity*, Hubbard et al.'s (2004) *Key Thinkers on Space and Place*, Massey's (2005) *For Space* Thrift's *Spatial Formations* (1996) and *Non-representational Theory* (2007).

Note: Full details of the above can be found in the reference list below.

NOTES

1   This is a little bit unfair. The exceptions to this rule include work by those who have been interested in the history of cartography and navigation, such as the late Eva Taylor (see Taylor, 1930).

2   I could have chosen many other examples, such as the history of the surveying of Britain or the mapping of Switzerland (Gugerli, 1998) or India.

3   This move also underlines how often space is bodied in various ways. And, if space is bodied, then it will, for example, be actively gendered. Therefore, to return to the beginning of this chapter, space has numerous sexual dimensions (see Pile and Nast, 2000).

References

Abrams, J. and Hall, P. (eds) (2006) *Else/Where: Mapping. New Cartographies of Networks and Territories*. Minneapolis, MN: University of Minnesota Press.

Allen, J. (2002) *Hidden Geographies of Power*. Cambridge: Polity Press.

Anderson, K., Domosh, M., Pile, S. and Thrift, N.J. (eds) (2003) *The Cultural Geography Handbook*. London: Sage.

Barcan, R. and Buchanan, I. (eds) (1999) *Imagining Australian Space: Cultural Studies and Spatial Inquiry*. Nedlands, WA: University of Western Australia Press.

Blunt, A. and Wills, J. (2000) *Dissident Geographies*. London: Longman.

Castells, M. (2000) *The Information Age* (3 Vols). Oxford: Blackwell.

Crang, M. and Thrift, N.J. (eds) (2000) *Thinking Space*. London: Routledge.

Cresswell, T. (2006) *On the Move*. London: Routledge.

Gregory, D. (1994) *Geographical Imaginations*. Oxford: Blackwell.

Grosz, E. (2001) *Architecture from the Outside: Essays on Virtual and Real Space*. Cambridge, MA: MIT Press.

Guedj, D. (2001) *The Measure of the World*. Chicago, IL: University of Chicago Press.

Gugerli, D. (1998) 'Politics on the topographer's table: the Helvetic triangulation of cartography, politics and representation', in T. Lenoir (ed.) *Inscribing Science: Scientific Texts and the Materiality of Communication*. Stanford, CA: Stanford University Press, pp. 91–118.

Harvey, D. (1989) *The Condition of Postmodernity*. Oxford: Blackwell.

Hubbard, P., Kitchin R. and Valentine, G. (eds) (2004) *Key Thinkers on Space and Place*. London: Sage.

Katz, J. and Aakhus, M. (eds) (2002) *Perpetual Contact: Mobile Communication, Private Talk, Public Communication*. Cambridge: Cambridge University Press.

Latour, B. (1997) 'Trains of thought: Piaget, formalism and the fifth dimension', *Common Knowledge*, 6: 170–91.

Latour, B. (1998) 'How to be iconophilic in art, science and religion?', in C.A. Jones and P. Galison (eds) *Picturing Science, Producing Art*. New York: Routledge, pp. 418–40.

Laurier, E. and Philo, C. (2006) 'Cold shoulders and napkins handed: gestures of responsibility', *Transactions, Institute of British Geographers*, NS 31: 193–207.

Lefebvre, H. (2004) *Rhythmanalysis: Space, Time, and Everyday Life*. London: Continuum.

Leyshon, A., Matless, D. and Revill, G. (eds) (2000) *The Place of Music*. New York: Guilford Press.

Massey, D. (2005) *For Space*. London: Sage.

McCarthy, A. (2001) *Ambient Television*. Durham, NC: Duke University Press.

Pile, S. and Nast, H. (eds) (2000) *Places Outside the Body*. London: Routledge.

Rose, G. (2001) *Visual Methodologies*. London: Sage.

Sheller, M. and Urry, J. (2006) Special Issue on Mobilities and Materialities. *Environment and Planning A*, 38(2).

Taylor, E.G.R. (1930) *Tudor Geography 1485–1583*. London: Methuen.

Thrift, N.J. (1996) *Spatial Formations*. London: Sage.

Thrift, N.J. (2005) 'But malice aforethought: cities and the natural history of hatred', *Transactions, Institute of British Geographers*, NS 30: 133–50.

Thrift, N.J. (2006) 'Space', *Theory Culture and Society*, 23: 139–55.

Thrift, N.J. (2007) *Non-representational Theory: Space, Politics, Affect*. London: Routledge

Urry, J. (2000) *Sociology beyond Societies: Mobilities for the Twenty-First Century*. London: Routledge.

6

Space: Making Room for Space in Physical Geography

Martin Kent

D efinition

Geographers are poor at defining space. The *Oxford English Dictionary* defines space in two ways: (1) 'A continuous extension viewed with or without reference to the existence of objects within it'; and (2) 'The interval between points or objects viewed as having one, two or three dimensions'. The geographer's prime interest is in the objects within the space and their relative position, which involves the description, explanation and prediction of the distribution of phenomena. The relationships between objects in space is at the core of geography.

INTRODUCTION

The importance of the concept of space in geography has always been controversial (Gatrell, 1983; Unwin, 1992; Holt-Jensen, 1999) and whether geography and geographers should primarily focus on, or at the very least give some recognition to, the importance of space remains a fundamental question for the discipline. This chapter examines the concept of space in the context of physical geography (see Chapter 5 of this volume for a human geography perspective on space). The chapter begins with the suggestion that physical geographers have neglected the vital spatial dimension of their subject over the past few decades and explores the possible reasons for this. The spatial units and approaches to mapping that are recognized by the major subdisciplines are then examined and, in turn, the relatively poor spatial synthesis across physical geography is considered. Finally, it is argued that new technologies and developments in related disciplines, such as ecology, present physical geographers with exciting possibilities for stimulating a new awareness of space. Recent developments in biogeography provide early indications of this 'spatial reawakening' in physical geography.

SPACE NEGLECTED

There is no doubt that physical geographers have generally been less willing or able than their human counterparts to adopt a predominantly spatial emphasis to their studies and research. A search through the most recent texts on the nature and philosophy of (physical) geography (e.g. Haines-Young and Petch, 1986; Stoddart, 1986; Haggett, 1990; Rogers et al., 1992; Unwin, 1992; Rhoads and Thorn, 1996; Slaymaker and Spencer, 1998; Holt-Jensen, 1999; Gregory, 2000) reveals that remarkably little is written on the importance of space as a concept in physical geography. Gregory lists 11 'Tenets for Physical Geography', of which the first is 'Emphasise the spatial perspective' (2000: 287) yet, paradoxically, the word 'space' is not mentioned in the index and there is just one reference to 'spatial analysis'.

Spatial analysis provided a unifying theme for geography during the 1960s (Unwin, 1992) and emphasized that geographers should 'pay attention to the spatial arrangement of the phenomena in an area and not so much to the phenomena themselves' (Schaefer, 1953: 228). Most developments took place in human geography rather than in physical, although some attempts were made to demonstrate the potential significance of spatial analysis to the physical side of the subject, notably by Doornkamp and King (1971) and Chorley (1972). Within spatial analysis, concepts of distance are fundamental, which traditionally in the three-dimensional Euclidean view of space is expressed as a straight line between any two points or a three-dimensional volume that may be occupied by any phenomena. Gatrell (1983) stresses that this is the concept of *absolute space*. Human geographers quickly developed new ideas about space and distance – for example, 'time-distance', 'cost or economic distance', 'cognitive or perceptual distance' and 'social distance' (Unwin, 1992) – all of which treat space as a relative phenomenon (*relative space*). Ultimately, these developments have led to widespread rejection of spatial analysis by human geographers, many of whom now regard space as a social construct.

In comparison, within physical geography, concepts of distance have remained strongly absolute and defined in terms of Euclidean and metric space. Perhaps physical geographers, with their ongoing emphasis on absolute space, could have been expected to take up the challenge and to focus more on spatial analysis. However, they have generally also chosen to reject this pathway and downgrade or neglect spatial analysis. This downgrading of space by physical geographers can be linked to three alternative views that, rather than space, human–environment relationships, process and the time dimension, are at the core of the discipline.

Thus, the first widely held view is that physical geography is primarily about human relationships with and impact on the *environment* and the resulting need for environmental management (Briggs, 1981; Pacione, 1999). While there is absolutely no doubt concerning the impact of human activity on all physical and environmental systems, to place this at the core of geography raises some interesting questions about the difference between physical geography and the much more recently evolved discipline of environmental science. Is environmental science simply physical (and some human) geography without a spatial emphasis? To be provocative, are physical geographers really geographers at all if they downgrade or neglect a spatial emphasis to the subject?

It is interesting that one of the most widely sold textbooks in physical geography since its publication in 1985 (Briggs and Smithson's *Fundamentals of Physical Geography*) reappeared in 1997 and 2002 as a second and third edition but under the revised title, *Fundamentals of the Physical Environment* (Briggs et al., 1997). Other core texts aimed at and written by geographers are now predominantly presented as 'environmental' rather than 'geographical'. While this is partly attributable to the wiliness of both publishers and authors in exploiting the market within both geography and environmental science degrees, it also can perhaps be seen as emphasizing the ambivalence of many physical geographers towards placing space and spatial analysis at the centre of physical geography and their tendency simply to equate geography with the environment.

Second, a commonly quoted idea is that physical geographers study 'pattern and *process*'. The word 'pattern' clearly has spatial connotations and is closely linked to ideas of geographers being concerned with the spatial description of phenomena. The word 'process', on the other hand, is linked with an explanation of (spatial) patterns and distributions and relates to the understanding of how patterns are derived, models of their underlying functionality and how patterns may change over time (Stoddart, 1997). Recently, physical geographers have tended to overemphasize the importance of process and hence function and explanation at the expense of pattern and the spatial approach.

This point is accentuated further when one considers the third alternative view and the recent emphasis that has also been placed on environmental change and the importance of *time* by many physical geographers. Time and process are inextricably linked. An excellent introduction to the general approach is presented in Slaymaker and Spencer (1998), who demonstrate the relevance of an understanding of past climatic and environmental variability and the increasing impact of human activity to key present and future issues of environmental management and climate change. Quaternary science and the study of palaeoenvironments have come to the core of physical geography over the past 30 years and studies of environmental change over the Holocene period have achieved particular significance (Roberts, 1998). Although it is implicit that environmental change is expressed in terms of change in spatial patterns of the various components of physical geography through time, the main emphasis has been primarily on temporal changes and understanding the processes that underlie them. Such research is vital in informing the future predictions of climate and environmental change in response to human forcing. This emphasis on time, important though it is, has perhaps been at the expense of space, and as long ago as 1987, Clark et al. (1987: 384), in the conclusion to their edited volume on physical geography, concluded that '[s]patial concepts are certainly of importance, but seem at the moment to be subsidiary to the elucidation of temporal behaviour'.

Space is, of course, intimately linked to time. The positive relationship between scales in space and time is widely acknowledged (see also Chapter 11 on scale in physical geography). Geographical studies over small areas can usually be completed over short periods of time, while those involving much larges areas of space require much longer periods of time (Schumm and Lichty, 1965; Harrison and Warren, 1970; Lane, 2001). Massey (1999: 273), writing on the relationships between physical and human geography, also concurs with the general view that space has been comparatively ignored:

> 'in contrast to the prominence of time and historicity in the debates that I have explored so far, space has had a very low profile. It is denigrated as a simple absence of history and/or not accorded the same depth of intellectual treatment as time.

Completely to divorce space and time is impossible, and any attempts to do so are artificial. The problem, however, remains that the time dimension has tended to receive greater attention than space in much of physical geography.

To summarize, general disinterest in spatial analysis, with the particular exceptions discussed below, and the rise of environment, process and time as core foci in physical geography mean that physical geographers have not worried greatly about the relevance and importance of the spatial basis of their discipline.

CONCEPTS OF SPACE AND SPATIAL UNITS IN PHYSICAL GEOGRAPHY

It is instructive to review the conventional spatial units and approaches to mapping that are recognized by the major subdisciplines of physical geography (Figure 6.1). The problem of scale is, however, immediately important since, for all subdisciplines, spatial units and approaches to mapping for research and management can be defined and completed at a range of scales (see Chapter 11 on scale). Spatial units are also often nested within each other across this range of scales. Of the various components of physical geography, perhaps biogeography can be shown to have identified most strongly with the spatial aspects of physical geography over the past 30 years.

Biogeography

The Ecosystem and Plant Community Concepts

Two of the major paradigms of biogeography are, first, the ecosystem concept, originally described by Tansley (1935). The spatial expression of the ecosystem concept is, however, linked to the second key paradigm, which is that of the plant community or, more generally, the habitat. In unmodified terrestrial environments, vegetation or habitat is 'the outward and visible sign' of a particular ecosystem type because it is within or underneath vegetation or habitat that all higher trophic levels live, feed and reproduce. Thus vegetation or habitat usually represent the key defining spatial concepts for ecosystems, and mapping or spatial representation is

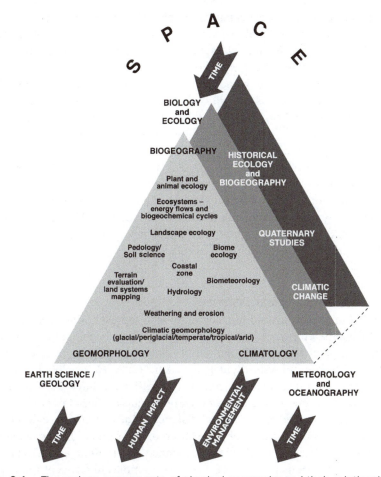

Figure 6.1 The various components of physical geography and their relationships in space and time

frequently completed on the basis of both physiognomic and/or floristic variability (Kent and Coker, 1992; Huggett, 2004). The question of whether plant community and ecosystem types can be mapped and thus given spatial expression at all and whether they are distributed as a continuum rather than as a set of mappable types is still heavily debated by both ecologists and biogeographers, and the idea that types can even be considered as 'concrete' mappable types, rather than 'abstract' categories, remains controversial (Kent and Coker, 1992; Kent et al., 1997). The completion of the National Vegetation Classification (NVC) for Britain (Rodwell, 1991–2000), with all major vegetation types now being recognized, demonstrates this problem clearly, with the main classification being essentially 'abstract' but with many examples now emerging of 'concrete' mapping of NVC types in different parts of Britain (e.g. Dargie, 1998). The nature of boundaries, transitions or ecotones between different plant community/vegetation types within a given region has tended to be conveniently ignored by both biogeographers and ecologists. Similarly, the application of multivariate analysis using techniques of numerical classification and ordination and the 'fitting' of vegetation samples that are 'transitional' between types, often using the MATCH and TABLEFIT computer programs of Malloch (1991) and Hill (1991), are also still highly contentious issues (Kent et al., 1997).

The process of defining and then mapping plant communities or habitats and hence ecosystems is thus fraught with difficulty. Often the level of detail that can be incorporated is low, and vegetation categories are reduced to generalized habitats and definition of boundaries between habitats is often arbitrary. The phase 1 habitat mapping programme for the mapping of plant communities in Britain during the 1980s and 1990s provides an excellent example. The level of generalization was often so high as to render the final maps as of comparatively little value, and for various practical reasons, the accuracy of many of the maps has subsequently been questioned (Cherrill and McClean, 1999). This problem of boundary recognition is highlighted as a major new potential research area for physical geography in the conclusion to this chapter.

The Concept of 'Ecological Distance'

Plant and animal ecology and biogeography also represent one area of physical geography where alternative concepts of distance have been applied. The search for pattern in plant and animal community data routinely involves the application of multivariate analysis (Kent and Coker, 1992; Waite, 2000). As Gatrell (1983) points out, numerous (dis)similarity coefficients have been devised by ecologists to assess the degree of matching between samples of community composition, often described as 'ecological distance'. Many of these are described as having 'metric' properties (i.e. they are based on concepts of Euclidean space) – not least the coefficient of Euclidean distance itself (Kent and Coker, 1992; Waite, 2000). Numerous methods of 'mapping' samples in terms of their species composition expressed as 'ecological distances' have been devised under the general heading of 'ordination techniques' – for example, non-metric multidimensional scaling, principal components analysis and various forms of correspondence analysis. Since all samples of species composition are collected in space, it is possible to relate true spatial distance (absolute space) to change in species composition expressed as various forms of 'ecological distance' (relative space). However, despite their convenience, problems of spatial distortion abound in such analyses (Kent and Coker, 1992). An interesting twist in river ecology is the potential importance of measuring absolute distance along river network paths, not 'as the crow flies'. For example, Ganio et al. (2005) demonstrate the use of a geostatistical tool that uses point-to-point distances measured in this way to evaluate the distribution of salmonid fishes in Oregon stream networks.

Mapping of Individual Species

The other key form of spatial expression in biogeography is the mapping of individual species distributions, often also with a view to understanding their tolerance ranges and environmental controls. In the case of rare and endangered species, changing species distributions on these maps are a vital tool for biological conservation. In Britain, the distribution mapping of individual species for most groups of organisms has now been completed on a 10 km² grid (Perring and Walters, 1982; Countryside Survey, 2008; Preston et al., 2002), and similar mapping schemes are increasingly available elsewhere in Europe and across the globe. However, terrestrial animal communities and species assemblages of higher trophic levels, as opposed to individual species, although recognized, are rarely mapped as such since they exist within the habitat provided by vegetation. More recent developments, linked to the description and analysis of species distributions and ranges, under the general heading of 'macroecology' are described towards the end of this chapter.

Changing Spatial Distributions Through Time – Quaternary and Historical Biogeography

Mapping of changes in species and community distributions through time introduces the fields of Quaternary science and historical ecology/biogeography (Figure 6.1). Most studies in this area again provide 'abstract' concepts of changes in both individual and community distributions through time rather than 'concrete' spatial mapping of those changes, in the sense that general changes affecting vegetation and ecosystems are described but the actual spatial changes in species ranges are not mapped in any detail. Also, as suggested above, most interpretation and inference concentrates on temporal rather than spatial variation. The current trend towards fine-resolution description and analysis of microfossils and environmental indicators in cores from peat, lakes, sediments and the oceans, which assists in linking rapid short-term ecological changes to longer-term geological time, emphasizes this point further (Bennett, 1997).

The 'Ergodic Hypothesis' – Substitution of Space for Time

Biogeographers and ecologists also often link space and time through the 'ergodic hypothesis' whereby, as an expedient research strategy, different areas in space are taken to represent different stages in time (Chorley and Kennedy, 1971; Bennett and Chorley, 1978). Classic examples are in the study of vegetation successional processes, where differing locations or spatial units within a local area at one point in time are taken to represent the sequences of species changes that would occur in that area over long periods of time. Similar approaches have been applied in geomorphology in the context of landform development (Chorley and Kennedy, 1971).

Soils and Pedology

Soils and pedology are traditionally seen as a part of biogeography but are, of course, also closely linked to geomorphology. In soil mapping, similar problems to those of mapping vegetation exist in that a set of soil 'types' is typically described in the 'abstract' sense for any region but precise mapping of those types in 'concrete' terms is often difficult (Gerrard, 2000; White, 2006). The problems of soil classification to define and recognize types or series are very similar to those for vegetation classification, and use of multivariate analysis and numerical classification and ordination techniques in this context is still debated (Webster, 1977, 1985). Problems of boundary recognition and definition are thus once again emphasized.

Landscape Ecology – a New Spatial Science?

A major trend within the ecological side of biogeography over the past decade has been to adopt the approach and methodology of landscape ecology (Forman and Godron, 1986; Forman, 1995; Kupfer, 1995; Kent et al., 1997; Turner et al., 2001; Burel and Baudry, 2003; Turner, 2005; Wiens and Moss, 2005; Kent, 2007a). Landscape ecology is concerned with the description, analysis and explanation of the spatial patterns of plant community and land-use types within a given landscape or region (see also Chapter 10). Landscapes are composed of 'patches' that are distributed as a mosaic within any local area. The mosaic of patches at the patch scale within a landscape can be aggregated to give the landscape scale, and both higher and lower scales may be identified above the landscape scale and below the patch scale (Figure 6.2). These individual patches have varying degrees of 'naturalness', and a spectrum of patch types with varying intensity of human impact and modification, known as land-cover types, may be identified. Linear patches and features, such as river courses and hedgerows, are known as corridors and again boundaries between patches and corridors of different type are of considerable interest. Indeed, aquatic ecologists and biogeographers have begun to apply the principles of landscape ecology to understanding the distribution of organisms within river networks – so called 'fluvial landscape ecology' (see Wiens, 2002 and other papers in that journal issue).

The methodology of landscape ecology is closely linked to recent developments in both remote sensing and geographic information systems (GIS) (Haines-Young et al., 1993; Bissonette, 1997; Klopatek and Gardner, 1999; Farina, 2000, 2006; Turner et al., 2001; Wiens and Moss, 2005; Kent, 2007a). Recognition of patches first requires classification of patches based on their ecological and land-use attributes into land-cover types. Once patches of a particular type have been defined, they can be examined in terms of numerous parameters relating to size, shape and fragmentation. Various computer packages for spatial description and analysis of landscape patches are available, of which the best known is FRAGSTATS, based on the GIS programs, Arc-Info, Arc-View and ArcGIS (ESRI, 2001; McGarigal and Marks, 2002).

This advent of landscape ecology, GIS and remote sensing has provided a valuable tool for habitat mapping in biogeography, with one of the best examples being the UK Countryside Surveys carried out between 1978 and 2000 (Department of Environment, 1993; Department of Environment, Transport and the Regions, 2000). Using remote sensing, field survey and geographic information systems, the Countryside Survey is now repeated every eight to ten years, thus representing a spatial census of habitat and land-cover change across

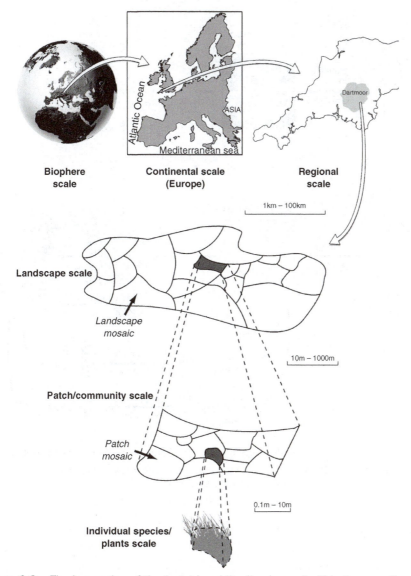

Figure 6.2   The key scales of the 'patch' and the 'landscape' within the overall spatial hierarchy of landscape ecology

Britain. The latest survey is nearing comptetion and the results are available from the website (Countryside Survey, 2008).

In the Countryside Information System (CIS) (Figure 6.3), the main categories derived from remote sensing are of generalized habitat types, principally because only major differences in vegetation and hence ecosystem physiognomy, rather than detailed variations in floristic composition, are picked out by remote-sensing imagery. However, this is counter-balanced by extensive repeat field surveys on the ground within all main landscape/habitat types, yielding data not only on changes in habitat extent but also in species richness and diversity of plants and freshwater organisms at ten-year intervals.

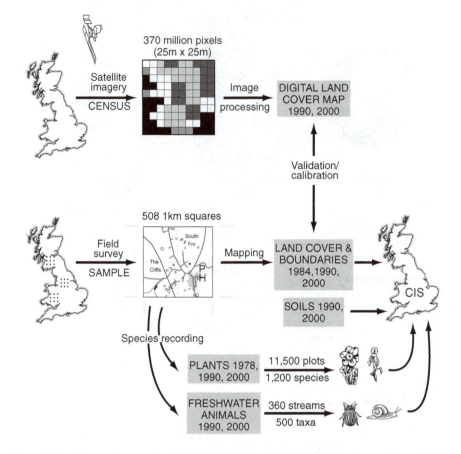

Figure 6.3   The methodological approach of the Countryside Surveys of Great Britain

Source: Adapted from Department of Environment (1993); Department of Environment, Transport and the Regions/Centre for Ecology and Hydrology (2000)

Climatology

Air Masses and Fronts

As arguably dealing with the most dynamic part of the environment, the spatial units of climatology are represented by air masses and their numerous different characteristics and the fronts that characterize the boundaries between them. Both air masses and fronts are constantly changing in space and time. Nevertheless, the major air mass and frontal zones of the planet are widely recognized (Briggs et al., 1997; Goudie, 2001; Smithson et al., 2002; Barry, 2003). Similarly, climate types are recognized and classified (Oliver, 1991). Within major air masses, further dynamic variability and interaction occur at more detailed scales in response to relief and to relative proximity in space of land and sea. Any weather map or sequence of weather maps through time constitutes a representation of the changing spatial patterns of air masses and the nature of the fronts between them. Spatial and temporal patterns of extreme events are now a vital part of climatology and have significant social, economic and even political consequences (Perry, 1995). Magnitude and frequency studies of extreme climatic events and their spatial expression are now a key input to hazard studies in the whole of physical geography.

Meteorology

The processes behind the patterns described in climatology are studied in the subject of meteorology, and the prediction of changes in air mass and frontal patterns at a wide range of spatial and temporal scales provides the essence of weather and climate forecasting, which is central to the discipline. However, the brevity of this section serves to emphasize the point that the part of physical geography where spatial concepts are probably most difficult to work with is climatology, primarily because of the exceptionally dynamic nature of atmospheric processes.

Geomorphology and Hydrology

The Drainage Basin as a Spatial Unit in Geomorphology and Hydrology

Over the past 30 years, perhaps the most important spatial unit for geomorphology and hydrology has been the drainage basin, catchment or watershed, linked to the functioning systems model of the land phase of the hydrological cycle (Chorley, 1969; Gregory and Walling, 1973; Likens et al., 1977; Newson, 1995). In terms of hydrology, the analysis of both quantity and quality of water and changes through time, particularly in response to human impact, has been facilitated by the drainage basin approach, and much practical management is now based on the catchment concept (Gower, 1980; Newson, 1995). Studies of ecosystem biogeochemistry and nutrient cycling have also been primarily based on the catchment/watershed–ecosystem approach (Likens et al., 1977).

The catchment has also represented a spatial unit for comparison, with morphological, hydrological and geomorphological properties of different sets of drainage basins being compared (Doornkamp and King, 1971; Chorley, 1972). Stream ordering and the nesting of catchments are a particularly important concept that links across different spatial scales (Likens et al., 1977; Newson, 1995). The catchment is also a key unit for those physical geographers studying environmental change, with lakes and their sediments representing a valuable record of past hydrological, geomorphological and anthropogenic land-use change in upstream catchments (O'Sullivan, 1979; Lau and Lane, 2001).

(Geo)Morphological Mapping

Evans (1990: 97) expresses the comparative lack of interest in spatial aspects of geomorphology when he states:

> The excitement of work on process mechanisms, and the decline of the spatial tradition in geography, have pushed mapping to the periphery of academic concern, though it is important everywhere in applied work. The production of geomorphological maps has a very low priority in Britain and North America, but many field-based publications contain maps of selected features relevant to their themes; these features may be the dominant landforms within the limited areas involved.

Morphological mapping in geomorphology developed from the work of Waters (1958) and Savigear (1965). Spatial units are defined using symbols for breaks of slope, and particular geomorphological features are shown using specific symbols (Bakker, 1963). Mapping may be completed in the field as well as from aerial photography (Verstappen, 1983). Further sophistication may be included, for example, by information on the genesis and nature of materials (Demek et al., 1972). However, problems occur in areas where breaks of slope are not obvious and slopes are smooth. The techniques have been extensively applied, for example, in the work of Sissons (1967, 1974), when mapping features of glacial geomorphology in Scotland. Some countries have produced national geomorphological maps (e.g. France, Belgium and Hungary, with maps at 1:25,000 and 1:50,000) but detail is often poor at this scale.

The advent of geographical positioning systems (GPS) and geographical information systems (GIS) has greatly improved accuracy and presentation of results (Raper, 2000; DeMers, 2005; Longley et al., 2005). Many geomorphological maps are produced for consultancy reports and in applied projects. These are often site- and problem-specific and may often not be in the public domain. Digital terrain modelling using GIS is now widely applied and is an important tool for geomorphologists (DeMers, 2005). Again GPS has proved invaluable in this respect (Longley et al., 2005).

As discussed below, geomorphological mapping, digital terrain modelling and GIS also provide the basis for the discipline of terrain evaluation and land-systems mapping, which provides a strong integrative spatial theme for the whole of physical geography.

SPATIAL SYNTHESIS IN PHYSICAL GEOGRAPHY

The problem of the increasing fragmentation of physical geography has been highlighted by many authors (Unwin, 1992; Holt-Jensen, 1999; Gregory, 2000), but geography is also meant to be a subject of synthesis and some notable attempts at spatial synthesis in physical geography exist. Examples of spatial synthesis occur at two scales: first, at the landscape unit–ecosystem scale, in the disciplines of terrain evaluation and mapping (Townshend, 1981; Vink, 1983; Mitchell, 1991), which are linked to the more recent developments in landscape ecology discussed above (Forman and Godron, 1986; Forman, 1995; Farina, 2000, 2006; Turner et al., 2001); and, second, at the biome scale, in what can be described as regional physical geography, of which the best contemporary examples are probably Briggs et al. (1997) and Goudie (2001).

Terrain evaluation and land-systems mapping
Terrain evaluation and land-systems mapping evolved in those parts of the globe where assessments of land resources over very large areas were required, often for resource exploitation – for example, the Commonwealth Scientific and Industrial Research Organization (CSIRO) in Australia and the Canadian Government Land Surveys (Environment Canada) (Townshend, 1981; Vink, 1983;

Mitchell, 1991). The importance of such surveys lay in their ideas on synthesis and that strong relationships existed among geology, geomorphology and slope form, overlying soils and vegetation (ecosystems) and, finally, land-use, where human modification was important. Particular types of slope form (slope facets) were linked to certain soil and vegetation (ecosystem) types and these in turn determined land-use activity, resulting in a form of habitat/ecosystem-type mapping. Warren (2001) called for a re-evaluation of these ideas in relation to valley-side slope processes to provide a basis for conservation management. Aerial photography, remote sensing, GIS and digital terrain modelling are now an essential part of such evaluation and yet again problems of boundary delimitation between different terrain types or slope facets represent a major research focus for the subject (Townshend, 1981; Burrough and Frank, 1996; Burrough and McDonnell, 1998).

Regional physical geography

The other potential area of spatial synthesis in physical geography is that of 'regional physical geography'. Although such ideas have taken a back seat over the past 30 years due to the 'process revolution' and fragmentation of subdisciplines in physical geography, there is no doubt that the possibility still exists. In theory, such spatial regional synthesis is possible at a range of scales, starting at the biome scale and progressively becoming more detailed at the regional and then local scales (Figure 6.2). Examples of physical synthesis at the biome scale still exist, most notably in the work of Goudie (2001), who takes a global perspective to physical geography, but then in the second part of his text examines the physical geography of the 'major world zones', as he describes them. Regional and local studies are, however, virtually non-existent.

THE FUTURE: BIOGEOGRAPHY AS AN INDICATOR OF A 'SPATIAL REAWAKENING' IN PHYSICAL GEOGRAPHY

Where does the future lie in terms of the awareness and degree of emphasis of the concept of space within physical geography? At the present time, the one area where a re-emergence of space as a key concept is clearly occurring is in biogeography and also within its related disciplines of biology and ecology. Fascinatingly, biologists and ecologists have become increasingly aware of the importance of space over the past decade. This increasing awareness is expressed in three areas: metapopulation ecology, macroecology, and the application of geostatistics and spatial analysis in ecology.

Metapopulation ecology

The ecological subdiscipline of 'metapopulation ecology', which is concerned with research into species populations in 'spatially patchy' environments, emerged during the 1990s. The following quotation from the concluding chapter of Hanski (1999: 261) demonstrates this new focus:

Ecologists are used to thinking about assemblages of interacting entities, such as individuals in populations, but usually without an explicit reference to space. The novelty of spatial ecology is in the claim that the spatial locations of individuals, populations and communities can have equally significant consequences on dynamics as birth and death rates, competition and predation. Spatial ecology is one of the most visible developments in ecology and population biology in recent years; some regard it as a new paradigm arising towards the end of the twentieth century. New paradigm or not, it is true that never before has space been considered to be so pivotal to so many biological phenomena as today.

Metapopulation ecology has evolved from and is now replacing one of the few examples of spatial theory in physical geography, namely, the theory of island biogeography, originally proposed by MacArthur and Wilson (1963, 1967). The theory has been widely interpreted as having important implications for biological conservation (Diamond, 1975; Kent, 1987; Shafer, 1990), but more recently has been seriously questioned, particularly the underlying assumptions of equilibrium (Stott, 1998; Whittaker, 1998, 2000).

Macroecology

In relation to the study of single species distributions, a whole new area called 'macroecology' is emerging (Gaston and Blackburn, 2000; Gaston, 2003, 2006; Storch and Gaston, 2004; Kent, 2005, 2007b), which is described as a distinctive approach within ecology, separate from biogeography, which not only examines patterns of species distributions but also aims to uncover natural laws and unifying principles that underlie the natural structure and function of ecological systems. Macroecology and biogeography overlap to a considerable extent, but macroecology claims to go beyond the simple description and explanation of species patterns to the discovery of fundamental relationships between species ranges, abundance, diversity (richness), body size and environmental correlates such as temperature and ecosystem energetics. As the name 'macroecology' suggests, the aim is to get away from the reductionist ecology of the past 30 years, representing the 'bottom-up' approach and change to a 'top-down' approach, working from species spatial patterns at regional and continental scales to explore numerous distributional and environmental relationships and their underlying processes.

Kent (2005; 2007b) summarizes much recent research in macroecology. To give one example, many studies have now been completed to test the validity of Rapoport's rule (that geographic range size of species increases from equator to poles). Arita et al. (2005) confirmed the relationship for the mammals of North America and similar results were found by Morin and Chuine (2006) and Lane (2007) for northern temperate/boreal trees. However, Ribas and Schoereder (2006) tested the Rapoport effect via null models and simulated data against real data from the literature, finding inconclusive results, a similar result to that of Weiser et al. (2007) working with New World woody plants. This mixed set of outcomes for different organism groups in differing locations is typical of many areas of macroecology, where the complexities of macroecological patterns and the confounding effects of anthropogenic activity make the search for clear and unambiguous relationships and the understanding of processes a difficult and demanding task. There is no doubt, however, that this is a rapidly growing and important area of physical geography with a markedly spatial emphasis (Kent, 2007b).

Geostatistics, spatial autocorrelation and spatial analysis in ecology

This new focus on space is expressed elsewhere in ecology. Dale (1999) and Fortin and Dale (2005) have reinvigorated and revitalized the methodology of spatial analysis for ecologists. Problems of boundary definition and mapping are assuming a new importance related to recent developments in GIS (Burrough and Frank, 1996). Kent et al. (1997) have reviewed the concepts of the plant community and the problems of defining and mapping plant community boundaries and gradients. Once again in this area, ecologists and mathematicians, rather than biogeographers, have made major contributions to boundary delimitation and mapping through new advances in geostatistics, including applications of fuzzy classification and crisp and fuzzy boundary detection (e.g. Fortin et al., 2000; Jacquez et al., 2000; Kent et al., 2006). As an example, a user-friendly software package for boundary detection and spatial rate of change analysis in spatially distributed ecological data – BoundarySeer (TerraSeer, 2001) – has recently been released (Figure 6.4), based on the work of Jacquez and Fortin, neither of whom, incidentally, are geographers. Many of these methods are also available within standard GIS applications such as ArcInfo, ArcView and ArcGIS (McGarigal and Marks, 2002; ESRI, 2001) and there is no doubt that use of GIS as a tool of spatial description and analysis within all parts of physical geography will increase even further over the next decade (Burrough and Frank, 1996; Longley et al., 2005).

A new awareness of the importance of understanding and accounting for spatial autocorrelation using geostatistics within ecological and biogeographical data is also emerging (Chappell, 2003; Kent et al., 2006). Kent (2006) has summarized the implications of spatial autocorrelation for multivariate analysis using numerical classification and ordination methods.

CONCLUSION

As yet, these rapidly developing trends in biology and ecology are only just beginning to percolate into biogeography and, ironically, it seems almost as if everyone *but* geographers and biogeographers are at the forefront of this new revolution. Whether any similar revitalization of spatial concepts and spatial analysis is likely in the rest of physical geography remains unclear, although the use of geospatial statistics and advanced techniques, including wavelet analysis, is gaining popularity in geomorphology (e.g. Torgersen et al., in press). While many physical geographers will always argue that space and a spatial emphasis are implicit in their work, perhaps now is an appropriate moment for them to re-examine the importance of space and spatial concepts and analysis in their various researches and

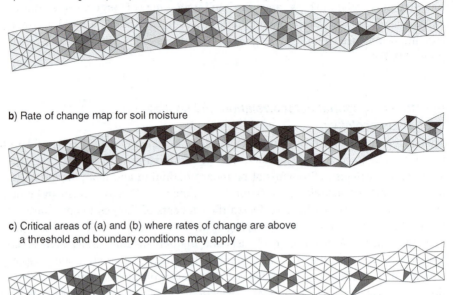

a) Rate of change map for plant community species composition

b) Rate of change map for soil moisture

c) Critical areas of (a) and (b) where rates of change are above
    a threshold and boundary conditions may apply

Sea   ←——— 2 km ———→   Inland

Figure 6.4   Examples of mapping and boundary detection of plant communities and
an environmental variable (soil moisture) using the 'BoundarySeer' package of Jacquez
et al. (2000) and TerraSeer (2001). Data are from a 2 km × 200 m 'tranome' (a 2-D
transect) across the machair sand dunes of South Uist in the Outer Hebrides of
Scotland. The first map (a) shows rate of change in plant community composition
derived from 'triangulation wombling' of scores from the first axis of an ordination of
the species data from quadrats located at each of the 217 points in the triangular
grid. The darker the tone, the greater the rate of change. The second map (b) presents
rates of change of soil moisture values on the same grid, while the third map (c)
shows critical areas where rates of change are above a given threshold and thus
boundary conditions may apply
Reproduced with the kind permission of Catherine Reid, University of Plymouth

in their teaching. The increasing importance of GIS and remote sensing within
physical geography and related new developments in spatial analysis almost certainly represent a significant way forward in addition to the major trends in the
subject identified at the outset of this chapter. The understanding of process is
only as useful as the ability to apply it to explain the highly spatially variable
nature of the earth's surface. Thus defining the geographic variation of the key
input parameters and variables of all process models is essential, as is the geographic variability of their outputs. The above technologies can assist greatly
with doing this. While the emphasis of physical geographers on processes and
time will undoubtedly and rightly continue, in the future physical geographers
may not be able to ignore the concept of space quite as easily as perhaps they
have done over the past 30 years.

SUMMARY

- Spatial concepts have been neglected in physical geography over the past 30 years.

- Physical geographers are poor at explicitly defining and explaining spatial concepts in their subject.

- Physical geographers have perhaps emphasized process at the expense of pattern (space) in recent years.

- The time dimension in physical geography has received greater attention than the spatial dimension.

- Spatial units, such as the plant community, the air mass and the drainage basin, are fundamental to physical geography.

- Spatial synthesis across physical geography is poorly developed.

- The development of geographical information systems (GIS) and remote sensing offers exciting new possibilities for spatial analysis in physical geography.

- Recent developments in biology and ecology may stimulate a new awareness of the importance of space in physical geography through the three new areas of metapopulation ecology, macroecology and the application of spatial analysis in biogeography using geostatistics.

Further Reading

Gatrell's (1983) *Distance and Space: A Geographical Perspective* still provides an excellent introduction to concepts of space in geography. A valuable overview of the nature of physical geography is provided in *The Changing Nature of Physical Geography* (Gregory, 2000). The time element and time–space relationships are well covered in Slaymaker and Spencer's (1998) *Physical Geography and Global Environmental Change*, and the article by Schumm and Lichty (1965) is still essential reading for all physical geographers. Forman and Godron's (1986) *Landscape Ecology* remains the best introduction to landscape ecology as a truly spatial-based approach to biogeography, with more recent statements in Turner et al. (2001), Burel and Baudry (2003), Turner (2005) and Farina (2006). For a really challenging read on the awakening of spatial analysis in biology and ecology rather than physical geography, look at (but don't try to read in detail!) Hanski's (1999)

(Continued)

(Continued)

Metapopulation Ecology; Dale's (1999) **Spatial Pattern Analysis in Plant Ecology** and most of all Fortin and Dale's (2005) **Spatial Analysis: A Guide for Ecologists**. Geographical Information Systems (GIS) are still insufficiently integrated into physical geography yet will be an even more vital tool in spatial description and analysis for the subject in the future. Burrough and McDonnell's (1998) **Principles of Geographical Information Systems** and Longley et al.'s (2005) **Geographic Information Systems and Science** provide the best introductions to the subject. For an introduction to ideas of spatial autocorrelation and geostatistics, read the chapter by Chappell (2003) in the companion volume to this book.

Note: Full details of the above can be found in the references list below.

References

Arita, H.T., Rodriguez, P. and Vázquez-Domínguez, E. (2005) 'Continental and regional ranges of North American mammals: Rapoport's rule in real and null worlds', *Journal of Biogeography*, 32: 960–71.

Bakker, J.P. (1963) 'Different types of geomorphological maps, problems of geomorphological mapping', *Geographical Studies*, 46: 13–31.

Barry, R. (2003) *Atmosphere Weather and Climate* (8th edn). London: Routledge.

Bennett, K.D. (1997) *Evolution and Ecology: The Pace of Life*. Cambridge: Cambridge University Press.

Bennett, R.J. and Chorley, R.J. (1978) *Environmental Systems: Philosophy, Analysis and Control*. London: Methuen.

Bissonette, J.A. (ed.) (1997) *Wildlife and Landscape Ecology: Effects of Pattern and Scale*. New York: Springer.

Briggs, D.J. (1981) 'Editorial: the principles and practice of applied geography', *Applied Geography*, 1: 1–8.

Briggs, D.J. and Smithson, P. (1985) *Fundamentals of Physical Geography*. London: Hutchinson/Routledge.

Briggs, D.J. Smithson, P., Addison, K. and Atkinson, K. (1997) *Fundamentals of the Physical Environment*. London: Routledge.

Burel, F. and Baudry, J. (2003) *Landscape Ecology: Concepts, Methods and Applications*. Enfield, NH: Science Publishers Inc.

Burrough, P.A. and Frank, A.U. (eds) (1996) *Geographic Objects with Indeterminate Boundaries*. GISDATA 2. London: Taylor & Francis.

Burrough, P.A. and McDonnell, R.A. (1998) *Principles of Geographical Information Systems* (2nd edn). Oxford: Clarendon Press.

Chappell, A. (2003) 'An introduction to geostatistics', in N.J. Clifford and G. Valentine (eds) *Key Methods in Geography*. London: Sage, pp. 383–407.

Cherrill, A. and McClean, C. (1999) 'The reliability of "Phase 1" habitat mapping in the UK: the extent and types of observer bias', *Landscape and Urban Planning*, 45: 131–44.

Chorley, R.J. (1969) 'The drainage basin as a fundamental geomorphic unit', in R.J. Chorley (ed.) *Water, Earth and Man*. London: Methuen, pp. 77–100.

Chorley, R.J. (ed.) (1972) *Spatial Analysis in Geomorphology*. London: Methuen.

Chorley, R.J. and Kennedy, B.A. (1971) *Physical Geography: A Systems Approach*. London: Prentice Hall.

Clark, M.J., Gregory, K.J. and Gurnell, A.M. (eds) (1987) *Horizons in Physical Geography*. Basingstoke: Macmillan.

Countryside Survey (2008) Countryside Survey website,http://www.countrysidesurvey.org.uk (accessed 20.5.2008).

Dale, M.R.T. (1999) *Spatial Pattern Analysis in Plant Ecology*. Cambridge: Cambridge University Press.

Dargie, T.C.D. (1998) *Sand Dune Vegetation Survey of Scotland: Western Isles. Scottish Natural Heritage Research, Survey and Monitoring Report 96* (3 vols). Battleby, Perth: Scottish Natural Heritage.

Demek, J., Embleton, C., Gellert, J.F. and Verstappen, H. (1972) *Manual of Detailed Geomorphological Mapping*. Prague: Academia.

DeMers, M. (2005) *Fundamentals of Geographic Information Systems* (3rd edn). London: Wiley.

Department of Environment (1993) *Countryside Survey 1990 – Summary Report*. London: HMSO/Department of the Environment.

Department of Environment, Transport and the Regions/Centre for Ecology and Hydrology (2000) *Accounting for Nature: Assessing Habitats in the UK Countryside*. London: Department of Environment, Transport and the Regions.

Diamond, J.M. (1975) 'The island dilemma: lessons of modern biogeographical studies for the design of nature reserves', *Biological Conservation*, 7: 129–46.

Doornkamp, J.C. and King, C.A.M. (1971) *Numerical Analysis in Geomorphology*. London: Edward Arnold.

ESRI (Environmental Systems Research Institute) (2001) *ArcGIS 8.1 Software*. Redlands, CA: ESRI Inc.

Evans, I.S. (1990) 'Cartographic techniques in geomorphology', in A. Goudie (ed.) *Geomorphological Techniques* (2nd edn). London: Unwin Hyman, pp. 97–108.

Farina, A. (2000) *Landscape Ecology in Action*. Dordrecht: Elsevier.

Farina, A. (2006) *Principles and Methods in Landscape Ecology* (2nd edn). Dordrecht: Kluwer Academic.

Forman, R.T.T. (1995) *Landscape Mosaics*. Cambridge: Cambridge University Press.

Forman, R.T.T. and Godron, M. (1986) *Landscape Ecology*. Chichester: Wiley.

Fortin, M.J. and Dale, M. (2005) *Spatial Analysis: A Guide for Ecologists*. Cambridge: Cambridge University Press.

Fortin, M.J., Olson, R.J., Ferson, S., Iverson, L., Hunsaker, C., Edwards, G., Levine, D., Butera, K. and Klemas, V. (2000) 'Issues related to the detection of boundaries', *Landscape Ecology*, 15: 453–6.

Ganio, L.M., Torgersen, C.E. and Gresswell, R.E. (2005) 'A geostatistical approach for describing spatial pattern in stream networks', *Frontiers in Ecology and Environment*, 3: 138–44.

Gaston, K.J. (2003) *The Structure and Dynamics of Geographic Ranges*. Oxford: Oxford University Press.

Gaston, K.J. (2006) 'Biodiversity and extinction: macroecological patterns and people', *Progress in Physical Geography*, 30: 258–69.

Gaston, K.J. and Blackburn, T.M. (2000) *Pattern and Process in Macroecology*. Oxford: Blackwell Science.

Gatrell, A.C. (1983) *Distance and Space: A Geographical Perspective*. Oxford: Clarendon Press.

Gerrard, J. (2000) *Fundamentals of Soils*. London: Routledge.

Goudie, A. (2001) *The Nature of the Environment* (4th edn). Oxford: Blackwell.

Gower, A.M. (ed.) (1980) *Water Quality in Catchment Ecosystems*. Chichester: Wiley.

Gregory, K.J. (2000) *The Changing Nature of Physical Geography*. London: Edward Arnold.

Gregory, K.J. and Walling, D.E. (1973) *Drainage Basin Form and Process*. London: Edward Arnold.

Haggett, P. (1990) *The Geographer's Art*. Oxford: Blackwell.

Haines-Young, R., Green, D.R. and Cousins, S. (eds) (1993) *Landscape Ecology and Geographic Information Systems*. London: Taylor & Francis.

Haines-Young, R. and Petch, J. (1986) *Physical Geography: Its Nature and Methods*. London: Harper & Row.

Hanski, I. (1999) *Metapopulation Ecology*. Oxford: Oxford University Press.

Harrison, C.M. and Warren, A. (1970) 'Conservation, stability and management', *Area*, 2: 26–32.

Hill, M.O. (1991) *TABLEFIT – for Identification of Vegetation Types. Version 0.0*. Monks Wood, Huntingdon/UK: Institute of Terrestrial Ecology.

Holt-Jensen, A. (1999) *Geography: History and Concepts* (3rd edn). London: Sage.

Huggett, R.J. (2004) *Fundamentals of Biogeography* (2nd edn). London: Routledge.

Jacquez, G.M., Maruca, S.L. and Fortin, M.-J. (2000) 'From fields to objects: a review of geographic boundary analysis', *Journal of Geographical Systems*, 2: 221–41.

Kent, M. (1987) 'Island biogeography and habitat conservation', *Progress in Physical Geography*, 11: 91–102.

Kent, M. (2005) 'Biogeography and macroecology', *Progress in Physical Geography*, 29: 256–64.

Kent, M. (2006) 'Numerical classification and ordination methods in biogeography', *Progress in Physical Geography*, 30: 399–408.

Kent, M. (2007a) 'Biogeography and landscape ecology', *Progress in Physical Geography*, 31: 227–38.

Kent, M. (2007b) 'Biogeography and macroecology: now a significant component of physical geography', *Progress in Physical Geography*, 31: 643–57.

Kent, M. and Coker, P. (1992) *Vegetation Description and Analysis: A Practical Approach*. Chichester: Wiley.

Kent, M., Gill, W.J., Weaver, R.E. and Armitage, R.E. (1997) 'Landscape and plant community boundaries in biogeography', *Progress in Physical Geography*, 21: 315–53.

Kent, M., Moyeed, R.A., Reid, C.L., Pakeman, R. and Weaver, R. (2006) 'Geostatistics, spatial rate of change analysis and boundary detection in plant ecology and biogeography', *Progress in Physical Geography*, 30: 201–31.

Klopatek, J.M. and Gardner, R.H. (eds) (1999) *Landscape Ecological Analysis: Issues and Applications*. New York: Springer.

Kupfer, J.A. (1995) 'Landscape ecology and biogeography', *Progress in Physical Geography*, 19: 18–34.

Lane, C.S. (2007) 'Latitudinal range variation of trees in the United States: a reanalysis of the applicability of Rapoport's rule', *The Professional Geographer*, 59: 115–30.

Lane, S. (2001) 'Constructive comments on D. Massey "Space-time, 'science' and the relationship between physical geography and human geography"', *Transactions, Institute of British Geographers*, 26: 243–5.

Lau, S.S.S. and Lane, S.N. (2001) 'Continuity and change in environmental systems – the case of shallow lake ecosystems', *Progress in Physical Geography*, 25: 178–202.

Likens, G.E., Bormann, F.H., Pierce, R.S., Eaton, J.S. and Johnson, N.M. (1977) *Biogeochemistry of a Forested Ecosystem*. Berlin: Springer-Verlag.

Longley, P.A., Goodchild, M.F., Maguire, D.J. and Rhind, D.W. (2005) *Geographic Information Systems and Science* (2nd edn). Chichester: Wiley.

MacArthur, R.H. and Wilson, E.O. (1963) 'An equilibrium theory for insular zoogeography', *Evolution*, 17: 372–87.

MacArthur, R.H. and Wilson, E.O. (1967) *The Theory of Island Biogeography*. Princeton, NJ: Princeton University Press.

Malloch, J.C. (1991) *MATCH (VERSION 1.3) – a Computer Program to Aid the Assignment of Vegetation Data to the Communities and the Sub-communities of the National Vegetation Classification*. Lancaster: University of Lancaster.

Massey, D. (1999) 'Space-time, "science" and the relationship between physical geography and human geography', *Transactions, Institute of British Geographers*, 24: 261–7.

McGarigal, K. and Marks, B.J. (2002) *FRAGSTATS: Spatial Pattern Analysis Program for Quantifying Landscape Structure. Version 3*. Corvallis, OR: Oregon State University (available at http://www.innovativegis.con/fragstatsarc/manual/manpref.htm *or* www.umass.edu/landeco/research/fragstats/fragstats.html).

Mitchell, C.W. (1991) *Terrain Evaluation: An Introductory Handbook to the History, Principles and Methods of Practical Terrain Assessment*. Harlow: Longman Scientific and Technical.

Morin, X. and Chuine, L. (2006) 'Niche breadth, competitive strength and range size of tree species: a trade-off based framework to understand species distribution', *Ecology Letters*, 9: 185–95.

Newson, M. (1995) *Hydrology and the River Environment*. Oxford: Clarendon Press.

Oliver, J. (1991) 'The history, status and future of climatic classification', *Physical Geography*, 12: 231–51.

O'Sullivan, P. (1979) 'The ecosystem–watershed concept in the environmental sciences: a review', *Journal of Environmental Studies*, 13: 273–81.

Pacione, M. (1999) *Applied Geography: Principles and Practice*. London: Routledge.

Perring, F.H. and Walters, S.M. (1982) *Atlas of the British Flora* (3rd edn). Wakefield: E.P. Publishing.

Perry, A.W. (1995) 'New climatologists for a new climatology', *Progress in Physical Geography*, 19: 280–5.

Preston, C.D., Pearman, D.A. and Dines, T.D. (2002) *New Atlas of the British and Irish Flora*. Oxford: Oxford University Press.

Raper, J. (2000) *Multidimensional Geographic Information Science*. London: Taylor & Francis.

Rhoads, B.L. and Thorn, C.E. (1996) *The Scientific Nature of Geomorphology*. Chichester: Wiley.

Ribas, C.R. and Schoereder, J.H. (2006) 'Is the Rapoport effect widespread? Null models revisited', *Global Ecology and Biogeography*, 15: 614–24.

Roberts, N. (1998) *The Holocene: An Environmental History* (2nd edn). Oxford: Blackwell.

Rodwell, J.S. (ed.) (1991–2000) *British Plant Communities* (Vols 1–5). Cambridge: Cambridge University Press.

Rogers, A., Viles, H. and Goudie, A. (1992) *The Student's Companion to Geography*. Oxford: Blackwell.

Savigear, R.A.G. (1965) 'A technique of morphological mapping', *Annals of the Association of American Geographers*, 55: 514–38.

Schaefer, F.K. (1953) 'Exceptionalism in geography: a methodological examination', *Annals of the Association of American Geographers*, 43: 226–49.

Schumm, S.A. and Lichty, R.W. (1965) 'Time, space and causality in geomorphology', *American Journal of Science*, 263: 110–19.

Shafer, C.L. (1990) *Nature Reserves: Island Theory and Conservation Practice*. Washington, DC: Smithsonian Institute Press.

Sissons, J.B. (1967) *The Evolution of Scotland's Scenery*. Edinburgh: Oliver & Boyd.

Sissons, J.B. (1974) 'A late glacial ice cap in the central Grampians, Scotland', *Transactions, Institute of British Geographers*, OS 62: 95–114.

Slaymaker, O. and Spencer, T. (1998) *Physical Geography and Global Environmental Change*. Harlow: Addison Wesley Longman.

Smithson, P., Addison, K., Atkinson, K. and Briggs, D. (2002) *Fundamentals of the Physical Environment*. London: Routledge.

Stoddart, D.R. (1986) *On Geography*. Oxford: Blackwell.

Stoddart, D.R. (ed.) (1997) *Process and Form in Geomorphology*. London: Routledge.

Storch, D. and Gaston, K.J. (2004) 'Untangling ecological complexity on different scales of space and time', *Basic and Applied Ecology*, 5: 389–400.

Stott, P. (1998) 'Biogeography and ecology in crisis: the urgent need for a new metalanguage', *Journal of Biogeography*, 25: 1–2.

Tansley, A.G. (1935) 'The use and abuse of vegetational concepts and terms', *Ecology*, 16: 284–307.

TerraSeer (2001) *BoundarySeer – Software for Geographic Boundary Analysis*. Ann Arbor, MI: Biomedware.

Torgersen, C.E., Gresswell, R.E., Bateman, D.S. and Burnett, K.M. (in press) 'Spatial identification of tributary impacts in river networks', in S.P. Rice, A. Roy and B.L. Rhoads (eds) *River Confluences, Tributaries and the Fluvial Network*. Chichester: Wiley, pp. 159–81.

Townshend, J.R.G. (ed.) (1981) *Terrain Analysis and Remote Sensing*. London: George Allen & Unwin.

Turner, M.G. (2005) 'Landscape ecology in North America: past, present and future', *Ecology*, 86: 1967–74.

Turner, M.G., Gardner, R.H. and O'Neill, R.V. (2001) *Landscape Ecology in Theory and Practice*. New York: Springer.

Unwin, T. (1992) *The Place of Geography*. Harlow: Longman.

Verstappen, H.T. (1983) *Applied Geomorphology: Geomorphological Surveys for Environmental Development*. Amsterdam: Elsevier.

Vink, A.P.A. (1983) *Landscape Ecology and Land Use* (trans. and edited D.A. Davidson). London: Longman.

Waite, S. (2000) *Statistical Ecology in Practice*. London: Prentice-Hall.

Warren, A. (2001) 'Valley-side slopes', in A. Warren and J.R. French (eds) *Habitat Conservation: Managing the Physical Environment*. Chichester: Wiley, pp. 39–66.

Waters, R.S. (1958) 'Morphological mapping', *Geography*, 43: 10–17.

Webster, R. (1977) *Quantitative and Numerical Methods in Soil Classification and Survey*. Oxford: Oxford University Press.

Webster, R. (1985) 'Quantitative spatial analysis of soil in the field', *Advances in Soil Science*, 3: 1–10.

Weiser, M.D., Enquist, B.J., Boyle, B., Killeen, T.J., Jorgensen, P.M., Fonseca, G., Jennings, M.D., Kerkhoff, A.J., Lacher, T.E. and Monteagudo, A. (2007) 'Latitudinal patterns of range size and species richness of New World woody plants', *Global Ecology and Biogeography*, 16: 679–88.

White, R.E. (2006) *Principles and Practice of Soil Science: The Soil as a Natural Resource* (3rd edn). Oxford: Blackwell Science.

Whittaker, R.J. (1998) *Island Biogeography*. Oxford: Oxford University Press.

Whittaker, R.J. (2000) 'Scale, succession and complexity in island biogeography: are we asking the right questions?', *Global Ecology and Biogeography*, 9: 75–85.

Wiens, J.A. (2002). 'Riverine landscapes: taking landscape ecology into the water', *Freshwater Biology*, 47: 501–15.

Wiens, J.A. and Moss, M. (2005) *Issues and Perspectives in Landscape Ecology*. Cambridge: Cambridge University Press.

ACKNOWLEDGEMENTS

Martin Kent would like to thank Cheryl Hayward, Marzuki Haji, Brian Rogers and Tim Absalom of the Cartography Unit of the School of Geography, University of Plymouth, for drawing the figures.

7

Time: Change and Stability in Environmental Systems

John B. Thornes

Definition

Time is a framework in which geomorphological events are often placed to infer cause-and-effect relationships. Historic geomorphology uses this framework to reconstruct the past. Evolutionary geomorphology attempts to argue from process deductions about how landforms develop in time – what are the trajectories through time? Dynamical geomorphology attempts to explain this evolution through time in terms of non-linear behaviour. Time is not a variable, nor does it explain outcomes.

INTRODUCTION

Since the appearance of *Geomorphology and Time* (Thornes and Brunsden, 1977), a number of key developments across the broad field of physical sciences have had a fundamental impact on the ways we view time. These changes, like those of the quantitative revolution that profoundly affected physical geography in the 1960s and 1970s, parallel the shift from the 'old science and old thinking' to the 'new science and new thinking' (Marshall and Zohar, 1997). They involve the paradigms of dynamical systems, non-linearities, chaotic behaviour and panarchy. The purpose of this chapter is to provide a readable and digestible introduction to those ideas and the impacts they are likely to have on physical geography (see Chapter 8 on time in human geography). It is not the objective to provide a history of the subject itself that has been admirably accomplished by Gregory (1985).

Although many of these ideas took root long before the 1960s, they have taken time to mature and become accepted. The rapid progress of quantitative ecology under the influence of May in the UK and the Princeton School in the USA; of non-linear geomorphology by Schumm at Fort Collins and Favis-Mortlock at Oxford; and of dynamical climatology by Lorenz and Trenberth; have so much altered the view of how natural physical systems change in time that the moment has come for an overall reconstruction of what and how we research and teach in physical geography. This chapter attempts in some ways to provide a basis for this much needed review. It is consciously non-mathematical in approach. It reflects heavily the influences and originality of the author's own teachers: W.E.H. Culling, F.K. Hare and D.R. Harris.

A VERY BRIEF HISTORY OF THE TIME PERSPECTIVE

Physical geography comprises geomorphology, biogeography and climatology – all of those subjects that study the character of the face of the earth, its form, shape, variety and origins, its history and evolution. Because geomorphology took its roots in American geology (and before that, from the stratigraphers who roamed the British countryside), it had, from the outset, a heavily historical and evolutionary conceptual basis, as illustrated in the works of W.M. Davis. This was typified by Davis's model of the cycle of erosion (1899). In this model Davis characterized the landform sequence that would occur with the passage of time, following an initial disturbance in the form of a relatively sudden uplift. Later, geomorphologists used these landform characteristics to give a relative age to areas of country, notably in southeast England (Wooldridge and Linton, 1955) and in the Appalachians, just as geologists use fossils to date rocks and petrography to characterize the conditions in which rock formations were developed and therefore to infer a time sequence of the history of events. In the 1960s, the interest and emphasis lay in using these stratigraphic events to infer the historical sequence of events in the Quaternary period. Dating methods ranged from the simple rule of superimposition for sedimentary rocks (deepest are oldest) to radio-isotope dating. As the methods of dating improved, so the precision of the historical sequence could also be improved. Emphasis on climatic dating (such as glaciations and warm and cold periods), coupled with pollen stratigraphy, confused the issue by shifting the emphasis to dating landforms and away from the study of processes, sometimes leading to circularity of argument through a lack of clarity of objectives.

The coupling of environmental change and historical dating methods flowered in the middle of the twentieth century when techniques were invented, developed, improved and applied to environmental reconstruction. Among the most important of these were radiocarbon dating (Libby, 1955) and oxygen-isotope temperature estimations and dating (Shackleton, 1977). These illustrate the point made by Harré (1969) that scientific progress is often made by technical innovation. Of course, the greatest example of this lies with the mid-century development and proliferation of computers, closely followed by the great accessibility of remotely sensed images and processing capacity. The first ensured that early analytical models in geomorphology, ecology and climatology could be solved by digital numerical methods as illustrated in the third example towards the end of the chapter. The second, mainly through the openness and generosity of the US government, enabled serious and purposeful monitoring of the earth's surface at progressively higher frequencies. Data to test theories of change through time suddenly became abundant, and the numerical capacity to model these changes mathematically ushered in the expansion of studies of global and regional climate changes and their impacts that were the precursors of the present preoccupation with global warming and its actual and potential circumstances for the earth's population.

Throughout most of the last century, the concept of succession prevailed in biogeographical thought. This involved the idea that, following disturbances, there was a progressive change leading to a new equilibrium state called the 'climax', in which the vegetation achieved its full potential, given the prevailing soil, climate and socio-economic conditions. This was able to rest empirically on the description and characterization of the vegetation across the globe that had been successively accomplished by the earliest physical geographers (notably von Humboldt) in their exploration of the globe. Classification of climates had also been practised for some decades before it became central to the problems of global change.

Already by 1920, ecology was experiencing the innovative development of the exploration of single-species population models and models of competition between competing species that would herald the eventual dismissal of the simplified conceptual basis of the theory of succession. By the 1940s, the climatologist Edward Lorenz had stumbled, through computational problems, on to the essential difficulties of non-linearity and chaos in climatic modelling, the so-called sensitivity-to-initial-conditions (see below). Schumm (1979) had upset the geomorphological apple-cart by pointing out that, contrary to the idea of smooth progressive change in geomorphological behaviour, many processes proceed by a series of episodes of intensive change as 'intrinsic' thresholds were crossed. This brought about a major rethink in geomorphology when it was realized that geomorphological history depended not only on gradually changing extrinsic forces (mainly climatic and tectonic), but also on the robustness (resilience) of geomorphological systems to absorb the impacts of these changes. It was only by the end of the century that Phillips (1999) collected together the material and consolidated the view of non-linear physical geography.

Another important landmark appeared about the same time. This is the *panarchy paradigm* that builds on the original work of Holling (1973) in developing the metaphor of stability and resilience, admitting the existence of multiple stable states and applying it to understanding transformations in human and natural systems. In particular, Holling and Gunderson (2002) developed the concept of adaptive cycles following disturbance. Together with newly developed ideas about spatial ecology (Tilman and Karieva, 1997; Dieckmann et al., 2000), these intellectual advances seem set to bring about a restructuring of the biogeography component of physical geography in the near future.

The main thrust of the quantitative revolution of the 1960s and 1970s was with empirical analysis of large datasets. As well as the spatial data, attentions also focused on time-structured data and the empirical analysis of time-series for both physical environmental series (such as water quality data) or social series (such as employment, demographic and epidemiological data). The work largely comprised the decomposition of time-series into trend, periodic, persistence and noise components (for further details, see Thornes and Brunsden, 1977: 81–4). The methodology was essentially inductive, though epidemiological work also had to come to grips with modelling the wave-like behaviour of infectious outbreaks, and ecologists were already deducing non-linear competitive behaviour using difference and differential equations. Scheidegger (1960) and Kirkby (1972) opened up geomorphology to mathematical modelling through imaginative applications of standard mathematical procedures in hillslope morphology closely linked to the newly developed hillslope hydrology. By the early 1980s, the stage was set for the development of a dynamical geomorphology called 'evolutionary geomorphology', in which the emphasis is on the trajectories of behaviour through time rather than the description of the sequence of (dated) events in the historical approach we outlined earlier.

By the 1990s, the *Economic Ecology* method, an outgrowth of the entomological work of Gutierrez and Baumgartner used an evolutionary approach to the interactions of biophysical and economic systems that appears to provide a possible bridge between physical and human geography that has been the Holy Grail of geography since the middle of the 1930s, and of environmental sciences since the 1980s. This approach is exemplified in the fourth and final section of this chapter. Even more recently, Ibanez et al. (2007) have developed a General Desertification Model that does precisely that bridge-building, using a dynamical system approach. A series of generic equations are used to model processes ranging from soil erosion to groundwater extraction and the impacts of human activity on the outcomes. The gap between human and physical geography has thus been closed quite dramatically.

THE NON-LINEAR PARADIGM

Until the mid-twentieth century, scientific thinking was dominated by change as a linear phenomenon in which small forces produce proportionately small responses. Doubling the force doubled the response. Moreover, the phenomena were assumed to be kept in check by negative feedbacks that tended to subdue strong departures from the average in time. Natural systems were perceived to be well behaved, analysable, predictable and controllable. As Marshall and Zohar (1997: 248) describe it: 'Linear science complemented the dreams of a nineteenth century society that was rule-bound, reliable, predictable and unlikely to shock, a society that believed in eternal progress through the manipulation and exploitation of natural and human resources.'

In physical geography, the linear world is represented by the idea of convergence on an equilibrium state. Equilibrium is a mathematical concept in which the rate of change is zero. It is generally taken to indicate a non-changing state of the system under observation. Equilibrium thinking has been dominated by Le Chatalier's principle that systems react to changes in ways that tend to

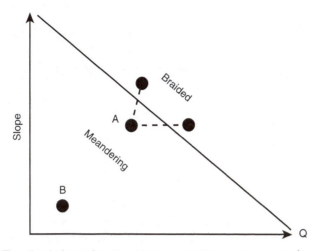
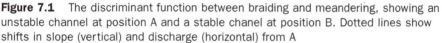

Figure 7.1   The discriminant function between braiding and meandering, showing an unstable channel at position A and a stable chanel at position B. Dotted lines show shifts in slope (vertical) and discharge (horizontal) from A

Source: Based on Schumm (1968)

minimize the effect of the initial disturbance. This has been implicit for almost 100 years in the concepts of vegetation succession and climatic climax, usually attributed to F.E. Clements (1928). We shall see later in this chapter that this cardinal role of the concept has now to be rejected. Similarly, geomorphology was preoccupied by progression towards stable equilibrium states of alluvial channel hydraulic geometry, stable network patterns and 'characteristic' slope forms. In climatology, the global circulation, notably the circumpolar Westerlies and their wave characteristics, was thought of as a stable equilibrium form of the earth's atmospheric heat engine, brought into being by advection of heat and momentum from the Equator to the poles.

      In non-linear systems, change is rapid and unexpected, and can be triggered by small events. The response is usually out of all proportion to the causative event. This is usually named the 'Butterfly Effect', following the work of climatologist, Edward Lorenz (1963). It expresses the idea that the global circulation is so non-linear that a butterfly fluttering its wings on one side of the world would perturb the circulation and that, in a non-linear system, these perturbations might be amplified to interfere with the general circulation in another part of the world.

      Big effects, then, do not necessarily have big causes. Brunsden and Thornes (1979) developed this theme in geomorphology on the basis of Schumm's earlier work. They suggested that a key feature of geomorphological systems was their sensitivity to small perturbations. This is illustrated by Schumm's example of the shift from the meandering state to a braided state in rivers. It had been shown that meandering and braided channels can be separated by a line on a graph of slope and discharge (Figure 7.1). When the line is crossed as a result of a change in either slope or discharge (or both), the system changes from one river morphology to another. In other words, it is not simply a

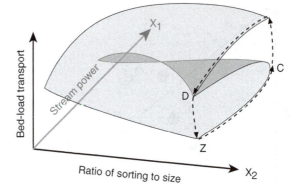

Figure 7.2   Catastrophe theory representation of bed-load sediment transport. At C, there is suden high entrainment of bed material as stream power reaches a threshold value, leading to high bed-load transport. At D, there is general deposition as stream power falls. The dotted line represents the passage of a flood wave in an ephemeral channel

Source: from Thornes (1980)

linear response but the whole character of the system changes. If a point (e.g. A) is near to the line, only tiny changes in the controlling variables (slope or discharge) will push it across the threshold – for example, from a meandering to a braided habit, as shown by the arrows in Figure 7.1. At point B, the system is very stable because large shifts would be needed in either slope or discharge before any change in state could result.

Although this example illustrates the case of two controlling variables (slope and discharge), Figure 7.2 illustrates the problem of three controlling variables. In this case, the vertical axis is the response (here the sediment transport in a river). The horizontal axes are the stream power (x_1) and the sorting of bed material (x_2). In all cases (represented by the shaded curved surface), sediment transport increases as stream power increases (going from front to back of the figure, parallel to x_1). At low values of x_2 (poorly sorted), there is always some sediment that matches the available stream power and which, therefore, can be entrained (picked up), so the curve is relatively smooth. With better sorting (further along the x_2 axis) the grain size becomes more concentrated at one size group. Only when the stream power is sufficient to entrain that grain size can the sediment load increase. The threshold of entrainment (the value at which transport of a given particle size begins) increases as the ratio of mean grain size/sorting (x_2) increases. This produces a jump in the amount of sediment transported after the tiny increase in stream power that crosses the threshold. This jump is called a catastrophe (French for step), not to be confused with catastrophic, meaning disaster. The discovery of non-linear catastrophes has led to a whole area of non-linear science called catastrophe theory with its own specialized body of mathematics. This example is considered at length for ephemeral stream behaviour in Thornes (1980), and catastrophe theory has been the subject of application in human geography and ecology in Wilson's excellent text (1981).

Sometimes the system is completely transformed and a new system re-emerges from the change. This may occur where an ecosystem is completely ravaged by fire, climatic change or cultivation. After a fire, for example, Mediterranean scrub re-emerges as woodland after 12–15 years (Obando, 2002). After oil spills, the ecosystem reorganizes itself. Emergent phenomena form another special field of non-linear behaviour that attempts to explain how patterns emerge after a major disturbance. An example is the emergence of a drainage pattern in a former lake shore or an ancient sea plain after uplift. Rebeiro-Hargrave (1999) developed a model to account for the gully-dominated drainage system that has emerged since the middle Tertiary in the Guided and Baza sedimentary basins in southeast Spain, east of Granada. A technique called cellular automata was used to model the evolution of the drainage of the basin under rules of gully extension on a sedimentary plane undergoing tectonic tilting. The emergent pattern produced in this model is very similar to that which exists today, and which emerged during the Quaternary era.

Non-linearity has become a way of life. It can be summarized as highly complicated responses to simple changes in inputs that may appear insignificantly small, but lead the systems across thresholds that unexpectedly and rapidly shift the behaviour. This has been a major threat to historical sciences such as geology and geomorphology as they existed in the middle of last century. Large and sudden changes, such as the onset of glaciation, had been assumed to be produced by large and sudden changes in the controlling systems (Oerlermans and van der Veen, 1984). It is now recognized that the Dansgaard–Oeschlager events (large flows of cold meltwater into the North Atlantic) and even the onset of the Quaternary glaciations themselves are the manifestations of the non-linear behaviour of the ice–land–ocean–atmosphere system of interaction. Many of the phenomena that were attributed in the 1960s to climate change are now attributed to hidden non-linearities in natural systems.

THE PANARCHY PARADIGM

The apparent importance of understanding the behaviour of environmental systems, particularly their non-linear behaviour, is that it appears to offer comfort to environmental managers. If the threshold of stream plan–form behaviour (meandering vs. braided) can be exactly identified, then, for a given (engineered) discharge, it is theoretically possible to train a river to avoid the slopes across which instability is likely to be generated. By lengthening or shortening a channel between two heights, the slope can be changed. Even better, if the stable condition for pool and riffle and meander wavelength can be identified and if this attractor can be reached, the river system will find it hard to shift away from this equilibrium. This is the essential basis of river restoration (Thornes and Rowntree, 2006) which first identifies the potential stable equilibria, subsequently engineers the river to one of these equilibrium states, and thereafter, keeps it there by minor engineering and geomorphological modifications (such as artificial bars and pools).

A natural system that can be managed in this way is said to be resilient – that is, it responds to management perturbations by adopting a new equilibrium. There are abundant examples of this kind of philosophy, but there are two major

types of uncertainty that undermine it. The first is that extreme events can oblit-
erate the good intentions (a great flood can simply remove all the gravel bars that
represent equilibrium conditions). Second, there are unknown and unforeseen
complications (including hidden thresholds) that make an oversimplified view of
the system exceptionally dangerous.

Such unforeseen complications are revealed in the concept of 'carrying
capacity'. Carrying capacity is the number of animals that can graze grassland
without damage to the rangeland ecology, without the consequent soil erosion and
land degradation. Of course, a pig is not the same as a cow, so the environmental
managers designed 'animal units' as the basis for agronomists to estimate the carrying capacity, and this could act as a management target that would enable them
to keep the system close to the equilibrium state with respect to ecology and land
degradation. Unfortunately, this approach fails to accept that the target grass cover
(the desired equilibrium to prevent erosion) and the perturbations that could
shift the system away from the target equilibrium are unforeseeable (the magnitude and frequency of wetter or drier periods and their impact on grassland are
discussed in a later section of this chapter). Even in the very short term, the internal dynamics of the competition between erosion and grass for water leads to
unstable oscillations of the vegetation cover (see Brandt and Thornes, 1993).
Therefore, the push to develop 'carrying capacity' as an indicator of desertification in Mediterranean environments, long after it has been abandoned elsewhere,
is based on false premises. A further problem is that this approach engenders a
false sense of security because of its apparent basis in science and engineering and
the attachment to hard targets. In an easily read and refreshingly honest appraisal
of the problems of understanding transformations in human and natural systems,
Gunderson and Holling (2002: xxii) attack the simple prescriptions of the concept
of environmental systems control because '[t]hey seem to replace inherent uncertainty with the spurious certainty of ideology, precise numbers or action. The theories implicit in these examples ignore multiple stable states.' They note the
problem imposed by non-linearities of the types described above. They go on to
say that: 'The theories ignore the possibility that the slow erosion (meaning
change) of key controlling processes can abruptly flip an ecosystem or economy
into a different state that might be effectively irreversible.'

This is an absolutely crucial point in the evolutionary view of systems. It
throws into serious doubt the belief that complex system management could (or
should) be based on a few key indicators. There is an obsession on the part of the
environmental agencies in Europe and the USA to develop single-figure indicators to monitor environmental management targets.

Holling et al. (2002) translate the discussion of equilibria into 'caricatures
of nature' as four myths, as illustrated by Figure 7.3. The caricatures are as
follows:

1   *Nature flat* – a system in which there are few or no forces affecting stability.
2   *Nature balanced* – a system at or near an equilibrium condition, the system at
    the bottom of a cup.
3   *Nature anarchic* – a view of the nature as globally unstable – dominated by
    hyperbolic processes of growth and collapse.

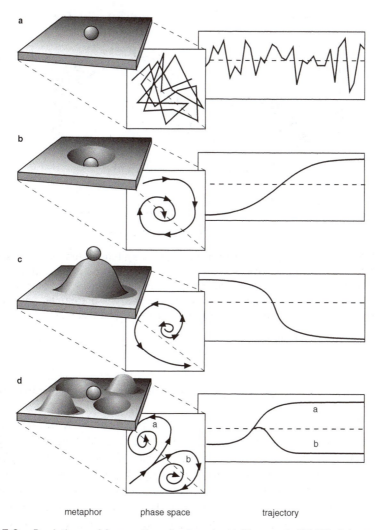

metaphor   phase space   trajectory

Figure 7.3   Depictions of four myths of nature by Holling et al. (2002): (a) nature flat; (b) nature balanced; (c) nature anarchic; (d) nature resilient (multiple stable states)

Source: from Holling et al. (2002)

4   *Nature resilient* – a nature of multi-stable states, some of which become irreversible traps while others become natural alternating states that are experienced as part of the internal dynamics.

This caricature recognizes that instabilities may organize system behaviour as much as stabilities do. (This is especially relevant in considering the dynamical systems that prevail in physical geography in which non-linearities dominate.) Holling has argued this case since 1973. Holling et al. (2002: 5) are seeking a theory of adaptive change 'to help us to understand the changes occurring globally'. They use the term panarchy to 'describe the cross-scale, interdisciplinary and dynamic nature of the theory whose essential focus is to rationalize the

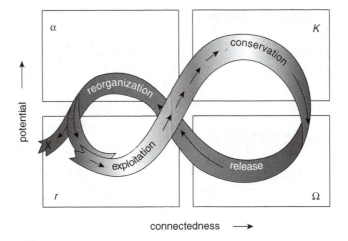

Figure 7.4 A stylized representation of the four ecosystem states as an ecosystem adapts following change (Holling et al., 2002). The vertical axis is the capacity or potential that is inherent in the accumulated resource of biomass and nutrients. The horizontal axis is the connectedness. The cycle starts at the bottom left and the main (light) curve from *r* to *k* is the logistic curve discussed in the text

Source: from Holling et al. (2002)

interplay between change and persistence, between the predictable and the unpredictable'.

Their general theory emerges as a model of the cycle of resilience and adaptation following change. This is envisaged as a Mobius strip trajectory through time, as shown in Figure 7.4. If we start following a major perturbation (say a shock) at the bottom left of the strip, the new situation is characterized by reorganization and the transient appearance or expansion of organisms – the pioneer stage (or *r*-phase in ecology). This leads to steady consolidation leading towards the relatively stable *k*-phase, which is characterized by conservative behaviour at equilibrium (again, see Obando, 2002). The system's connectedness increases until the system collapses, perhaps as a result of some small change leading to the crossing of an internal threshold. This, in turn, leads to what the authors call the release, and eventually to a reorganization that starts the adaptive cycle again. The cycle, so described, is reminiscent of Davis's cycle of erosion (1899). Uplift leads to a new initiation of a drainage system on the raised surface (close to the end-surface of Penck) and this is consolidated through its control on runoff and sediment yield and feedback that lead to self-adaptation of the drainage network, as modelled by Rodriguez-Iturbe (1994). There is, however, no direct analogy with the 'release' phase of the equilibrium in the Davis model. Thornes (1990) showed that, through the competition between erosion and vegetation, there appeared a boom-and-bust cycle in Mediterranean *matorral* that was simulated by Brandt and Thornes (1993) and which shared a regular cycle with the adaptive behaviour of the Holling et al. model.

The adaptive cycle model is still in its infancy and only time will tell if it is sufficiently robust (see, for example, the discussion by Stafford Smith and Reynolds, 2002). The weakness at the moment seems to be in the explanation of the release (collapse) phase, which appeals more to external variables (wind, fire and drought) than to intrinsic internal dynamics. Periodic flips from one stable state to another are 'mediated by changes in the slow variables (the horizontal axis variables in the catastrophe theory) that suddenly trigger fast variable response, or escape' (Holling et al., 2002: 35). There seems little doubt that the panarchy paradigm will provide the focus for thinking and exploration of the systems control in environmental management philosophy in the future. If it overturns the idea that systems are generally in a stable equilibrium state and hence resilient in the engineering sense, this will represent a significant forward step.

THREE TIME-BASED PROBLEMS

In this section several examples have been chosen to illustrate the ideas expressed so far in this chapter.

Global changes and vegetation shifts

The concept of a progressive move towards a potential has been widely explored in ecology and biogeography. Here the upper limit of biomass or species richness became fixed in the concept of succession. This was viewed by Odum (1969) as: (1) an orderly process that is reasonably directional and therefore predictable; (2) resulting from modification of the physical environment by the evolving plant community (for example, a soil develops); and (3) culminating in a stable (climax, mature) ecosystem with homeostatic properties. The upper limit is reached when all the resources available to the community (water, radiation and nutrients) are consumed in producing biomass. Likewise, a lower stable limit would be fixed by any limitation on resources – Liebig's Law. Any one of them might be 'limiting' and, in the 1960s and 1970s, they were reinterpreted in the 'new ecology' as the capacity or potential. Thus, in semi-arid environments, water provides the limit to growth. The rate of growth increased or decreased according to the difference between the current state of the system and the capacity. This is modelled by the logistic growth equation. Because any of water, radiation and nutrients can be limiting, the succession concept of Clements (1928) has been interpreted as leading to the climate climax vegetation. The assumption that the vegetation reaches the capacity determined by the limits of growth is implicit in the climatic climax vegetation and the concept of succession. It forms the basis of the interpretation of historical evidence of vegetation cover in climatic terms (for example, in palynology) and the reconstruction of past vegetation shifts resulting from global climate change. Sometimes it is used to predict the potential of future climate changes.

As an example, the shifts in the grass–bush boundary in the Eastern Cape of South Africa were modelled by assuming logistic vegetation growth at a rate proportional to rainfall (see Thornes, 2005). The rainfall was simulated using the mean, variance and periodicities of the last 50 years and adding some random noise. As precipitation changes, the biomass responds, usually with a time-lag. The vegetation cover is said to 'track' the precipitation. This tracking is of the boom-and-bust variety referred to above. In periods of heavy rain, vegetation 'overshoots' the potential then dies back to the equilibrium or capacity value. In the dry periods, the vegetation dries back to a very sparse cover. Because rainfall is modelled as a random walk, the vegetation cover tracks it also as a random walk. Because South African rainfall is dominated by the ENSO oscillations (see below), wet and dry periods succeed each other. The lags in vegetation modelled in this way can be up to 50 years after the causal extreme rainfall or drought (see Thornes, 2005).

It is well known that these reactions to climate change are complicated by grazing activities (Noy-Meir, 1982) and that the impacts of grazing are non-linear, perhaps even chaotic in nature, with small effects pushing the system into new trajectories. The consequence is that there are two stable states, one with no vegetation, the other with the maximum possible amount. Small shifts of rainfall can move the vegetation system to one or another of them. Once a change is set in train, it moves towards one of the stable states. This also supports the arguments of Holmgren and Scheffer (2001) that, in managing the vegetation cover, it might be best to wait until the turns in rainfall create the conditions that will lead towards a new desired stable state. It also follows that the interpretation of shifts in geomorphological process (notably soil erosion) as a direct, linear consequence of climate change needs to be treated with caution, given the non-linear nature of vegetation responses.

Circulation dynamics and complexity

Much of the chaotic dynamics paradigm, arising from Edward Lorenz's discovery of the Butterfly Effect, came from his attempt to model atmospheric circulation. We should therefore not be surprised to find that the modelling of circulation change, at the very core of the debate on the impacts of global warming, is fraught with difficulties created by non-linearities that lead to chaotic outcomes. There are several aspects to this:

- Non-linearities mean that complicated and unexpected responses may be expected from quite small changes in inputs. Similarly, very complicated behaviour can be the outcome of simple causes. We do not have to seek complicated inputs to get complicated outputs from complex systems.
- Because of this chaotic behaviour, time-series of outputs, as discussed above, should be explored for indications of deterministic chaos (meaning oscillations caused by unstable numerical behaviour).
- Prediction becomes even more difficult as a result of chaos so that global climate change modelling is subject to even greater uncertainty. Not only do impacts of climate change reflect non-linearities, but the inputs themselves (climate scenarios) are also the outcomes of chaotic dynamics.

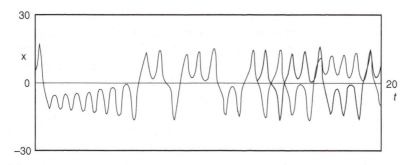

Figure 7.5   Numerical solution of the Lorenz equations for $r = 28$ and with $(x, y, z) =$ $(x_0, 5, 5)$ at $t = 0$. The bold curve is for $x_0 = 5.000$, and the lighter curve for $x_0 = 5.005$. The two diverge after about 13 oscillations, showing sensitivity in the outcome to small differences in initial conditions (x_0)

Source: Acheson (1997), reprinted with permission of Oxford University Press

Lorenz (1963) produced a drastically oversimplified model of thermal convection in a layer of fluid. The motion was driven by the temperature difference between the bottom of the fluid layer and the top. Subsequently, the equations used by Lorenz to describe the speed of the moving air have attracted attention purely on their own mathematical merits, as demonstrated by Acheson (1997). Given three differential equations, Acheson demonstrates that there are three equilibrium states. He also shows that if the van der Pol equations are solved for the evolution through time, they can be seen to exhibit chaotic behaviour (Figure 7.5). This is shown by their irregular behaviour and their extreme sensitivity to initial conditions (Acheson, 1997). With an initial difference of just one part in 1,000, the oscillations diverge as the difference becomes greater than about 13. The bold curve is for the starting value of x with $(x_0) = 5.000$ and the lighter curve with $x_0 = 5.005$. Even when the difference is reduced to a factor of 100, to just one part in 10,000, the outcomes stay together for only a little longer, until $t =$ about 16. Lorenz realized that this extreme behaviour had profound implications. He concluded in 1963 that, '[i]n view of the inevitable inaccuracy and incompleteness of weather observations, precise very long range forecasting would seem to be non-existent' (cited in Acheson, 1997: 159). Here 'long' means greater than three days!

Although some of these problems persist at the scale of global circulation, they are less acute. Prediction of climate (rather than weather) is more feasible because average conditions, not day-to-day variations, are being considered (Schneider, 1992). Imagine a billiard table with many balls. It would be difficult to predict the behaviour of a single ball if many are set in motion. Nevertheless, the average number of balls reaching a pocket in an average amount of time may be rather easier to predict. This statistical mechanics approach makes very helpful statements about the probability distribution of a large number of atoms in a gas (the ensemble) compared with the behaviour of individual atoms. Moreover, at the space and time scales of climate, the atmospheric changes are forced by other changes that respond only very slowly and by their own feedback systems, such as the effect of oceanic temperature. Generally, the degree of predictability is related to the period of amplitude and forcing. Schneider goes on to point out that, after a long time period, the global-scale climate system will move to a unique equilibrium after a transient adjustment for these longer times, assuming the existence of such unique and stable equilibria.

The periodically changing climate response that illustrates internal working of the climate system is exemplified in the ENSO (El Ninõ Southern Oscillation) circulation of the eastern Pacific and Australo-Indonesian regions (Cane, 1997). El Niño is a marked warming of the coastal waters off Peru and Ecuador that is accompanied by torrential rain and catastrophic flooding. El Niño events have been known since 1726, and there is evidence indicating occurrences for at least a millennium before that (Quinn et al., 1987). The Southern Oscillation describes a difference in the sea-surface pressures between the southeast tropical Pacific and the Australo-Indonesian regions (between the stations of Tahiti and Darwin). When the waters in the eastern tropical Pacific are abnormally warm, sea-level pressure drops in the eastern Pacific and rises in the west. The ENSO reveals characteristic features of simple non-linear dynamics – regular oscillations and periodic switching between alternate quasi-stable states.

The first signs appear in the Northern Hemisphere spring, build to a peak at the end of the year and, by the following summer, the warm event is usually over. The phenomenon occurs every four years, and modellers have been able to predict it very successfully on the basis of Bjerknes's (1969) hypothesis of the interactions among the Kelvin waves, the sea temperature and the Rossby waves of the tropical Pacific. According to Cane (1997: 586, emphasis added):

> ENSO is the premium example of variability, *stemming entirely from the internal* working of the climate system. Attempts to model and predict ENSO test our ability to model and predict larger time-scale variations and do it in a context with more observational data to verify against. It is the most advanced example of modelling two-way interactions between the atmosphere and ocean.

Ecological economics and complexity

As natural ecosystems come under increasing exploitation by human activity, the very stability of those systems is undermined. There are many studies in historical geography of the collapse of natural systems due to excessive pressures and failure to recognize the stability issues involved over short, medium and long time-scales. The breakdown of fishery and whaling communities, of tropical hunter-gatherers (Watt, 1973) and semi-arid grazers are but a few examples. This has been reflected in the preoccupation of geographers with the problems of sustainability of resources over time and the problem of the Tragedy of the Commons (see Chapter 14).

Ecological economics seeks to better understand the interaction of biophysical and socio-economic systems by examining the conditions under which

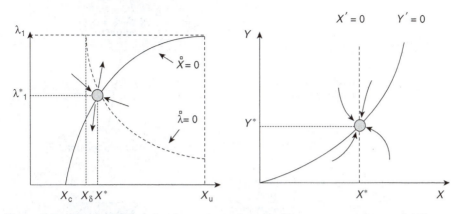

Figure 7.6   Regev–Gutierrez's phase space for gain (*Y*) versus resource availability
(*X*). The right-hand case is for societal co-operation and the equilibrium point is stable.
The left-hand figure is for free competition and the equilibrium point is an unstable
saddle
Source: Gutirrez et al. (2005). Copyright Elsevier, 2006 and reproduced with permis-
sion from Thornes (2007)

environmental systems remain stable when subjected to perturbations by external forces. This approach was extensively explored by Regev et al. (1998) in
a heavily mathematical approach using the methodology of systems engineering.
A more accessible outline of the approach is contained in Gutierrez and
Regev (2002).

The essential background to this model-building is the competition for
resources that underlies the Lotka–Volterra equations and the concept of the
trophic web (food chain). It is claimed that the model applies to all trophic levels
in a food chain, including human harvesting of renewable resources. Analogies
between the economies of humans and other species are used to develop a formulation of the goals that steer the benefits from energy allocation. This provides
a system of equations that can be tested for stability, using mathematical techniques and expressing the results in phase space (Figure 7.6) in which the pathways to stable and unstable solutions can be described.

Figure 7.6 shows the relationship between gain (profit) on the *Y* axis and
available resources on the *X* axis according to their model formulation. In the
left-hand graph, the objective function is for maximum individual gain under
conditions of free competition. In the right-hand graph, the objective function is
for maximum long-term community gain with co-operation between individual
harvesters. These graphs represent the outcomes of the model that incorporates
the controls on the available resources (such as climate) as well as economic controls on the gain (such as interest rates and technological advances).

The solid and dashed lines are isoclines (curves along which the two
variables – gain and resources – are in equilibrium). In the left-hand graph, the
solid line is the curve along which the resource availability is in equilibrium
with the profit. A high resource availability is in equilibrium with a high gain.

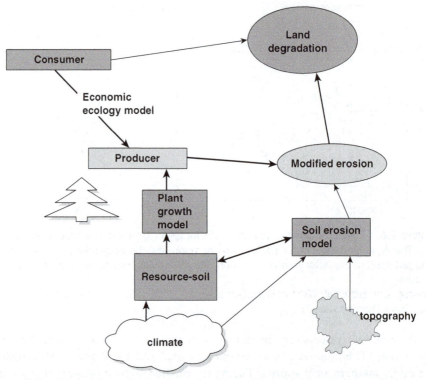

Figure 7.7   Schematic representation of a resource competition model applied to land degradation. For further explanation see text

As the resource availability decreases, so the equilibrium gain decreases. The heavy point marks the intersection of the two isoclines where both variables are in equilibrium with each other. The arrows around the heavy point show how the system will evolve in time after perturbations. In the right-hand graph, all the arrows are moving towards the equilibrium, whereas in the left, two are moving away.

Perturbations to the societal case (right-hand graph) lead the system back to the joint equilibrium. By contrast, in the free competition (left-hand case) perturbations in Y lead to instability, with the system moving to unacceptably high gains or to zero gains (bankruptcy). Figure 7.7 shows how the approach might be schematically applied to land degradation. Here the critical level is the vegetation production level. This is driven by rainfall and sunshine (the resource level), and reduced by harvesting in the form of grazing, cropping and deforestation. Another level above this would be the human demand for the livestock. At each level, differential equations describe the mass or population change and they have to be solved simultaneously. Growth or decline at each level is controlled by the ratio of supply (from below) to demand (from the level above). This is the Baumgartner–Gutierrez equation (Gutierrez and Baumgartner, 1984) and the units of production and consumption have to be standardized (e.g. as energy or mass) and therefore have to be converted between levels. As yet, the model is formulated for one predator (harvester) level. With more than one, the stability analysis may be too difficult and then the trajectories have to be solved using digital simulation.

In the land degradation case, we can easily envisage several predators on the vegetation, such as grazing, fire and firewood collecting. It should be possible to identify what harvest would lead to the unstable breakdown of the system with no net gain to the harvesters (farmers).

CONCLUSIONS

In the last 25 years, it has become obvious that the adoption of classical scientific reductionist approaches in physical geographical investigations and other environmental studies may not necessarily provide a basis for explanation. The understanding of physical processes requires that we accept that, instead of smooth linear behaviour over time, the real world is characterized by suddenly changing oscillatory behaviour in which the systems tend to switch between stable states. In the past, there was a tendency to seek externally varying forces to produce changes in a somewhat simple cause-and-effect mentality. It was as if every geomorphological change in the stratigraphical or morphological record should have a climate change explanation. So, for example, the debate over the origin of Dartmoor tors became a heated argument about alternative climatic explanations rather than a discussion of relative rates of weathering and erosion, where it rightfully belongs. The more the stratigraphy appeared to be periodic, the greater was the tendency to seek external forcing in the form of external controls that themselves vary periodically.

Decomposition of time-series can be misleading and oscillatory behaviour that persists for a long period in one form or another should lead us to seek evidence of sensitivity to initial conditions and evidence of chaotic dynamics intrinsic to the system. The critical question when there is evidence of periodic behaviour should be 'Does it indicate dynamical instability?'

This chapter is not a call for the abandonment of historical studies in physical geography, but rather a caution about the interpretation of unexpected behaviour. Moreover, it has to be recognized that the systems engineering paradigm described in this chapter is not without its critics: 'There has been a growing sense that traditional scientific approaches are not working and indeed make the problem worse' (Ludwig et al., 1993, cited in Westley et al., 2002). It is claimed that:

'Rigid technological approaches fail because they tend to focus on the wrong types of uncertainty and on narrow types of scientific practice. Many formal techniques of assessment and policy analysis presume a system near equilibrium, with a constancy of relationships and uncertainties that arise not from errors in tools or models, but from lack of appropriate information. (Westley et al., 2002)

SUMMARY

- In the last 25 years the importance of non-linearities, chaos, complexity and issues of the stability of equilibria have dominated the study of phenomena through time.

- Conventional time-series analysis is empirically based. Special techniques are needed to identify the non-linearities.

- Non-linearities are characterized by periodic behaviour, sensitivity to initial conditions and switching between main behaviours. The switching may be caused by external forcing, crossing internal thresholds or by small random perturbations near to critical points.

- There are families of new models to be used in the case of non-linear behaviour.

- All future physical geography will have to accept, identify and incorporate non-linear behaviour and identify its effects on the time-series of outputs from environmental systems.

Further Reading

Acheson's (1997) *From Calculus to Chaos* is a mathematical treatment of most of the dynamical systems concepts, pitched just within the reach of A-level mathematics. Brunsden and Thornes' (1979) paper, illustrates the basic equilibrium approach to 1960s geomorphology, and suggests how the sensitivity of geomorphological systems to external perturbations can be identified and quantified through the transient-form ratio. The chapter by Chorley (1965) on Davis's cyclical model, broke the mould of mid-twentieth-century historical geomorphology (denudation chronology) and opened the way to process studies and model-building. Bennett and Chorley's (1978) *Environmental Systems* is the classical approach to systems analysis by two leading geographers, and the first to recognize the stability problems that are discussed in this chapter. It is demanding reading. Gunderson and Holling's edited volume (2002), *Panarchy*, provides a refreshing, up-to-date appraisal of oversimplified approaches to environmental management. It stresses the need for understanding the complexity of environmental behaviour and proposes new concepts to deal with it. Marshall and Zohar's (1997) *Who's Afraid of Schröedinger's Cat?* is a remarkably readable collection of items on the new science from the arrow of time, through fuzzy logic, to the wave equation with complexity and chaos on the way. Introductory chapters are an absolute must. The most recent comprehensive review of complex behaviour in geomorphological systems is provided by Phillip's (1999) **Earth Surface Systems.** Roetzheim's (1994) **Enter the Complexity Lab** is a fun book that clearly explains the main concepts of complexity, stability and chaos. It includes a disk of computer programs (games?) that illustrate the principles, using 'birds', ants and even mice in mazes. Chaos, cellular automata, and life models are all here. Although 30 years old, Wilson's (1981) **Catastrophe Theory and Bifurcation** is still the clearest and most accessible source for catastrophe, bifurcations and other aspects of non-linear behaviour in geographical systems.

Note: Full details of the above can be found in the references list below.

References

Acheson, D. (1997) *From Calculus to Chaos: An Introduction to Dynamics.* Oxford: Oxford University Press.

Bennett, R.J. and Chorley, R.J. (1978) *Environmental Systems: Philosophy, Analysis and Control.* London: Methuen.

Bjerknes, J. (1969) 'Atmospheric teleconnections from the equatorial Pacific', *Monthly Weather Review*, 97: 163–72.

Brandt, C.J. and Thornes, J.B. (1993) 'Erosion–vegetation competition in an environment undergoing climatic change with stochastic rainfall variations', in A.C. Millington and K.T. Pye (eds) *Environmental Change in the Drylands: Biogeographical and Geomorphological Responses.* Chichester: Wiley, pp. 306–20.

Brunsden, D. and Thornes, J.B. (1979) 'Landscape sensitivity and change', *Transactions, Institute of British Geography*, 4: 463–84.

Cane, M.A. (1997) 'Tropical Pacific ENSO models: ENSO – a model of the coupled system', in K.E. Trenberth (ed.) *Climate System Modelling.* Cambridge: Cambridge University Press, pp. 583–616.

Chorley, R.J. (1965) 'A re-evaluation of the geomorphic system of W.M. Davis', in R.J. Chorley and P. Haggett (eds) *Frontiers in Geographical Teaching.* London: Methuen, pp. 21–38.

Clements, F.E. (1928) *Plant Succession and Indicators: A Definitive Edition of Plant Succession and Plant Indicators.* New York: H.W. Wilson.

Davis, W.M. (1899) 'The geographical cycle', *Geographical Journal*, XIV: 481–504.

Dieckmann, U., Law, R. and Metz, J.A.J. (2000) *The Geometry of Ecological Interactions: Simplifying Spatial Complexity.* Cambridge: Cambridge University Press.

Gregory, K.J. (1985) *The Nature of Physical Geography.* London: Edward Arnold.

Gunderson, L.H. and Holling, C.S. (2002) 'Preface', in L.H. Gunderson and C.S. Holling (eds) *Panarchy: Understanding Transformations in Human and Natural Systems.* Washington, DC: Island Press, p. xxii.

Gutierrez, A.P. and Baumgartner, J.U. (1984) 'Multitrophic level models of plant-herbivore-parasitoid-predator interactions', *Canadian Entomologist*, 116: 933–49.

Gutierrez, A.P. and Regev, U. (2005) 'The bioeconomics of tritrophic systems: applications to invasive species', *Ecological Economics*, 52: 383–96.

Harré, R. (1969) *Scientific Thought 1900–1960.* Oxford: Clarendon Press.

Holling, C.S. (1973) 'Resilience and stability of ecological systems', *Annual Review of Ecology and Systematics*, 4: 1–24.

Holling, C.S. and Gunderson, L.H. (2002) 'Resilience and adaptive cycles', in L.H. Gunderson and C.S. Holling (eds) *Panarchy: Understanding Transformations in Human and Natural Systems*. Washington, DC: Island Press, pp. 25–63.

Holling, C.S., Gunderson, L.H. and Ludwig, D. (2002) 'In quest of a theory of adaptive change', in L.H. Gunderson and C.S. Holling (eds) *Panarchy: Understanding Transformations in Human and Natural Systems*. Washington, DC: Island Press, pp. 3–24.

Holmgren, M. and Scheffer, M. (2001) 'El Niño as a window of opportunity for the restoration of degraded arid ecosystems', *Ecosystems*, 4: 141–9.

Ibanez, J., Martinez, J. and Schnabel, S. (2007) 'Desertification due to overgrazing in a dynamic commercial livestock-grass soil system', *Ecological Modelling*, 205: 277–88.

Kirkby, M.J. (1972) 'Characteristic slope forms', in D. Brunsden (compiler) *Hillslope Form and Process*. London: IBG Special Publication, pp. 15–31.

Libby, W.F. (1955) *Radiocarbon Dating* (2nd edn). Chicago, IL: Chicago University Press.

Lorenz, E.N. (1963) *The Essence of Chaos*. London: University College Press.

Ludwig, D., Hillborn, R. and Walters, C.S. (1993) 'Uncertainty, resource exploitation and conservation: lessons from history', *Science*, 260: 17–36.

Marshall, I. and Zohar, D. (1997) *Who's Afraid of Schroedinger's Cat?* London: Bloomsbury Press.

Noy-Meir, I. (1982) 'Stability of plant herbivore models and possible application to savanna', in B.J. Huntley and B.J. Walker (eds) *Ecology of Tropical Savannas*. Berlin: Springer-Verlag, pp. 591–609.

Obando, J.A. (2002) 'The impact of land abandonment on regeneration of semi-arid vegetation: a case study from the Guadalentin', in N.A. Geeson et al. (eds) *Mediterranean Desertification: A Mosaic of Processes*. Chichester: Wiley, pp. 247–68.

Odum, E.P. (1969) 'Generalization of successional–climax paradigm with overshoot', *Science*, 164: 262.

Oerlermans, J. and van der Veen, C.J. (1984) *Glacial Fluctuations and Climate Change*. Dordrecht: Kluwer Academic.

Phillips, J.D. (1999) *Earth Surface Systems: Complexity, Order and Scale*. Oxford: Blackwell.

Quinn, W., Dopf, D., Short, K. and Kuo-Yang, R.W. (1987) 'Historical trends and statistics of southern oscillations, El Niño and Indonesian droughts', *Fisheries Bulletin,* 76: 663–78.

Rebeiro-Hargrave, A. (1999) 'Large-scale modelling of drainage evolution in tectonically active asymmetric intermontane basins, using cellular automata', *Zeitschrift für Geomorphologie*, Suppl. Bd, 118: 121–34.

Regev, U., Gutierrez, A.P., Schreiber, S. and Zilberman, D. (1998) 'Biological and economic foundation of renewable resource exploitation', *Ecological Economics*, 26: 227–42 .

Rodriguez-Iturbe, I. (1994) 'The geomorphological unit hydrograph', in K. Bevan and M.J. Kirkby (eds) *Channel Network Hydrology*. Chichester: Wiley, pp. 43–69.

Roetzheim, W.H. (1994) *Enter the Complexity Lab: Where Chaos Meets Complexity*. Indianapolis, IN: Sam's Publishing/Prentice-Hall.

Scheidegger, A.E. (1960) *Theoretical Geomorphology*. Berlin: Springer-Verlag.

Schneider, S.H. (1992) 'Introduction to climate modelling', in K.E. Trenberth (ed.) *Climate System Modelling*. Cambridge: Cambridge University Press, pp. 3–26.

Schumm, S.A. (1968) 'River adjustment to altered hydrologic regime: Murrumbridge River and palaeochannels, Australia', *United States Geological Survey Professional Paper*, 598: 68 pp.

Schumm, S.A. (1979) 'Thresholds in geomorphology', *Transactions, Institute of British Geographers*, 4: 485–515.

Shackleton, N.J. (1977) 'The oxygen isotope record of the late Pleistocene', *Philosophical Transactions of the Royal Society*, B280: 169–82.

Stafford Smith, D.M. and Reynolds, J.F. (2002) 'Desertification: a new paradigm for an old problem', in J.F. Reynolds and D.M. Stafford Smith (eds) *Global Desertification*. Berlin: Dahlem University Press, pp. 403–25.

Thornes, J.B. (1980) 'Structural instability and ephemeral channel behaviour', *Zeitschrift für Geomorphologie*, Suppl. Bd, 39: 136–52.

Thornes, J.B. (1990) 'The interaction of erosional and vegetational dynamics in land degradation: spatial outcomes', in J.B. Thornes (ed.) *Vegetation and Erosion*. Chichester: Wiley, pp. 41–53.

Thornes, J.B. (2005) 'The extremeness of extreme events', *Geografica Polonika*, 76: 157–75.

Thornes, J.B. (2007) 'Modelling soil erosion by grazing: recent developments and new approaches', *Geographical Research*.

Thornes, J.B. and Brunsden, D. (eds) (1977) *Geomorphology and Time*. London: Methuen.

Thornes, J.B. and Rowntree, K.M. (2006) 'Integrated catchment management in semi-arid environments in the context of the European Water Framework Directive', *Land Degradation and Development*, 17: 355–64.

Tilman, D. and Kareiva, P. (eds) (1997) *Spatial Ecology: The Role of Space in Population Dynamics and Interspecific Interactions*. Monographs in Population Biology 30. Princeton, NJ: Princeton University Press.

Watt, K.K. (1973) *Principles of Environmental Science*. New York: McGraw-Hill.

Westley, F., Carpenter, S.R., Brock, W.A., Holling, C.S. and Gundersen, L.H. (2002) 'Why systems of people and nature are not just social and ecological systems', in L.H. Gundersen and C.S. Holling (eds) *Panarchy: Understanding Transformations in Human and Natural Systems*. Washington, DC: Island Press, pp. 103–20.

Wilson, A.G. (1981) *Catastrophe Theory and Bifurcation*. London: Croom Helm.

Wooldridge, S.W. and Linton, D.L. (1955) *Structure, Surface and Drainage in South East England* (2nd edn). London: George Philip & Son.

8

Time: From Hegemonic Change to Everyday Life

Peter J. Taylor

Definition

Time and space form the basic physical dimensions of the universe. As such, time is used to measure change, including societal change. Social time indicates time with content: human phenomena in the process of change. Social time is invariably linked to social space as 'time-space'.

INTRODUCTION: THE MEANINGS OF TIME IN HUMAN GEOGRAPHY

In human geography, time has been conceptualized in two distinct ways (see Chapter 7 on time in physical geography). One view of time is as a physical dimension, something that can be measured precisely. Thus a geographer might add a 'time dimension' to a model of, say, settlements to show how patterns have developed over a given period. Such models are called *dynamic models* (in contrast to 'static models' that describe a situation at one point in time) and rely on *time-series data and analysis*. The second view of time is as social change, where the emphasis is upon the 'content of time'. Thus a geographer might study the evolution of a particular settlement pattern as an outcome of industrialization. Such an approach focuses upon *social processes*: industrialization is a bundle of such processes relating to shifting work practices with many concomitant economic, political and cultural changes. The form that such study takes depends upon the *social theory* that is used to define the nature of social change.

This chapter will focus on this second view of time. The argument consists of three parts: first, the ways in which geographers have studied time are described; second, different patterns of time are discussed; and, finally, we interpret how we use our ideas about time to interpret the present. We cannot begin the argument, however, without introducing two basic relations: time and space, and time and modernity.

For geographers, time cannot be studied independently of space. Like time, space can be viewed as either 'physical' space, which we think of as *geometry*, or 'social' space, which we think of as *place*, that is a space with content (see Chapters 5 and 9 for more the multiple definitions of space and place in human geography). In the former case, space is viewed as three-dimensional so that time becomes the 'fourth dimension'. In the latter case, social processes are studied *in situ* so that we study the composition of places as in regional geography. Because space and time are so indelibly linked, in much of the discussion below there will be reference to *'time-space'* phenomenon. In short, throughout their concern to understand time – whether measuring temporal trends or interpreting compositional changes – geographers will be interested in locations, from local to global, wherein the trends or changes occur.

We live in a world we call 'modern'. The idea of modern is a very time-laden one. Being modern is to 'move with the times', to be 'up to date', to be a user of the newest gadgets or ideas, to be a follower of the latest fashions in clothes, furnishings or games. Collectively, a society of modern people defines a state of modernity. There is a sense that when either an individual or a society does not possess the newest artefacts, then they are deemed to be somehow 'behind' in time. Note that this is pre-eminently a 'social time' idea; it makes no physical sense for contemporary individuals or societies to be in front or behind in 'real' time. However, a consequence of concerns for this social time in modernity is that the latter has come to define a society that experiences rapid and ceaseless social change. Thus the concept of time is central to the meaning of modernity and of geographers' attempts to understand this condition.

HOW TIME HAS BEEN STUDIED BY HUMAN GEOGRAPHERS

The study of time in human geography has long been the domain of historical geographers. Their traditional concern for the development of landscapes and regions was challenged through the upheavals of the discipline in the 1960s. It is from this period that concepts of time began to be incorporated more generally throughout human geography in new systematic ways (Carlstein et al., 1978a, 1978b and 1978c). Five temporal models and concepts can be identified.

The first of these is *time-space convergence*. Through noting that the time it takes to transverse distances has fallen dramatically in recent centuries, it is sometimes suggested that the world is becoming somehow 'smaller'. This idea of a 'shrinking world' is not unique to geography but its precise measurement as time-space convergence is. This concept was devised by Janelle (1969). Drawing on data for the time it took to travel between particular pairs of towns from using

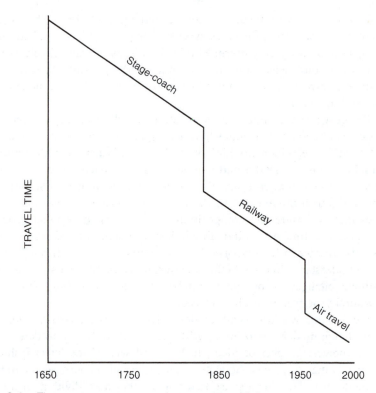

Figure 8.1   Time-space convergence between two cities

stage-coaches to flying aeroplanes, Janelle constructed graphs showing the decline
in length of time from the seventeenth to the twentieth century (Figure 8.1).
As a particular example, he measures travel times between Edinburgh and
London, showing how they declined from taking weeks to complete the journey
to just a few hours. More precisely, for the period 1776–1966, Janelle calculated
an average 'convergence rate' between the two cities of 29 minutes per year.
Subsequent researchers have used travel times between several cities to create
new 'time maps' wherein physical distances between locations are replaced by
'time-distances' (Forer, 1978).

Hägerstrand's (1973) *time-geography*, the second of the concepts identified
here, is arguably the most original contribution by human geography to the study
of time. Using a two-dimensional space as a base map to which time was added
as a vertical dimension, Hägerstrand attempted to trace the time-space paths of
individuals 'upwards' and sideways through this three-dimensional diagram as
they carried out everyday tasks (Figure 8.2). For instance, during a single day a
person would start on the base map at his or her residence, would then travel
upwards (through time) and sideways (through space) to a workplace, followed
by further movements on upwards through time and across the map to a
lunchtime meeting, and so on until returning at the 'top' of the diagram to the
initial spatial starting point, the home. For each individual, depending on their
access to travel facilities, there is a time-space prism that defines the boundaries
of what activities are possible from their home base. This defines possible

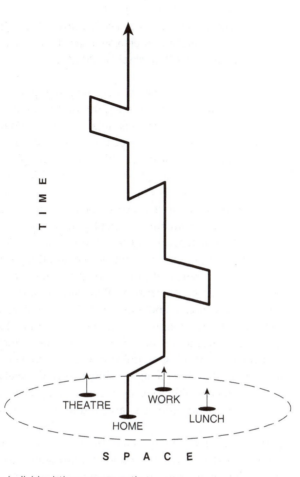

Figure 8.2 An individual time-space path

resources available to an individual: for instance, local schools but not multilingual international schools, or local general shops but not highly specialized retail outlets. In addition, movement through the prism is further constrained by necessary interactions with others, as at a work meeting or collecting a child from daycare, and where there is competition for time-space with others carrying out their everyday tasks. The classic case of the latter is a typical city 'rush hour' where commuters' time-space paths converge to create a time-space population concentration. This framework illustrates how space and time are resources routinely deployed in everyday life. The historical research of Pred (1986) is the exemplar for using the time-geography model for understanding how everyday life is altered through social changes.

For some, time-geography now has a dated image (witness here our modernist desire for the new), but other authors argue that shortcomings in the approach have been overcome:

> While some may think that time-geography only concentrates on constraints, is
> based on Cartesian space and Newtonian time and characterized by underdeveloped

notions of power and agency (cf. Giddens, 1984; Harvey, 1990; Rose, 1993), Hägerstrand and co-workers sought to overcome many of these criticisms in the 1980s and 1990s (Lenntorp, 1999). (Schwanen, 2007: 9)

Indeed, Schwanen's (2007) own research on material geographies combines time-geography with (post) actor-network theory (a more recent set of ideas shaped by the influential writings of Latour, see for example, Latour, 2005) to elucidate the ways that 'things' are implicated in time-space routines, in his example by exploring the roles played by mundane objects (childseats, mobile phones, notebooks, toys) as working parents live out the time-space complexities of their everyday caring responsibilities (Schwanen, 2007).

Harvey's (1990) concept of *time-space compression* is undoubtedly the most influential contribution of human geography to the recent study of time. Fascinated by the nineteenth-century notion of the 'annihilation of space by time', Harvey develops an argument that links space and time to both economic necessities and cultural expressions. Starting with the rationalization of space in the cartographies of the sixteenth-century Renaissance and the eighteenth-century Enlightenment that expressed both the (spatial) power of state and property, Harvey argues that in order to 'conquer space' new space has to be produced, notably in transport and communications. This is the same reduction of 'spatial barriers' that produces the shrinking world of time-space convergence, but here it is treated as much more than the changing dimensions of society. Within modern capitalist relations this shrinking world is part of a cyclical process of 'creative destruction' as new investments are required to resolve crises of over-production. These new spaces create a general feeling of the world 'speeding up', leading to an 'overwhelming sense of compression'. It is the latter experience that is reflected in how the world is represented as cultural elites attempt to harness the maelstrom of incessant change that is modernity (Berman, 1982). The great artistic movements of modernism in the late nineteenth and early twentieth centuries, and postmodernism in the late twentieth century, are both interpreted as cultural reactions to this 'speeding up' of time. Time-space compression is, therefore, a powerful organizing concept that connects economy and culture.

World-systems analysis as developed by Wallerstein (1988, 1991) offers an alternative materialist interpretation of the relations between time and space in human geography. To emphasize the indissolubility of the two concepts he coins the word *TimeSpace* to describe his particular integration. Starting with Braudel's (1980) identification of three categories of social time – the short term, the medium term, and the long term – Wallerstein adds a geographical component by attaching a spatial scale to each temporal span. Braudel's first time is episodic time, the time of traditional history that traces change through events, affairs, occasions and happenings. Wallerstein identifies geopolitical space, the immediate locales of events, as the spatial equivalent thus creating episodic-geopolitical TimeSpace. Braudel's second span is a more general patterned time (trends and cycles) to which Wallerstein adds an ideological space, locales of different divisions of the world, such as East/West during the Cold War. This produces a cyclical-ideological TimeSpace. Braudel's third span is structural time, which deals with the long-term slow movement of everyday life that underpins society. Wallerstein interprets this as the structures of historical systems, such as the historical rise and demise of the great ancient empires. The structural space of this time is the boundary expansion of the system and its reproduction through core–periphery structures. This all defines a structural TimeSpace. Wallerstein's research focuses on the modern world-system – our 'historical system' – whose structural TimeSpace evolved from a European capitalist world-economy in the sixteenth century to a global world-economy in the twentieth century, and which was reproduced through a cyclical-ideological TimeSpace of distinctive periods that developed through the episodic-geopolitical TimeSpace of events. World-systems integration of time and space is adapted to regional geography as 'historical regions' by Taylor (1988, 1991).

Wallerstein's TimeSpace is not the only use of Braudel's (1980) social times in human geography. Historical geography has developed into a multifarious enterprise and among these different themes there has been an engagement with Braudel's work (Baker, 1984). There is a revived concern for understanding the major patterns of social change in time and space (e.g. Thrift, 1990). Dodgshon (1998) has developed a *geographical perspective on change* wherein he provides an overview of social change in different spaces. He includes a taxonomy of change which compares mechanisms of change (e.g. system-feedback), sources of change (e.g. external), products of change (e.g. expansion) and morphologies of change (e.g. non-linear change). The latter encompasses the continuity-discontinuity debate that defines a basic division in the treatment of time as social change: is change a relatively smooth phenomenon or are their major discontinuities in the nature of changes? The former position is represented by theories of progress that see social change as a 'forward march' of humanity; in contrast, many geographers identify basic disruptions where there are 'accelerations' in the degree of social change, the 'industrial revolution' being the classic case. With disruptions there is a time divide between patterns of social relations (e.g. pre-industrial and industrial society or between medieval 'feudal' Europe and early-modern 'capitalist' Europe). The remainder of this chapter will delve more deeply into these time morphologies.

MORPHOLOGIES OF TIME

The condition of modernity is generally defined by social pressure for incessant change. Berman (1982: 15) has famously described this condition as 'a maelstrom of perpetual disintegration and renewal, of struggle and contradiction, of ambiguity and anguish'. Change can be both very exciting and very stressful: modern men and women (us) experience modernity as creating new opportunities while simultaneously destroying old cherished ways. The trick is to ensure that we are not mere objects of the vicissitudes of modernity but rather that we become active subjects, participants in the processes of change. Controlling change is the job of *planning*, the archetypal modern activity that focuses upon time.

A planning exercise can be defined as any project that attempts to control social change over a specified time horizon. As modern people, we live our lives through many individual and group projects as we plan our short-, medium- and long-term futures. Modern institutions operate through various planning instruments. For instance, during parts of the twentieth century all of the following 'grand planning' activities prospered for at least a period: the urban planning of cities, the corporate planning of firms, the military planning of the Cold War military–industrial complexes, the Keynsian[1] economic planning of welfare states, the development planning of third world states, and the five-year industrial planning of communist states. The last example reminds us that planning does not always work. In fact, all the surviving examples in the list have been latterly modified to become a much more 'flexible' version of their former selves. Flexible means, of course, much less control on change, an admittance that modern change cannot be simply tamed through the application of planning. Modernity is much too complex to be 'planned'.

In hindsight we can see a common process in action here. Planning is a reaction to a social problem. For instance, urban planning developed in the early twentieth century as a reaction to the legacy of poor living conditions in the Victorian industrial city. The modernist movement in planning set about clearing the 'slums' and relocating people into new, clean, high-rise blocks of housing. All planning provides a solution to its problem as defined at a given time. But modernity is perpetual change so that as soon as grand planning projects begin they progressively get out of date. Planning is condemned to solve yesterday's problems. The classic example is the Soviet Union, which produced the greatest nineteenth-century industrial state in history; the only problem was that it was created in the twentieth century – the modern world had moved on. This is equally true of urban planning wherein high-rise housing became 'slums in the sky' and have had to be abandoned as 'urban solutions'.

There is one example of failed planning that we need to consider further because it is associated with a time-based terminology that continues to distort our thinking. From the 1950s onwards the poorer countries of the world have been encouraged to embark on 'development planning' (see Chapter 21 on development). Such planning typically involved harnessing investment in order to pass through various stages in a path that culminates in something termed 'development'. In the most famous such model, states were deemed to pass through five stages from 'traditional society' to 'high mass consumption' (Rostow, 1960). Clearly, the last term shows that what was happening here is that poor countries were being advised to mimic the growth patterns of rich countries. All countries, it seems, were expected to proceed along identical parallel paths separated only by time (Taylor, 1989). Put another way, poor countries are merely lagging behind and, given the right policies, they will soon catch up. Hence the coining of two terms: 'developed countries' for the rich states, since they have already reached the goal to be 'developed', and 'developing countries' for the poor countries still on their way to 'development'.

Now let's return to the reality that is modernity. In hindsight, we can see that economic development in the rich countries is not actually an end-point, economic change in these countries has proceeded in leaps and bounds with new high-tech industries. In contrast, most of the erstwhile 'developing countries' are economically falling further and further behind. What are termed 'developed countries', implying an 'end-state', are developing more and more, and what are termed 'developing countries' are, to a large degree, simply not developing, certainly not in the way envisaged by development planning. The terms 'developing countries' and 'developed countries' have become topsy-turvy euphemisms for simply 'poor states' and 'rich states' derived from an optimistic time when it was thought planning could steer modernity over all the world.

Does this all mean that we are condemned to live as objects of modernity with little or no control on our futures, either individually or collectively? Fortunately not. All that planning failures illustrate is that modernity cannot be packaged into controlled, timed segments. There are many success stories within modernity and they occur through what has been called *surreptitious modes of change* rather than overt 'top-down' planning (Taylor, 2000). These modes of activity involve a multitude of decisions by ordinary people as part of their everyday lives. Where such a social movement captures an ongoing trend and recreates it as a major form of social change, then we have subjects successfully making modernity for their own needs and wants. The rise of the suburb as the dominant urban form is the classic case a surreptitious mode of change. Ridiculed by architects and planners as 'urban sprawl', ordinary people voted with their feet (actually cars) to create large swathes of single-family dwellings with gardens through the twentieth century (Fishman, 1987; Hall, 1996; Hayden, 2003). The very opposite of state planning, here was a popular and commercially profitable mode of change which – despite the problems it sometimes caused by inscribing a classed, gendered and racialized division of space on the landscape (Duncan and Duncan, 2004; Blunt and Dowling, 2006) – continues to be popular with many heterosexual nuclear families, and hence property developers, today.

The rise of the suburb was very much a US-led process of social change and represents an element of what is sometimes called the American hegemonic cycle. This is part of a Wallersteinian cyclical-ideological TimeSpace. In world-systems analysis, the modern world-system has developed through three hegemonic cycles each based upon the economic successes of a world hegemonic state. These hegemonic states are defined in terms of their economic world leadership in production, trade and finance, which generates concomitant political and cultural leadership as well. The first example of a state achieving this status was the Dutch Republic in the seventeenth century with its mercantile hegemonic cycle. In the nineteenth century, Britain was predominant with its industrial hegemonic cycle, and in the twentieth century the American hegemonic cycle was based on mass consumption (Taylor, 1996). Each cyclical phase reached its fruition when a basic enabling breakthrough was consolidated by cutting-edge economic activity. For the Dutch, it was innovations in shipbuilding that led to them being the leading traders of their era. For the British, it was the steam engine that enabled factory textiles to rule the world market. And for the Americans, it has been the communications and computer technologies, and their eventual integration, that has enabled the development of a vast advertising industry that produces the necessary demand for consumption.

This cyclical model is directly implicated in the *creation* of modernities (Taylor, 1999). As previously noted, the condition of modernity is experienced as incessant change and the economic upheavals caused by the three hegemons – new mercantilism, new industrialism, new consumerism – intensified the degree of social change. Worlds of new opportunities and dangers were created to which ordinary people were forced to respond. Within each hegemonic state the heightened pressure of change led to new forms of everyday life that has been called *ordinary modernity* (Taylor, 1999). In effect, the traditional household was the invention of the modernist ideal of the home as a haven from the turmoils occurring outside. This is a private world of comfort, an idealized place where the family relaxes away from the stresses of social change. Multitudes of modern people aspired to this ideal place of comfort, but not all succeeded: havens can also be cages, places of hidden violence (Taylor, 2000). Moreover, as second-wave feminism highlighted, the home can be both a space of isolation and labour for women, as well as a site of resistance (Blunt and Dowling, 2006).

Under the leadership of American hegemony everyday life has been centred on historically unprecedented consumption by masses of ordinary people. The suburb is the archetypal landscape of modern consumption. This is our contemporary world of suburban living based upon machines for access (the motor car), machines for domestic work (washing machines, microwaves, vacuum cleaners, dishwashers), and machines for entertainment (TVs, video players, music centres, game players). These large, general and necessary items are supplemented by many individual items of consumption, such as furniture, clothing and toys, that make each home different and distinctive. This is a world pioneered in the USA in the first half of the twentieth century by such 'household names' as Ford, Hoover and General Electric, and diffused to the other parts of the world in the rest of the twentieth century. But this everyday consumer modernity did not arise from nowhere. Before the Americans it was the nineteenth-century British who developed the nuclear, family-centred everyday life in their new industrial world. The Victorian home may have lacked the 'modern conveniences' that we expect, but it was still, in its own way, an unprecedented zone of comfort. Famously individual in character – Victorian homes were cluttered with family 'knick-knacks' – this was when identifiable modern (comfortable) furniture first becomes widespread. But it was the Dutch in the seventeenth century who invented the modern home itself (Rybczynski, 1986). Before them, dwellings were relatively public places where business, entertainment, eating and sleeping were mixed. In seventeenth-century Dutch houses the upstairs became a separate private area for family members and invited friends only. This is the crucial separation of home from paid work and business, a hallmark of modernity. The new Dutch homes were furnished and decorated for and by the family to reflect their individuality, and this included children – the Dutch are said to have invented childhood. Thus our modern everyday life has a time trajectory that goes from the houses of Dutch burghers to contemporary suburbia, reflecting the hegemonic cycles of economic change.

Hegemonic cycles with their associated everyday modernities are a classic example of a morphology of social time. As a cyclical model of change they counter simple progress models that assume a linear pattern of advancement. However, this is not a case of discontinuity over continuity. The nature of social change in the modern world-system is too complex to be captured by such either/or models. Obviously, there must be discontinuities in cyclical models – the worlds of mercantilism, industrialism and consumerism are different – but there can also be continuity. Alongside the differences there are enough similarities for us to identify a generic ordinary modernity of everyday life that has its own trajectory of development culminating in contemporary mass consumption. Social change in our modern world is a complex mix of cycles and trends as social times. Ultimately, however, it may be that it is the trends that are all-important to our futures.

CONCLUSION: TIME TODAY

Contemporary globalization is a classic example of how the concepts of time and space are linked together. The idea of globalization has dominated much thinking in human geography and beyond in the last decade ago or so (see Chapter 19 on globalization and human geography). It is self-evidently a spatial term since it references and announces a specific geographical scale of activity, the 'global'. But this spatial-scale reference makes sense only in relation also to time. Consider the titles of three classic books on globalization: *Global Formation* (Chase-Dunn, 1989), *Global Shift* (Dicken, 1998) and *Global Transformations* (Held et al., 1999). Each one links global with a particular process of social change: the three different terms describing change reflect alternative social-theoretical bases behind each book's argument. The point is that globalization is studied because it represents an important element of contemporary social change, so important in fact that it is sometimes said to define a new historical era. Globalization is a time-space concept *par excellence*.

The new communication technologies that make possible instant world-wide connections are the basic enabling mechanism of contemporary globalization. This has created a new relationship between time and space: information, knowledge, ideas and instructions can be electronically transmitted instantaneously around the world. It has been said to denote the 'end of geography' (O'Brien, 1991). More realistically, it marks another change in the relations between time and space. The most influential writer on this topic is Castells (1996). For him, social space is materially produced as a means to facilitate meetings – he calls them 'time-sharing practices' – between social agents. Social space has traditionally been organized so as to bring together people simultaneously so that they can interact as social beings. Such simultaneous practices had always relied upon spatial contiguity. Today, there is a global space of flows that enable social practices to occur across large distances: with electronic technology spatial contiguity and temporal simultaneity have been physically separated (Castells, 1996: 411). This does not presage the 'end of geography' but rather points towards exciting new geographies with, for instance, the development of a world city network to simultaneously facilitate global social practices within the new spaces of flows (Beaverstock et al., 2001; Taylor et al., 2006).

One of the prime characteristics of our 'globalizing' times is that we are very self-conscious of intensive social change. The human geography and social sciences literatures are awash with descriptions of things being new. This is indicated by the many terms that proclaim the passing of a recent past: postcolonialism, post-industrialism, postFordism, postdevelopment, post-Marxism, post-structuralism and, of course, postmodernism are most common. And there are identifications of many associated processes that are supposedly changing our world: restructurings, new orders, new identities, transitions and crises are the most common. The sheer number of 'posts' and related processes indicate that the modern maelstrom of incessant social change remains very much with us. In fact, the cacophony of such time-laden concerns brings up the question as to whether contemporary times are indeed a special time of change.

This is where the morphology of social time is crucial. For those who broadly follow a progressive linear social time model, then our times are but a stepping stone to more modern technological breakthroughs leading to a more advanced society. Those for whom cycles are part of the morphology have to ask whether the conditions are right for the creation of a new cycle. Who follows the USA? Given that the last hegemon has led us to mass consumption, the question is asked whether this is sustainable – is the earth big enough for ever-growing and never-ending mass consumption? Ultimately, time in human geography and the social sciences reduces to a question of social justice across generations (de Shalit, 1995). Taking a progressive position means that we should push on so that as yet unborn generations can cumulatively reap the benefits of technological advance. On the other hand, if modern consumption is not sustainable, it behoves us to make sure we bequeath to future generations a quality of environment on our planet at least as good as the one we inherited from previous generations.

SUMMARY

- Social time is a 'time with content', social process.

- Social time is inherently linked to social space or place.

- We live in modern times where change in incessant.

- Geographers have used several time-space concepts to understand modern times: space-time convergence, time-geography, space-time compression, TimeSpace and a geographical perspective on change.

- Planning is the 'top-down' modern practice for controlling social change.

- Planning is condemned to apply yesterday's solutions to today's problems.

- There are surreptitious modes of change where everyday behaviour creates large-scale historical change.

- Successful surreptitious changes are associated with the hegemonic cycles of the Dutch, British and Americans.

- These changes create time-spaces of ordinary modernity where people find a haven from incessant social change.

- Contemporary globalization is a classic time-space concept based upon new communication technologies.

- Globalization is constituted by new electronic spaces of flows that separate temporal simultaneity from spatial contiguity.

- The bottom line for globalization is intergenerational justice: will we leave the planet in as healthy a physical condition as we inherited?

Further Reading

The best starting point for following up this chapter is Leyshon (1995), who provides a comprehensive discussion of the idea of a 'shrinking world'. In terms of communications, the collection of essays edited by Brunn and Leinbach (1991) **Collapsing Time and Space**, is useful and includes an update of his ideas by Janelle. The basic statements on time-space compression and TimeSpace are to be found in Harvey's (1990) **The Condition of Postmodernity** and Wallerstein's (1991) **Unthinking Social Science** respectively, both are difficult reads but well worth the effort. Dodgshon's (1998) **Society in Time and Space** provides a valuable recent historical geography contribution to studying social change, one which is well versed in social theory. Finally, the links between time and modernity are developed further in Taylor's (1999) **Modernities: A Geohistorical Perspective**.

Note: Full details of the above can be found in the reference list below.

NOTE

1 Keynsian economic planning is named after the economist John Maynard Keynes, who devised a theory and practice of 'demand management' that dominated economic policy in the mid-twentieth century.

References

Baker, A.R.H. (1984) 'Reflections on the relations of historical geography and the *Annales* school of history', in A.H.R. Baker, and M. Billinge (eds) *Explorations in Historical Geography*. Cambridge: Cambridge University Press, pp. 1–24.

Beaverstock, J.V., Smith, R.G. and Taylor, P.J. (2001) 'World–city network: a new metageography?', *Annals, Association of American Geographers*, 90: 123–34.

Berman, M. (1982) *All That is Solid Melts into Air*. New York: Simon & Schuster.

Blunt, A. and Dowling, R. (2006) *Home*. London: Routledge.

Braudel, F. (1980) *On History*. London: Weidenfeld & Nicolson.

Brunn, S.D. and Leinbach, T.R. (eds) (1991) *Collapsing Space and Time*. London: Harper Collins.

Carlstein, C., Parkes, D. and Thrift, N. (eds) (1978a) *Making Sense of Time*. London: Edward Arnold.

Carlstein, C., Parkes, D. and Thrift, N. (eds) (1978b) *Human Activity and Time Geography*. London: Edward Arnold.

Carlstein, C., Parkes, D. and Thrift, N. (eds) (1978c) *Time and Regional Dynamics*. London: Edward Arnold.

Castells, M. (1996) *The Rise of Network Society*. Oxford: Blackwell.

Chase-Dunn, C. (1989) *Global Formation*. Oxford: Blackwell.

Dicken, P. (1998) *Global Shift*. London: Paul Chapman.

Dodgshon, R.A. (1998) *Society in Time and Space*. Cambridge: Cambridge University Press.

Duncan, J.S. and Duncan, N.G. (2004) *Landscapes of Privilege: The Politics of Aesthetic in an American Suburb*. London: Routledge.

Fishman, R. (1987) *Bourgeois Utopias*. New York: Basic Books.

Forer, P. (1978) 'A place for plastic space?', *Progress in Human Geography*, 2: 230–67.

Giddens, A. (1984) *The Constitution of Society: Outline of the Theory of Structuration*. Berkeley, CA: University of California Press.

Hägerstrand, T. (1973) 'The domain of human geography', in R.J. Chorley (ed.) *Directions in Geography*. London: Methuen, pp. 67–87.

Hall, P. (1996) *Cities of Tomorrow*. Oxford: Blackwell.

Harvey, D. (1990) *The Condition of Postmodernity*. Oxford: Blackwell.

Hayden, D. (2003) *Building Suburbia: Green Fields and Urban Growth, 1820–2000*. New York: Pantheon.

Held, D., McGrew, A., Goldblatt, D. and Perraton, J. (1999) *Global Transformations*. Cambridge: Polity Press.

Janelle, D. (1969) 'Spatial reorganisation: a model and a concept', *Annals, Association of American Geographers*, 59: 348–64.

Latour, B. (2005) *Reassembling the Social: An Introduction to Actor-network Theory*. Oxford: Oxford University Press.

Lenntorp, B. (1999) 'Time-geography – at the end of its beginning', *Geo-journal*, 48(3): 155–8.

Leyshon, A. (1995) 'Annihilating space?: the speed-up of communications', in J. Allen and C. Hamnett (eds) *A Shrinking World?* Oxford: Oxford University Press, pp. 11–54.

O'Brien, R. (1991) *Global Financial Integration: The End of Geography*. London: Pinter.

Pred, A. (1986) *Place, Practice and Structure: Place and Society in Southern Sweden, 1750–1850*. Cambridge: Cambridge University Press.

Rose, G. (1993) *Geography and Feminism*. Cambridge: Polity Press.

Rostow, W.W. (1960) *Stages of Economic Growth*. Cambridge: Cambridge University Press.

Rybczynski, W. (1986) *Home: A Short History of an Idea*. London: Penguin.

Schwanen, T. (2007) 'Matter(s) of interest: artefacts, spacing and timing', *Geografiska Annaler*, 89(B), 1: 9–22.

Shalit, de A. (1995) *Why Posterity Matters*. London: Routledge.

Taylor, P.J. (1988) 'World-systems analysis and regional geography', *Professional Geographer*, 40: 259–65.

Taylor, P.J. (1989) 'The error of developmentalism in human geography', in D. Gregory and R. Walford (eds) *Horizons in Human Geography*. London: Macmillan, pp. 303–19.

Taylor, P.J. (1991) 'The theory and practice of regions: Europes', *Environment and Planning D: Society and Space*, 9: 183–95.

Taylor, P.J. (1996) *The Way the Modern World Works*. London: Wiley.

Taylor, P.J. (1999) *Modernities: A Geohistorical Perspective*. Cambridge: Polity Press.

Taylor, P.J. (2000) 'Havens and cages: reinventing states and households in the modern world-system', *Journal of World-Systems Research*, 6: 544–62.

Taylor, P.J., Derudder, B., Saey, P. and Witlow, F. (eds) (2006) *Cities in Globalization: Practices, Policies and Theories*. London: Routledge.

Thrift, N. (1990) 'Transport and communication, 1730–1914', in R. Dodgson and R.A. Butlin (eds) *An Historical Geography of England and Wales*. London: Academic Press, pp. 453–86.

Wallerstein, I. (1988) 'The inventions of TimeSpace realities', *Geography*, 73: 7–23.

Wallerstein, I. (1991) *Unthinking Social Science*. Cambridge: Polity Press.

9

Place: Connections and Boundaries in an Interdependent World

Noel Castree

Definition

Place is among the most complex of geographical ideas. In human geography it has three meanings: a point on the earth's surface; the locus of individual and group identity; and the scale of everyday life. Until recently, all three meanings were framed by a 'mosaic' metaphor that implied that different places were discrete and singular. However, in the wake of globalization, it has become necessary for human geographers to rethink their ideas about place. This is not to imply that places are becoming the same, as if globalization is an homogenizing process. Rather, the challenge has been to conceptualize place difference and place interdependence simultaneously. The metaphors of 'switching points' and 'nodes' enable us to see places as at once unique and connected. The chapter shows how these metaphors have been applied to the three definitions of place identified in the chapter. It ends with a brief discussion of new place-related research frontiers being explored by human geographers.

INTRODUCTION: THE END OF PLACE? THE END OF GEOGRAPHY?

Geography is concerned to provide accurate, orderly and rational description and interpretation of the variable character of the earth surface. (Hartshorne, 1939: viii)

The fundamental fact is that ... places ... become diluted and diffused in the ... [new] logic of a space of flows. (Castells, 1996: 12)

Places are not what they used to be. Consider the two quotations above. Writing over six decades ago, Hartshorne, one of the most influential geographers of his generation, famously argued that geography's principal aim was the study of 'areal differentiation'. The world, he argued in *The Nature of Geography* (1939), was a rich and fascinating mosaic of places, and the geographer's task was to describe and explain this 'variable character' in both its human and physical dimensions. Writing on the cusp of a new millennium, the sociologist-cum-geographer Castells saw things very differently. The globalization of production, trade, finance, politics and culture, themselves facilitated by remarkable advances in transport and telecommunications, has made the world a 'global village'. For Castells, globalization thus signals the end of place. In our brave new world, he argues, a 'space of flows' – flows of people, information and goods – is increasingly breaking down the barriers that have hitherto rendered places distinct and different. The contrast between this argument and Hartshorne's is striking. If Castells is right, the twenty-first century arguably entails something Hartshorne could scarcely have anticipated: namely, 'the end of geography' (O'Brien, 1992). In other words, if areal differentiation is diminishing, if places are becoming 'diluted and diffused', geography as a subject arguably loses one of its *raisons d'être*. Globalization, it seems, forments a crisis of disciplinary identity.

Or does it? In this chapter I want to argue that far from signalling the end of place, the global interconnections to which Castells refers have resulted in an exciting and innovative redefinition of what place means. Accordingly, the discipline of geography is still very much about the study of the world's variable character – and thus still very much alive and well. The point, though, as we'll see, is that this variation can no longer be accounted for by treating places as relatively bounded and separate. This 'mosaic view' of the world was already outliving its usefulness in Hartshorne's time. By the 1940s it was becoming clear that places were no longer isolated, a fact that posed a challenge to Hartshorne's idea of 'areal differentiation'. Over 60 years later, places worldwide are, as Castells argues, more intimately interlinked than ever before. However, as we will see in this chapter, contemporary human geographers argue that this does not result in the diminution of place differences. Their challenge is to explain an apparent paradox: how can places remain different at a time when they're more interconnected – indeed interdependent – than ever before? Surely, the globalization of trade, finance and the like to which Castells points signals a more homogeneous world? This paradox, as we shall see, is indeed apparent rather than real: for contemporary geographers have argued that a concept of place fit for our times is one that sees *place differences as both cause and effect of place connections*. Far from heralding the end of place, the argument is that globalization is coincident with new forms of place differentiation. This, if you like, is Harthorne's areal differentiation resurrected but with an important new twist. In the twenty-first century, the geographical study of place cannot afford to remain caught in the conceptual straitjacket of a mosaic view of the world. But neither should it buy into Castells's exaggerated vision of a placeless planet where geographical sameness is replacing geographical difference.

In what follows I want to explain how human geographers have fashioned a concept of place that is appropriate for this era of globalization (see

Chapter 19 on globalization and Chapter 10 on place and physical geography).
Moreover, I want also to explain why this concept matters – both for geography
as a discipline and for people living in the interdependent world geographers
study. First, though, we need to look a little more closely at what place means,
how geographers have defined it in the past, and its importance as a concept to
geography as a discipline.

THE 'PLACE' OF GEOGRAPHY

> ...the significance of place has been reconstituted rather than undermined.
> (McDowell, 1997: 67)

The term place, as geographer Tim Cresswell (1999: 226) has observed, 'eludes
easy definition'. My *Concise Oxford Dictionary* identifies 20 meanings of the term,
and this semantic elusiveness is compounded by the fact that human geographers
have used it in a variety of ways throughout the discipline's history. John Agnew
(1987), writing many years ago, cut through this complexity to identify three
principal meanings of the term in geographical discourse. These meanings
arguably remain in force today:

1 *Place as location* – a specific point on the earth's surface.
2 *A sense of place* – the subjective feelings people have about places, including
   the role of place in their individual and group identity.
3 *Place as locale* – a setting and scale for people's daily actions and interactions.

In the following sections of this chapter I want to explore these three
meanings of place in more detail. In each case my overarching concern is to
explain how contemporary geographers have reckoned with the fact of the
increasing interconnections among places while still insisting that places are not
somehow becoming more alike (see Figure 9.1). For now, though, I simply want
to describe how this triad of approaches to place has emerged, waxed and waned
in the years before and since Hartshorne's plenary statement about areal differentiation and the nature of geography. As we'll see in the chronology that follows,
the second and third definitions of place emerged to challenge the first in the
1970s, 1980s and 1990s, since when attempts have been made to synthesize and
update them.

Beginnings

Hartshorne used the term 'place' rather imprecisely, often conflating it with
the equally complex term 'region'. This fact notwithstanding, it's probably fair to
say that Hartshorne viewed place as location – the first and oldest meaning
identified by Agnew – and places as distinct points on the earth's surface. Indeed,
in the five decades or so that geography had been a university subject in
western Europe and North America up to 1939, the normal expectation was that

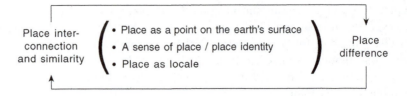

Figure 9.1   Approaches to place in contemporary human geography

professional geographers would study particular places, in both their human and environmental aspects, in great detail and publish articles and books on them (see Chapters 1 and 3 on the strength of regional geography). Classic examples include *Tableau de la Géographie de la France* (1917), by French geographer Paul Vidal de la Blache, and H.J. Fleure's *Wales and Her People* (1926). For Hartshorne, these types of study were what made geography special among academic disciplines. In *The Nature of Geography* (1939), he distinguished among 'systematic', 'chronological' and 'idiographic' subjects. The former take just one main aspect of reality and study it in detail – thus economics studies the economy and chemistry the world's chemical elements, and so on. Chronological disciplines study change over time – as history and geology do. However, Hartshorne argued that few disciplines look at how multiple different processes and events come together in the real world in specific places. Geography, he insisted, is precisely this 'synthetic' or integrative discipline. Moreover, because economic, social, political, hydrological, topographic and all manner of other factors never relate in quite the same way in any two places, he argued that geography studies the unique rather than the general. This, for Hartshorne, is what made it an idiographic discipline: it was about accounting for difference rather than sameness.

In truth, Hartshorne exaggerated the importance of place study to geography's disciplinary identity. Others had for decades seen geography less as the study of place and more the study of 'man–land [*sic*] relationships'. Indeed, after Oxford University's first professional geography appointment – Halford Mackinder – had famously defined geography as a 'bridging subject' between the human and natural sciences in 1887, many geographers had devoted their energies to studying these relationships right up to the continental and global scales. Moreover, the 'nature' of the geography Hartshorne sought to define and defend was to change almost as soon as his widely read book was published. There were three reasons why. First, many professional geographers were drafted into the armed forces during the Second World War and soon found that they lacked the technical skills required to undertake military and intelligence activities. The problem was many of the place studies geographers undertook were broad and largely descriptive. Geographers were trained to be jacks of all trades – to know a bit about a lot of things in given places – but masters of none. Second, this gave the subject a 'dilettantish image', as historian of geography David Livingstone (1992: 311) has put it, which served it poorly in a postwar educational environment where specialization was the norm. When the Geography Department in America's most prestigious university – Harvard – was closed down in 1951, many geographers keenly felt the need to make the discipline more rigorous and respectable. Finally, the mosaic view of place that seemed common sense to Hartshorne and his predecessors started to look highly unrealistic, both during and after the war. As one geographical critic of the time put it: 'We are no longer dealing with a world of neatly articulated entities. ... Our suspicion ... [is] that ... geographers may perhaps be trying to put boundaries that do not exist around areas that do not matter' (Kimble, 1951/1996: 500, 499).

Dis-placements

Consequently, the concept and study of place fell into disuse for almost three decades after Hartshorne's tome was published. In the immediate postwar era a new generation of geographers instigated what one of them called a 'scientific and quantitative revolution' (Burton, 1963). Learning and applying the tools of mathematics and statistics, this new generation sought to make geography a science (see Chapter 3 on geography and the social science tradition). This entailed specialization – including the increasing separation of human and physical geography and the subdivision of each[1] – and the attempt to develop testable theories, models and laws. Rather than looking for the unique, the different and the particular, geographers sought to mimic the physical sciences by looking for similarity, generality and pattern. In the words of Hartshorne's great rival, the geographer Fredrick Schaefer (1953: 227): '... Geography has to be conceived as the science concerned with the formulation of the laws governing the spatial distribution of certain features on the surface of the earth.' The keynote titles of the period said it all and were a far cry from the regional monographs of the prewar era: *Theoretical Geography* (Bunge, 1962), *Models in Geography* (Chorley and Haggett, 1967), *Locational Analysis in Human Geography* (Haggett, 1965) and *Explanation in Geography* (Harvey, 1969). To the extent that Hartshorne's vision of place figured at all, it was when events or things in one place were shown to be 'a particular realisation of the laws governing all similar events and things' (Rogers, 1992: 244). So much for place difference and uniqueness. Geography was now to be a 'spatial science', devoted to searching for geographical order at a variety of scales, to measuring numerically both people and things, and to testing rigorously hypotheses and models so as to develop generally applicable laws, rules and theories.

   Mid-century geography therefore survived quite happily without place as a central, organizing concept. By the early 1970s, however, it started to become clear that scientific geography was not to everyone's liking. Specifically, a cohort of human geographers wondered whether people's activities could and should be studied 'scientifically'. Within a decade this critique of spatial science, as this chapter now goes on to explain, led to what Rogers (1992) described as 'the rediscovery of place'.

The return of the repressed

This critique and rediscovery came in two phases. To begin with, a set of so-called 'humanistic geographers' argued that spatial science was 'in-human' (see Chapter 4 on geography and the humanities tradition). By treating people as 'little more than dots on a map or integers in an equation' (Goodwin, 1999: 38), it ignored the subjective, qualitative and emotional aspects of human existence and amounted to a 'Geography without man [*sic*]' (Ley, 1980/1996). Consequently, the attempt to rehumanize human geography took the form of close and careful studies of individual and group 'lifeworlds'. Two classic examples were David Ley's (1974) exploration of gang 'turf' rivalries in poor inner-city neighbourhoods in Philadelphia and Graham Rowles's (1978) detailed analysis of a group of old people's attachment to their home-place. In effect, what Ley, Rowles and other humanistic geographers were doing was resurrecting the importance of place. However, in the humanistic lexicon places were not, *pace* Hartshorne, conceived as objective points on the earth's surface. Rather, the aim was to recover people's varying *sense of place* (the second definition of place identified by Agnew): that is, how different individuals and groups, within and between places, both interpret and develop meaningful attachments to those specific areas where they live out their lives.

This concern with geographical experience was a vital corrective to the passionless, placeless grids of spatial scientific analysis. But it was not the only alternative to scientific human geography. From the early 1970s humanistic geographers were both accompanied and challenged by another group of dissenters from spatial science: Marxist geographers. Led by David Harvey, a former darling of geography's scientific establishment, these politically left-wing geographers argued that spatial science did little to address pressing real-world problems, like poverty, famine and environmental degradation. Moreover, they argued that by hiding behind a mask of 'objectivity' spatial science was dishonest about its own conservative, '*status quo*' political commitments. As Harvey made clear in human geography's first overtly Marxist book, *Social Justice and the City* (1973), a radical geography should be focused on non-trivial issues and should be geared to changing the world rather than simply understanding it (see Chapter 25 on relevance and human geography). What has all this got to do with place? A good deal as it turns out. Despite their common disdain for spatial science, tensions developed between Marxist and humanistic geography – and it was, among other things, over the question of place. For Harvey and his Marxist colleagues, the humanistic concern for a sense of place was worthy but ultimately problematic, for it tended to treat people and places in isolation and was obsessed with the minutiae of local attachments and local experiences. Against this, the Marxists – pointing to the development of a truly global economy by the early 1970s – argued that places were increasingly not only interconnected but also interdependent. That is, places were not only related to one another but related in ways that meant that what happened in one place could have serious consequences for another place many thousands of miles away. Harvey's (1982) *The Limits to Capital* was a major attempt to explain and criticize the nature and consequences of these global interconnections: namely, those specific to capitalism.

Overcoming dualisms

This brings me to the second phase in human geography's rediscovery of place. Though the Marxists were right to argue that human geographers needed an objective understanding of what places had in common, they were, by the early 1980s, as guilty as the spatial scientists had been of failing to pay sufficient attention to place difference. They also tended to give far more attention to the global economic and other processes that supposedly 'structured', and even, it was sometimes said, 'determined', the thoughts and actions of people in specific places (Duncan and Ley, 1982). That is to say, the Marxists were preoccupied with interplace connections more than specific place differences. By the same token, though humanistic geographers were right to emphasize the particularity of place experience, their concern with difference and lifeworlds arguably blinded them to the common processes linking places worldwide – 'stretched out' processes that could change the 'objective' nature of place and, thereby, locals' 'subjective' sense of place. Likewise, they tended to over-emphasize the degree to which people in place could control their own lives since Marxists like Harvey argued that global systems (like capitalism) constrain people's 'agency' in their home-places. How, then, to connect 'local worlds' with 'global worlds'? This was the challenge taken up by a set of British and American geographers from the mid-1980s. What inspired these geographers' efforts was a mixture of dramatic real-world changes and new theoretical developments.

During the previous decade, Britain and the USA, like many other countries, had seen their human geography literally remade by the ravages of a sustained economic crisis. The geography of people and places in the two countries was being restructured in the face of global economic competition and neoliberal governments (led by Thatcher and Reagan) intent on creating a new Britain and a new America. But the point, as Doreen Massey showed in her germinal book, *Spatial Divisions of Labour* (1984), is that the *same* processes of economic competition were having *varying* effects across the face of these and other countries. In other words, the global interconnections that meant that British and American cities and towns could not be analysed in isolation were producing not *geographical similarity* but *geographical difference*. The task of the so-called 'localities projects' which followed Massey's study (and which involved UK human geographers undertaking detailed studies of different British towns and cities) was to explain how global forces could have such variable local effects. Concurrent with the writings of Massey and the localities researchers were those inspired by new theoretical developments from outside geography. In a series of books, the now famous sociologist Anthony Giddens had developed 'structuration theory' in order to overcome the impasse between structural (or determinist) explanations of people's actions and free-will (or voluntarist) explanations. How, Giddens asked, could one combine a focus on 'big social systems' with a focus on individual and group action? (1984). In geography the impasse was represented by the Marxist obsession with global socio-economic processes and the humanistic geographers' concern with locally variable place experiences and actions. The geographers Derek Gregory and Allan Pred sought to spatialize Giddens's thinking (and to answer his question) in their innovative books, *Regional Transformation and Industrial Revolution* (Gregory, 1982) and *Place, Practice and Structure* (Pred, 1986) – books which used historical examples to show how previously isolated places became embroiled in translocal forces. What Gregory and Pred demonstrated is that social structure and social agency come together differently in different places such that they *mutually* determine one another.

Conducted in the wake of the stand-off between Marxist and humanistic geography, the localities research projects and structuration theory-inspired work

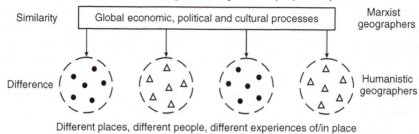

Figure 9.2   Marxist and humanistic geographers' approach to place

of Gregory and Pred sought to find a middle ground between two dualistic and untenable positions: that is, that places are *either* all the same *or* all different and that people in places are *either* free agents – able to develop their own singular attachments to, and practices in, a place – *or* the victims of overwhelming global social forces. The result was a conception of *place as locale* – the third meaning of place identified by Agnew. For Massey, Gregory, Pred and their fellow travellers, a locale was the scale at which people's daily life was typically lived. It was at once the objective arena for everyday action and face-to-face interaction *and* the subjective setting in which people developed and expressed themselves emotionally. It was at once intensely local *and yet* insistently non-local to the extent that 'outside' forces intruded into the objective and subjective aspects of local life in an interdependent world. And every locale was at once unique and particular *and yet* shared features in common with the myriad other locales worldwide to which it was connected (see Figure 9.2).

To summarize, after fading into mid-century obscurity, place is once again 'one of the central terms in ... geography' (Cresswell, 1999: 226). Over the last decade human geographers have extended and enriched the return to place pioneered by those writing in the 1970s and 1980s. In the remaining sections of the chapter I want to take each of the three approaches to place discussed here and illustrate briefly, using examples, how contemporary geographers have shown that place interconnection and interdependence in the modern world mark not the end but what Neil Smith (1990: 221) once called 'the beginning of geography' (see Chapter 19 on globalization and human geography). In terms of our three definitions of place, we can ask three key questions – namely, how can places be unique and yet subject to similar global forces? How is people's sense of place intensely local and yet (implicitly or explicitly) extroverted? And how can human actions be place-based, unpredictable and variable and yet considerably constrained by extra-local forces hailing from far away? In the last few years human geographers have offered innovative answers to all these questions. It's to these answers that I now turn.

RETHINKING PLACE AS LOCATION: POROUS PLACES

People and things are increasingly out of place. (Clifford, 1988: 6)

I've already called into question the mosaic view of place. Globalization entails the 'stretching' of social relationships across space such that the boundaries between the 'inside' of a place and the 'outside' are rendered porous. Today, we must appreciate the openness of places; that is, we need what Massey (1994: 51) calls 'a global sense of the local'. It's not just that today more and more places are interlinked and interdependent. It's also that the *intensity* of these global connections has increased: we live in an age of what Peter Dicken (2000: 316) calls 'deep integration'. In sum, the world is no longer a vast mosaic of places. At this point it might be tempting to join Castells and declare 'the end of place'. But this would be to confuse the redundancy of *a particular conception of place* with the disappearance of place as such. As I said in the introduction, places are not what they used to be. But places still undoubtedly exist. For instance, Manchester, where I live, is not the same as and remains far distant from, say, Manila – even though the two cities might be directly connected by relations of finance, trade or immigration. As Massey (1995: 54) puts it, 'we ... [therefore] need to rethink our idea of places...' because 'place has been transformed...' (Agnew, 1989: 12).

In metaphorical terms, this rethinking can be evoked as follows. Since the mosaic view conceptualizes places as distinct points in space – which is, today, largely unrealistic – it is perhaps better to see them as *switching points in a larger global system* or else *nodes in translocal networks* (Crang, 1999) (see Figure 9.3). These metaphors, as I'll now explain, allow us to think of places as inextricably interconnected – indeed interdependent – *and* as different and unique. Let us take each half of this metaphorical equation in turn.

Places in the contemporary world are, clearly, no longer separate. For instance, the bank where I this morning deposited a cheque is but one local fragment of a global financial system, while the apple I just consumed in front of my Taiwanese computer implicated me in a production network stretching back to an orchard in New Zealand (from whence the apple came). Moreover, with interconnection also comes *interdependence*. For instance, barely a day passes without newspaper reports of job losses and job creation in places as diverse as Chicago, Calcutta or Cairo. Often, though not always, these changing local employment situations can be explained with reference to interplace competition for investment and markets. For example, if Calcuttan workers can make auto-parts more cheaply than labourers in Chicago, a firm like Ford might favour an Indian auto-parts supplier for its vehicles. In short, what happens *then and there* can have sharp consequences in the *here and now*.

But if places are no longer separate, the more difficult argument to understand is that they somehow remain unique. No two places are quite the same, even in this era of globalization – or so several geographers, disagreeing with Castells, have argued. Notice that I use the word unique and not singular. In Hartshorne's worldview, places were *singular*: that is, they were all so obviously or subtly different from one another as to be absolute one-offs. The same combination of human and environmental factors, the argument went, was never found twice. However, if we see places as *unique*, we can argue that they are different *and* that they have something in common in an interdependent world (just as we are all unique as people in terms of looks and personalities and yet share the same biological make-up). This is the argument made by Ron Johnston (1984)

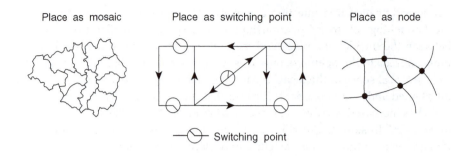

Figure 9.3   Metaphors for understanding place

in 'The world is our oyster' and by Doreen Massey (1995) in 'The conceptualisation of place'.

The question therefore arises: how can places continue to differ in a world of increasingly intimate global interrelationships? There are five answers. Together these answers explain why the metaphors of switching points and nodes are apt: for both evoke the idea that different places are 'plugged in' to different sets of global relations with different degrees of power over those relations. First, and most obviously, while globalization brings places closer together in terms of the reduced time taken to cross the space between them, the fact of geographical distance still remains. Thus, to return to the example of Manchester and Manila, while the two cities are *relatively* closer together, their *absolute* locational differences endure. Second, globalization has not unfolded across a homogeneous space. Rather, it has linked places *because* they are different. For instance, precisely because Boeing is a leading aircraft manufacturer, places without the capacity to produce aircraft have imported Boeing products all the way from its base in Seattle. Third, even though many places are subject to the same global forces, they react to and mould them differently. An aerospace company like Boeing, for example, has a number of choices as to how to respond to foreign competition. It can close its factories in cities like Seattle altogether, lay off some but not all workers or retrain these workers and sell new products to new places. More radically, it could shift production operations to cheaper or more efficient sites outside the USA. Likewise, McDonald's – sometimes held up as a potent symbol of cultural globalization and homogenization (Ritzer, 1996) – means different things in different places. In Maputo it might be a 'trendy' sign of all that's modern or new, in Tangiers it might be symbol of soulless American commerce, while here in Manchester it's but a familiar and rather banal marker of consumer culture. Fourth, even today all or most social relationships are not global in reach. Many remain insistently local – like the one I enjoy with my family or my local football club team-mates. Finally, we should not forget that not all places in the world are equally 'wired in'. Globalization, as Dicken (2000) notes, can take the form of 'shallow' as well as deep integration. Thus many places in sub-Saharan Africa, for example, remain partially cut off from the rest of the world or else subject to very one-sided relationships that exacerbate poverty – the kind of 'difference' that places in the developing world certainly do *not* want to preserve.

For instance, it may surprise you to discover that Ethiopia – one of the world's poorest places – continued to produce large quantities of food throughout the horrific famines of the mid-1980s and 1990s. How and why? Because wealthy landowners were producing export crops for European and North American markets rather than food crops for their own people. For the five reasons mentioned above, it is simply misconceived to think that globalization equals sameness and homogeneity. On the contrary, human geographers have shown that the more linked places become, the more place differences endure and are remade. In Swyngedouw's (1989) apt neologism, we need to talk less about globalization and more about an uneven process of 'glocalization'.

RETHINKING A SENSE OF PLACE: 'GLOCAL' IDENTITIES

> ... even local identities are completely caught up in a web of global interdependence. (Mitchell, 2000: 274)

In the previous section we considered places, implicitly, in terms of their objective properties – that is, as material and physical locations – and how to conceptualize them. But what of the subjective questions of how people interpret their home-places and those of others? As we've seen, humanistic geographers were among the first to take the subjective aspects of place existence seriously. As these geographers were right to argue, the thoughts and feelings that people have towards places are every bit as real and material as the places themselves. Disclosing people's 'sense of place' requires 'empathetic' enquiries into the realms of feelings, emotions and values. Some three decades after the likes of Ley launched this so-called 'hermeneutic approach' in human geography, it's clear that subjective attachments to, or interpretations of, place matter as much as ever. Global interdependency notwithstanding, most people live their lives within just a few square kilometres. Moreover, at certain times of life, people can be *highly* confined to specific places, as with children and many elderly people. So place remains a crucial locus for daily experience. Think about yourself: which places matter to you and why? Your answer will probably involve just a few places, and one of them will almost certainly be your home-place(s). You will have a highly personal sense of place that's bound up with specific events in your life, involving not just your perception of place(s) but your feeling about place(s). So apart from their physical dimensions, there's an imaginative and affective dimension to places too. This need not always be positive (some of us associate particular places with suffering or unhappiness), but rarely are any of us agnostic about those places that had formative impacts on our sense of self.

How, though, to understand these non-physical realms of thought and feeling? The humanistic desire to disclose people's sense of place will no longer suffice, for two reasons. First, cultural geographers have argued that place is linked to the formation of personal and group *identities* (Keith and Pile, 1993). People have more than just a sense of place: additionally, place is written into their very characters. Think, for example, of how we tend to characterize people – often stereotypically – by their place of origin (e.g. in Britain there are 'Cockneys' and

'Geordies', in North America rural 'rednecks' and an inner-city 'underclass'). And think about how your very sense of self, as a person, is intimately linked to the place you are from. For instance, though I live in south Manchester I'm originally from a town in north Manchester and both my accent and my character still carry the traces, 20 years since I left, of my upbringing. So place runs deep. Second, there was an implication in humanistic writing that there was one ultimately 'real' or 'authentic' sense of place for people. The Canadian geographer Edward Relph (1976), for example, complained about the 'placelessness' of so many modern towns with their high-rise towers and bland, serial suburbs. He believed that the spread of faceless modern architecture and planning was 'dehumanizing' place experience such that people's senses of place were being thinned out and rendered uniform. The problems with this kind of argument are manifold. To begin with it's rather conservative in nature, seeing 'outside' influences as a 'threat' to the supposedly 'authentic' nature of places. It's almost as if Relph lamented the fact that places were increasingly interlinked rather than different pieces in a mosaic. As problematically, it underestimates the sheer *variety* of place attachment and identities that people can and do develop in the *same* places. There is ultimately no one sense of place or place identity (think of how a poor immigrant woman in Hackney, London, might view that place as opposed to a wealthy young male professional) but many. Finally, geographers like Relph underestimate how different senses of place and place identity could persist not despite but *because of* 'external' influences hailing from other places.

This last comment brings us to the important insight that different local identities might result from, or be expressed because of, similar global connections. Identities are not natural. They are, rather, socially fabricated over people's life course. People tend, when considering the place element of identity, to conjure up the image of a settled community – literally, a home-place. But in a globalizing world, most places are anything but settled. They are subject to ongoing change, both physically (the factory that shuts down or the new shopping centre that opens) and socially (the foreign immigrants that move in or the older generation who die off) – and much of this change is, as we saw in the previous section, about local changes resulting from global/extra-local processes. So we must recognize that while identities are, today, still formed in places (they are place-based) they are not place-bound – that is, the result of *purely* local experiences. Rather, locally variable identities partially arise from 'outside' influences, paradoxical though this may seem.

Contemporary human geographers have illustrated this 'glocal' nature of identity in two ways. First, there are those cases where identities *seem* to be purely local but where human geographers have shown that they are in fact not so. For instance, in mid-2001 a set of serious 'race' riots erupted in the poor, old industrial towns of Bradford, Burnley and Oldham in the north of England. These towns, like so many multicultural places in western Europe, have had large immigrant populations from the Indian subcontinent for over three decades. Yet extreme right-wing political groups – like the National Front – want to expel them, thereby 'purifying' these places and returning them to their purportedly 'true' character as white and English. The irony, of course, is that this attempt to define and defend a 'local' identity from unwanted 'foreign' influences arises precisely in and through the presence of those 'outside' influences! A further irony is that the Indians and Pakistanis being discriminated against consider themselves to be very much *a part of* these three places – and rightly so, having lived there for over two generations. So *seemingly* local identities that attempt to shut out non-local influences – in the three places mentioned, influences of international immigration – are, in the modern world, not straightforwardly local at all (Harvey, 1995).

Second, human geographers are also showing that many 'local' identities are overtly and explicitly 'extra-local'. There are two main cases to consider here. The first is where people who are not indigenous to a place characterize it in a way that both reflects their own worldview and which therefore takes on a certain reality – even though it might be a far cry from the local residents' view of that place. The best example here is modern tourism, which serves up the world as a set of idealized places, each with a specific image that is marketed to potential tourists. For example, the Caribbean is usually thought of as a peaceful, paradisical place, full of exotic resorts; what tourists rarely see behind this 'imagined geography' are the slums and poverty that are endemic to most Caribbean towns and cities (Cater, 1995; see also Torres and Momsen, 2005). In addition, and rather differently, geographers have shown that many place-based identities today are openly 'extroverted' and outward-looking – in effect explicitly incorporating 'non-local' influences (unlike the National Front in Bradford, Burnley and Oldham). The best examples come from so-called 'transnational communities' – that is, communities that are spread out among different places but which remain connected. In Vancouver, Canada, for example, there are many Chinese residents who are from Hong Kong and who maintain strong familial and cultural links with this former British colony. So their identity as Vancouverites, living in a western Canadian city abutting the USA, is complemented by their identity as Hong Kong Chinese. Theirs is an avowedly *hybrid* identity, such that even though they live physically in one place their place loyalties are plural and transnational (Mitchell, 1993) (see Figure 9.4). In sum, in many places in the contemporary world the identities of people who live in those places are rarely local in the 'mosaic' sense of the word. As Massey (1998) insists, we need to look not for the *roots* of people's identity but the *routes*. That is, we need to trace how 'local' identities are built from the way people internalize a whole array of 'non-local' influences as the latter converge on different places.

RETHINKING PLACE AS LOCALE: GLOBAL FORCES, LOCAL RESPONSES

Life chances are materially affected by the lottery of location. (Crang, 1999: 24)

We live in a highly uneven world. Global interconnection and interdependency have been coincident with inequality and uneven development rather than homogeneity. Since the first incursions of Marxism into human geography, geographers have argued that local inequalities are *caused by* global interlinkages, not merely correlated with them. If we take the example of Ethiopian famines cited

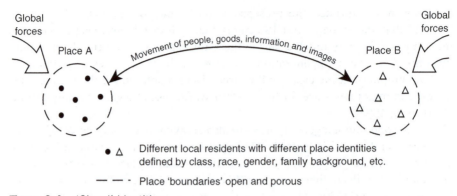

Figure 9.4   'Glocal' identities

earlier, it's clear that these local tragedies were a direct outcome of colonial and trade ties to Europe and beyond. But the traffic is not all one way. People acting in places are not simply marionettes whose actions and life chances are dictated by movements of the world economy and global politics. In other words, people acting in place have a degree of 'agency' to control their destinies and those of the places they reside in. So local action cannot only *react to* global pressures but also *act back on them*. Since Gregory, Pred and others, following Giddens, first made this argument in the 1980s, human geographers have not only shown the nature and limits of place-based agency, but also how it varies from place to place. This geographically variable interaction between global and international structures and people's place-based agency is the process of what Giddens, as we saw earlier, famously called 'structuration'. In Doreen Massey's terms, there is a global 'power geometry' to which actors in such places get to 'call the shots' (or not) for other actors near and far.

This uneven geography of structuration can be illustrated, in simple terms, by the following interpretation of a recent, little-known but fascinating event: the attempt by the small, central American country Costa Rica to make money by selling off it's 'genetic resources'. Like other central American counties, Costa Rica is relatively poor in global terms and classed as a 'developing country'. Its principal means of income is the export of coffee beans and bananas. However, large, western transnational pharmaceutical companies have, in recent years, become very interested in tropical countries – such as Costa Rica – that are so-called 'genetic hotspots'. The tropics contain the bulk of the world's plant, animal, insect and bacterial species, and it's estimated that some 50% of these species are yet to be discovered. Transnationals like Monsanto, Pfizer and Smith-Kline-Beecham are now actively 'prospecting' for these species, hoping that their physical and genetic properties might some day be usable in the development of pharmaceutical products, such as drugs or cosmetics. Among developing countries, Costa Rica has been at the forefront of this 'merchandizing' of currently unowned and undiscovered tropical species and, in 1991, set up an organization – INBio [the National Institute of Biology] – to collect species samples and sell them to interested western companies. Thus far InBio has made over US $5 million selling Costa Rica's genetic resources.

In this case, the 'structure' that both conditioned the decision to sell Costa Rica's genetic heritage and led to the establishment of INBio was the world economy, an economy in which Costa Rica has become overly reliant on two staple exports, coffee and bananas. The 'agency' at work here, embodied in INBio's everyday operations in the country's capital, San José, has yielded Costa Rica 5 million valuable dollars. However, this agency has been unequally distributed within Costa Rica. Historically, Costa Rica was widely populated by indigenous or so-called 'First Nations' peoples. These peoples were displaced during the Spanish conquests of the sixteenth and seventeenth centuries and, today, some 30,000 of them live in small, poor 'native reserves' located in out-of-the-way rural areas. Many of these peoples have a unique knowledge of local environmental resources and, more generally, have legitimate claims to the Costa Rican genetic inheritance being sold off by scientists and bureaucrats at INBio. However, there's little evidence that any of the $5 million earned through INBio has made its way into Costa Rica's native reserves. The country's indigenous peoples are locked in a political structure that offers them little power or opportunity, and their exclusion from INBio's operation illustrates this graphically. On top of this, their physical location in places distant from the capital city, the centre of political authority in Costa Rica, makes it doubly difficult to be heard.

NEW QUESTIONS ABOUT PLACE

> 'Place' … is not the sole property of human geographers. We are, however, in a unique position to … examine … the concept … in everyday life. (Cresswell, 2004: 123)

Before I conclude this chapter, let me point to some of the new place-related research frontiers being opened up by geographers. The first relates to cutting-edge information technologies. In my introduction I mentioned globalization and the idea of 'the end of geography', before going on to question the apparent equation between global interconnectivity and the erasure of spatial difference. More than other media of communication, new information technologies seem precisely to signal the placeless 'space of flows' of which Castells speaks. So do we, in effect, have two co-existing worlds: one a virtual, head-of-a-pin world ('cyberspace') in which peoples' location is irrelevant; and the other a 'real' world of place connectivity and difference I have presented in this chapter? The answer is yes and no. While information technologies do indeed 'annihilate space' (to borrow Karl Marx's evocative term), 'cybergeographers' like Martin Dodge, Rob Kitchin, Ken Hillis and Paul Adams suggest their geographies are rather more complex. First, information technologies require a physical-technical infrastructure in order to operate: networks of machines, satellites, fibre-optic cables and the like. This infrastructure itself has an uneven geography of production and distribution, linking myriad places to different degrees and with different consequences. The infrastructure is anything but uniform and placeless. Second, a good deal of virtual interaction between geographically separated IT users has an imagined – but very real – locational element to it. Think of 'chat rooms' or 'cyber cafés' where the physical metaphors used actively structure the modes of interaction between interlocutors, in some cases facilitating fictional identities and providing outlets of social behaviour impossible in 'real' locations.

Finally, it is easy to forget that much of the world does not enjoy the benefits of email, the worldwide web or video-conferencing. For all its supposed 'placeless' qualities, modern information technologies can serve to further marginalize certain places in economic and other terms, even as others thrive through their use.

A second research frontier relates to place and morality. With few exceptions, geographers interrogating place have focused on so-called 'cognitive' issues (i.e. those concerning description, explanation and/or evocation). But it has long been clear that questions of place connectivity and difference have a profound ethical dimension in a world marked by uneven geographical development. One ethical issue is what David Harvey (1996: 325) has called 'the right to geographical difference', by which he means the universal right of people to create and maintain places as they see fit. A second, related ethical issue is the responsibilities that places have to those they are connected to. For instance, should wealthy places actively assist poorer ones and how? The British geographer David Smith is one of a very few geographers to address both these ethical issues together, but others are now following his lead and thinking hard about the moral aspects of place interrelations and difference (see Sack, 2003).

A final, and related research frontier concerns the politics of place, by which I mean conscious actions undertaken to maintain a particular locality or else transform it. Going back some years, geographers have undertaken important research into purely local actions by locally-based actors: workers, neighbourhood organizations or small businesses, say. However, one feature of the modern world – closely linked with globalization – is the so-called 'up-scaling' of place politics (see Chapter 12 on scale and human geography). By this I mean the enrolment of distant others in local campaigns or activism. Many good examples can be found in a newish research field called 'labour geography'. What labour geographers like Andrew Herod and Jane Wills are showing in their research is how transnational actions are increasingly being undertaken to defend or enhance the interests of particular workforces. This kind of 'borderless solidarity' often works by using workers abroad to affect the operations of a transnational firm whose operations are being challenged in one particular place. In other words, workers overseas are asked for help by one local labour force because they are strategically well placed to disrupt the firm's extraction, production, distribution or marketing activities. Though this sort of up-scaling of place-politics may seem to transcend place difference through acts of long-distance co-operation, the reality is that the place differences matter. In most translocal worker campaigns, the location and nature of the workers enrolled in common struggle is important in determining the struggle's likely success.

CONCLUSION: THE MATTER OF PLACE

> ... the significance of place depends on the issue under consideration and the sets of social relationships that are relevant to the issue. (McDowell, 1997: 4)

Place matters and its importance is multifaceted. Some three decades after spatial science reached its zenith, difference is back on the geographical agenda.

The discipline is once again concerned with the idiographic, but in a very different and much wider sense than Hartshorne could ever have imagined. Place difference, both objective and subjective, is now understood in terms of uniqueness rather than singularity. We again have a style of human geography that is integrative and synthetic rather than analytical and place-blind. But it must reckon with a world where places are infinitely more complex and changing than they were during geography's first engagements with place in the early twentieth century. In addition, we must also acknowledge that place matters in a very profound and very worldly sense, which is why other subjects – like sociology, anthropology, communications studies and economics – are now very interested in the difference that place makes. We need to understand the variable nature of places not just out of sheer curiosity (though that's reason enough). More than this, as the bloody struggles over place in Israel, Northern Ireland, the Basque country, Eritrea, Sri Lanka, the former Yugoslavia and elsewhere show so tragically, local attachments and differences remain fundamental aspects of the human condition. In short, the renewed study of place is too important to be left to geographers alone. This is why Massey (1993) argues that geographers need to advocate a 'progressive sense of place' to people in the world at large. What she means is that geographers have a moral obligation to show people that their place-based actions and understandings make no sense without acknowledging all those things impinging on place from the outside. What's 'progressive' about this, for Massey, is that it encourages an openness to the wider world, not a defensive putting up of barriers. We must, she says, live with the incontrovertible fact that the global is *in* the local and vice versa. This is more than a merely academic observation. In a world of place difference, stressing what connects places has real practical and political relevance. It can make all the difference between a world of inward-looking local rivalries and a cosmopolitan world where place differences are respected and place connections celebrated.

SUMMARY

- Place is a complex concept with three principal meanings in modern human geography.

- As the world has changed, so too have human geographers' conceptions of place.

- Human geographers have tried to rethink place in a way that respects place differences while acknowledging heightened place interconnections and interdependencies. That is, places are conceived as being unique rather than singular.

- This rethinking has taken human geographers away from older 'mosaic' metaphors of place to newer notions of 'switching points' and 'nodes'.

- Using these notions, we can rethink all three definitions of place in order to show how local and non-local events and relations intertwine.

- New human geography research on place looks at information technologies, ethics and transnational forms of solidarity

- The importance of a place concept that stresses how 'outside' processes impact on the 'inside' of places is that it challenges the idea that places and the peoples in them can ever thrive by defensively putting up barriers against non-local forces.

Further Reading

A good place to start is with the following entries in the most recent edition of **The Dictionary of Human Geography** (Johnston et al., 2008): place, locale, locality, local–global relations, localization, sense of place, placelessness, globalization and boundary. Cresswell's (2004) **Place** is good, but does not conform entirely to my own presentation of 'place'. Likewise, the chapters on 'Agency–Structure', 'Local–Global' and 'Space–Place' in Cloke and Johnston's (2005) **Spaces of Geographical Thought** make for interesting reading. The introduction and Chapter 1 of Hannerz's (1997) **Transnational Connections** provide a good general introduction to the meaning of place in the contemporary world. A comprehensive introduction to the place concept in geography is provided by Holloway and Hubbard (2000) in **People and Place**, while Massey (1995) and Allen and Hamnett (1995) offer first-rate general introductions to conceptualizing place in an era of globalization. McDowell's (1997) edited book, **Undoing Place?,** showcases the best writing on place in geography and cognate fields. In relation to the three meanings of place explored in this chapter, see the following: on the local, the global, difference and sameness, Crang (1999) and Allen and Hamnett (1995); on 'glocal' identity, Cloke (1999) and Driver (1999); on local action and global processes, Meegan (1995) and Bebbington (2000). Finally, some examples of research on the three research frontiers mentioned are as follows: Adams (1998), Massey (2004) and Herod (2003).

Note: Full details of the above can be found in the references list below.

NOTE

1 And this book reflects these enduring divisions, with almost every key concept given separate treatment by a human and a physical geographer.

References

Adams, P. (1998) 'Network topologies and virtual place', *Annals of the Association of American Geographers*, 88(1): 88–106.

Agnew, J. (1987) *Place and Politics*. Boston, MA: Allen & Unwin.

Agnew, J. (1989) 'The devaluation of place in social science', in J. Agnew and J. Duncan (eds) *The Power of Place*. Boston, MA: Allen & Unwin, pp. 9–30.

Allen, J. and Hamnett, C. (1995) 'Uneven worlds', in J. Allen and C. Hamnett (eds) *A Shrinking World?* Oxford: Oxford University Press, pp. 233–54.

Bebbington, A. (2000) 'Reencountering development: livelihood transitions and place transformations in the Andes', *Annals of the Association of American Geographers*, 90(3): 495–520.

Bunge, W. (1962) *Theoretical Geography*. Lund: Kleerup.

Burton, I. (1963) 'The quantitative revolution and theoretical geography', *The Canadian Geographer*, 7: 151–62.

Castells, M. (1996) *The Rise of the Network Society*. Oxford: Blackwell.

Cater, E. (1995) 'Consuming spaces: global tourism', in J. Allen and C. Hamnett (eds) *A Shrinking World?* Oxford: Oxford University Press, pp. 183–222.

Chorley, R. and Haggett, P. (eds) (1967) *Models in Geography*. London: Methuen.

Clifford, J. (1988) *The Predicament of Culture: Twentieth-Century Ethnography, Literature*. Cambridge, MA: Harvard University Press.

Cloke, P. (1999) 'Self–other', in P. Cloke et al. (eds) *Introducing Human Geographies*. London: Edward Arnold, pp. 43–53.

Cloke, P. and Johnston, R.J. (eds) (2005) *Spaces of Geographical Thought*. London: Sage.

Crang, P. (1999) 'Local–global', in P. Cloke et al. (eds) *Introducing Human Geographies*. London: Edward Arnold, pp. 24–34.

Cresswell, T. (1999) 'Place', in P. Cloke et al. (eds) *Introducing Human Geographies*. London: Edward Arnold, pp. 226–34.

Cresswell, T. (2004) *Place*. Oxford: Blackwell.

Dicken, P. (2000) 'Globalisation', in R.J. Johnston et al. (eds) *The Dictionary of Human Geography* (4th edn). Oxford: Blackwell, pp. 315–16.

Driver, F. (1999) 'Imaginative geographies', in P. Cloke et al. (eds) *Introducing Human Geographies*. London: Edward Arnold, pp. 209–17.

Duncan, J. and Ley, D. (1982) 'Structural Marxism and human geography', *Annals of the Association of American Geographers*, 72: 30–59.

Fleure, H.J. (1926) *Wales and her People*. Wrexham: Hughes and Son.

Giddens, A. (1984) *The Constitution of Society*. Combridge: Polity Press.

Goodwin, M. (1999) 'Structure–agency', in P. Cloke et al. (eds) *Introducing Human Geographies*. London: Edward Arnold, pp. 35–42.

Gregory, D. (1982) *Regional Transformation and Industrial Revolution*. London: Macmillan.

Haggett, P. (1965) *Locational Analysis in Human Geography*. London: Edward Arnold.

Hannerz, U. (1997) *Transnational Connections*. London: Routledge.

Hartshorne, R. (1939) *The Nature of Geography*. Lancaster, PA: Association of American Geographers.

Harvey, D. (1969) *Explanation in Geography*. London: Edward Arnold.

Harvey, D. (1973) *Social Justice and the City*. London: Edward Arnold.

Harvey, D. (1982) *The Limits to Capital*. Oxford: Blackwell.

Harvey, D. (1995) 'Militant particularism and global ambition', *Social Text*, 42(1): 69–98.

Harvey, D. (1996) *Justice, Nature and the Geography of Difference*. Oxford: Blackwell.

Herod, A. (2003) 'Workers, space and labor geography', *International Labor and Working Class History*, 64: 112–38.

Holloway, L. and Hubbard, P. (2000) *People and Place*. Harlow: Prentice-Hall.

Johnston, R.J. (1984) 'The world is our oyster', *Transactions, Institute of British Geographers*, 9: 443–59.

Johnston, R.J., Gregory, D., Pratt, G. and Watts, M. (eds) (2008) *The Dictionary of Human Geography* (5th edn). Oxford: Blackwell.

Keith, M. and Pile, S. (eds) (1993) *Place and the Politics of Identity*. London: Routledge.

Kimble, G. (1951/1996) 'The inadequacy of the regional concept', in J. Agnew et al. (eds) *Human Geography: An Essential Anthology*. Oxford: Blackwell, pp. 492–512.

Ley, D. (1974) *The Black Inner City as Frontier Outpost*. Washington, DC: Association of American Geographers.

Ley, D. (1980/1996) 'Geography without man', in J. Agnew et al. (eds) *Human Geography: An Essential Anthology*. Oxford: Blackwell, pp. 192–210.

Livingstone, D. (1992) *The Geographical Tradition*. Oxford: Blackwell.

Massey, D. (1984) *Spatial Division of Labour*. London: Macmillan.

Massey, D. (1993) 'Power geometry and a progressive sense of place', in J. Bird et al. (eds) *Mapping the Futures*. London: Routledge, pp. 62–8.

Massey, D. (1994) *Space, Place and Gender*. Oxford: Polity Press.

Massey, D. (1995) 'The conceptualisation of place', in D. Massey and P. Jess (eds) *A Place in the World?* Oxford: Oxford University Press, pp. 46–79.

Massey, D. (1998) 'The spatial construction of youth cultures', in T. Skelton and G. Valentine (eds) *Cool Places*. London: Routledge, pp. 121–9.

Massey, D. (2004) 'Geographies of responsibility', *Geografiska Annaler*, B86: 5–18.

McDowell, L. (ed.) (1997) *Undoing Place?* London: Edward Arnold.

Meegan, R. (1995) 'Local worlds', in J. Allen and D. Massey (eds) *Geographical Worlds*. Oxford: Oxford University Press, pp. 53–104.

Mitchell, D. (2000) *Cultural Geography*. Oxford: Blackwell.

Mitchell, K. (1993) 'Multiculturalism, or the united colors of capitalism?', *Antipode*, 25: 263–94.

O'Brien, R. (1992) *Global Financial Integration: The End of Geography?* London: Pinter.

Pred, A. (1986) *Place, Practice and Structure*. Cambridge: Polity Press.

Relph, E. (1976) *Place and Placelessness*. London: Pion.

Ritzer, G. (1996) *The McDonaldsization of Society*. Thousand Oaks, CA: Pine Forge Press.

Rogers, A. (1992) 'Key themes and debates', in A. Rogers et al. (eds) *The Student's Companion to Geography*. Oxford: Blackwell, pp. 233–54.

Rowles, G. (1978) *The Prisoners of Space?* Boulder, CO: Westview Press.

Sack, R. (2003) *The Geographical Guide to the Real and the Good*. New York: Routledge.

Schaefer, F. (1953) 'Exceptionalism in geography', *Annals of the Assoication of American Geographers*, 43(3): 226–49.

Smith, N. (1990) *Uneven Development* (2nd edn). Oxford: Blackwell.

Swyngedouw, E. (1989) 'The heart of a place', *Geografiska Annaler*, B71: 31–42.

Torres, R.M. and Momsen, J. (2005) 'Gringolandia: the construction of a new tourist space in Mexico', *Annals of the Association of American Geographers*, 95(2): 314–35.

Vidal de la Blache, P. (1917) *Tableau de la Géographie de la France*. Paris: Hachette.

10

Place: The Management of Sustainable Physical Environments

Ken Gregory

D **efinition**

Place has not explicitly been a primary focus for physical geographers although it has been implicit in much of the development of physical geography for more than a century. The description of places was essential as environments were explored; such descriptions were then compared, leading to systems of categorization of places, so that places could subsequently be evaluated against the background of general models. As physical geography now extends to environmental management, place warrants greater explicit attention by physical geographers in relation to the management of sustainable physical environments as exemplified by urban places.

INTRODUCTION: PLACE LOCATED

The aphorism 'Geography is about maps, but biography is about chaps' (Bentley, 1905) encapsulates much public perception of 'geography' as the study of places, with the word 'geographer' still connoting someone who not only knows where places are but what they are like. Paradoxically, physical geographers have given comparatively little explicit attention to place, although it will be argued here that whereas physical geography was *implicitly* concerned with place for much of the twentieth century, the theme is now becoming more *explicit* in the twenty-first century. To the physical geographer, place is the particular part of space occupied by organisms or possessing physical environmental characteristics. Place is associated with a number of related terms, including environment,

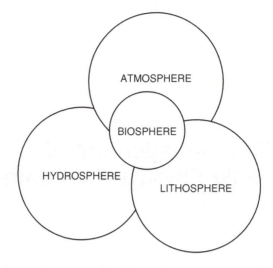

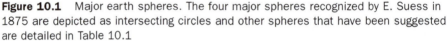

Figure 10.1   Major earth spheres. The four major spheres recognized by E. Suess in 1875 are depicted as intersecting circles and other spheres that have been suggested are detailed in Table 10.1

landscape and nature. The range of terms used indicates how place is not exclusive to any one discipline because others, including ecology, geology, other environmental sciences and landscape architecture, also focus upon place, using their own terms and approach. It has been suggested (Rolston, 1997) that six words model the world we view: nature, environment, wilderness, science, earth, value. In addition to descriptions by physical geographers, our appreciation of the physical character of place is gradually established from a variety of images provided by literature, art, mathematics, science, language and various forms of media. The distinctiveness of places is shown by the fact that in the Russian language there are words for unique types of valley, and Finnish and Swedish vocabulary includes words for aspects of winter which do not occur in other countries (Mead and Smeds, 1967). Initially, attempts were made to restrict definitions of place to natural conditions, prior to human activity and cultural influence, until it became appreciated that the impact of human activity is such that there are now few, if any, really natural places, environments or landscapes. Most recently, it has been realized that even physical environment is culturally determined: do people from different cultures see physical landscape in the same way?

Just as the definition of place has varied over time, most recently reflecting perception of physical environments by particular cultures, so the places studied by physical geographers have changed over time. Places that are the focus of attention for physical geographers are located within the spheres in the envelope from about 200 km above the earth's surface to the centre of the earth. Since 1875, when the Austrian geologist Suess invented the terms hydrosphere, lithosphere and biosphere to complement atmosphere, which had been used since the 1700s (Figure 10.1), earth and life scientists have gone sphere crazy (Huggett, 1995). No one discipline can study all the major spheres now identified (Table 10.1); it is by focusing research upon a particular blend of spheres that the basis is provided for the study of places in a discipline such as physical geography.

However, some spheres have not always been studied by physical geographers. Thus the ocean sphere was included in physical geography in the first half of the twentieth century but became the exclusive province of oceanography in the second half, when the physical geographer made increasingly substantial contributions to the terrestrial hydrosphere in the study of hydrology. Whereas other spheres, or combinations of them (Table 10.1), are the province of other disciplines in the way that the biosphere is dominantly the realm of the biological sciences, there is a physical geography perspective which is distinctive in the way in which it links the several spheres together. In focusing upon a particular interaction of spheres, the physical geographer has to locate places, to describe them both in terms of their characteristics and their dynamics, to explain them and their development in relation to adjacent places, and to evaluate them for particular purposes.

This chapter outlines how the study of places by physical geographers evolved, how it stimulated integrated approaches, and how opportunities and challenges, illustrated by urban areas, are now available for a physical geography perspective.

PLACE LOCATED AND FRAGMENTED

Since Arnold Guyot (1850) wrote that physical geography 'should compare, it should interpret, it should rise to the how and the wherefore of the phenomena which it describes', it has been appreciated that more than mere description is required. Over the subsequent 150 years, phases of exploration and audit, of classification, and of categorization were necessary to establish the fundamental knowledge of the physical environment. In the twentieth century, exploration furnished the maps that provided the basic information, and only in the latter half of that century were many of the characteristics of place inventoried in maps or in detailed surveys greatly accelerated by the advent of remote sensing. Once sufficient information had been amassed, it was necessary to classify it, a process often requiring major schemes of mapping of types of geomorphology, of climate, of soils and of vegetation, and thence of regions. Significant debates occurred about how classification should be attempted and whether it should be based upon static or dynamic characteristics for example. It was not until the 1960s, with the arrival of the quantitative revolution, that physical geographers could embark upon the quantitative description of place, greatly enhanced in the late twentieth century with advances in information technology. A methodological change could then occur as the idiographic focus upon the description of the unique character of distinctive places, which had featured in the first half-century, was succeeded by a more nomothetic, law-giving approach seeking more general explanation and models in the second half of the twentieth century (Gregory, 2000).

As physical geography developed, it became subdivided into branches, broadly corresponding with the spheres that had been identified (Figure 10.1; Table 10.1), with each branch adopting a basic unit of resolution for a particular place in physical environment. Such units could be the primary basis for characterization

Table 10.1   Earth spheres

The spheres included are those of greatest relevance to the physical geographer. Other spheres identified include: magnetosphere; celestial sphere; cosmosphere; exosphere; thermosphere; ionosphere; mesosphere; startosphere; heterosphere; homosphere; relief sphere; landscape; anthroposphere; asthenosphere; barysphere; centrosphere.

Sphere	Source	Interpretation
Atmosphere	Used since end of seventeenth century for the gaseous envelope of air surrounding the earth up to 200 km. Includes **troposphere** (0–10 km); **stratosphere** (10–25 km); **ozonosphere** (25–60 km); **mesosphere** (60–100 km); and **thermosphere, ionosphere** and **exosphere** (100–500 km)	The envelope of air surrounding earth
Troposphere		The lowest 12 km of the atmosphere
Geosphere	Used to signify lithosphere, or lithosphere + hydrosphere + atmosphere, or any of the terrestrial spheres or shells (Huggett, 1995). Huggett applies to core, mantle and all layers of the crust. (a) lithosphere or (b) lithosphere + hydrosphere + atmosphere or (c) any of so-called spheres or layers of earth (Bates and Jackson, 1980)	Lithosphere + hydrosphere + atmosphere Vink (1983) used geosphere as the zone of interaction on or near the earth's surface of atmosphere, hydrosphere, biosphere, lithosphere, pedosphere and noosphere or anthroposphere
Geoecosphere	The landscape sphere (Vink, 1983) termed geoecosphere (Huggett, 1995) for the sphere in which other spheres (biosphere, toposphere, atmosphere, pedosphere and hydrosphere) interact	The sphere in which other spheres (biosphere, toposphere, atmosphere, pedosphere and hydrosphere) interact
Pedosphere	Used by Mattson (1938). Shell or layer of the earth in which soil-forming processes occur (Bates and Jackson, 1980). The sphere of the regolith affected by soil forming processes called **edaphosphere**, and the remainder of the pedosphere of weathered rock and unconsolidated material called **debrissphere** by Huggett (1995)	Layer of the earth in which soil-forming processes occur
Cryosphere	Snow, ice, frozen ground, and sea ice. The part of the earth's surface that is perennially frozen, the zone of the earth where ice and frozen ground are formed (Bates and Jackson, 1980)	The part of the earth's surface that is perennially frozen and includes snow, ice, frozen ground, and sea ice
Hydrosphere	Water, both fresh and saline, in liquid, solid or gaseous state close to or on the surface of the earth. 95% in oceans and seas; 2% in glaciers and permanent snow; 2.5% is freshwater, of which 70% is in ice and snow and 30% is groundwater. Less than 3% is freshwater in lakes and rivers and 50% of that is in Great Lakes, Lake Baikal, and African Rift lakes	Water, both freshwater and saline water in lakes and seas, in liquid, solid or gaseous state close to or on the surface of the earth

Table 10.1   *(Continued)*

Sphere	Source	Interpretation
Toposphere	**Relief sphere** used by Budel (1982) for the totality of the earth's topography. Huggett (1995) used toposphere for the sphere at the interface of the pedosphere and atmosphere and of pedosphere and hydrosphere	The sphere at the interface of the pedosphere and atmosphere and of pedosphere and hydrosphere
Biosphere	Coined by E. Suess in 1875 but not given strict definition. Now used in more than one way (Hutchinson, 1970): one is as zone or surface envelope of the earth which is naturally capable of supporting life; another is synonymous with biota as sum of living creatures on the earth Where life exists and where its influence extends Bates and Jackson (1980) have two meanings: (a) all area occupied or favourable for occupation by living organisms; (b) all living organisms of the earth and its atmosphere There are three meanings: (a) totality of all living things dwelling on earth (Teilhard de Chardin, 1969; Huggett, 1995); (b) space occupied by living things (Vernadsky, 1945); (c) part of the earth in which life exists (Hutchinson, 1970).	Sphere naturally capable of supporting life from few hundred metres above and below the land/water surface and so overlaps with atmosphere, hydrosphere and lithosphere
Ecosphere	Life and the inorganic environment that sustains it (Cole, 1958) Global sum of life and life-support systems (Huggett, 1995)	Life and the inorganic life-support systems
Noosphere	The realm of human consciousness in nature. Developed by Pierre Teilhard de Chardin (1881–1955) Vernadsky (1863–1945) began to use the term in the mid-1930s for the transformation of the biosphere under the influence of humankind	The thinking layer arising from the transformation of the biosphere under the influence of human activity
Lithosphere	The solid earth The earth's crust and portion of the upper mantle	The earth's crust and portion of the upper mantle
Asthenosphere	Two uses are either solid portion of the earth – the rocks – or the outer shell of the earth	Zone in the upper mantle extending from a depth of 50–300 km below surface to a depth of about 700 km

and monitoring studies. In respect of the atmosphere, the site conditions were carefully specified for monitoring weather data to ensure that the weather station was typical of the surrounding area. The basic unit for soil study was the vertical section through all the constituent horizons of a soil, recognized as the soil profile since the time of Dokuchaev and other early Russian soil scientists in the late nineteenth century. As the profile is two-dimensional, the pedon was introduced as the smallest unit or volume of soil that represents, or exemplifies, all the horizons of the soil profile: it is a vertical slice of soil profile of sufficient thickness and width to include all the features that characterize each horizon (Wild, 1993), is usually a horizontal, more or less hexagonal area of 1 m² but may be larger (Bates and Jackson, 1980), and is an integral part of many soil survey classification systems.

In biogeography, basic places were recognized because all organisms live in niches; either a fundamental niche which an individual may occupy in the absence of competition with other species, or a realized niche which is the actual niche occupied when competition is in progress (Watts, 1971). The niche has subsequently been defined as the habitat in which the organism lives, but also the periods of time during which it occurs and is active, and the resources it obtains there. Other terms have therefore been developed for the organisms, for their place or habitat, and for the combination of organisms and habitat. Microhabitat is a precise location within a habitat where an individual species is normally found; a biotope is the smallest space occupied by a single life-form, as when fungi grow on biotopes found in the hollows of uneven tree trunks. Habitat, a term employed in various ways according to when and in which branch of science it has been used (e.g. Morrison, 1999), can also designate the living place of an organism or a community, applying to a range of scales from the microscale relating to organisms of microscopic or sub-microscopic size through to the macroscale at continental or subcontinental scale. In ecology, habitat has come to signify a description of where an organism is found, whereas niche is a complete description of how the organism relates to its physical and biological environment. The duality of place implicit in definitions that involve organisms as well as the environment of the organisms was reflected in biogeocoenosis, a Russian term equivalent to the western term ecosystem, and involving both the biocoenosis, a term introduced by Mobius in 1877 for a mixed community of plants and animals, together with its physical environment (ecotope). Biotope was defined as an area of uniform ecology and organic adaptation, although it was subsequently thought of as a habitat of a biocoenosis or a microhabitat within a biocoenosis. In relation to the land surface, the relief sphere or toposphere (Table 10.1), the basic place unit for the geomorphologist is the morphological unit; the undivided flat or slope is the basic unit of relief characterized by Linton (1951) as the electrons and protons of which physical landscapes are built.

The pedon, biotope and morphological unit are all examples of bases for specifying basic characteristics of place which emerged during the course of study of the physical environment, increasingly concentrated upon particular sub-branches of physical geography, including climatology, geomorphology and biogeography with pedology. After acquisition of information to characterize the physical environment of such basic units, several further requirements had to be fulfilled:

- Parameters were needed to describe the character of place – these were often chosen from the sub-branches of physical geography (associated with one of the spheres of Table 10.1). In climatology, mean annual temperature or mean annual precipitation were extensively used, but in other branches of physical geography there were insufficient basic data available so that mapping schemes devised to characterize place included systems of morphological, geomorphological or vegetational mapping.

- Mapping schemes provided detailed field survey scale information concerning the spatial units of physical environment necessary for the study of climate, geomorphology (landforms), soils (soil profile) or vegetation (plant community) and had to be related to a range of other scales up to global. This is illustrated by 16,000 different kinds of soil recognized within the USA (Buol, 1999), which had to be hierarchically arranged as part of a soil classification system to the global level. Detailed maps of landforms, local climates, soil series, or ecosystems could be amalgamated to larger scales and eventually to the world scale.

- Classification was achieved by reference to some spatial or temporal framework in the way that rivers could be categorized in terms of stream order, and landforms could be dated according to age where underlying datable material was expected to be contemporaneous in age with the landform.

- Because basic data and many classifications were expressed in terms of single aspects of environment, of physical form/slope/geomorphology, of climate parameters, of soil series, and of plant or animal communities, ways of integrating the different aspects were needed. The greatest limitation of regional geography was that the so-called integration was left to the reader to accomplish. Hence parameters were sought which described the dynamics or process of place and which reflected the way in which place functioned. With the advent of hydrology, this was exemplified by the way in which output parameters of water and sediment delivery reflected the integrated characteristics of the catchment areas upstream. Four systems incumbent in the systems approach (Chorley and Kennedy, 1971) provided particular ways of characterizing place.

- Place characterized by using single parameters, frequently expressed in terms of average values, was seen to be insufficient in physical geography. The climate at a point location expressed in terms of mean annual temperature or mean annual precipitation did not reflect the extremes of temperature, the incidence of extreme rainfalls or droughts – possibly inducing climatic hazards, or the variation from year to year or from decade to decade that was more pertinent to environmental processes and indeed to human activity.

- Not all sciences investigate places to the same level of analysis or resolution. Whereas the basic sciences, such as physics, chemistry and biology, may investigate sections of the earth spheres at a microscopic or sub-atomic scale, physical geography and other environmental sciences, conceived as composite sciences (Osterkamp and Hupp, 1996), adopt a broader scale of resolution. Although the explicit quantitative description of landform development has lagged behind the understanding of processes, recent technical developments in terrain monitoring now apply at the microscale, where information can be gathered at the total station level; at the mesoscale aided by global positioning systems (GPS); and at the macroscale facilitated by remote sensing and aided by digital elevation models (Lane et al., 1998).

Place is not exclusive to any one discipline but, as the physical geographer and other scientists initially characterized environment in terms of single characteristics, it was necessary to develop more integrated methods of depiction.

INTEGRATED PLACES

Advances in the more integrated depiction of place involved more than that of climate, soil, vegetation and surface form, and were achieved in three ways: by recognition of integrated units; by production of system models that depended upon the relationship of basic units; and by integrated classification of places into a hierarchy of levels culminating in world regions or types.

As an example of an *integrated unit*, singularities (short seasonal episodes, lasting as little as just a few days, commonly occurring at specific dates of the year) are a way of describing the climate experienced at a place, based upon analysis of meteorological or climatic records over periods of time. Thus, based upon analysis of 50 years of daily weather maps (1898–1947), Lamb (1964) demonstrated the existence of five seasons in the British Isles within which there were 22 singularities, many well known in folklore by terms such as April showers. In some places it is hazards, naturally occurring, and often extreme geophysical conditions threatening life or property, that are especially influential. Natural hazards in the hydrosphere, biosphere, lithosphere and atmosphere include the risk of extreme events which differ substantially from their mean values (Alexander, 1999). For southwestern Ontario it was shown (Hewitt and Burton, 1971) that in a 50-year period there would be one severe drought, two major windstorms, five severe snowstorms, eight severe hurricanes, 10 severe glaze storms, 16 severe floods, 25 severe hailstorms, and 39 tornadoes. It was therefore possible to express the hazardousness of a place as the complex of conditions that define the hazardous part of a region's environment. Hazards can be simple, including a single damaging element such as wind, rain, floodwater or earth tremor; compound, involving several elements acting together above their respective damage thresholds, illustrated by the wind, hail and lightning of a severe storm; and multiple, when elements of different kinds coincide or follow one another sequentially as when a hurricane is succeeded by landslides and floods.

In relation to land-surface relief and morphology, places can be characterized as landforms, employing terms such as eskers, limestone pavement, or type of river channel pattern. However, more generally, the undivided flat or slope, the morphological unit, is a basic unit of relief (Linton, 1951) having much in common with the *site* originally described as 'an area which appears for all practical purposes to provide throughout its extent similar conditions as to climate, physiography, geology, soil' (Bourne, 1931). Indeed, the site was a primary feature in soil investigations because it was the area over which soil profiles were investigated. Soil profiles and sites were then grouped, or classified, into the fundamental soil mapping units which might be a series, defined as groups of soils with similar profiles formed on lithologically similar parent materials.

In biogeography, as in ecology, it had been appreciated that a habitat plus the community it contains are a single working system; the term 'ecosystem' was invented by Tansley (1935) for a community of organisms plus its environment as one unit, therefore embodying the community in a place together with the environmental characteristics, relief, soil, rock type (the habitat) that influence the community. Ecosystems can vary in size from one to thousands of hectares and can be a pond upstream of a debris dam or a large section of the Russian steppe. As examples of integrated descriptions of place, ecosystems represented the more open-system thinking developed in physical geography, whereby place in physical geography was described by types of system. The ecosystem depends upon a dynamic relationship between a community of organisms and its environment.

Other ways of focusing upon dynamic relationships in order to characterize place included the drainage basin, implicitly employed in hydrology since the nineteenth century when the integrity of the water balance equation was first appreciated (Gregory, 1976a), and subsequently proposed as the fundamental geomorphic unit (Chorley, 1969). The drainage basin, the area drained by a particular stream or drainage network and delimited by a watershed, is an integrated unit that may be described in terms of drainage basin characteristics, including rock type, soil, vegetation and land-use, and relief characteristics. It is also a dynamic response unit from which outputs of water, sediment and solutes reflect the characteristics of the drainage basin which acts as the transfer function. The drainage basin unit is definable in relation to streams and rivers of a great range of sizes, can be a basis for spatial variations (see Chapter 6 of this volume), and is often employed in relation to environmental management (see below).

Just as there are ways of characterizing individual places in physical environments, there are also ways in which adjacent places are interrelated, expressed in *system models*. Such relationships may arise as drainage basins forming part of a nested hierarchy, or they may be linked by energy flows in the way that cascading systems have been recognized as structures with output from one subsystem forming input to the next (Chorley and Kennedy, 1971). Sequences of the character of places can be identified for climate (climosequences), relief (toposequences), lithology (lithosequences), ecology (biosequences), and time (chronosequences). In geomorphology, a nine-unit hypothetical land-surface model (Dalrymple et al., 1967) showed how nine particular slope components could occur in land-surface slopes anywhere in the world (Figure 10.2a). Each component was associated with a particular assemblage of processes so that it was possible to predict how slopes could occur under different morphogenetic conditions. A similar approach was applied to pedogeomorphic research (Conacher and Dalrymple, 1977) although a simple five-unit slope may be sufficient (e.g. Birkeland, 1984). Recurrent patterns of spatial variation have included the catena concept (Milne, 1935), expressing the way in which a topographic sequence of soils of the same age, and usually on the same parent material, can occur in landscape usually reflecting differences in relief/slope and drainage (Figure 10.2b), an arrangement which others have described as a toposequence (Bates and Jackson, 1980). Most catenas or hillslope models are thought of in two dimensions but combined with the drainage basin concept can produce a three-dimensional model of a small watershed (e.g. Huggett, 1975) embracing the manner in which water moves through the surface layers and over the surface leading to the creation of soil profiles, so providing an integrated understanding of the dynamics of place. Further scope remains to elaborate the catena idea and, in southeast Australia, sequences of erosion and deposition over time periods were related to soil profile development in a series of K cycles (Butler, 1959). Each period of time called a K cycle is composed of an unstable phase when erosion or deposition

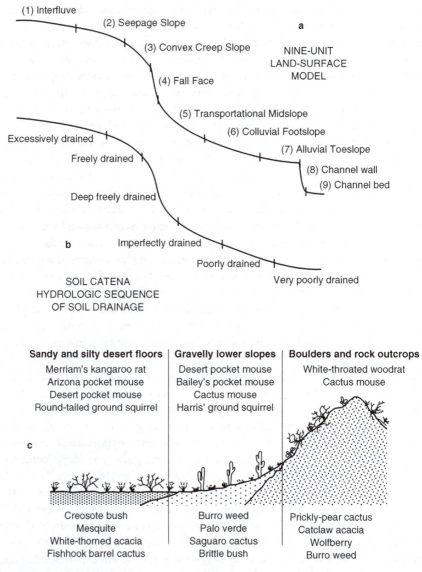

Figure 10.2   Sequences of places in the physical environment: (a) the components of the nine-unit land-surface model (based on Dalrymple et al, 1967); (b) the drainage categories found in a hydrological sequence of soils which makes up a catena; (c) is a particular example showing how communities vary with contrasting types of place in North American deserts

Source: After Vaughan (1978), as developed by Huggett (1995)

may occur, and a stable phase when soil profile development occurs. Evidence for up to 8 K cycles has been found preserved in some Australian soil landscapes. Toposequences of vegetation associations (Figure 10.2c) have been demonstrated showing how slope, soil drainage and vegetation vary empathetically in a repeated sequence across a landscape.

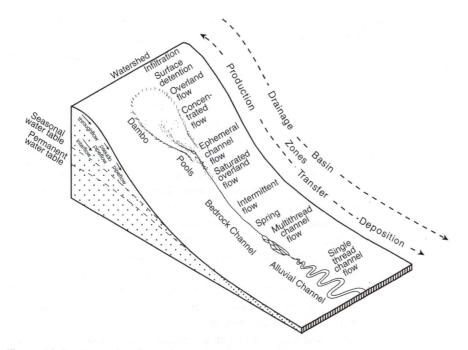

Figure 10.3   Hierarchy of places. The drainage basin is one scale of place, and at the other extreme is study at a point in the river channel. In between these two extremes are major zones such as the production, transfer and deposition zones indicated, within which it is possible to identify segments (bedrock, alluvial or colluvial) of valleys, within each of which there can be different types of reach (e.g. pool-riffle), within which there are channel units such as pools and specific aquatic communities. Such a hierarchy of scales of place, rather like a series of Russian dolls, exists in a physical environment which is dynamically changing according to the sequence of hydrological events. The diagram indicates the way in which the drainage basin network changes dynamically

The search for recurrent patterns of place is really a step towards the search for *integrated classifications* of physical environment. As spatial patterns were explored and related to controlling factors such as climate, the characteristics of soil and slope catenas (Ollier, 1976), of hydrological slope models (Kirkby, 1976) or of drainage networks (Gregory, 1976b) were exemplars of the relationships established between geomorphology and climate (Derbyshire, 1976). It was also necessary to relate places at a range of different scales such as that from a point in a river channel to a drainage basin (Figure 10.3). A particular viewpoint can be adopted, and two particular visions, at completely opposite scales, were the global and the local. The global approach started from a world distribution and sub-divided it, whereas the local vision described places in detail and amalgamated them to show how they fitted into broader regional, national and even world patterns.

The global vision had been evident in the classification of climates, of vegetation, of soils and of relief. Some interrelationships had been found between such world patterns, for example one scheme of classification by Köppen attempted to fit climatic values to world vegetation distribution patterns. However, such approaches did not really yield integrated approaches to the physical environment so that climatic-based schemes were sought to relate to world distributions in climatic geomorphology. These included the scheme of nine morphogenetic systems (Peltier, 1950, 1975), each distinguished by a characteristic assemblage of geomorphic processes; 13 morphoclimatic zones (Tricart, 1957) related to climates, geomorphic processes and also to soils and vegetation; and five, later increased to eight, climato-morphogenetic zones (Budel, 1977), each characterized by particular landscape-forming processes and by relief features related in a distinctive way to past landscape development (Figure 10.4). Although climatic factors are undoubtedly of significance in affecting the global pattern of physical environments, other factors can be equally significant (Twidale and Lageat, 1994); there are dangers in adopting a very simplistic approach so that climatic geomorphology was viewed as not new, not well established and premature (Stoddart, 1968), and climate impacts may have been overestimated (Twidale and Lageat, 1994).

Perhaps most successfully, or at least most industriously, pursued were Russian efforts to develop the energy budget approach of Budyko (1958), which attempted a physico-geographical zonation of the earth and provided the foundation for energy and moisture regimes to be related to vegetation types (Grigoryev, 1961), to genetic soil types (Gerasimov, 1961) and to geographic zonality (Ye Grishankov, 1973). Such schemes could be extended to link with the approach to integrated systems based upon field survey.

A contrasted local vision arose from attempts to describe the physical environment in an integrated way at the level of field survey. Russian studies recognized the *urochischa* as a basic physical-geographical unit of landscape with uniform bedrock, hydrological conditions, microclimate, soil and mesorelief (Ye Grishankov, 1973). Reconnaissance investigations in Australia, Africa and New Guinea, aided by the rapidly growing use of air photograph data, involved the recognition of land-systems, defined as areas with a recurring pattern of topography, soils and vegetation (Christian and Stewart, 1953). The land-systems approach employed for resource evaluation in particular areas of the world utilized landscape ecology and land evaluation. These two approaches developed from soil survey sometimes founded upon land capability analysis, a grouping of kinds of soil into classes according to their potential use, and the treatments required for their sustained use; and upon suitability analysis, defined as the fitness of a given tract of land for a defined use. Landscape ecology, a term first used by Troll to connote the interaction between landscape and ecology, provides an approach that interprets landscape as supporting interrelated natural and cultural systems (Vink, 1983; see also, Chapter 6 in this volume). The central focus of landscape ecology, which is the study of pattern and process at the landscape scale (Forman, 1995), involves the interrelationship between landscape structure, the spatial patterning of ecosystems, and landscape functioning, the interactions of flows of energy, matter and species within and among component ecosystems (Kupfer, 1995). Whereas landscape ecology focuses on what the systems in the landscape can generally be used for, landscape evaluation is the estimation of the potential of land for particular kinds of use which can include productive uses such as arable farming, livestock production and forestry, together with uses that provide services or other benefits, such as water catchment areas, recreation, tourism and wildlife conservation (Dent and Young, 1981). Such approaches have been refined with the advent of information systems (Cocks and Walker, 1987), advanced

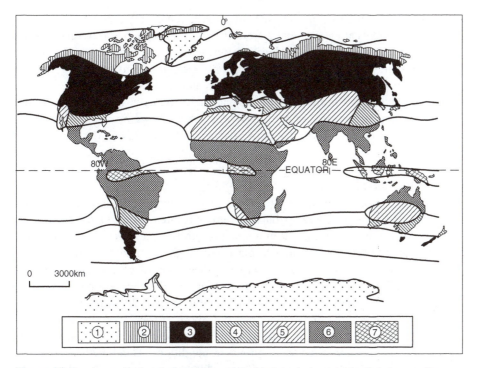

Figure 10.4   An example of places in a global integrated context: climato-genetic zones (after Budel, 1969). Many attempts have been made to subdivide the earth's surface according to the physical character of places. One integrated approach involved recognition of climato-genetic zones distinguished according to contemporary environmental processes and the past development of the physical environment. The zones were described as: (1) glacier zone; (2) subpolar zone of excessive valley formation; (3) extratropical zone of former valley formation (later called ectropic zone); (4) subtropical zone of mixed relief formation; (5) arid zone; (6) peritropical zone of excessive planation; (7) inner tropical zone of partial planation

developments in remote sensing and development of geographical information systems (e.g. Heywood et al., 1998). The land-system approach provided an applied approach to physical environment greatly assisted by geographical information systems (GIS) which liberated multiscale approaches to data acquisition and analysis. Further progress was aided by GPS facilities and by real-time analysis so that the way in which it is now possible to perceive place in the physical environment has been completely revolutionized as a result of advances in data acquisition and analysis techniques. Such advances have meant that the initially independent global and local approaches can now be amalgamated by remote sensing and GIS.

PLACE REDISCOVERED? OPPORTUNITIES AND CHALLENGES

Until recently, place has not received such explicit interest from physical geographers as that accorded by human geographers. However, attention was given implicitly to what place means, how it is described, and how it is related to conceptual models at various scales from the local to the global. Interrelationships between the significance of the physical environment of place (landscape ecology) and the way in which place may be used (landscape evaluation) are just two strands that are fundamental to the move towards greater understanding of the capacity and potential of environment. However, place now assumes a greater significance in physical geography in relation to five themes: how do we place it in its environmental context or position it, sustain it, manage it, restore it, and design it?

Positioning place

To 'position' place we need to move towards more general models which can cope with non-linear and complex spatial and temporal interactions. Older models had a dependence upon linear models of interaction (and of space) and thus engendered a lack of attention to the specifics of place. Although the movement towards general conceptual models (e.g. Figure 10.2) was a necessary development, it has been argued (Phillips, 2001) that historical and spatial contingency are responsible for the character of places. Historical contingency means that the state of a system or environment is partially dependent on one or more process states or upon events in the past. It arises from inheritance, conditionality and instability. Inheritance relates to features inherited from previous conditions. Conditionality is when development might occur by two or more different pathways according to the intensity of a particular phenomenon, for example whether a threshold is exceeded to instigate different trajectories of development. Instability refers to dynamical instabilities whereby small perturbations or variations in initial conditions vary or grow over time, giving divergent evolution. Spatial contingency occurs where the state of an earth surface system is dependent on local conditions which relate to local histories, landscape spatial patterns and scale contingency. The local history of a place independent of variations in environmental controls can contribute to spatial variation in earth surface systems. Landscape patterning refers to landscape scale processes that create and modify spatial structures including processes leading to spatial persistence or contagion (Phillips, 2001), such as catenary relationships (Figure 10.2). Scale contingency relates to the tendency for controls over process–response relationships to vary with spatial extent or resolution encompassing relationship to scales of measurement and observation. Quantitative spatial analytic approaches in landscape ecology and pedology have increasingly focused on modelling local spatial variability rather than a search for global laws. Predictions of, and generalizations about, human impacts on the environment, must be adapted to local or regional or environment-specific conditions. The recognition and confrontation of spatial and historical contingency does not entail abandonment of theory or rigour, but rather the adaptation of scientific and spatial enquiry to the real world (Phillips, 2001).

Sustaining place

How we sustain it requires consideration of *sustainable development*, which originated from conservation ideas in North America and Europe, came to the fore with the 1992 Rio United Nations Conference, and has subsequently been thought of as 'development that respects the life quality of future generations and that is accomplished through support for the viability of the Earth's resources and ecosystems' (Saunier, 1999). Instead of thinking about ecosystems (and place) as physical objects, it is possible to visualize them in terms of attributes with value for people as natural assets or 'natural capital' – an approach employed by Haines-Young (2000) to combine the scientific and cultural traditions of landscape ecology in managing landscapes. This challenges the view that the goal of sustainable development was to maintain the quantity and quality of ecological resources at approximate steady state so as to ensure that they are not depleted at rates exceeding their renewal (Bartell, 1996), because characterization of environmental management in terms of the 'natural capital' paradigm is fundamentally different. It provides an understanding of how the physical and biological processes associated with landscapes have value in an economic and cultural context, and so it is the study of natural capital from a dynamic, evolutionary and landscape perspective. It is not a steady state that we seek but rather a sustainable trajectory for ecosystems and landscapes so that equilibrium models rarely apply with no single sustainable state but a whole set of landscapes that are more or less sustainable (Haines-Young, 2000) (see also Chapters 18 and 22 of this volume on nature and sustainability).

Managing place

In the management of place, a more cultural approach to physical environment (Gregory, 2000, 2006) is required: two of the major approaches available are *conservation* and *holistic dynamic management*. Conservation has a long tradition in relation to natural resources, but the conservation of areas or place as nature reserves, sites of special scientific interest (SSSIs), or wilderness areas is more recent and is one approach to management. In the UK, Regionally Important Geological and Geomorphological Sites (RIGS) were established in 1990 to provide regional complements to SSSIs. However, it may be more pragmatic to adopt an approach, often styled environmental, which acknowledges the dynamics of environmental systems, requires a sustainable solution, and recognizes the totality of all the aspects of the physical environment – to which the term holistic has often been applied. The drainage basin or catchment has been extensively used as a management planning unit, capable of being employed for both physical management of environment and for administrative purposes. However, management has not always covered the entire breadth of environmental characteristics and may not be as holistic as it could or should be, so that further progress remains to be made (National Research Council, 1999; Downs and Gregory, 2004).

Restoring place

A range of disciplines is currently engaged with *restoration* of places because wetlands, prairies, lakes, wildlife habitats and other areas have been damaged by mineral extraction and many other forms of human impact. Restoration ecology is concerned with the restoration of badly damaged ecosystems to some pre-disturbance condition (Cairns, 1989), although, as we pursue environmental restoration as an addition to conservation (Berger, 1990), the question arises to what do we restore the original 'place': the condition that would have developed if disturbance had not occurred, or one that appears natural? In which case, what is natural? In river management, restoration is now a major theme (Downs and Gregory, 2004) and changing public opinion is becoming at least as important as gaining new scientific knowledge (Douglas, 2000). Yi Fu Tuan wrote his book *Topophilia* (1974: 1–2) 'out of the need to sort and order in some way the wide variety of attitudes and values relating to man's physical environment'. An awareness of such cultural strands is now more evident in physical geography (Gregory, 2000, 2004).

Designing Place

Renewed interest in place by physical geographers and environmental scientists raises the question as to whether physical geographers should be engaged in *environmental design* (Gregory, 2000, 2004, 2006) and, if so, how? Ian McHarg, in his book *Design with Nature* (1969), proposed ways of approaching natural environment that were of wider importance than landscape architecture itself. When the book was subsequently revised, McHarg (1992) commented that in 1969 'scientists had not yet discovered the environment, the mandarins were molecular biologists and physicists concerned with sub-atomic particles' but hoped that his book 'provided a method whereby environmental data could be incorporated into the planning process, by interpreting ecological studies to include the full panoply of the environmental sciences, with the subject of values being crucial to the environmental movement'. Although the full range of environmental sciences was embraced, McHarg (1992) identified social systems to be the one significant omission; in 1969 the influence of economics was antithetical to ecology, and other social sciences were then, unlike now, oblivious to the environment. The vision in McHarg's book was characterized by Lewis Mumford (1999) as one of 'organic exuberance and human delight, which ecology and ecological design promise to open up for us, McHarg revives the hope for a better world'. In river channel management (Downs and Gregory, 2004), design can be involved at the preliminary stage when an approach is evolved, at the implementation stage when alternatives are reviewed, at the stage when the design is effected and at the post-project consideration stage when the managed channel is kept under review (Gregory, 2006).

These five themes are exemplified by the physical geography of urban areas. Indeed, McHarg had developed ideas of design with nature for the application to the expansion of urban areas, defined by the World Health Organization (WHO) in 1990 as a 'man-made environment encroaching and replacing a natural setting and having a relatively high concentration of people whose economic activity is largely non-agricultural', although the US Census Bureau definition is areas with more than 386 people per km^2. Although urban physical geography, is not always identified as a field of physical geography, as only 2% of the world's land surface is built up, it is estimated that urban areas could include 60% of world population by 2025. The 'ecological footprint' represents the environmental impact of cities in terms of the amount of land required to sustain them and, if the earth's population consumed resources at the same rate as does a typical resident of Los Angeles, it would require at least three planet earths to provide all the energy they required.

Urban environments have distinctive urban climates above roof level, forming the urban boundary layer (UBL), and below roof level is the urban canopy layer (UCL). Urban heat islands are created and a large city is typically 1.3°C warmer annually than the surrounding area, although the heat island varies diurnally, with up to 10°C difference near midnight. Hydrologically, the extensive impervious areas of urban environments reduce surface storage so that infiltration is not possible and evapotranspiration is much less than in rural areas. Increased amounts of surface runoff are complemented by the surface water runoff system which collects water from roads and roofs. Stream discharges from urban areas tend to have higher peak flows and lower base flows, and the flood frequencies of rivers draining urban areas will be significantly changed from the time before the urban area existed. Urban areas also generate characteristic water quality with water temperatures often higher than those of rural areas, higher solute concentrations reflecting additional sources, including pollutants, and suspended sediment concentrations high during building activity but lower after urbanization when sources are no longer exposed. Urban ecology is changed as a result of exterminations and introductions: although once thought of as ecological deserts, urban environments are now known to support a variety of plant and animal species, including songbirds, deer and fox, and an increasing number of North American metropolitan areas now have to contend with large predators including alligators, coyotes, pumas and bears. Cities have been described as a karst topography (Bunge, 1973), with sewers performing precisely the function of limestone caves in Yugoslavia, which causes a parched physical environment, especially in city centres, or as ecosystems (Douglas, 1983) which embrace population ecology, system ecology, the city as a habitat, and energy and material transfer within cities. The European Commission (18 February 2004) revealed plans to improve environmental aspects of towns and cities with a new EU-wide strategy which aims to provide a 'best practice' style approach, with successful projects implemented on a widespread basis across the Union.

Many physical models of cities and urban areas have been constructed, for example, for urban climate and for urban hydrology, but, in common with all places, each urban area is affected by historical and spatial contingency so that it is difficult to predict how an urban area will affect physical processes or environmental characteristics without reference to characteristics of the particular area. With this in mind, Table 10.2 suggests some of the ways in which urban physical geography places present challenges and opportunities for the five themes identified. At first it may be thought that these themes transgress into other disciplines, but in fact there is a singular contribution that geographers can make through concern with place. To position urban places in relation to contingency it is first necessary to see how to relate to certain characteristics of climate, geomorphology, hydrology, soils and ecology. Then for each sub-branch of physical geography it is possible to indicate some of the techniques that can be utilized in urban environments to meet the challenges of sustainability, management, restoration and design. For example, in the case of urban drainage, the original intention – to remove stormwater from the urban area as rapidly as possible – is why channelization was often undertaken, whereas more recent management methods have been devised to meet at least five objectives: concerned with actual location of the urban area or of urban expansion; retention of precipitation; delay of runoff; management of the effects of runoff in the urban area; and planning for downstream consequences. Particular approaches to urban stormwater management now necessarily transcend several of these themes, especially as urban drainage has moved away from the conventional thinking of designing for flooding towards balancing the impact of urban drainage on flood control, quality management and amenity (CIRIA, 2001). Thus Sustainable Drainage Systems (SUDS) are more sustainable than conventional drainage methods because they manage runoff flow rates (reducing the impact of urbanization on flooding), protect or enhance water quality, are sympathetic to the environmental setting and the needs of the local community, provide habitat for wildlife in urban watercourses and, where appropriate, encourage natural groundwater recharge (Herrington Consulting, 2006). Low-Impact Development (LID) techniques provide an approach to environmentally friendly land-use planning, following the approach of Ian McHarg (McHarg, 1969, 1992), including a suite of landscaping and design techniques that attempt to maintain the natural, pre-developed ability of a site to manage rainfall, also embracing the idea that stormwater is not merely a waste product to be disposed of, but is a resource (Massachusetts Government, 2006). Similarly, smart growth approaches minimize the runoff impact of urbanization (Tang et al., 2005). Associated with these approaches, a range of innovative measures has been evolved, often with sustainable benefits (see Table 10.2).

The five themes necessarily overlap which is why several of the columns in Table 10.2 are combined. In addition, it could be argued that, at a time when a more holistic view is being taken of place, the climate, geomorphology, hydrology, soils and ecology should not be considered separately. Hence the final line represents the holistic approach to urban areas – an approach which resonates through many of the environmental sciences at present. It also requires a more cultural approach because a range of cultural issues (Gregory, 2006) can affect the way in which the river channel, or the environment more generally, is perceived and managed.

CONCLUSION

Physical geographers can now be involved in the design of physical environment or of place, rural or urban, building upon involvement in the sustainability agenda, a greater awareness of different cultural attitudes to place, the need to restore damaged places, and the possibility of using understanding of the dynamics and evolution of place in physical environment as the basis for sustainable environmental design. As physical geography refreshingly progresses to become more concerned with the holistic design of particular places, we still have to remember that our existing knowledge, models and theories have been

Table 10.2 Opportunities and challenges for the physical geographer in relation to urban places

Branch of physical geography	Environmental context: position	Opportunity and challenge			
		Sustainability	Management and planning	Restoration	Design
	Physical characteristics of urban area in relation to historical and spatial contingency	Visualize ways in which urban developments can embrace characteristics which contribute to sustainability	Conservation of physical environment in urban areas and holistic dynamic approach Planning has to be specific to the particular urban environment	Identify where restoration is possible and/or necessary	Suggest pattern of physical characteristics of new or expanded urban area or modification of existing one
Climate	Prevailing weather systems, singularities, atmospheric hazards	Control programmes for atmospheric emissions Tree planting (to augment tree biomass and diversity and influence atmospheric environment)			Estimate how urban area will generate urban heat island and climate Vary air conditioning according to urban heat island; use other measures to ease discomfort during high temperature periods Weather sensitivity analysis
Geomorphology	Topography, environmental hazards, proximity to environmental thresholds	Minimize loading effects of buildings Minimize exposure of bare soil (temporary ground cover – geotextiles, mulches, plastic sheeting) Control water and sediment on building sites Landslide protection measures Insulating procedures in permafrost areas		Reclaim waste land and contour appropriately for area. Design to reduce channel velocities, increase residence time in channels and accommodate or delay pollutant loads with possible changes, including permeable revetment, swales (*shallow vegetated channels*), excavation of pools, plunge pools, set backs from the channel filter strips (*drain water from impermeable areas and filter out silt*), sediment traps in channels Restore or rehabilitate channelized rivers with daylighting (*excavation of culverted or buried streams*) where possible	

(Continued)

Table 10.2 (Continued)

Branch of physical geography	Environmental context: position	Opportunity and challenge			
		Sustainability	Management and planning	Restoration	Design
Hydrology	Flow and sediment regime; position in drainage basin, drainage basin characteristics	Urban drainage systems may need flow velocity reduction; channelization; floodplain levees. Reduction of flood hazards may require minimizing connections between impervious surfaces, encouraging local storage such as rain gardens in each garden/backyard, using diversion channels for water from construction sites, or considering use of land-use zoning, recharge of aquifers, land-use regulation, flood insurance. Overall may employ SUDS (sustainable drainage systems – to militate against flooding and pollution), and/or LID (low-impact development techniques)		Separation of foul water and stormwater systems Restoration and diversification of urban drainage system may include use of underground storage reservoirs (*slow release of stormwater*; collection of water on roof gardens, brown roof, green roof; downpipes on to pavements and roads (*not directly connected to stormwater drainage system*); soakaways; filter drains (*linear trenches of permeable material*); permeable pavement; detention ponds; balancing ponds; infiltration basins, bioretention areas, infiltration trenches; water conservation structures Restoration of baseflows (*groundwater cultivation by construction that facilitates infiltration*)
Soils	Susceptibility to soil erosion, salt weathering	Protection against salt weathering in drylands Slope stabilization and design Integrated planning for river corridors Include soil as integral component of park planning and management		Measures to control or reduce gully erosion include construction of layered, filtering check dams to control erosion and trap sediment, planting long-rooted grasses on gully walls, reducing seepage scarps. On eroding slopes distribute coarse material on bare soils to reduce speed of water flow	
Ecology	Vegetation and faunal regions population ecology	View the city as a habitat, recalling that typically 30% of the surface of urban areas remains vegetated Ensure presence of botanical gardens, urban nature reserves, public parks and conservation areas where feasible Maintain natural areas for survival of resident plants and animals Manage sustainable ecology associated with other areas, including golf courses		Remove undesirable species; reduce pollution (to encourage organisms to return); reduce use of fertilizers and pesticides where feasible Preserve wetlands, floodplains and tree cover when feasible (*increases infiltration and reduces storm runoff*) Encourage street tree planting; create urban nature reserves; establish conservation areas and natural reserves; maintain or introduce botanical gardens and parks Integrate golf courses, public gardens, backyards into the urban ecology, encourage character of window-boxes and personal gardens to be consistent with local environment conditions	

Table 10.2 *(Continued)*

Branch of physical geography	Environmental context: position	Opportunity and challenge		
		Sustainability	Management and planning	Restoration	Design
Holistic	New attitude to environment with scientific and political concern about ecological and cultural sustainability of monocultural resource systems	Energy and material transfer within cities Urban areas should be not only diverse functionally but also self-perpetuating ecosystems that require minimal maintenance and are sustainable Respond to public views in relation to management and sustainability		Identify subjects for restoration and propose plans that involve rehabilitation Determine what is character of place that restoration will produce	Managing with nature whenever possible involving establishing what is appropriate for the place, what alternatives are feasible, which alternative is the basis for the most appropriate design, and after the design is implemented how post-project appraisal can be effective

built upon experience of Anglo-American places, of some physical environments more than others, and influenced by dramatic, eye-catching examples such as the Grand Canyon, the Amazon rainforest, or the Yosemite National Park. In the development of physical geography we can discern threads running through science reflecting the investigation of particular types of place. When physical geography encompassed the quantitative revolution, spatial analysis tended to concentrate upon a number of approaches which were essentially morphological in character, such as network classification, and even systems approaches did not allow sufficient attention to be given to the inherent physical distinctiveness of places. As we have become more holistic, having progressed from an idiographic to a nomothetic approach, are physical geographers sufficiently objective in their attitudes to place so that they can now profit from a postmodern idiographic approach, concerned with the character and design of individual places? This could be realized by building on the achievements of general spatial models to locate place in both space and time, including recognition of the importance of cultural identity. Knowing how physical geographers have reacted to place in the past assists in becoming more holistic in our future attitudes to urban as well as rural places, by progressively adapting scientific and spatial enquiry to the real world, which is made up of actual places with specific characteristics.

SUMMARY

- Place is central to physical geography but until recently has not been given explicit attention.

- Places studied by physical geographers can be located in relation to spheres of study (Figure 10.1 and Table 10.1) of the environment. Physical geographers focus their research upon a particular combination of these spheres.

- In each branch of physical geography basic units are required to characterize place and examples are the morphological unit, the landform, the soil profile, or the niche.

- Emphasis upon separate aspects of place, in terms of climate, soil, ecology and geomorphology, meant that more integrated relationships were sought, such as the land system

- Place now receives more explicit attention by physical geographers in relation to the way in which it is positioned in historical and spatial contingency; and then according to sustainability, management, restoration and design of physical environments.

- Urban places exemplify how these five themes are evident (Table 10.2), revealing significant opportunities and challenges for the physical geographer.

Further Reading

Useful background is provided by the **Encyclopedia of Environmental Science** (Alexander and Fairbridge, 1999), and the development of physical geography, including cultural physical geography, is reviewed in **The Changing Nature of Physical Geography** (Gregory, 2000). An integrated approach to physical geography in the context of environmental change is given in **Physical Geography and Global Environmental Change** (Slaymaker and Spencer, 1998), and **Landform Monitoring, Modelling and Analysis** (Lane et al. 1998) shows how current techniques have advanced description of landform developments – an example of place in physical geography. **Design with Nature** (McHarg, 1969) is a salutary read because it shows how a landscape architect appreciated environmental challenges before they were acknowledged in physical geography. However, **Geoecology: An Evolutionary Approach** (Huggett, 1995) and the article by Haines-Young (2000) indicate some ways in which physical geography can progress. Ways in which river channel management is being achieved is demonstrated in **River Channel Management** (Downs and Gregory, 2004). **Topophilia** (Yi Fu Tuan, 1974) is a typically thought-provoking product of a stimulating writer who began research as a physical geographer, and the article by Phillips (2001) provides a stimulating position statement.

Note: Full details of the above can be found in the references list below.

References

Alexander, D.E. (1999) 'Natural hazards', in D.E. Alexander and R.W. Fairbridge (eds) *Encyclopedia of Environmental Science*. Dordrecht: Kluwer Academic Publishers, pp. 421–5.

Alexander, D.E. and Fairbridge, R.W. (eds) (1999) *Encylopedia of Environmental Science*. Dordrecht: Kluwer Academic Publishers.

Bartell, S.M. (1996) 'Ecological risk assessment and ecosystem variation', in R.D. Simpson and N.L. Christensen (eds) *Ecosystem Function and Human Activity*. New York: Chapman and Hall, pp. 45–70.

Bates, R.L. and Jackson, J.A. (1980) *Glossary of Geology*. Falls Church, VA: American Geological Institute.

Bentley, E.C. (1905) *Biography for Beginners*. London: T. Werner Laurie.

Berger, J. (ed.) (1990) *Environmental Restoration*. Washington, D.C: Island Press.

Birkeland, P.W. (1984) *Pedology, Weathering and Geomorphological Research*. New York: Oxford University Press.

Bourne, R. (1931) 'Regional survey and its relation to stocktaking of the agricultural and forest resources of the British Empire', *Oxford Forestry Memoir*, 13.

Budel, J. (1969) 'Das System der klima-genetischen Geomorphologie', *Erdkunde*, 23: 165–82.

Budel, J. (1977) *Klima-Geomorphologie*. Berlin/Stuttgart: Borntraeger.

Budel, J. (1982) *Climatic Geomorphology* (trans. L. Fischer and D. Busche). Princeton, NJ: Princeton University Press.

Budyko, M.I. (1958) *The Heat Balance of the Earth's Surface* (Trans. N. Steepanova from original dated 1956). Washington, DC: Weather Bureau.

Bunge, W. (1973) 'The geography', *Professional Geographer*, 25: 331-7.

Buol, S.W. (1999) 'Soil', in D.E. Alexander and R.W. Fairbridge (eds) *Encyclopedia of Environmental Science*. Dordrecht: Kluwer Academic Publishers, pp. 563-4.

Butler, B.E. (1959) *Periodic Phenomena in Landscapes as a Basis for Soil Studies*. Soil Publication 14. Melbourne: Commonwealth Scientific and Industrial Research Organization (CSIRO).

Cairns, J. (1989) 'Restoring damaged ecosystems: is pre-disturbance condition a viable option?', *The Environmental Professional*, 11: 152-9.

Chorley, R.J. (1969) 'The drainage basin as the fundamental geomorphic unit', in R.J. Chorley, (ed.) *Water, Earth and Man*. London: Methuen, pp. 59-96.

Chorley, R.J. and Kennedy, B.A. (1971) *Physical Geography: A Systems Approach*. London: Prentice-Hall.

Christian, C.S. and Stewart, G.A. (1953) *Survey of the Katherine – Darwin Region 1946*. Land Research Series 1. Melbourne, Vic.: CSIRO.

CIRIA (2001) *Sustainable Urban Drainage Systems: Best Practice Manual for England, Scotland, Wales and Northern Ireland*. CIRIA C523. London:

Cocks, K.D. and Walker, P.A. (1987) 'Using the Australian Resources Information System to describe extensive regions', *Applied Geography*, 7: 17-27.

Connacher, A.J. and Dalrymple, J.B. (1977) 'The nine-unit land-surface model: an approach to pedogeomorphic research', *Geoderma*, 18: 1-154.

Dalrymple, J.B., Conacher, A.J. and Blong, R.J. (1967) 'A nine-unit hypothetical land-surface model', *Zeitschrift für Geomorphologie*, 12: 60-76.

Dent, D. and Young, A. (1981) *Soils and Land Use Planning*. London: Allen & Unwin.

Derbyshire, E. (ed.) (1976) *Geomorphology and Climate*. Chichester: John Wiley & Sons.

Douglas, I. (1983) *The Urban Environment*. London: Edward Arnold.

Douglas, I. (2000) 'Fluvial geomorphology and river management', *Australia Geographical Studies*, 38: 253-62.

Downs, P.W. and Gregory, K.J. (2004) *River Channel Management*. London: Edward Arnold.

Forman, R.T.T. (1995) *Land Mosaics: The Ecology of Landscapes and Regions*. New York: Cambridge University Press.

Gerasimov, I.P. (1961) 'The moisture and heat factors of soil formation', *Soviet Geography*, 2: 3-12.

Gregory, K.J. (1976a) 'Changing drainage basins', *Geographical Journal*, 142: 237-47.

Gregory, K.J. (1976b) 'Drainage networks and climate', in E. Derbyshire (ed.) *Geomorphology and Climate*. Chichester: John Wiley & Sons, pp. 289-318.

Gregory, K.J. (2000) *The Changing Nature of Physical Geography*. London: Edward Arnold.

Gregory, K.J. (2004) 'Human activity transforming and designing river landscapes: a review perspective', *Geographica Polonica*, 77: 5-20.

Gregory, K.J. (2006) 'The human role in changing river channels', *Geomorphology*, 79: 172-91.

Grigoryev, A.Z. (1961) 'The heat and moisture regions and geographic zonality', *Soviet Geography*, 2: 3-16.

Guyot, A. (1850) *The Earth and Man: Lectures on Comparative Physical Geography in its Relation to the History of Mankind*. New York: Scribners.

Haines-Young, R. (2000) 'Sustainable development and sustainable landscapes: defining a new paradigm for landscape ecology', *Fennia* 178: 7-14.

Herrington Consulting (2006) http://www.herringtonconsulting.co.uk/SUDS.htm

Hewitt, K. and Burton, I. (1971) *The Hazardousness of a Place: A Regional Ecology of Damaging Events*. Toronto: University of Toronto Press.

Heywood, I., Cornelius, S. and Carver, S. (1998) *An Introduction to Geographical Information Systems*. Harlow: Longman.

Huggett, R.J. (1975) 'Soil landscape systems: a model of soil genesis', *Geoderma*, 13: 1-22.

Huggett, R.J. (1980) *Systems Analysis in Geography*. Oxford: Clarendon Press.

Huggett, R.J. (1995) *Geoecology: An Evolutionary Approach*. London: Routledge.

Hutchinson, G.E. (1970) 'The biosphere', *Scientific American*, 223: 45-53.

Kirkby, M.J. (1976) 'Hydrological slope models: the influence of climate', in E. Derbyshire (ed.) *Geomorphology and Climate*. Chichester: John Wiley & Sons, pp. 247-68.

Kupfer, J.A. (1995) 'Landscape ecology and biogeography', *Progress in Physical Geography*, 19: 18–34.

Lamb, H.H. (1964) *The English Climate*. London: English University Press.

Lane, S.N., Richards, K.S. and Chandler, J.H. (eds) (1998) *Landform Monitoring, Modelling and Analysis*. Chichester: John Wiley & Sons.

Linton, D.L. (1951) 'The delimitation of morphological regions', in L.D. Stamp, and S.W. Wooldridge (eds) *London Essays in Geography*. London: London School of Economics, pp. 199–218.

Massachusetts Government (2006) http://www.mass.gv/envir/smart_growth_toolkit/pages/glossary.html

Mattson, S. (1938) 'The constitution of the pedosphere', *Annals of the Agricultural College of Sweden*, 5: 261–76.

McHarg, I.L. (1969) *Design with Nature*. New York: Natural History Press.

McHarg, I.L. (1992) *Design with Nature*. Chichester: John Wiley & Sons.

Mead, W.R. and Smeds, H. (1967) *Winter in Finland*. London: Hugh Evelyn.

Milne, G. (1935) 'Some suggested units of classification and mapping, particularly for East African soils', *Soil Research*, 4: 183–98.

Mobius, K. (1877) *Die Auster und die Austernwirtschaft*. Berlin: Wiegundt, Hampel and Parey.

Morrison, M.L. (1999) 'Habitat and habitat destruction', in D.E. Alexander, and R.W. Fairbridge (eds) *Encyclopedia of Environmental Science*. Dordrecht: Kluwer Academic Publishers, pp. 308–9.

Mumford, L. (1992) 'Introduction' in I.L. McHarg (ed.) *Design with Nature*. Chichester: Wiley, pp. vii-viii.

National Research Council (NRC) (1999) *New Strategies for America's Watersheds*. Washington, DC: National Academy Press.

Ollier, C.D. (1976) 'Catenas in different climates', in E. Derbyshire (ed.) *Geomorphology and Climate*. Chichester: John Wiley & Sons, pp. 137–70.

Osterkamp, W.R. and Hupp, C.R. (1996) 'The evolution of geomorphology, ecology and other composite sciences', in B.L. Rhoads and C.E. Thorn (eds) *The Scientific Nature of Geomorphology*. Chichester: John Wiley & Sons, pp. 415–41.

Peltier, L.C. (1950) 'The geographic cycle in periglacial regions as it is related to climatic geomorphology', *Annals Association of American Geographers*, 40: 214–36.

Peltier, L.C. (1975) 'The concept of climatic geomorphology', in W.N. Melhorn and R.C. Flemal (eds) *Theories of Landform Development*. Binghamton, NY: State University of New York Press.

Phillips, J.D. (2001) 'Human impacts on the environment: unpredictability and the primacy of place', *Physical Geography*, 22: 321–32.

Rolston, III, H. (1997) 'Nature for real: is nature a social construct?', in T.D.J. Chappell (ed.) *The Philosophy of the Environment*. Edinburgh: Edinburgh University Press, pp. 38–64.

Saunier, R.E. (1999) 'Sustainable development, global sustainability', in D.E. Alexander and R.W. Fairbridge (eds) *Encyclopedia of Environmental Science*. Dordrecht: Kluwer Academic Publishers, pp. 587–92.

Slaymaker, O. and Spencer, T. (1998) *Physical Geography and Global Environmental Change*. Harlow: Addison Wesley Longman.

Stoddart, D.R. (1968) 'Climatic geomorphology: review and assessment', *Progress in Geography*, 1: 160–222.

Tang, Z., Engel, B.A., Lim, K.J., Pijanowski, B.C. and Harbor, J. (2005) 'Minimizing the impact of urbanization on long term runoff', *Journal of the American Water Resources Association*, 41: 1347–59.

Tansley, A.G. (1935) 'The use and abuse of vegetational concepts and terms', *Ecology*, 16: 284–307.

Teilhard de Chardin, P. (1959) *The Phenomenon of Man*. London: Collins.

Tricart, J. (1957) 'Application du concept de zonalité a la géomorphologie', *Tijdschrift van het Koninklijk Nederlandsch Aarddrijiskundig Geomootschap*, 422–34.

Troll, C. (1950) 'Die geographische Landschaft und ihre Erforschung', *Studium Generale* 3: 163–81.

Tuan, Yi Fu (1974) *Topophilia: A Study of Environmental Perception, Attitudes and Values.* Englewood Cliffs, NJ: Prentice-Hall.

Twidale, C.R. and Lageat, Y. (1994) 'Climatic geomorphology: a critique', *Progress in Physical Geography*, 18: 319–34.

Vaughan, T.A. (1978) *Mammalogy.* Philadelphia, PA: W.B. Saunders.

Vink, A.P.A. (1983) *Landscape Ecology and Land Use* (Trans. and edited by D.A. Davidson from the original Dutch). London: Longman.

Watts, D.A. (1971) *Principles of Biogeography: An Introduction to the Functional Mechanisms of Ecosystems.* London: McGraw-Hill.

Wild, A. (1993) *Soils and the Environment.* Cambridge: Cambridge University Press.

Ye Grishankov, G. (1973) 'The landscape levels of continents and geographic zonality', *Soviet Geography*, 14: 61–77.

11

Scale: Resolution, Analysis and Synthesis in Physical Geography

Tim Burt

Definition

Physical geographers study the world across a wide range of scales, from the molecular to the global. In any investigation, we will pay attention to some things and ignore others. In terms of spatial scale, the resolution of any study indicates the level at which we focus on a particular item of interest. However, geographers have never confined themselves to one scale alone: on the one hand, they may narrow their perspective in order to focus on the detailed way in which a system operates; on the other hand, they may wish to extrapolate their findings at one scale to the wider region.

INTRODUCTION: SCALE AND RESOLUTION

> All this time the guard was looking at her, first through a telescope, then through
> a microscope, and then through an opera glass. (Lewis Carroll, 1872)

Somewhere between the atom and the universe lies the geographical field of inquiry. The *Penguin Dictionary* defines geography as the science that describes the earth's surface. This clearly provides plenty of scope for the physical geographer and, as Figure 11.1 shows, a scale of interest that ranges over many orders of magnitude (see Chapter 12 on scale and human geography). A number of related issues have always confronted the geographer regardless of the specific topic of interest. This chapter engages with two of these key questions: at what scale should the inquiry be focused? How can findings at one scale be related to

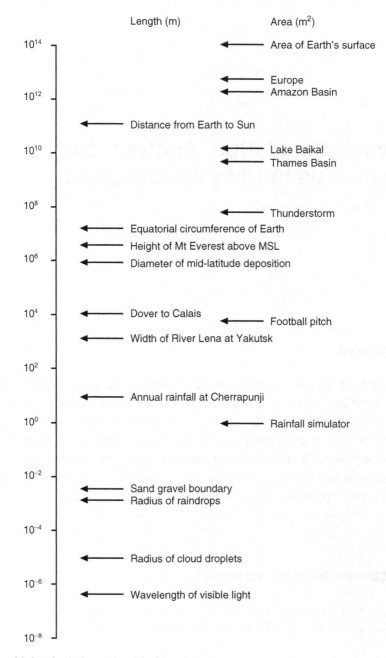

Figure 11.1   Scale in geographical inquiry

another? These two issues are considered in an integrated manner in the main part of the chapter, but initially it is useful to clarify their meanings separately.

The first question relates to the *resolution* of the study. One meaning of resolution is breaking into parts and, in our context, resolution indicates the spatial scale at which we observe individual items of interest and, by inference, the scale below which we do not seek to focus our studies. Thus, a study of the particle size of a soil sample might focus on the fractions of sand, silt and clay, but need not extend down to electrochemical bonding mechanisms at the molecular scale. Resolution is a familiar notion in remote-sensing studies where spatial resolution indicates the greatest resolving power of the instrument in question; for example, on early Landsat images, individual pixels (or picture elements) were approximately 80 m × 80 m and features much smaller than this would not be detectable. Spectral resolution indicates both the discreteness of wavebands measured by the satellite's sensor and its ability to measure radiation intensity levels. You could think of this as the numbers of different colours that the sensor can measure. What should we choose – a few primary colours or a larger number of shades? A simple analogy in grain-size analysis would be the number of sieves used and thus the number of categories of particle size that can be distinguished.

The second question relates to the need to apply the results of one scale of analysis at different scales. This may involve *upscaling* of results from smaller to larger areas – for example, extending results from small catchment studies to large river basins. It may also in some circumstances involve *downscaling* – for example, applying the results of general circulation models (global scale) to particular regions. It has long been known that generalizations made at one level do not necessarily hold at another, and that conclusions derived at one scale may be invalid at another (Haggett, 1965). For example, consider the recolonization of the Indonesian volcano Krakatau after its famous 1883 eruption. At one scale, today's plant distribution on the island reflects the local ecological niches – shoreline, slope, wetland, and other habitats. However, at the larger scale, the general biodiversity is dependent on other controlling factors. How far is the island from the mainland (this controls the likelihood of species migration)? What is the biodiversity of the source area (the new community will obviously mirror the biodiversity of its close neighbours)? Clearly, as we change scale, the questions – and the answers – are different.

There is, of course, a strong relationship between the spatial and temporal scales of study. In general, as spatial scale increases, so does the timescale of interest. Thus, Knighton (1984) suggests the approximate linkage between scales in studies of fluvial landforms, as shown in Table 11.1. The correlation between spatial and temporal scales of analysis does not always hold but, in general terms, short-term studies tend to focus on process dynamics whereas longer-term studies are more likely to involve statistical analysis of form and structure. In geomorphology this shift in scale can be equated with the contrast between functional analysis of dynamic systems as opposed to historical studies of landform evolution. This distinction is less clear today than it used to be because increasingly powerful computers can run sophisticated simulation models that allow the application of detailed process understanding over long timescales and large spatial scales.

There is also a link between scale and causality, as pointed out in a seminal paper by Schumm and Lichty (1965). At the shortest timescale, processes operate within an essentially fixed environment. For example, in a river channel, patterns of water flow are controlled by the channel morphology: form controls

Table 11.1   Approximate linkage between scales in studies of fluvial landforms

	Length scale (m)	Timescale (years)
Bed forms in sandy channels	0.1–10	0.1–1
Cross-sectional form	1–100	1–50
Bed forms in gravel-bed channels	1–100	5–100
Meander wavelength	10–500	10–1,000
Reach gradient	100–1,000	50–5,000
Long profile analysis	1000–100,000	100–10,000

process at this scale and the system is said to be in 'static equilibrium'. However, over the longer term, properties that were fixed at the shorter timescale themselves become variable. Process now controls form and a 'steady-state equilibrium' may be identified. Large events may perturb the system but there is then recovery to a characteristic form. At the longest timescales, even characteristics like the long profile of the river valley will eventually change. A 'dynamic equilibrium' involves progressive evolution of landform as a response to ongoing erosion. This is also the timescale at which major changes in climate can affect landforms at the regional scale – for example, advances and retreats of ice sheets over entire continents. Chorley et al. (1984) provide further discussion of the scale issue in geomorphology, while time is discussed in more detail by Thornes (Chapter 7). Burt (2003) provides examples of the use of computer simulation models to show how landforms evolve over time under the operation of a constant set of processes.

UPSCALING

Chorley (1978) has drawn the distinction between functional and realist approaches in geomorphology. Functional theories involve statistical generalizations whereas a focus on process dynamics implies greater realism of understanding. Two questions arise: can the results of small-scale studies, whether functional or realist, be applied to the larger scale? And to what extent are local studies representative of the wider region or merely unique case studies?

Filtering

In time-series analysis, techniques are applied to remove small-scale noise (rapid fluctuations in system output) in order to emphasize larger-scale patterns such as periodic cycles, trends, major perturbations and threshold changes. Trend-surface analysis has been applied in spatial studies too, generalizing regional patterns from point-scale data. In both cases, the filter passes components of certain frequencies while excluding others. Chorley and Haggett (1965) provide a detailed discussion and many examples of trend-surface mapping. One example included there, and discussed in more detail in Chorley et al. (1966), concerns grain size variations of soils within the Breckland, a distinctive region of sandy deposits in

East Anglia. A nested sampling design was employed to allow variations at different scales to be identified. Systematic sampling at 2 km intervals indicated a coarsening towards the northeast, the presumed source of the wind-blown material. At the smallest scale, a peak in variability at around 8 m spacing showed the effect of sediment sorting under periglacial conditions when classic patterned ground features, polygons and stripes, developed. Residual analysis indicated features at scales of 125–1000 m, possibly related to dune formation during a very dry epoch.

More recent developments in the field of geostatistics include spatial autocorrelation techniques and methods such as kriging, originally developed to map variation in mineral concentrations within ore bodies. Kriging helps identify the scales over which certain processes remain important and how different phenomena show continuity in space. Such methods provide a real improvement on traditional mapping techniques – in the old days, when faced with a scatter of observations in space, one used a mixture of knowledge, judgement and pure guesswork to draw the isolines on a map. Modern geostatistical methods introduce more rigour into such interpolation exercises. A recent example of the use of such techniques in hillslope geomorphology can be found in Burt and Park (1999), who used linear kriging to model the spatial distribution of soil properties across a hillslope. Soil samples were collected from 64 soil pits based mainly on a 25 m sampling grid. Typical results are shown in Figure 11.2. The map for total exchangeable bases (TEB) shows that there is a reasonably uniform distribution across the slope. This is because cations like calcium, magnesium and potassium are important nutrients for plant growth and their distribution reflects a tight soil–vegetation system that minimizes leaching losses. Such nutrients never pass through a soil without biological involvement and they tend therefore to be present in topsoil regardless of location. The map for manganese (Mn) shows a much clearer catenary distribution, with strong leaching in more acid soils upslope and deposition in more neutral footslope soils. This example illustrates the way in which we upscale from point data (soil profile) to the hillslope scale, and additionally shows how the questions asked change subtly as we refocus from vertical processes at the soil profile scale (e.g. leaching, nutrient cycling) to horizontal processes at the hillslope scale (subsurface runoff, solutional denudation).

Sampling at the regional scale

In some cases, sampling has been sufficiently widespread for analyses to be conducted over very wide areas. For example, Hewlett et al. (1977) used data from a large number of stream gauging stations to categorize the stormflow response of rivers in the eastern USA. It is critical for river basin managers to understand the flood response of a river basin following rainfall. The approach differs from trend-surface analysis in that no use is made of regression-type techniques, but both methods allow regional responses to be mapped. The flashiness of a river basin may be judged by calculating the ratio of stormflow volume (Q) to storm rainfall (P). Figure 11.3 maps the mean stormflow response for all rainfall events over 25 mm. For the eastern USA as a whole, the mean response is 0.2 – in other words, typically about 20% of rainfall becomes stormflow. However, the

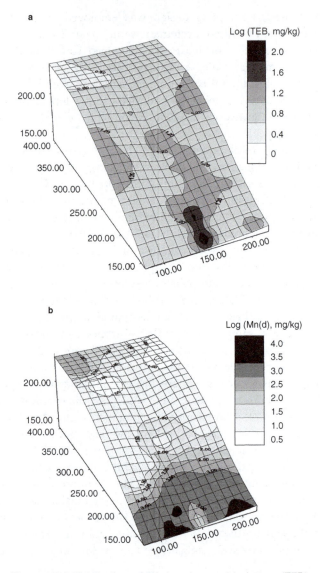

Figure 11.2   The spatial distribution of total exchangeable bases (TEB); and manganese (Mn) in topsoil (0–10 cm) for a hillslope section at Bicknoller Combe, Quantock Hills, England

stormflow response varies widely through space, exceeding 0.4 in the Louisville basin where soils are thin and basin storage low. By contrast, because of deep, permeable, sandy soils, in the Sand Hills of the Piedmont the ratio is as low as 0.04. Note that at the regional scale, stormflow response is strongly associated with geology, whereas more locally (i.e. at the scale of individual fields) spatial variations in infiltration capacity become more important. The approach allows estimates to be made for ungauged basins, an important benefit since estimating runoff response in the absence of stream discharge records remains one of the

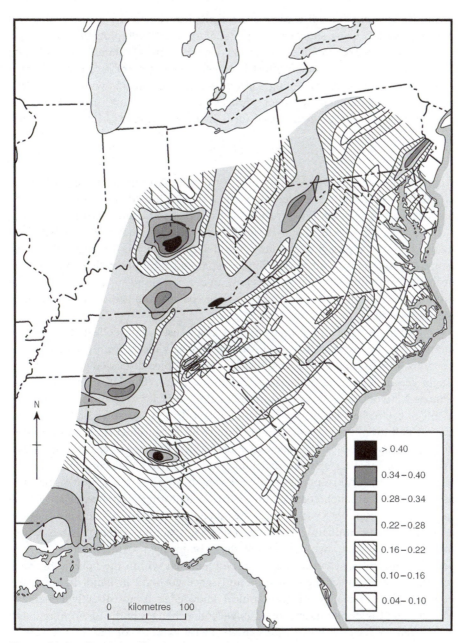

Figure 11.3   Flood runoff response, expressed as the ratio of stormflow to rainfall, for the southeastern USA

Source: Based on Hewlett (1982)

biggest problems in hydrology. It is also a good example of the application of geo-graphical knowledge to a related sphere of interest, namely flood forecasting in hydrology.

We can develop the idea that different variables are dominant at different scales by considering the Universal Soil Loss Equation (USLE), which was developed in the USA to predict soil erosion from agricultural land. By co-ordinating soil erosion research using standardized plots, more than 8,000 plot-years of erosion research data were compiled from 36 locations in 21 states (Mitchell and Bubenzer, 1980). While much effort has been devoted more recently to finding alternative methods of modelling soil loss, the USLE remains widely used. It is essentially a functional approach: process mechanisms are implied but not explicitly modelled. The generalized form of the equation is as follows:

$$A = f(R\ K\ L\ S\ C\ P)$$

where A the soil loss is a function of f, R the rainfall erosivity factor, K the soil erodibility factor, L the slope length factor, S the slope gradient factor, C the crop management factor and P the erosion control practice factor. As with the map of stormflow response, USLE allows soil loss at unknown sites to be predicted. In many ways, the approach is similar to trend-surface analysis, the main difference being the collation of data from widely separated locations rather than the co-ordinated sampling strategy that usually underpins trend-surface analysis. The difficulty of upscaling plot erosion data to the basin scale is discussed further in the next section.

Both the stormflow analysis and USLE used data collected as part of monitoring programmes. One problem with observations from a single site is knowing how to generalize across space. The examples discussed so far relate to single variables such as soil loss or runoff response. The UK's Environmental Change Network (ECN) was established in 1992 in order to co-ordinate measurements of ecological change, including the major driving variables such as climate and air pollution (Burt, 1994). A wide variety of measurements are made within ECN: physical variables like solar radiation and stream chemistry that need measuring regularly, other characteristics like soil properties that vary more slowly, plus many ecological responses from birds and butterflies to frog spawn. Because of their high cost, such complex and varied monitoring programmes are necessarily few in number, but at least by having a network of stations there is a chance of discriminating regional responses from local effects. It is worth mentioning too that, by maintaining the monitoring programme over several decades, subtle long-term changes can be identified within the noisy short-term record. Once again, changes in time and space go hand in hand. In this case, however, the application of geographical knowledge is not directly beneficial. Monitoring provides an indication of unwelcome changes in the environment and only when we have such evidence can we devise ways to combat the problem.

Nested experiments

A common problem in geography is how to generalize the results of small-scale experiments at the larger scale. The Breckland example showed the advantage of detailed spatial sampling but this may not always be possible, for reasons of time or the sheer cost of process studies. Of course, results from a single study – in effect, a sample of one – must always be treated with great caution, and this has led to some uncertainty about the wider value of field experiments.

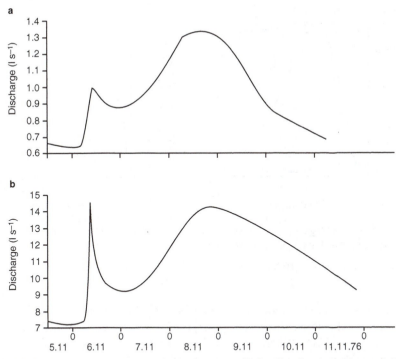

Figure 11.4 Double-peaked storm hydrographs at Bicknoller Combe, Quantock Hills, England, for (a) a hillslope section; and (b) the entire catchment area

To counter this, a number of studies have adopted a 'nested' approach to field experiments – each one designed at a scale to fit inside the next scale of attention. For example, the CHASM (Catchment Hydrology And Sustainable Management) research initiative in the UK has nested mesoscale (100 km²), miniscale (10 km²) and microscale (1 km²) basins in order to bridge the gap between small-scale studies and the relevant scale for integrated river-basin management. Even smaller 'patch' or plot-scale studies will provide the most detailed data for model calibration and process rate measurements. The whole approach recognizes, as its focus, the need to upscale from detailed local-scale models to larger-scale models with lower spatial resolution but greater spatial coverage.

As an example of this sort of approach, Anderson and Burt (1978) presented results from a two-level nested experiment, comparing the hydrological response of a single section of hillslope (3 ha – the same slope as mapped in Figure 11.2) with that of a small first-order catchment (60 ha). This shift of scale is not great enough to lose the dominant control of hillslope processes on the catchment response (Figure 11.4). Even when the shift is larger still, to a 5th order basin (2300 ha), the double-peak hydrographs, characteristic of a delayed throughflow response, remain very obvious (Burt, 1989). It is only in very much larger basins that the detail of the headwater response is lost, subsumed within an aggregate hydrograph that reflects travel time of water through the channel network and the total volume of flood runoff produced, rather than

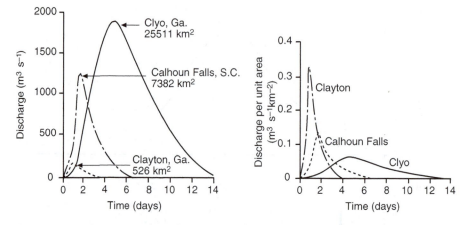

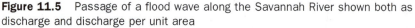

Figure 11.5   Passage of a flood wave along the Savannah River shown both as discharge and discharge per unit area

Source: Based on Hewlett (1982)

the peak discharge in individual tributary basins. This is illustrated in Figure 11.5 (Hewlett, 1982) which shows the passage of a flood wave down the Savannah River, Georgia, USA. When expressed in terms of discharge per unit area the flow at Clayton looks impressive but it is the total discharge at Clyo, further downstream, that is much more likely to be problematic. This is apparent in the discharge plot that shows how the peak discharges upstream are lost in the volumes of water that create the downstream flood.

    When we think about the soil properties shown in Figure 11.2 and tie this in with what we know about the hydrology of the same hillslope (Figure 11.4), we begin to see further links between local processes and patterns at the landscape scale. Ecologists have been much interested recently in 'patch dynamics'. One way of viewing an area is as a mosaic of patches, each patch with its individual assemblage of plants and processes. This is a largely 'vertical' view of ecosystems in that most of the important transfers of energy and matter move up and down, from atmosphere to root zone or vice versa. However, the distribution of patches in space is often orderly rather than random – when we upscale we see that patches are functionally related to one another. A more spatially extensive 'horizontal' approach encourages a structural, pattern-orientated, or geographical approach, emphasizing the spatial distribution of – and interactions among – ecological entities (Burt and Pinay, 2005). Thus, on the hillslope already considered, we see the following toposequence: podzolic soils and heathland vegetation on the interfluve, merging downslope to brown earth soils with grass and bracken – a typical soil catena, in other words (Gerrard, 1981). The important point is that pattern is related to process, with the distribution of soil and vegetation being intimately related to the hydrological and biogeochemical processes operating across the slope. When we add a third dimension, we find that some footslopes are wetter than others because water converges into hillslope hollows (Anderson and Burt, 1978). Soil saturation can allow different processes to occur: soils become anoxic and reduction processes like denitrification take over. In terms of stream-water quality, these footslope wetlands can provide a buffer between the upslope regions (where leaching is another of our vertical processes) and the stream (Cirmo and McDonnell, 1997; Burt and Pinay, 2005). For example, acid waters draining from upslope may become more neutral during their residence within the saturated zone downslope making the stream less vulnerable to acid pollution. The distribution of manganese reflects this, with leaching from acid soils upslope and deposition at the base of the slope where the pH is greater. In the same way, nitrate draining from farmland may be denitrified within a riparian buffer zone, thereby helping to keep river water below the legal limit for nitrate concentration (Haycock et al., 1997; Burt et al., 1999). Given the importance of such abrupt changes in biogeochemical environment, it is important that any field measurement programme reflects the landscape scale, but focuses particularly on the near-stream zone (Burt, 2005). At the landscape scale, it is not so much the individual patch dynamics that are important, more the way in which patches link together (Burt and Pinay, 2005).

Rainfall–runoff modelling

Discussion of the different controls on the flood hydrograph at increasing scale leads neatly into a brief consideration of rainfall–runoff modelling. There has been a great deal of effort devoted to this task over recent decades, aided by a dramatic increase in computer power. As a basic premise, model structure has tended to become more detailed in space and time as the scale of interest *decreases*. At the hillslope scale, models are likely to incorporate equations describing all the important surface and subsurface processes. Such models are also usually fully distributed, dividing up the catchment area into a large number of elements or grid squares and solving the equations for the state variable at each point. As the scale of interest increases, hydrological models are likely to be more conceptual in structure, formulated on the basis of a simple arrangement of a relatively small number of components, each of which is a simplified representation of one process element in the system being modelled. These conceptual models tend also to be 'lumped', treating the catchment as a single unit, with state variables that represent averages over the entire area (Beven, 2000). Empirical or black-box models have also proven reliable for making predictions at the catchment scale despite their lack of theoretical structure. Rather, they rely on establishing a statistical correspondence between input and output using data already collected. By definition, their resolution precludes any interest in within-catchment variation in runoff processes. Beven (2000) provides an extensive and thorough review of rainfall–runoff modelling while several different models are described in Anderson and Burt (1985). Beven concludes that physically-based distributed hydrological models are the state of the art but, despite much effort and great investment in computing resources, such models have often been restricted to research programmes with simpler models continuing to be preferred (both in terms of cost and accuracy) for practical purposes. Moreover, a good deal of uncertainty remains about just how well such models can replicate the real world: having a complex mathematical model is no guarantee of success and a simpler model may well yield more accurate predictions. Of course, hydrology straddles the divide between science and engineering. Geographers have been much involved over the years in the study of process hydrology (including modelling) but from the modelling point of view it is notable that, as scale increases, the spatial unit of interest gets larger and the process description becomes simpler. One of the great challenges in hydrological research therefore is how to upscale the results from the small scale to the large scale. No doubt, as we have already seen, as size of drainage basin increases, the important variables will change too.

Soil erosion and sediment delivery

A major problem arises from the difficulty of upscaling soil erosion plot results to the level of sediment delivery at the drainage-basin scale (Walling, 1983). As noted above, USLE measurements were obtained from standardized erosion plots (22 m long), but the delivery of eroded sediment to the river channel involves consideration of the whole landscape. An isolated plot may provide useful information on erosion itself, but its very isolation precludes a complete appreciation of the erosion problem. Real slopes will behave differently from a 22 m plot: for example, longer flow paths and convergence of water into hollows can both encourage gullies to develop, something ignored in USLE. There will also be deposition of eroded soil, for example, on shallow footslopes or behind hedges; and once in the channel, further storage is possible, in-channel or on floodplains. Typically, only a fraction of the eroded material makes it to the basin outlet – but what fraction exactly? There have been two research groups working on these overlapping but essentially different problems: those studying soil erosion *per se* and those studying sediment delivery systems. Not surprisingly, their conclusions have differed somewhat, especially when it comes to assessing the significance of soil loss. Catchment-scale processes can buffer the erosion process so that the pattern of in-field soil erosion can look very different from the pattern of sediment yield at the basin outlet. This difference between the field and landscape response is the reason why a vigorous debate arose following the publication of Trimble (1999; Table 11.2). His study of the Coon Creek basin, Wisconsin, showed that erosion rates in the period 1975–99 were much less than in the 1930s (as little as 6% if today's rate is compared with the *maximum* rate from the 1920s and 1930s), and he argued that soil loss is no longer a major concern. Moreover, because of lags within the system, sediment yield remained constant throughout the 140-year period of study; in other words, the dramatic soil erosion of the early twentieth century did not produce a signal at the basin outlet. A number of soil erosion experts condemned Trimble for daring to say that the soil erosion threat was less than (they) generally estimated. However, his data certainly do suggest a large reduction in 'upland' (i.e. field) sources of eroded soil. More importantly from our point of view, the complexity of the sediment transfer system shows that a local-scale viewpoint is not enough if we are to extrapolate to larger basins.

Diffuse pollution hydrology

From the examples already given, we can see that there is, not surprisingly, a close analogy between the study of soil erosion and sediment delivery, and the

Table 11.2   A sediment budget for Coon Creek, Wisconsin

	1853–1938	1938–75	1975–93
Sources			
Net upland sheet and rill erosion	326	114	76
Upland gullies	73	64	19
Tributaries	42	35	9
Upper main valley	0	27	13
Sinks			
Upland valleys	38	38	0
Tributary valleys	87	0	25
Upper main valley	71	27	4
Lower main valley	209	139	51
Total sources	441	240	117
Total *soil erosion* sources	399	178	95
Net export to upper main valley	316	175	79
Net export to lower main valley	245	175	88
Sediment yield to Mississippi	38	36	37

Source: Based on Trimble (1999)

study of nutrient leaching and the conveyance of dissolved load to the stream: both rely on a knowledge of runoff processes and hydrological pathways. The study of diffuse nutrient pollution builds on this geographical approach to hydrology. As point sources of water pollution have been increasingly dealt with (e.g. effluent from sewage treatment works), so attention has turned to non-point or diffuse sources. It has long been known that there is a greater risk of pollution from some parts of a catchment than from others (e.g. Burt and Arkell, 1987); for example, a heavily fertilized wheat field is more liable to be a source of nitrate if it is well connected to the river (e.g. on a floodplain) than if it is poorly connected (e.g. located near the interfluve). To ensure the most efficient deployment of any mitigation measures, it is important to focus on those parts of a catchment where most value will be gained from any investment. This implies that analysis must be undertaken at the catchment scale, so that all possible diffuse pollution sources can be judged with respect to one another in terms of their probable relative importance, in order to identify those farms and fields where resources should be focused. Lane et al. (2006) have developed a computer model (SCIMAP – the Sensitive Catchment Integrated Modelling Analysis Platform) which is based on traditional hydrological theory (Lane et al., 2004) but which uses a simple risk-based approach to prioritize critical zones within the river basin. The model identifies both source areas of pollution and the accumulated risk of pollution along flow paths (defined on the basis of catchment topography) from field to channel. It allows identification of those parts of the catchment that, in relative terms, are the most risky, that is most likely to be responsible for an observed problem. In their case, the problem relates to the inability of fish to spawn because of sediment, derived from fields upstream, clogging river gravels. The point is that the approach establishes a link between landscape-scale processes and in-channel conditions, by explicitly considering the transfer of water to the channel network from the catchment slopes. Given that topography controls the slope to channel linkage, then land-use impacts can only be understood with respect to position in the landscape. Moreover, by comparing all sites, this methodology is able to map relative pollution risk, focusing attention on the most hazardous sites. It is interesting that the analysis scales up from a detailed geographical knowledge of hillslope hydrology to a catchment-scale analysis of pollution risk, and then downscales to identify pollution 'hot-spots'. In this way, the example neatly anticipates the next section.

DOWNSCALING

Despite an obvious geographical need to generalize over wide areas, the focus of analysis often moves in the opposite direction. Such reductionism has been the hallmark of science, in some cases quite literally using the microscope to examine underlying structures and processes. Thus, the basis of Anderson and Burt's (1978) study was a detailed examination of soil moisture movement in soil, in order to better understand the way in which delayed throughflow hydrographs were generated. Downscaling has been an important activity in some areas of modelling too, notably in relation to general circulation models (GCMs) of global climate. It has been necessary to develop regional-scale models in order to show how forecast changes in global climate will manifest themselves at, say, the scale of North America or Europe. The resolution of GCMs is of the order of 500×500 km. Processes operating at finer spatial scales, such as cloud formation and precipitation, are 'parameterized', estimated by empirical or conceptually based relationships. Processes at the land surface operate at this smaller scale too and are similarly parameterized (Arnell, 2002). Clearly, treating a large area of land as a single column of homogeneous soil, vegetation and atmosphere with a uniform climate is very unrealistic. Models with improved spatial resolution are therefore used to help translate the results of GCM simulations to the regional scale. One example has been the Amazon basin where the interaction of global climate change and the impact of deforestation may have a significant impact on local rainfall and runoff. Arnell (2002) notes that, although different models vary in their predictions, a complete removal of Amazonian rainforest could reduce evaporation by up to 20% and thereby decrease rainfall by up to 30% across the basin as a whole (Nobre et al., 1991). Field studies reduce the spatial scale of investigation much further, of course, by several orders of magnitude; plot studies in the Amazon basin (Shuttleworth, 1988) have measured interception and evaporation rates to provide a basis for forecasting the effect of forest canopy removal.

ANALYSIS AND SYNTHESIS

There has always been a tension in physical geography between the need to understand the detail of process mechanisms set against the need to understand how broader, complex systems operate. Analysis encourages use of the microscope while synthesis requires the wide-angle lens! For much of the first half of the twentieth century, geography was dominated by regional studies (see Chapters 1 and 26, this volume). At the largest scale, Herbertson (1905) divided the earth into major natural regions, mainly on the basis of climate. Small areas were expected to show the same individuality. Thus, the focus was on areal differentiation, on the variable character of the earth's surface. However, by the 1950s, geographers had become disillusioned with the regional paradigm. There was a major reorientation in the nature of geographical research, involving the strengthening of systematic studies, attempts to develop law, theories and models, and application of mathematical and statistical procedures to facilitate the search for generalizations (Johnston, 1983). One almost inevitable outcome, at least in the short term, was a focus on small-scale process studies (Anderson and Burt, 1990).

In recent years there has been reaction to this focus and physical geographers have raised their sights somewhat, whether in looking once again at large-scale landforms (Sugden et al., 1997) or the impact of climate change and human impact on the world's major biomes (Goudie, 2000). The difference today is that knowledge of process mechanisms is fundamental, providing the basis for large-scale synthesis via modelling studies. However, as we have seen, the important variables change as we increase our scale of interest, and we must learn how to restructure our models as we move to increasingly larger scales, a point discussed by Beven (2000) in relation to hydrological modelling, for example. It may be that scale and complexity go hand in hand but the challenge is to find order and pattern in larger as well as smaller systems. Noble (1999: 297) nicely confronts this issue in relation to the ecological complexity of landscapes: 'We live in landscapes; we manage landscapes. We often describe the environment around us in terms of landscapes. Yet landscapes have long been a scientific blind spot.'

We cannot just add up the individual parts of a landscape to tell us what we have – the whole is, in that sense, greater than the sum of the parts because of functional linkage between different elements of the system. Nevertheless, because complex, large-scale systems do function, we have the opportunity to link pattern and process at that scale. Thus, for physical geographers, whatever their specialist interest, the challenge remains to understand this scale linkage, bridging the gap between studies of process dynamics and the application of such geographical knowledge to larger areas.

CONCLUSION

The scale of interest in physical geography is extremely wide, from molecular to global. All scientists aim to make generalized statements about the phenomena they study and, in so doing, there are choices to be made – what to focus on and what to ignore. For the physical geographer, this choice also involves a decision about the resolution of study – at what scale to focus the investigation. Whatever scale is selected, geographers have always realized that results obtained at one scale are not enough. On the one hand, they will wish to delve beneath their current level of interest to understand more about the process mechanics of the system being studied. And at the same time, they will want to demonstrate the relevance of their work at the larger scale and to apply knowledge gained from one study to speculate about other places.

SUMMARY

- When we approach any topic, we must define the scale at which we want to focus our attention, but we must also consider what other scales may be relevant to our work.

- Upscaling can be achieved in a number of ways: through the use of 'nested' field experiments, by applying appropriate statistical methods and through the use of computer simulation models.

- Downscaling is traditionally concerned with a reduction of our scale of interest in order to learn more about the dynamics of the system being studied. Much work today is focused at the global scale and the results invariably need translating to the regional scale – a rather different type of downscaling. At the smaller scale, process-based models of catchment hydrology allow areas at most risk of generating diffuse pollution to be identified.

- In practice, physical geographers work across a range of scales. As the scale changes, so do the questions being asked and the results obtained. Factors important at one scale may be less relevant at a different scale, so a flexible and perceptive approach is needed.

Further Reading

For those interested in computer modelling, Beven's (2000) *Rainfall–Runoff Modelling* is a purpose-built and up-to-date analysis of hydrological modelling and contains much of interest on the scale issue. Brunsden and Thornes's (1979) **'Landscape sensitivity and change'** is an important paper that introduces the concept of landscape sensitivity and demonstrates how scale and location are important factors in landscape change. Some of their ideas feed through into Burt and Pinay's (2005) discussion of biogeochemical landscape systems. Despite its age, Chorley and Kennedy's (1972) *Physical Geography: A Systems Approach* is still a very relevant and readable introduction to systems analysis in physical geography. Material covering a wide variety of scales and applications is included. *The Geographer's Art* (Haggett, 1990) is not a physical geography text but it is easy reading and very stimulating for any geographer; most of his case studies involve the scale issue. *Fluvial Forms and Processes* (Knighton, 1998) provides an accessible guide to fluvial geomorphology, the different scales of interest involved and how these couple together. Schumm and Lichty's (1965) **'Time, space and causality'** is a classic paper in geomorphology, explaining how causal variables differ with changing scale of interest, in both space and time. Not an easy read but a 'must' nevertheless!

Note: Full details of the above can be found in the references list below.

References

Anderson, M.G. and Burt, T.P. (1978) 'The role of topography in controlling throughflow generation', *Earth Surface Processes*, 3: 331–4.

Anderson, M.G. and Burt, T.P. (1985) 'Modelling strategies', in M.G. Anderson and T.P. Burt (eds) *Hydrological Forecasting*. Chichester: Wiley, pp. 1–13.

Anderson, M.G. and Burt, T.P. (1990) 'Geomorphological techniques. Part one. Introduction', in A.S. Goudie (ed.) *Geomorphological Techniques* (2nd edn). London: Unwin Hyman, pp. 1–29.

Arnell, N. (2002) *Hydrology and Global Environment Change*. London: Prentice Hall.

Beven, K.J. (2000) *Rainfall–Runoff Modelling*. Chichester: Wiley.

Brunsden, D. and Thornes, J.B. (1979) 'Landscape sensitivity and change', *Transactions, Institute of British Geographers*, 4: 463–84.

Burt, T.P. (1989) 'Storm runoff generation in small catchments in relation to the flood response of large basins', in K.J. Beven and P.A. Carling (eds) *Floods*. Chichester: Wiley, pp. 11–36.

Burt, T.P. (1994) 'Long-term study of the natural environment: perceptive science or mindless monitoring?,' *Progress in Physical Geography*, 18: 475–96.

Burt, T.P. (2003) 'Some observations on slope development in South Wales: Savigear and Kirkby revisited', *Progress in Physical Geography*, 27(4): 581–95.

Burt, T.P. (2005) 'A third paradox in catchment hydrology and biogeochemistry – decoupling in the riparian zone', *Hydrological Processes*, 19: 2087–9.

Burt, T.P. and Arkell, B.P. (1987) 'Temporal and spatial patterns of nitrate losses from an agricultural catchment', *Soil Use and Management*, 3: 138–43.

Burt, T.P., Matchett, L.S., Goulding, K.W.T., Webster, C.P. and Haycock, N.E. (1999) 'Denitrification in riparian buffer zones: the role of floodplain sediments', *Hydrological Processes*, 13: 1451–63.

Burt, T.P. and Park S.J. (1999) 'The distribution of solute processes on an acid hillslope and the delivery of solutes to a stream. I. Exchangeable bases', *Earth Surface Processes and Landforms*, 24: 781–97.

Burt, T.P. and Pinay, G. (2005) 'Linking hydrology and biogeochemistry in complex landscapes', *Progress in Physical Geography*, 29(3): 297–316.

Carroll, L. (1872) *Through the Looking Glass and what Alice Found There*. London: Macmillan & Co.

Chorley, R.J. (1978) 'Bases for theory in geomorphology', in C. Embleton et al. (eds) *Geomorphology: Present Problems and Future Prospects*. Oxford: Oxford University Press, pp. 1–13.

Chorley, R.J. and Haggett, P. (1965) 'Trend surface mapping in geographical research', *Transactions, Institute of British Geographers*, 37: 47–67.

Chorley, R.J. and Kennedy, B.A. (1972) *Physical Geography: A Systems Approach*. London: Prentice Hall.

Chorley, R.J., Schumm, S.A. and Sugden, D.E. (1984) *Geomorphology*. London: Methuen.

Chorley, R.J., Stoddart, D.R., Haggett, P. and Slaymaker, H.O. (1966) 'Regional and local components in the areal distribution of surface sand facies in the Breckland, eastern England', *Journal of Sedimentary Petrology*, 36: 209–20.

Cirmo, C.P. and McDonnell, J.J. (1997) 'Linking the hydrologic and biogeochemical controls of nitrogen transport in near-stream zones of temperate forested catchments: a review', *Journal of Hydrology*, 199: 88–120.

Gerrard, A.J. (1981) *Soils and Landforms*. London: George Allen & Unwin.

Goudie, A.S. (2000) *The Human Impact on the Natural Environment* (5th edn). Oxford: Blackwell.

Haggett, P. (1965) *Locational Analysis in Human Geography*. London: Edward Arnold.

Haggett, P. (1990) *The Geographer's Art*. Oxford: Blackwell.

Haycock, N.E., Burt, T.P., Goulding, K.W.T. and Pinay, G. (1997) *Buffer Zones: Their Processes and Potential in Water Protection*. Harpenden: Quest Environmental.

Herbertson, A.J. (1905) 'The major natural regions', *The Geographical Journal*, 25: 300–10.

Hewlett, J.D. (1982) *Principles of Forest Hydrology*. Athens, GA: University of Georgia Press.

Hewlett, J.D., Cunningham, G.B. and Troendle, C.A. (1977) 'Predicting stormflow and peakflow from small basins in humid areas by the R-index method', *Water Resources Bulletin*, 13: 231–53.

Johnston, R.J. (1983) *Geography and Geographers* (2nd edn). London: Edward Arnold.

Knighton, A.D. (1984) *Fluvial Forms and Processes*. London: Edward Arnold.

Knighton, A.D. (1998) *Fluvial Forms and Processes* (2nd edn). London: Edward Arnold.

Lane, S.N., Brookes, C.J., Heathwaite, A.L. and Reaney, S. (2006) 'Surveillant science: challenges for the management of rural environments emerging from the new generation diffuse pollution models', *Journal of Agricultural Economics*, 57(2): 239–57.

Lane. S.N., Brookes, C.J., Kirkby, M.J. and Holden, J. (2004) 'A network index based version of TOPMODEL for use with high resolution digital topographic data', *Hydrological Processes*, 18: 191–201.

Mitchell, J.K. and Bubenzer, G.D. (1980) 'Soil loss estimation', in M.J. Kirkby and R.P.C. Morgan (eds) *Soil Erosion*. Chichester: Wiley, pp. 17–62.

Noble, I.R. (1999) 'Effect of landscape fragmentation, disturbance, and succession on ecosystem function', in J.D. Tenhunen and P. Kabat (eds) *Integrating Hydrology, Ecosystem Dynamics and Biogeochemistry in Complex Landscapes*. Chichester: Wiley, pp. 297–312.

Nobre, C., Sellers, P.J. and Shukla, J. (1991) 'Amazonian deforestation and regional climate change', *Journal of Climatology*, 10: 957–88.

Schumm, S.A. and Lichty, R.W. (1965) 'Time, space and causality', *American Journal of Science*, 263: 110–19.

Shuttleworth, W.J. (1988) 'Evaporation from Amazonian rainforest', *Philosophical Transactions of the Royal Society of London*, B233: 321–46.

Sugden, D.E., Summerfield, M.A. and Burt, T.P. (1997) 'Editorial: linking short-term geomorphic processes to landscape evolution', *Earth Surface Processes and Landforms*, (special issue) 22: 193–4.

Trimble, S.W. (1999) 'Decreased rates of alluvial sediment storage in the Coon Creek Basin, Wisconsin, 1975–1993', *Science*, 285: 1245–7.

Walling, D.E. (1983) 'The sediment delivery problem', *Journal of Hydrology*, 65: 209–37.

12

Scale: The Local and the Global[1]

Andrew Herod

> **D**efinition
>
> Within human geography, scale is typically seen in one of two ways: either as a real, material thing which actually exists and is the result of political struggle and/or social processes, or as a way of framing our understanding of the world.

INTRODUCTION

Many commentators have argued that contemporary economic, political, cultural and social processes, such as that of globalization, are rescaling people's every-day lives across the planet in complex and contradictory ways. Thus, we have seen the creation of supranational political bodies such as the European Union at the same time that we have witnessed the devolution of some political power from member states to regional bodies. Equally, we appear to be witnessing an increased homogenization and 'Americanization' of global culture while, simultaneously, we are seeing the growth of localist tendencies in many parts of the world among those who have sought to defend traditional ways of life. Such examples of an apparent simultaneous globalization and localization of everyday life, together with myriad others like them, have raised important conceptual questions about this rescaling of people's lives and, particularly, about the relationship between what are often taken as the two extremes of our scaled lives, namely the 'global' and the 'local'. For instance, what does it really mean when we say that what started as a 'local' family business has now grown to become a 'global' transnational corporation? What exactly is the relationship between 'global' climate change and 'local' weather patterns? How is a 'global' language, such as English, 'localized' in different parts of the world, so that British English, American English, Australian English, Indian English, Nigerian English, and Singaporean English appear as quite distinct? What does it mean to talk about a war on 'global' terrorism if acts of violence occur in quite 'local' settings – particular streets in New York City, Ramallah, Belfast, or wherever?

In this chapter, then, I explore some issues related to how we use the concepts of the 'global' and the 'local' to make sense of the world around us. Specifically, I discuss three aspects of the local and the global, namely: (1) their ontological status;[2] (2) how the relationship between the global and the local has often been conceptualized within geographic writing; and (3) how the use of different metaphors concerning scales such as the global and the local can shape the ways in which we understand the scaled relationships between different places.

THE ONTOLOGICAL STATUS OF THE LOCAL AND THE GLOBAL

Although scale has long been considered one of geography's core concepts, until the 1980s it had largely been a taken-for-granted concept used for imposing organizational order on the world. Thus, while geographers – both physical and human – had frequently employed scales such as the 'regional' or the 'national' as frames for their research projects, looking at particular issues from a 'regional scale' or a 'national scale', they had spent very little time theorizing the nature of scale itself. However, following the publication of two articles by Peter Taylor (1981, 1982) and of Neil Smith's book *Uneven Development* (1984), the issue of what came to be known as the 'politics of scale' began to be hotly debated within human geography and continues to be so today, particularly as it relates to processes of globalization, localization, (re)regionalization, the hollowing-out of the nation-state, and several other current processes (for an extended discussion of these early works see Herod (2001), esp. pp. 37–46).

Although there were a number of issues at stake, one of the key ones concerned scale's ontological status, particularly whether scale is simply a mental device for categorizing and ordering the world or whether scales really exist as material social products. This debate reflected diverse epistemologies (theories of knowledge) upon which different geographers drew for understanding the world, particularly those epistemologies which were idealist and those which were materialist in their origins.[3] Hence, while geographers such as John Fraser Hart (1982: 21) drew upon Immanuel Kant's idealist philosophy to suggest that scales were no more than handy conceptual mechanisms for ordering the world ('subjective artistic devices ... shaped to fit the hand of the individual user'), others drew upon Marxist ideas of materialism to argue that scales were real social products (that is to say, that scales really exist in the world) and that, as such, there was a politics to their construction – Neil Smith, for instance, contended that the production of scales emerged out of contradictions within capital, showing, by way of example, how restructuring in the US economy during the 1980s was leading to a rescaling and reregionalization of its industrial landscape (Smith and Dennis, 1987; Smith, 1988).[4] In response to criticism that such a first cut at a theory of how scales are produced was somewhat capital-centric, by the early 1990s he had begun to explore how political struggles shape scale-making processes – such as how anti-gentrification activists in New York City sought to 'upscale' their endeavours, so that isolated actions in different neighbourhoods could be united into a citywide movement (Smith, 1989, 1993). Although much of this debate in the 1980s and early 1990s concerned whether economic regions really exist or not, and was tied up with efforts to develop a 'reconstructed regional geography' which focused upon processes of regional formation (for a discussion of this debate, see Pudup, 1988), it has also affected how we conceptualize the 'global' and the 'local'.

For those who draw their inspiration from Kantian idealism, then, the local and the global are seen as part of a pre-existing conceptual matrix of scales within which social life is lived. As such, they are simply mental contrivances for circumscribing and ordering processes and practices so that these may be distinguished and separated from one another as part of a hierarchy of spatial resolutions – a particular process or set of social practices can thus be considered to be 'local' whereas others are considered to be 'global' in scope. Within such an analytical framework, the 'global' is usually defined by the geologically given limits of the Earth, whereas the 'local' is seen as a spatial resolution useful for comprehending processes and practices which occur at geographical ranges smaller than the 'regional' scale, which in turn is seen to be anything which is smaller than the 'national' scale (which, for its part, is seen as being anything smaller than the 'global' scale).

For materialists, on the other hand, the key aspect of geographical scale is to understand that scales are socially produced through processes of struggle and compromise. Hence, for instance, the 'national' scale is not simply a scale which exists in a logical hierarchy between the global and the regional but, instead, is a scale that had to be actively created through economic and political processes which consolidated into larger nation-states the various duchies, principalities and fiefdoms that had been the major political units (at least in Europe) until the Middle Ages. This was a process that was formally recognized by the 1648 Peace of Westphalia (which legally established the notion of the territorial integrity of each nation-state, even those which had been defeated in war, and the primacy of the nation-state over its citizenry) and was not completed in some European countries, such as Germany and Italy, until the late nineteenth century. In contrast to idealists, then, materialists maintain that such scales as the local and the global are actively created through the practices of various social actors – scales do not just exist, waiting to be utilized, but must instead be brought into being. Thus, transnational corporations do not simply adapt their activities to a pre-made global scale defined by the Earth's geologically given limits but must, instead, actively build their own global scale of operation. They must, in effect, *become* 'global'.

The notion of 'becoming' and the focus on the politics of producing scales have been central to materialist arguments concerning the global scale. Thus, there has been much attention paid to how transnational corporations have 'gone global', how institutions of governance have 'become' supranational and how labour unions have sought to 'globalize' their operations to match those of an increasingly 'globalized' capital. In such an approach, the various scales at which social actors and processes operate cannot be conceived of as separate from the actors and processes that create them. In making such an argument, though, some materialists have tended to assume that while the global is actively forged through social practices, the local is somehow a more 'natural', less socially produced 'default' scale (i.e. that all social actors start as inherently 'local' actors and subsequently *become* regional, national and/or global). However, such an approach is problematic, for it privileges the local scale over all others, viewing it as a kind of foundation upon which all other scales are built. In contrast, others have argued that the local, too, is 'produced' and is no more a natural scale than is the global – social actors must work to become local in the same way that they must work to become global (or regional or national). The key issue, then, is to examine *how* social actors make themselves global and/or local, that is to say how they embed themselves locally or extend themselves globally. Thus manufacturers may have to 'localize' themselves by developing business linkages with local suppliers, by training a community's workforce to operate particular types of machinery used in their plant, by establishing credit with a community's banks and other financial institutions, and by building trust with politicians who represent the community in different levels of government. Through such activities, they actively *become* local in much the same way that other manufacturers may become global through establishing business relationships with suppliers, financial institutions, workforces and politicians in communities located throughout the world.

A useful attempt to think about such social practices theoretically has been Kevin Cox's (1998a: 2) distinction between what he calls 'spaces of dependence' ('those more-or-less localized social relations upon which we depend for the realization of essential interests ... for which there are no substitutes elsewhere [and which] define place-specific conditions for our material well being and our sense of significance') and 'spaces of engagement' (the spaces in and through which social actors construct associations with other actors located elsewhere). In making this distinction, Cox suggests that it is important to understand the production of scale as emerging from the interactions that link one particular actor's local spaces of dependence with those of other actors located at some distance as part of a strategy of engagement with them. Certainly, some writers (e.g. Jones, 1998; Judd, 1998) have questioned what exactly Cox means by terms such as 'localized social relations', while others (e.g. Herod, 2001) have suggested that 'spaces of dependence' may not necessarily be only 'local' (a transnational corporation, for instance, may be dependent upon resources found in several different places across the globe, such that its space of dependence may be argued to be 'global'). Nevertheless, Cox's approach does allow us to think about how different social actors may be dependent upon particular spaces yet may seek to engage with other social actors operating within their own, quite different, spaces of dependence. For Cox, then, the production of scale is conceptualized as emerging out of the ways in which actors build spaces of engagement to link the various spaces upon which they, or those with whom they must deal, are dependent, such that moving from the local to the global scale 'is not a movement from one discrete arena to another' (Cox, 1998a: 20), but a process of developing networks of associations that allow actors to shift between various spaces of engagement. Put another way, scales are conceived of in terms of a process rather than fixed entities (i.e. the global and the local are viewed not as static 'arenas' within which social life plays out but as constantly being made and remade by social actions).

In reviewing this growing body of theorizing about scale in the 1990s, Sallie Marston (2000) suggested that there was, however, a significant lacuna – most of the theorizing had focused upon issues of how scales are forged out of the struggles centred on capitalist production and had largely ignored issues of social reproduction and consumption. While she accepted that for Marxists, many of whom had led scalar theorizing in the 1980s and 1990s, social reproduction is implicated in relations of capitalist production – that is to say, capitalist commodity production and the accumulation of capital is also about reproducing capitalist social relations – Marston argued that such theorizing had failed to explore how non-capitalist social relations (specifically, patriarchy) had shaped the construction of one scale in particular, namely that of the household. By way of a corrective, she explored how nineteenth-century middle-class homes were constituted as particular spaces and subsequently 'utilized as a scale of social and political identity formation that eventually enabled American middle-class urban women to extend their influence beyond the home to other scales of social life, enabling them to influence issues of production, social reproduction and consumption in the process' (Marston, 2000: 235).

In an effort to address what he saw as a 'noticeable slippage in the literature between notions of geographical scale and other core geographical concepts, such as place, locality, territory and space' that had occurred since geographers first began to evaluate critically scale's ontological status, Neil Brenner (2001: 592) argued that much of the writing about what Smith (1984) had called the 'politics of scale' had tended to use the term 'scale' too broadly. Taking aim specifically at Marston's (2000) call for the scale of the household to be considered more centrally within the broader scalar lexicon, Brenner suggested that geographers were diluting the analytical power of the concept of the 'politics of scale' by using it to refer to things that could just as easily be deemed to be the 'politics of *place*' or the 'politics of *territory*' (e.g. political struggles taking place within particular self-enclosed spatial units, such as a household or a territory). To maintain its critical power, then, Brenner contended that the term 'politics of scale' should be applied in a more circumspect manner to refer specifically to 'the production, reconfiguration or contestation of particular differentiations, orderings and hierarchies *among* geographical scales', with geographical scale being understood primarily as a 'modality of *hierarchization* and *rehierarchization* through which processes of sociospatial differentiation unfold both materially and discursively' (Brenner, 2001: 600). For Brenner, in other words, lest its overuse dull its analytical edge, the concept of the 'politics of scale' should really be restricted to exploring processes of spatial scaling, while the longstanding geographical lexicon of 'locality', 'place', 'territoriality' and so forth could continue to be used to refer to the politics of sociospatial organization *within* relatively bounded geographical arenas.[5] Although Marston and Smith (2001)

responded by suggesting that what they saw as Brenner's effort to separate conceptually the notion of the 'production of space' from that of the 'production of scale'
made sense up to a point, and that the two should not be conflated, they argued
that it was impossible to separate entirely these two concepts, since the 'production of scale is integral to the production of space' (Marston and Smith, 2001: 616).

Whereas this debate within neo-Marxist human geography over terminology
and ontology significantly shaped conceptions of scale – and hence of the local
and the global – one of the most provocative recent interventions has been by
Marston et al. (2005: 422), in which they argue for 'expurgat[ing] scale from the geographic vocabulary' and call for the abandonment of notions of scale, which they
see as problematic – principally because, they maintain, such notions privilege
views of the world which see it as hierarchical, a perspective that tends to promote
one scale (often 'the global') over others. While they accept that scale may exist as
an epistemological ordering frame – naming something 'the national' or 'the global'
can dramatically shape how it is viewed – they contend that this is different from
claiming the landscape is organized into various nested hierarchies (i.e. 'scales'). In
place of approaches which assert that landscapes are materially structured into various hierarchies, Marston et al. make the case for what they call a 'flat ontology', distinguishing this from a 'horizontal ontology', which they see as merely replacing the
'up-down vertical imaginary' with a 'radiating (out from here) spatiality' (2005: 422).
Commentators have responded to this position in a number of ways – from claiming that Marston et al.'s formulation mirrors the propositions of actor-
network theory (Collinge, 2006) to averring they have misunderstood the distinction
between ontology and epistemology (Hoefle, 2006) and that theirs is little more than
a repackaged Kantianism, although unlike Hart (1982) – who accepts a nested
framework of scales ('the regional', 'the national', etc.), albeit one imposed by our
brains on the world to give it order – Marston et al. (2005: 420) argue against the
very notion that the world is hierarchically constituted 'as a nesting of "legal, juridi-
cal and organizational structures"'. For them, then, 'scale' is simply a representational trope, though one that may have material effects.

Although these more recent writings appear to have resurrected the
debate about scale's ontological status – does it really 'exist' or not? – after almost
two decades in which the Marxist materialist view had held sway in critical
human geography, all sides do agree that scalar language can be used as a powerful tool to frame political struggles. Kurtz's (2002) analysis of an environmental
dispute over the siting of a petrochemicals factory in Louisiana has been one
of the clearest exemplars of how discourses of scale can be used to 'frame'
particular social conflicts so as to make legitimate or illegitimate certain actors'
participation. Specifically, she shows how various sets of interests attempted to
characterize the dispute in different ways so as to make it appropriate or inappropriate for actors who were organized at different scales to participate in decision-
making processes – by arguing that the decision over whether to grant a permit
for the plant should be made by state economic development officials, for
instance, such officials attempted to exclude local parish-level officials and interest groups from being involved, while parish-level groups invoked both localism
and the national scale (by appealing to federal regulators) in their efforts to stop
the permit being granted.

Having considered, then, a number of debates concerning scale's ontological status, in the next section I want to highlight how discourses pertaining to scale (especially the global and the local) have been an important aspect of human geography during recent years.[6] Regardless of how scale is thought of ontologically, it is important to understand that the ways in which the global and the local – and especially the relationship between them – are presented rhetorically can fundamentally shape how we conceptualize the world and its social processes.

DISCOURSES OF THE GLOBAL AND THE LOCAL

As alluded to above, within debates over processes of globalization the local and the global have frequently been thought of as the two ends of the scalar spectrum, with the local being understood through its contrast with, and status as 'Other' to, the global, and vice versa.[7] Within this binary way of thinking, the global and the local have often been associated, respectively, with other sets of binaries – for instance, there exists a correspondence in much Western thinking between the global and the abstract, and between the local and the concrete, such that global activities are often perceived as somehow more abstract and less concrete than are local ones.

Within such binary thinking, Gibson-Graham (2002) has identified at least six ways in which the relationship between the local and the global is often viewed.[8] These six ways are:

1 The global and the local are seen not as things in and of themselves, but are viewed instead as interpretative frames for analysing situations. For example, when considering processes of economic restructuring, what is seen from a 'global perspective' (perhaps a worldwide economic slowdown) may appear different from what is seen from a 'local perspective' in particular places (some places may actually be experiencing economic expansion during such a global economic slowdown).

2 The global and the local each derive meaning from what they are not. Much like our conception of what a slave is only makes sense if we can contrast that with a conception of what constitutes a free person (and vice versa), the global and the local make sense only when contrasted with each other. Thus, drawing on Dirlik (1999: 4), Gibson-Graham suggests that in such a representation the global is 'something more than the national or regional ... anything other than the local'. In turn, the local is seen as the opposite of the global. This view, however, represents an important semantic shift, because whereas once the local 'derived its meaning from its contradiction to the national', now anything other than the global is often seen to be 'local' – for instance, in much rhetoric about globalization, nations and even entire extra-national regions (such as the European Union) are frequently referred to as 'local actors' within the process of globalization.

3 Whereas many writers have viewed spatial scales in terms of a hierarchy of fixed and separate arenas within which social life occurs (what could be called the 'process X occurs within global space but process Y occurs within local

spaces' approach), some, such as French social theorist Bruno Latour, have
tended to see the world as constituted through a series of networks which link
different places. In such an approach, both the local and the global 'offer
points of view on networks that are by nature neither local or global, but are
more or less long and more or less connected' (Latour, 1993: 122). Thus, much
like it would be impossible to describe a spider's web in hierarchical spatial
fashion – where does one part of the web end and another begin? – in a view
such as Latour's it is impossible to distinguish where the local ends and the
global (or other scales) begins. Instead, such a view sees the global and local
as simply different 'takes' on the same universe of networks, connections,
abstractness and concreteness. The global and the local are not so much opposite ends of a scalar spectrum, but are a terminology for contrasting shorter
and less-connected networks with longer and more-connected networks.

4   The global *is* local. In this perspective, Gibson-Graham argues, the global does
    not really exist, and if you scratch anything 'global' you will find locality. In
    such a view multinational firms, for instance, are actually 'multilocal' rather
    than 'global'.

5   The local *is* global, and place is a 'particular moment' in spatialized networks
    of social relations. Much like the tips of an octopus's legs might touch par-
    ticular locations on the seafloor while the body of the octopus floats above it,
    in such an approach the local appears as the location where glocal forces
    'touch down' on to the Earth's surface. In turn, the local is not a place but is
    an entry point to the world of global flows which encircle the planet.

6   The global and local are not locations but processes. Put another way, global-
    ization and localization produce all spaces as hybrids, as 'glocal' sites of both
    differentiation and integration (Dirlik, 1999: 20).[9] Thus, the local and the
    global are not fixed entities but are always in the process of being remade.
    Hence, local initiatives can be broadcast to the world and adopted in mul-
    tiple places across space, while global processes always involve localization
    (for instance, in the process of globalizing itself McDonald's tailors its prod-
    ucts to particular local tastes, serving beer in France, pineapple fritters in
    Hawaii and vegetarian 'hamburgers' in India).

In reviewing these different ways of articulating the relationship between the local
and the global, Gibson-Graham (2002) suggests that the history of this binary has
been one in which the power of the global has usually been assumed to be greater
than that of the local. Part of this results from the widely held view in Western
thought that greater size and extensiveness imply domination and superior power,
such that the local is often represented as 'small and relatively powerless, defined
and confined by the global' (Gibson-Graham, 2002: 27). In such a representation
'the global is a force, the local is its field of play ... the global is penetrating, the
local penetrated and transformed'. Thus, the global is conceived of as 'synonymous
with abstract space, the frictionless movement of money and commodities, the
expansiveness and inventiveness of capitalism and the market. But its Other, localism, is coded as place, community, defensiveness, bounded identity, *in situ* labor,
noncapitalism, the traditional'. Hence, Gibson-Graham (2002: 33) argues, what
emerges from the above review 'is an overriding sense that [power] is either

already distributed and possessed or able to be mobilized more successfully by "the global"' than by the local.

Paradoxically, such a vaunting of the power of the global has frequently been engaged in both by neoliberals and by many Marxists.[10] Thus neoliberals frequently use the rhetoric of globalization to undermine local opposition to their agenda (as in the myriad claims that national welfare states and local union work rules 'must' be given up 'due to the imperatives of global capitalism'), whereas many on the political left appear to have given up any hope of challenging capitalism globally and prefer instead to 'think globally' but to 'act locally' (for more on this argument, see Herod, 2001: 128). However, Gibson-Graham argues that such a discursive reification of the global and diminution of the power of the local restricts the possibilities for progressive political action, because it suggests that capital (which is usually seen as being global in its operation) can always outmanoeuvre its opponents (who, as Other to capital, consequently are usually viewed, by implication, as non-global or local in their operation).

Instead, Gibson-Graham has sought to deconstruct the local–global binary to suggest ways in which the local can be thought of as enabling political struggle and as a powerful basis for challenging (global) capital (in this vein see Herod, 2000, for an example of how a strike in a single community brought to a virtual standstill in North America the operations of global automobile giant General Motors). Relatedly, others (e.g. Wills, 1998; Herod, 2001) have shown, through examination of international trade union activities, that it is empirically incorrect to assume that workers and others opposed to global capital have not themselves been able to act globally, thus shaping the political and economic geographies of 'the global'.[11] Put another way, critics of the view which holds that the local is inherently weaker than the global as a scale of political action and that the global is a scale capable of construction only by capital have argued both that organizing locally can indeed be effective in particular circumstances and that we should not think of the global as a scale at which only capital can operate effectively. Nevertheless, the ability of neoliberal ideologues successfully to represent discursively the global as more powerful than the local can secure them significant advantages, for it may encourage their opponents to believe that organizing locally is doomed to failure. Likewise, corporate executives' success in representing the global as capital's domain may encourage labour unions and others not to attempt to organize globally because they may feel they have no hope of succeeding on 'capital's terrain'. At the same time, workers' viewing of a transnational corporation as 'multilocational' rather than as 'global' can significantly shape their political calculations – taking on a corporation that is conceived of as 'multilocational' is a different ballgame from taking on a 'global' entity. Clearly, then, how the terms 'local' and 'global' are deployed discursively can be very important for the politics of political struggle.

METAPHORS OF SCALE

Following from the above, it is clear that within human geography during the past decade or so a great deal of attention has been paid to the politics of discourse concerning scale, that is to say how the local and the global (and other scales) are talked about and represented. Centred upon the work of writers such as Jacques Derrida, Michel Foucault, Bruno Latour and others, much of this 'postpositivist' concern with language has focused upon the metaphors used to describe and make sense of the world, for metaphors can be powerful shapers of how we understand things.[12] When thinking about the use of metaphor, however, it is important to recognize that the choice of one metaphor over another is usually not made on the basis of which is empirically a 'more accurate' representation of something but, rather, on the basis of how someone is attempting to under-stand a particular phenomenon. Thus, for example, the Victorians used the metaphor of the steam engine to describe the functioning of the human body (muscles were described as the body's pistons, food as its fuel, lungs as the body's boilers, etc.) because that was a technology with which they were very familiar. Today, on the other hand, the body is often described using metaphors from the world of computing (the brain is seen as the 'central processing unit', for example). In thinking about the metaphors used to describe the rescaling of contemporary social life, then, it is important to realize that changing the metaphors we use to describe the world does not change the way the world actu-ally is, but it does change the ways in which we engage with the world. Hence, no one would (I hope) suggest that switching from thinking of the body in lan-guage redolent of the age of the steam engine to thinking of it using the language of computers actually changes the way bodies function, but such a switch in lan-guage does clearly alter fundamentally the ways in which we think about how they work.

In the case of metaphors used within geography to describe the relation-ship between the global and the local, several have been fairly commonplace, and each has different sets of implications associated with it. One of the most fre-quently used has been that of scale as a hierarchical ladder, with the implication being that one 'climbs' up the scalar hierarchy from local to regional to national to global, or down it from global to national to regional to local. In such a metaphor, the various scales are considered to be like the rungs on the ladder and there is a strict progression between them (see Figure 12.1). In using such a metaphor, the global – as the highest rung on the ladder – is seen to be 'above' the local and all other scales. At the same time, each scale is seen to be distinct from every other scale. Clearly, then, the use of such a metaphor leaves us with a particular way of conceptualizing the scalar relationships between places. However, if we choose a different popular metaphor to conceptualize scalar relationships – that of scales represented as a series of ever-larger concentric circles – then we get an understanding which is in some ways similar to, and in some ways quite different from, that derived from the ladder metaphor (see Figure 12.2). Thus, in this second metaphor the local is conceived of as a relatively small circle, with the regional as a larger circle encompassing it, while the national and the global scales are still larger circles encompassing the local and the regional. In some ways, this second metaphor has similarities to the first – scales are still seen as being quite separate entities, for instance (e.g. the circle representing the local is quite distinct from that representing the global). Yet, there are also some distinct differences. Whereas in the ladder metaphor the global was seen as being

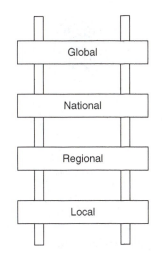

Figure 12.1   Scale as a ladder

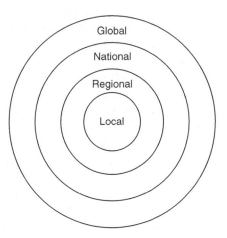

Figure 12.2   Scale as concentric circles

'above' other scales, this is not the case with the circle metaphor. Instead, the global is seen to encompass all other scales but is not necessarily seen as being 'above' them. In these metaphors, then, we have two varied understandings of the relationship between the global and the local.

Of course, the metaphor of the ladder and of concentric circles are not the only metaphors we could use to think about the relationship between the local and the global. Frequently, scales are also talked about as being part of a 'nested hierarchy' which can be thought of in terms similar to that of Russian Matryoshka ('nesting') dolls (see Figure 12.3). In such a representation, each doll (i.e. each scale) is separate and distinct and can be considered on its own. However, the piece as a whole is complete and can be comprehended in its totality only with each doll/scale sitting inside the one that is immediately bigger than itself, such that the dolls and scales fit together in one and only one way (that is to say, a larger doll/scale simply will not fit inside a smaller one). If we

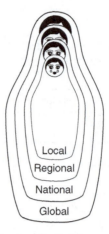

Figure 12.3   Scale as Matryoshka (nesting) dolls

think about scale in this way, then we have a situation in which there is no scale which is 'above' any other in the vertical sense that the ladder metaphor suggests. Likewise, the global scale (the outside doll) is viewed as being 'larger' than all the other 'smaller' scales, such that the global can contain other scales but this does not work the other way round (i.e. the local cannot contain the global). On the other hand, the Matryoshka doll metaphor implies much more forcefully than does either the circle or the ladder metaphor that there is a *nested* hierarchy of scales, with each scale fitting neatly together to provide a coherent whole. Significantly, then, if the ladder metaphor gives us a language of 'above/below' and the circle metaphor gives us one of 'larger/smaller' and 'encompassing/encompassed', the Matryoshka metaphor gives us a language of 'enveloping/enveloped' or, perhaps, 'containing/contained'.

Still another metaphor for thinking about scale, one popularized by French social theorist Bruno Latour, is that of a world of places which are 'networked' together. Thus, he argues (1996: 370), the world's complexity cannot be captured by 'notions of levels, layers, territories, [and] spheres', and should not be thought of as being made up of discreet levels (i.e. scales) of bounded spaces which fit together neatly. Rather than portraying scales as capable of somehow being stacked one above the other (as in the ladder metaphor), placed within one another (as in the circle metaphor), or fitted together like Matryoshka dolls, Latour maintains that we need to think about the world as being 'fibrous, thread-like, wiry, stringy, ropy, [and] capillary'. Clearly, using such a metaphor gives us yet another way of thinking about the scaled relationships between places – a topological rather than topographical one (Castree et al., 2007). Thus, drawing upon Latour's imagery, we might think of scale as more akin to a set of earthworm burrows or tree roots which are intertwined through different strata of soil (see Figures 12.4 and 12.5). Such metaphors leave us with an image of scale in which the global and local, together with other scales, are not separate from one another but are connected together in a single whole. Moreover, although it is possible to recognize different scales (much like it is possible to think of earthworm burrows or tree roots penetrating different strata of the soil, with some

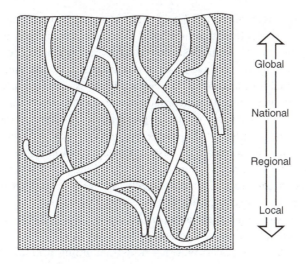

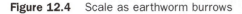

Figure 12.4   Scale as earthworm burrows

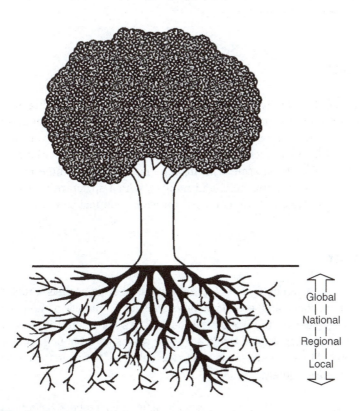

Figure 12.5   Scale as tree roots

going deeper than others), it is difficult to determine exactly where one scale ends and another begins and it is not necessarily possible to think in hierarchical terms about scale.[13] This is a quite different way, then, of conceptualizing scale than in the other metaphors discussed above. Hence, if we adopt Latour's representation, then we are no longer really talking of scale in terms of bounded spaces or the metrics of the clearly defined hierarchies of Euclidean space that were at the heart of the ladder, circles and Matryoska doll metaphors.[14] Such a different metaphor dramatically changes the way we think about scale and the relationship between the global and the local, leading us to question what exactly it would mean to talk of a 'larger' scale if we use the metaphor of earthworm burrows, or whether it makes sense to talk any more of the global scale as 'encompassing' the local if both are seen to be 'located' at the respective ends of such burrows.

Again, the significance of using different metaphors to talk about the scaled relationships between places is not to suggest that these different metaphors necessarily represent empirically different situations or that one is necessarily a better representation of the world and all its complexities than is another. Rather, such an appreciation of metaphor is important because it suggests that how we talk about scale impacts the ways in which we engage socially and politically with our scaled world – notwithstanding Marston et al.'s (2005) assertion that the world is, in scalar terms, fundamentally flat – and that this, in turn, may impact how we conduct our social, economic and political praxis and so make geographical landscapes.

CONCLUSION

In sum, there are clearly a number of important issues concerning how we think about and represent geographical scales in general, and the local and the global in particular. These issues go not only to the heart of what scale is ontologically, but also highlight the fact that how we conceptualize the relationship between the local and the global scale will, in large measure, determine how we understand the processes, both social and natural, which structure human and physical landscapes. This, in turn, can have immense political consequences.

SUMMARY

- There has been a debate in geography about whether scale is a real thing made through political and economic processes or is merely a mental device for imposing order on the world. This debate has considered scale in both topographical and topological terms – that is, scales as areal units and as parts of networks. More recently some have even called for the abandonment of the concept of scale entirely.

- There are at least six different ways in which geographers have thought about the relationship between the global and the local:

  1   The global and the local are not actually things but ways of framing situations.

  2   The global and the local each derive meaning from what they are not.

3   The global and the local simply offer different points of view on social networks.

4   The global *is* local; scratch anything global and you find locality. For instance, multinational firms are actually multilocal rather than global.

5   The local *is* global; the local is only where global processes 'touch down' on the Earth's surface.

6   All spaces are hybrids of the global and the local; they are glocal.

- Typically, in Western thought, the global has been thought of as more powerful and active than the local; the local is seen as small and relatively powerless. However, the local can serve as a powerful scale of political organization; the global is not a scale just controlled by capital, but those who challenge capital can also organize globally.

- At least five different metaphors have been popular in representing how the world is scaled. These are: a ladder; concentric circles; Matryoshka (nesting) dolls; earthworm burrows; and tree roots.

Further Reading

Taylor (1981) is the first article really to consider a political economy of scale. This was quickly followed by Smith's (1984) **Uneven Development**, which argues that scales are produced out of the dynamics of capitalist accumulation, and Smith and Dennis (1987), which shows how the traditional industrial core of the USA was rescaled by economic restructuring during the 1980s. In something of a corrective to his earlier capital-centric view of scale production, Smith (1992) examines how scale is produced out of political conflict between different social actors. With Smith contended, among other things, that much of the postmodern theorizing in the social sciences during the late 1980s/early 1990s had focused upon issues of difference, using a spatialized language of 'subject positionality', 'location', etc., yet had failed to think seriously about the production of material spaces and scales. Jonas (1994) is an early article which argues for understanding space as deeply scaled and for considering how the language of scale is an important aspect of subject identity. Swyngedouw (1997) explores issues of scale and power, particularly with regard to the globalization of the economy and of the nation-state, while Brenner (1998) suggests that scale emerges out of the contradictions between fixity and mobility in the circulation of capital and that this tension has shaped the politics of territorial organization since the late nineteenth century.

(Continued)

(Continued)

One of the first collections of papers to deal explicitly with the scale question can be found in a special issue of *Political Geography* (1997), Volume 16, issue 2. The papers focus upon a number of different empirical examples, including immigration politics in the European Union, labour union politics in the US shipping industry, Italian electoral politics and the politics of the US anti-nuclear movement. Cox (1998a) lays out a conceptual argument for understanding politics as concerned with how social actors negotiate the connections between their various spaces of dependence, in the process producing spaces of engagement. His edited book, **Spaces of Globalization** (1998b), contains nine chapters which examine how the local as a scale of political and economic organization is still important under processes of globalization. The chapters also challenge a number of popularly held views about globalization, particularly that it is solely capital which has engaged in processes of globalization. Marston (2000) argues that much of the writing during the 1980s and 1990s concerning the politics of scale focused almost exclusively on what Marxists would call the 'sphere of production'. Marston argues for a broader consideration of the politics of scale in what Marxists would call the 'sphere of consumption' and the 'sphere of social reproduction'. Herod's (2001) **Labor Geographies** provides an overview of neo-Marxist efforts to theorize the production of scale and focuses upon workers' efforts to construct scales of geographical organization as a way to secure their political and economic goals. Gibson-Graham (2002) presents six prevalent ways of thinking about the global and the local; they offer both intellectual and political strategies for undermining the presumed power of the global and liberating the transformative potential of the local. Herod and Wright's (2002a) **Geographies of Power: Placing Scale** contains 11 essays which explore three aspects of the politics of scale, namely: theorizing scale, rhetorics of scale and scalar praxis. Marston et al. (2005) calls for abandonment of the concept of scale.

Note: Full details of the above can be found in the references list below.

NOTES

1　Parts of this chapter draw upon Herod and Wright (2002b).
2　Ontology refers to what exists and the nature of being. It is a term often used in conjunction with 'epistemology', which refers to theories of knowledge (e.g. positivism, realism, Marxism, feminism) that allow us to make sense of the world.
3　Idealism and materialism are two of the most significant epistemologies to have shaped Western thought during the past 200 years. Exemplified by the writings of Immanuel Kant, idealism assumes that any pattern we see in the world is a result of how our mind categorizes and orders spaces. Indeed, Kant viewed space and time as simply mental frameworks for structuring existence, rather than

observable realities. Materialism, on the other hand, is exemplified by the writings of people such as Karl Marx, who argued that there was indeed a real, material order to the world, one which was imposed by the workings of economic and political processes (such as those under capitalism).

4 Smith (1984) argued that there were two contradictory tendencies within capital, those towards spatial differentiation and those towards spatial equalization. Thus, he argued that capital had to be sufficiently embedded in space so that accumulation could occur (a process which led to the differentiation of the economic landscape as some places received more investment than others) yet also remain sufficiently mobile that it could take advantage of opportunities for investment that opened up elsewhere (a process which led capital to try to equalize the rate of profit across the landscape). The geographical resolution of these opposing tendencies was seen in the landscape through the production of scales at various spatial resolutions (the urban scale, the regional scale, the national scale and the global scale).

5 Brenner's argument parallels the argument by some that if everything is considered 'political' then the notion of 'politics' loses its diagnostic bite.

6 The term 'discourse' relates to how particular issues are represented through language.

7 One could just as easily argue for other scalar extremes, such as the body on the one hand, and the extra-global on the other, given that our planet is now surrounded by myriad artificial satellites and that a geopolitics of space is being constructed as some nations lay claim to the resources of the Earth's upper atmosphere and beyond. However, it is the local and the global that appear to have caught the scalar imagination, at least with regard to understanding contemporary processes of economic restructuring.

8 The following draws extensively on Gibson-Graham's (2002) argument. For ease in reading, I have not included quotation marks except for quotations of authors cited directly by Gibson-Graham.

9 'Glocal' is a combination of the two terms 'global' and 'local'. It expresses the tension between the two scalar extremes, such as when social actors, like transnational corporations, have a global outlook but must tailor their products or practices to local situations and conditions.

10 Neoliberalism is the ideology of free trade. It is the belief that government should intervene in the market as little as possible.

11 The significance of discourse is readily revealed in such a statement. Thus, talk of how workers may be able to 'act globally' and of political and economic geographies of 'the global' assume that 'the global' is a separate and distinct scale that can be contrasted with others, such as 'the local'. If, on the other hand, we were to see 'the global' as Gibson-Graham have expressed it in point 4 (p. 224), then such talk would conceive of workers as engaging in 'multilocal' conflicts against firms which are themselves 'multilocally' organized.

12 Positivism was a popular epistemology within human geography in the 1950s and 1960s. It argued that the world was knowable in all its facets and that the application of 'science' would reveal this world to us. Positivists believe that the world can be understood 'objectively'. Postpositivism, on the other hand, suggests that it is never possible to know the world in its entirety and that our understandings of things will always be partial. Furthermore, many postpositivists argue that there are multiple truths to the world, and that different representations of the world are, to all intents and purposes, different truths about that world. Consequently, they argue that language is an important arena of political struggle, for how we represent things will shape how we understand them.

13 In Figures 12.4 and 12.5 the 'global' has been represented as being closest to the surface of the Earth, whereas the 'local' scale is seen as being deeper in the

soil. Such a view could be taken to indicate that the 'local' scale is somehow more embedded and runs deeper in people's everyday lives than do the other scales. However, both of these figures could just as easily have been drawn to indicate the 'global' scale stretching deepest into the soil, with the 'local' scale closest to the surface of the Earth (i.e. with the arrows to the side of each diagram flipped the other way up). In such a situation, it could be argued that the deeper 'global' scale represents a more extensive spread of worm burrows or tree roots (which can be taken as surrogates for social and political relations) whereas where the burrows or roots come together and break the surface of the ground could represent the 'local', the point where global social relationships and processes become visible to the eye. Clearly, the metaphor of the worm burrows or of the tree roots can, in fact, express two quite different situations, neither of which should be taken to be more 'accurate' in some absolute empirical sense. Such is the power of metaphor.

14   The term 'Euclidean space' refers to the belief that space is absolute and that it can be carved up into smaller and smaller parts of separate, self-contained units. This is in contrast to views which see space as relational, that is to say space as produced by social processes which link different actors and which, therefore, cannot be carved up neatly into infinitely divisible units.

References

Brenner, N. (1998) 'Between fixity and motion: accumulation, territorial organization and the historical geography of spatial scales', *Environment and Planning D: Society and Space*, 16: 459–81.

Brenner, N. (2001) 'The limits to scale? Methodological reflections on scalar structuration', *Progress in Human Geography*, 25: 591–614.

Castree, N., Featherstone, D. and Herod, A. (2007) 'Contrapuntal geographies: the politics of organising across socio-spatial difference', in K. Cox, M. Low and J. Robinson (eds) *Handbook of Political Geography*. London: Sage. pp. 305–321.

Collinge, C. (2006) 'Flat ontology and the deconstruction of scale: a response to Marston, Jones and Woodward', *Transactions, Institute of British Geographers*, NS 31: 244–51.

Cox, K.R. (1998a) 'Spaces of dependence, spaces of engagement and the politics of scale, or: looking for local politics', *Political Geography*, 17: 1–23.

Cox, K.R. (ed.) (1998b) *Spaces of Globalization: Reasserting the Power of the Local*. New York: Guilford Press.

Dirlik, A. (1999) 'Place-based imagination: globalism and the politics of place'. Unpublished manuscript, Department of History, Duke University, Durham, NC.

Gibson-Graham, J.K. (2002) 'Beyond global vs. local: economic politics outside the binary frame', in A. Herod and M.W. Wright (eds) *Geographies of Power: Placing Scale*. Oxford: Blackwell, pp. 25–60.

Hart, J.F. (1982) 'The highest form of the geographer's art', *Annals of the Association of American Geographers*, 72: 1–29.

Herod, A. (2000) 'Implications of just-in-time production for union strategy: lessons from the 1998 General Motors–United Auto Workers dispute', *Annals of the Association of American Geographers*, 90: 521–47.

Herod, A. (2001) *Labor Geographies: Workers and the Landscapes of Capitalism*. New York: Guilford Press.

Herod, A. and Wright, M.W. (eds) (2002a) *Geographies of Power: Placing Scale*. Oxford: Blackwell.

Herod, A. and Wright, M.W. (2002b) 'Placing scale: an introduction', in A. Herod and M.W. Wright (eds) *Geographies of Power: Placing Scale*. Oxford: Blackwell, pp. 1–14.

Hoefle, S.W. (2006) 'Eliminating scale and killing the goose that laid the golden egg?', *Transactions, Institute of British Geographers*, NS 31: 238–43.

Jonas, A.E.G. (1994) 'The scale politics of spatiality', *Environment and Planning D: Society and Space*, 12: 257–64.

Jones, K.T. (1998) 'Scale as epistemology', *Political Geography*, 17: 25–8.

Judd, D.R. (1998) 'The case of the missing scales: a commentary on Cox', *Political Geography*, 17: 29–34.

Kurtz, H. (2002) 'The politics of environmental justice as the politics of scale: St. James parish, Louisiana, and the Shintech siting controversy', in A. Herod and M.W. Wright (eds) *Geographies of Power: Placing Scale*. Oxford: Blackwell, pp. 249–73.

Latour, B. (1993) *We Have Never Been Modern* (trans. C. Porter). Cambridge, MA: Harvard University Press.

Latour, B. (1996) 'On actor-network theory: a few clarifications', *Soziale Welt*, 47: 369–81.

Marston, S.A. (2000) 'The social construction of scale', *Progress in Human Geography*, 24: 219–42.

Marston, S.A., Jones, J.P. III and Woodward, K. (2005) 'Human geography without scale', *Transactions, Institute of British Geographers*, NS 30: 416–32.

Marston, S.A. and Smith, N. (2001) 'States, scales and households: limits to scale thinking? A response to Brenner', *Progress in Human Geography*, 25: 615–19.

Pudup, M.B. (1988) 'Arguments within regional geography', *Progress in Human Geography*, 12: 369–90.

Smith, N. (1984) *Uneven Development: Nature, Capital and the Production of Space*. Oxford: Blackwell.

Smith, N. (1988) 'The region is dead! Long live the region!', *Political Geography Quarterly*, 7: 141–52.

Smith, N. (1989) 'Rents, riots and redskins', *Portable Lower East Side*, 6: 1–36.

Smith, N. (1992) 'Geography, difference and the politics of scale', in J. Doherty, E. Graham and M. Malek (eds) *Postmodernism and the Social Sciences*. New York: St Martin's Press, pp. 57–79.

Smith, N. (1993) 'Homeless/global: scaling places', in J. Bird et al. (eds) *Mapping the Futures: Local Cultures, Global Change*. London: Routledge, pp. 87–119.

Smith, N. and Dennis, W. (1987) 'The restructuring of geographical scale: coalescence and fragmentation of the northern core region', *Economic Geography*, 63: 160–82.

Swyngedouw, E. (1997) 'Excluding the other: the production of scale and scaled politics', in R. Lee and J. Wills (eds) *Geographies of Economies*. London: Edward Arnold, pp. 167–76.

Taylor, P.J. (1981) 'Geographical scales within the world-economy approach', *Review*, 5: 3–11.

Taylor, P.J. (1982) 'A materialist framework for political geography', *Transactions, Institute of British Geographers*, NS 7: 15–34.

Wills, J. (1998) 'Taking on the CosmoCorps? Experiments in transnational labor organization', *Economic Geography*, 74: 111–30.

13

Social Systems: Thinking about Society, Identity, Power and Resistance

Cindi Katz

Definition

Society is used to describe social relations at a range of spatial scales and even to conceptualize stretched out social networks that are not place based. Societies are often produced through a sense of sameness or shared identification. Identity is a complex and contested term that concerns how we understand who we are. Identities are relational, the Self is always defined in terms of the Other, that is in terms of difference, what it is not. Questions of identification and difference necessarily entail issues of power and resistance. Power is conceptualized as a diffuse and unlocatable force, that permeates all levels of society and that is reproduced in indirect and often erratic ways through multiple mediatory networks. In this way, all individuals are both in a position of simultaneously exercising power but also being subject to it. Resistance places the emphasis on the way individuals and groups find ways of reworking their situation in the face of power and oppression.

INTRODUCTION

Society is a deceptively simple term. Common usage defines society as a readily identifiable social group, the nature of which depends upon the temporal and geographical scale at which we address the question. In an historical sense society is defined by time periods. Thus we refer to ancient society, medieval society, contemporary society and so forth. Some social scientists also continue to speak of societies in terms of a sense of development or progress in which they refer to complex and simple societies, the status of which depends upon such things as geographic location, level of integration with other social groups and political economic processes, characteristic relations of production and their level of development, and internal coherence (see Chapters 8 and 21 for a critique of this mode of thinking). In a geographic sense we refer to broad social formations that encompass geographical or territorial divisions. When we talk about urban society, rural society, Texan society, Hungarian society, European society etc., we are imputing social relations at a host of different scales from the local, the state and regional scales, up to the national or even transnational scale. The term 'society' is also used to refer to those who inhabit common spaces either by choice or coercion for various periods of time, such as prison societies, college societies, or workplace societies. At the same time we also use society to describe social relations without propinquity, in other words, people can develop and maintain social ties and networks that have no geographical basis. Think, for example, of so-called cybersocieties – online communities of interest – or campaign groups and professional societies (the Society for the Prevention of Cruelty to Animals or the Society for Chartered Accountants) that may only get together face to face occasionally. And then, of course, we also use society as an all-encompassing term to describe the human race.

My brief introductory reflection on all the ways that we use society in everyday life illustrates how little thought is often given to the sorts of relations that might make these groups coherent or recognizable as societies, their internal differences, or the webs of connection that bind them to other social formations. In this chapter, I want to explore the notion of society in more detail, before considering the ideas about identity and difference, power and resistance, on which it rests. In doing so, I draw out the ways in which societies, identities and power constitute, are constituted within, and reconstitute space and place.

SOCIETY: SAME AND OTHER

Demarcating a given social group as a society presumes a sense of sameness – a shared sense of identification, interests or belonging at a particular geographic scale. To be American, Norwegian, or Thai is to be part of a nation. Membership of a particular society in this way helps to construct our identity even if what that means is quite individualistic, shifts over time, and is frequently hyphenated, as in Norwegian-Americans. Likewise, to be a member of voluntary or formally constructed societies, such as a philatelic society, the Royal Geographical Society, or an association of race car drivers, is not only to mark one's identity by choice but to seek out others who do the same so that specialized interests may be shared. Such societies offer arenas – where particular normative patterns of behaviour, practices and relations are produced – that enable people to simultaneously share common interests and construct their identities in particular ways that make them philatelists, geographers, or race car drivers. So, for example, if you regularly attend the Royal Geographical Society meetings, take part in email discussion lists run by its members or receive its newsletter, you may start to build up a particular shared knowledge or history with other members of the society, to use a common language or vocabulary, or even to start to dress like other geographers. In turn, this means that you might begin to feel part of this society, that you belong, that you identify as a geographer, while at the same time other people may start to recognize from the way that you talk, dress, and the things you do etc., that you are a geographer. In other words, there is a mutually constitutive relationship between the making of particular identities and the social groups that render them meaningful and set the norms and parameters of their production and reproduction.

Of course, however, even though societies are linked by common concerns or a shared sense of sameness, this does not necessarily mean that all their members share the same geographical space/territory, meet each other face to face, know one another, or would even like or actually identify with one another if they did encounter each other. Rather, societies are perhaps best conceptualized as 'imagined communities' (Anderson, 1983). Anderson (1983: 6) explains that nations are '*imagined* because the members of even the smallest nation will never know their fellow-members, meet them, or even hear of them, yet in the minds of each lives the image of their communion'. He further suggests that nations are imagined communities because fellow citizens usually have a deep sense of comradeship or identity with each other, even though, in practice, there may be differences, exploitation or inequality between them. In this way, the concept of 'imagined communities' both acknowledges the way members of societies identify and reproduce a sense of sameness with each other, while at the same time recognizing the limitations of this identification. As such, although Anderson (1983) first used this term to describe nations, it is an equally useful way of thinking about social relations based on sameness across other geographical scales. Rose (1990), for example, has used it to think about social relations in a working-class neighbourhood community, while Dwyer (1999) has employed it to understand how young British Asian Muslim women construct their identities across transnational communities.

Recognizing that social relationships are defined and worked out in and through place, space and nature, geographers, more than other social scientists, are interested in the ways that societies and geographies are mutually constitutive (e.g. Massey, 1984, 2005; Lefebvre, 1991; Harvey, 1992). To this end, research addresses how particular geographies sustain or weaken social relations and interactions, and how these social relations and power struggles work, in turn, to make and remake their geographies. Geographers use the term 'spatiality' to denote this understanding of the intertwining of the social and the spatial, which makes clear that society and space are simultaneously produced (Keith and Pile, 1993).

Societies, however, are not only produced through processes which foster a sense of sameness, but also those which emphasize a differentiation from others. Each society must have something to define itself against. The construction of any particular notion of society also rests on making boundaries – however porous – between those who belong, those who are the same, and those who do not, those who are different. To be human is not to be an animal, Asian society is not European society, rural society is not urban society. Yet, each of these societies is unintelligible without the other. In other words, all societies are relational in that they are always constructed and understood in terms of their sameness to, and difference from others. This is perhaps best understood by looking at the work of the Palestinian social theorist Edward Said. In his book *Orientialism* Said (1978) reflects on the way that the Orient or the East has been an object of fascination to the West for centuries. Drawing on the evidence of paintings, photographs and writings (novels, poetry, etc., as well as academic work) he highlights how Western visitors have represented the Orient as exotic, mysterious, decadent, corrupt, barbarous, and so on. In doing so, he argues that the traditions of thought and imagery which give the Orient a reality – in particular the discourse of the mysterious Orient – are in fact a European invention, a product of the European imagination. Moreover, he argues that, by containing and representing the Orient through this dominant framework, European culture has gained in strength and identity, setting itself off against the Orient in a series of asymmetrical relationships in which it is always seen in a favourable light (i.e. as civilized and ordered compared to the barbarism and corruption of the East, etc.) (Clifford, 1988). Thus Said argues that '[b]y dramatizing the distance and difference between what is close to and what is far away' imaginative geographies not only produce images of the Other but of the Self too. I will return to the notion of Self and Other in the following section when I focus more specifically on the question of identity.

IDENTITY

In recent years the notion of identity, and with it, identity politics have become quite popular within the social sciences. In this work identities such as gender, 'race' and sexuality are understood to be social constructions – a product of the way we understand and interpret our bodies – rather than natural or biological essences (for further discussion of essentialism and social constructionism see Women and Geography Study Group, 1997; Blunt and Wills, 2000; Valentine, 2001). By conceptualizing identity in this way, theorists have problematized understandings of subjectivity as fixed and ahistorical. Rather, identity is regarded not only as changeable over both time and space, but also as potentially voluntaristic in that individuals, in contemporary Western societies at least, have more freedom to make their own lifestyle choices than ever before.

Further, this understanding of identity as a social construction, has led to a greater recognition of the fact that people are multiply positioned in the world. We are not just male or female. We are all simultaneously classed, raced, gendered, have a sexuality and so on. For some theorists this recognition led to notions of fluidity or mobility wherein identity appeared almost as a sort of Brownian motion in which people were imagined to shuttle among their multiple identifications depending upon their circumstances. For instance, someone who is simultaneously Welsh, British, working-class, and a young man, might in these terms be understood to identify as Welsh when he is watching the national football team, British when he is travelling on holiday, working-class when he is at home in his neighbourhood community or as a teenager when he is out clubbing with his friends. Other writers, however, have pointed out that singular understandings of social identification along the lines of, say, class or gender are inadequate to explain and understand the complex power dynamics and injustices of social life. For example, black feminists have argued that it does not mean the same thing to be a black woman as it does to be a white woman. Identities are not merely something one opts into or out of at different moments, nor are they additive; rather, being black changes what it means to be a woman and these meanings are themselves historically and geographical contingent (Mohanty, 1988; McKittrick, 2006).

The ideas associated with identity politics have led in turn to debates about the relative importance of particular identities and forms of oppression. For some writers, particular forms of oppression are understood to be more important than others. For example, David Harvey (1993) has argued that a focus on different aspects of identity – gender, race, sexuality, and so on – has distracted us from the importance of class politics and contributed to undermining potential alliances and solidarities between different social groups rather than enhancing or developing them. He made his argument by comparing a fire at the Triangle Shirtwaist Company factory in New York in 1911, in which 146 employees died, with a similar disaster at a chicken processing plant in a small town in North Carolina, USA, in 1991, in which 25 people died. Harvey highlighted the way the Triangle Shirtwaist Company fire in 1911 became a *cause celebre* with over 100,000 people taking to the streets to demand better workplace protection. In contrast, he argued that the fire at the chicken processing plant in 1991 passed virtually unnoticed, occurring as it did at the time of controversy surrounding the potential appointment of Clarence Thomas – a black judge who had been accused of sexual harassment – to the Supreme Court. He pointed out how despite the fact that it was mainly women who died in the fire, women's organizations were too pre-occupied trying to prevent Clarence Thomas being appointed to engage with what had happened in North Carolina. As such, Harvey (1993: 47) claimed that the fragmentation of what he termed 'progressive politics around special issues and the rise of the so-called new social movements focusing on gender, race, ethnicity, ecology, multiculturalism' were obscuring the need to pursue more universal notions of social justice – which he defined in terms of class oppression – and causing political paralysis.

Other writers have argued that focusing on hierarchies of oppression – assuming that particular social relations matter more than others – has instead instigated what has threatened to become a new form of essentialism, valorizing particular social locations, such as class or gender, at the heart of multiple oppressions rather than theorizing carefully which differences matter under what conditions. More productively, others still have developed the notion of intersectionality as a means of understanding multiple identifications (Crenshaw, 1995). Rather than a celebration of difference or a compounding of oppressions, Crenshaw's analysis carefully examined how particular social relations of oppression and exploitation associated with different identifications intersected under specific circumstances and might in the process be compounded, work at cross purposes, be contradicted or otherwise altered (see Bondi, 1993; see also Valentine, 2007).

These debates aside, geographers have particularly sought to think about identities in terms of how they are formed, maintained, and contested in and through spatial relations, among other things. Space works in a number of ways to maintain the various inequalities, uneven power relations, and social injustices associated with contemporary forms of domination and exploitation, notably capitalism, sexism, racism, imperialism, and heterosexism (see Keith and Pile 1993; *Professional Geographer*, 2002). Questions of identity are often at the nexus of exclusionary practices in space. Sibley (1995) has used psychoanalytic understandings of Self and Other to theorize the tendency of powerful or hegemonic social groups to purify and dominate space. He defines the purification of space as the rejection of difference and the securing of boundaries to maintain homogeneity. Such practices take many forms. For instance, through various modes of residential segregation that result from covenants, bank lending policies, gating, or informal means such as the results of income disparities or certain cultural practices, some groups may be excluded from particular residential neighbourhoods. Smith's (1987) work on the UK housing market, Davis's (1990) study of Los Angeles, and Duncan and Duncan's (2004) research on an American suburb all provide examples of different ways that residential neighbourhoods become segregated by race and class.

The purification of space is also evident in non-residential spaces. Public environments, such as particular streets or parks, may have normative codes of behaviour or be policed by public or private security services (see Fyfe and Bannister, 1998) in ways which serve to exclude groups or individuals who are regarded as undesirable Others. For example, Valentine (1996) has shown how young people often have their access to spaces such as shopping malls restricted by the police and public security guards because they are regarded as disorderly or disruptive in public space. Likewise, Parr's (1997) research shows how people with mental ill-health can be denied the freedom enjoyed by other citizens to use and occupy everyday spaces such as city centres, while Don Mitchell (1997) has shown how anti-homeless laws are being used to 'clean up' US parks and streets. Quite apart from purification, these issues are addressed in the literature and politics concerned with 'the right to the city' (Mitchell, 2003).

Just as we can imagine spaces associated with, excluding or embracing particular identities, we can also see identity as spatially formed and enacted. The idea of diasporic identity is one such idea (see Anderson, 1991; Gilroy, 1993). Diaspora literally refers to the dispersal or scattering of a population from an original homeland, but it is now used more loosely to capture the complex sense of belonging that people can have over several places, all of which they might think of as home (Clifford, 1994; see Mavroudi, 2007, for more on different theorizations of diaspora). Paul Gilroy (1993), for instance, speaks of the Black Atlantic to frame a diasporic African population that shares cultural and political-economic histories that are drawn on differently across specific geographies. Kay Anderson (1991) and Peter Kwong (1987) have examined the Chinese diaspora in the production of Vancouver's and New York City's Chinatowns. But as these works and others make clear, diasporic identities are not homogenized across space and time. Rather, they are altered by particularities of place and the unevenness of the ways the social relations of production and reproduction are played out in different locations and at different scales. In other words, if spatiality confers a certain notion of fluidity and the social constructedness of identity against earlier ideas of fixed identity rooted in a given essence, it also makes impossible notions of identity as infinitely malleable. In the simultaneous working out of spatial and social life, particular identifications inevitably rub up against and confront others, calling into question which differences matter under what circumstances (K. Mitchell, 1997a). For example, in Katharyne Mitchell's (1997b) work on Chinese immigrants to Vancouver, British Columbia, the question of class insistently disturbs the construction of homogenized identities of race or nation, while the differences produced around national identity often undermine potential class solidarities among the bourgeoisie of the area.

Questions of identity should, then, raise questions of politics. Why are particular identifications mobilized, under what circumstances, and by whom? When identity is understood in more fixed terms, such as class, gender, or race, it provides a more transparent basis for political organizing. However, given the by now well-recognized truism that all political actors are multiply positioned, attempts to organize around more singular notions of identity may founder or produce problems of their own, on the one hand by eclipsing important differences among those mobilized, and on the other by failing to recognize potential solidarities people identified in one way might have with others who are differently identified. Effective politics rarely arises from pigeon-holes. The challenge for political organizing, then, is to recognize which differences matter at particular historical and geographical conjunctures and to develop the sort of politics that can work the intersections of several modes of identification, such as class, race, sexual orientation and gender (see Haraway, 1985; Bondi, 1993; Crenshaw, 1995).

POWER AND DIFFERENCE

As these discussions of society and identity suggest, one of the ways that *power* operates and is felt is through the production and reproduction of *difference*. The French theorist Foucault (1977) has been particularly influential in shaping the way geographers think about power. Foucault argues that power is not something that one person or group of people holds over another or others, or that operates in a unidimensional way. It is not something that can be won or lost. Rather, he conceptualizes power as a diffuse and unlocatable force, that permeates all levels of society and that is reproduced in an indirect and often erratic way through multiple mediatory networks. In this way, all individuals are both in a position of simultaneously exercising power but also being subject to it. Disciplinary power, as Foucault called it, is witnessed in its diffuse, capillary effects as people conform to norms.

This way of thinking about power is perhaps best understood through an example. Taken at face value, prisons might be conceptualized as institutions where power is unidirectional; in other words, where the warders have power over the inmates. However, studies of prison life demonstrate that in practice power is much more fluid and diffuse. For example, within prisons there is often an informal economy with inmates trading commodities they have bought or own (phone cards, food, personal possessions, etc.) for commodities that are scarce and therefore precious, such as cigarettes, or those that are illegal, such as goods which have been smuggled into the prison like drugs and alcohol. Inmates who build up stockpiles of these valuable commodities – who are often known as barons – can exert power over others: bullying and blackmailing inmates who want these commodities into paying extortionate amounts for them or working for them within the prison. Indeed, sometimes the prison authorities allow such illegal trading, and the bullying and violence that goes with it, to take place because they may receive overtime payments if they have to carry out extra duties that result from illegal activities, or because the inmates may use their contacts outside the institution to threaten and intimidate the officers' families. Likewise, different barons within the prison will also try to advance their own power within the institution by keeping the influence of others in check. In this way, prison officers can benefit from these struggles because order is indirectly maintained in the prison. In other words, power is not something that officers just have and hold over inmates; rather, from this brief example it is apparent that officers and inmates are linked together through various webs of power that involves intimate alliances and collusions, such that prison society represents an endless play of dominations in which the distribution of power between officers and inmates, and between different individuals and groups of inmates, is dynamic, relational and fluid albeit wildly uneven (Valentine and Longstaff, 1999). (For other examples of entanglements of power, see Sharp et al., 2000.) Nevertheless it is crucial to remember that prisons as institutions and social settings exist to maintain social order – in and outside of them – and their practices of control are linked firmly to capital accumulation and dispossession (Foucault, 1977; Gilmore, 2007).

While it can be illuminating to imagine power in this way, and indeed Foucault was most informative in demonstrating how the diffuse effects of power were expressed and internalized by people and institutions, it remains important to name names so people can – as film-maker Spike Lee exhorts – 'fight the power'. Under contemporary conditions, the most formidable sources of political-economic power remain capitalist relations of production, patriarchy and racism. All of these social relations of domination interact – often, but not always, reinforcing each other – and produce difference. The differences associated with capitalist relations of production are those of class and, as discussed below, the geographical differences produced in the course of uneven development, including the differences of nation associated with imperialism. With patriarchal relations of domination the key differences are those of gender and sexuality. Finally, racist relations of domination produce differences around race and ethnicity. It is important for social theory to take into account each of these relations of domination and the kinds of differences they produce, not only to account for the ways these relations converge and diverge and reveal the ways they work with and against one another under particular circumstances, but also because each provides a broad and important arena for resistance and opposition.

In a particularly geographical vein, theories of uneven development have made clear that capitalist accumulation works in and through the production of difference in space – not only the development of certain areas and resources at the expense of others, but the deliberate development of particular places and regions while others are *underdeveloped* (Harvey, 1982; Smith, 1984; see Chapter 21 on development). The production of difference works at different scales. Indeed, geographic scale is a means of organizing spatial difference (Smith, 1992; Swyngedouw, 1997; Marston, 2000; see also Chapter 12). For instance, at the global scale, the underdevelopment of the global South is bound directly with the development of the northern industrialized nations, many of which are former imperial powers. At the national and regional scales, the processes of industrial development and decline are best understood within the framework of capitalist uneven development. Finally, at the urban scale, gentrification is predicated on long-term patterns of investment and disinvestment in urban neighbourhoods (Smith, 1996; Hamnett, 2003; Butler, 2007). The spatiality of uneven development is witnessed in the co-existence of environments of value whose landscapes reflect wealth, investment and development, and deteriorating natural and built environments which, under particular political economic circumstance, become targets for reinvestment and development. These produced differences, which reflect the peculiar dynamics of capitalist accumulation, are often naturalized by observers and analysts. For example, difference is explained as a result of such things as environmental conditions, social inadequacies, historical accident, or explained away with recourse to an evolutionary understanding of development which imagines various places or regions at different stages in the process.

The differences produced by patriarchal relations of power are spatialized as well. One of the most significant forms of spatial difference associated with patriarchy is that between private and public space. As feminist and other theorists have made clear, this commonly naturalized distinction is a political artefact and not so easily maintained; the two spaces and the spheres they represent blur and intersect. Nevertheless, maintaining the distinction in theory and practice has worked to limit women's place and sphere of influence to the home and thereby diminish their exercise of power (McDowell, 1999; Pratt, 2004). Racism also produces difference spatially through, among other things, formal and informal patterns of residential segregation, controls on international migration, and the making and maintenance of racialized environments through such things as fear and violence that produce places that are simultaneously exclusionary and inclusive depending upon one's racialization and experience (Anderson, 1991, 2007; Pred, 2000; *Professional Geographer*, 2002; Hubbard, 2005; McKittrick, 2006).

RESISTANCE

In the last decades of the twentieth century, geographers, like other social scientists, addressed the production of difference and its spatial consequences, complicating the assumed relationships between place and identity and making clear that all places were simultaneously unique and differentiated. The production of difference has been looked at with reference to the increasingly porous boundaries of the nation-state with the globalization of capitalist production and cultural forms and practices, as well as with reference to the production of difference through and in place. To this end, geographers have examined the contemporary romance with place as an often reactionary means to create a bulwark against the homogenizing effects of globalization (see also Chapter 9 and 19). Research on the geographies of difference demonstrates not only the ways that power works to produce difference, but how deployments of difference in the production of space, place and nature can counter and redirect power. Thus geographies of difference often suggest geographies of resistance.

If part of the operation of society is to set and maintain appropriate norms of behaviour and the social institutions that uphold them, and these norms, values and social relations generally – and not coincidently – work to ensure the endurance and reproduction of the society, we can expect *resistance* to the inequalities and uneven power relations fostered by them. Many geographers have looked at the ways that social power is produced and reinforced spatially, and more recently some have addressed the numerous means by which uneven power relations are contested and resisted. In various ways this work examines how hegemony is spatially secured – and thus might be interrupted, compromised, and undone on spatial as well as social and political grounds.

While geographers have, until recently, been more interested in analysing the operations of power in and through space, scholars in cultural studies and other fields have made the material social practices of resistance their focus since the 1970s (Hall and Jefferson, 1976; Willis, 1977; Scott, 1990). The burgeoning interest in resistance during this period seems to have led to its dilution as a political practice and analytic category. In all too many texts it seemed that any 'independent initiative', to use Gramsci's (1971) phrase, no matter how small, was understood as resistance to the social relations of domination. Meanwhile, in this same period, capitalist production has gone global; the disparities between rich and poor nations and between rich and poor people within nations have increased almost everywhere; violence has spread and intensified at all scales from the personal and domestic to the national and global; indicators of poverty remain stubbornly high; and a large and growing number of young people in rich as well as poor countries seemed likely to be denied the promises of modernity. For these reasons, it is important to parse resistance to develop more subtle and workable distinctions to understand its effects.

In my research on social reproduction and global economic restructuring in rural Sudan, I have distinguished between resilience, reworking and resistance as ways of understanding the material social practices of everyday life through which people confronted and coped with shifting relations of domination associated with, among other things, global economic restructuring, civil war and the local intrusion of a state-sponsored agricultural development project (Katz, 2001, 2004). In the village where I worked, deforestation and pasture deterioration, along with the restrictions on arable land associated with the agricultural project and its effects, compromised residents' ability to stay in the village and continue working the mix of agriculture, animal husbandry and forestry that had long sustained their village. I anticipated that as young adults came of age they would have to leave for urban areas, where they were ill-prepared to find sustaining work. Instead, the local population adjusted to these rather dramatic shifts by radically expanding the terrain of their work. In less than two decades the area they drew on to cut wood, produce charcoal, graze animals and cultivate had grown exponentially. I called this phenomenon time–space expansion. Through this spatial strategy of resilience, young people were able to remain in the village even if their work trajectories included occasional stints in nearby towns. In another vein, as adults in the village witnessed these changes and the ways their children were potentially deskilled in the process, they effected a series of self-help initiatives designed to increase school attendance, and in particular to educate girls. Through the installation of stand-pipes that saved much of the time girls spent fetching water, the construction of separate classrooms for girls with female teachers hired to teach them, and finally the construction of a secondary school in the village, school attendance was increased dramatically. These efforts of reworking demonstrate the flexibility of people's responses to the shifting conditions of their everyday lives. Finally, there were some events in the village and more broadly that called forth practices that might be understood as resistance, in that they were consciously directed at altering a condition that people recognized as oppressive. For instance, the Ministry of Agriculture initially required all tenants in the agricultural project to cultivate cotton and groundnuts exclusively, forbidding the cultivation of the staple dietary crop, sorghum. During the first decade of the project's operation the price of sorghum increased 2000% while cotton and groundnut prices remained relatively stagnant. Tenants organized and sent representatives to the Ministry of Agriculture to fight for the right to cultivate sorghum on project lands. They eventually won the right to devote their groundnut allotment to sorghum. As this brief summary might suggest, distinguishing between strategies that enable people to survive under difficult circumstances, those that rework particular conditions that compromise the conditions of their existence, and those that resist relations of power that are exploitative or oppressive to them can render clearer the political effects of varied material social practices exercised in the face of power. More nuanced understandings of these effects can be of use in the development of political strategies to redress social injustices and economic exploitation.

CONCLUSION

In this chapter I have focused on the countless relations of identification and difference, power and resistance that make up our individual everyday social relationships and that are implicated in the operation of social relations across a range of scales from the local to the global. In doing so I have sought to demonstrate how the social and the spatial are always mutually constituted. The complexity of the concepts – identity, difference, power and resistance – presented in this chapter also hints at the way understandings of them have changed over time and how their meanings and importance continue to be contested. At the same time, the chapter demonstrates the importance of continuing to refine these ideas if we are to develop a better understanding of how society works and therefore to find ways of working towards developing more just social relations and practices.

SUMMARY

- All societies are relational in that they are produced through people's collective imaginings of their sameness to, and difference from, others.

- Identities are social constructions rather than natural or biological essences. This understanding of identity problematizes subjectivity understood as fixed and ahistorical.

- A recognition that we are multiply positioned has undermined nations positing hierarchies of oppression (which forms of oppression are most important) and to diverse ways of theorizing multiple oppressions.

- Power is conceptualized as a diffuse and unlocatable force, that permeates all levels of society and that is reproduced in an indirect and often erratic way through multiple mediatory networks. In this way, all individuals are both in a position of exercising power but also subject to it, even these relations are uneven and contingent.

- Resistance is a broad category that until recently has been used by social scientists to describe any form of independent initiative. It is now being unpacked to emphasize elements such as the way people are *resilient* in the face of power and *rework* situations of oppression, domination and exploitation.

Further Reading

Said's classic book **Orientalism** (1978) is a good place to start when thinking about questions of sameness and difference. **Place and the Politics of Identity**, edited by Keith and Pile (1993), is a collection of essays that illustrates and debates the ways that identities are spatially as well socially constituted. Another edited book, **Mapping the Subject** (Pile and Thrift, 1995), focuses more explicitly on how we define our subjectivity in terms of our sense of Self and Other. There are too many books and special issues of journals that deal with specific forms of identification, such as class, gender, 'race' or sexuality, to list, but see, for example, Browne et al. (2007), Gough et al. (2006), **Professional Geographer** (2002), Women and Geography Study Group (1997). Valentine (2007) provides a useful route in to how we might think about intersectionality in geography and offers a case study of intersectionality as lived experience to show how it might be researched in practice. McKittrick (2006) offers a stunning syudy of the operations of multiple oppressions on bodies in space. In **Geographies of Exclusion**, Sibley

(Continued)

(Continued)

(1995) theorizes the way powerful groups can purify and dominate space, marginalizing those society defines as others. ***Entanglements of Power*** (Sharp et al., 2000) provides a collection of essays, drawing on a diverse range of case studies, that explore the ways practices of domination and resistance cannot be separated and are integral to all workings of power. In a similar collection of essays, ***Geographies of Resistance***, Pile and Keith (1997) focus more closely on the concept of resistance. While Katz's (2004) ***Growing Up Global*** unpacks this term into different components. Most of these terms are also considered in general human geography textbooks such as Blunt and Wills' (2000) ***Dissident Geographies***, Panelli's (2004) ***Social Geographies*** Valentine's (2001) ***Social Geographies: Space and Society*** and Knox and Marston's (2006) ***Human Geography: Places and Regions in Global Context.***

Note: Full details of the above can be found in the reference list below

---

References

Anderson, B. (1983) *Imagined Communities: Reflections on the Origin and Spread of Nationalism.* London: Verso.

Anderson, K. (1991) *Vancouver's Chinatown: Racial Discourse in Canada 1875–1980.* Montreal: McGill-Queen's University Press.

Anderson, K. (2007) *Race and the Crisis of Humanism.* London and New York: Routledge.

Blunt, A. and Wills, J. (2000) *Dissident Geographies.* Harlow: Pearson Education.

Bondi, L. (1993) 'Locating identity politics', in S. Pile and M. Keith (eds) *Place and the Politics of Identity.* London and New York: Routledge, pp. 84–101.

Browne, K., Lim, J. and Brown, G. (2007) *Geographies of Sexualities: Theory, Practices and Politics.* Aldershot: Ashgate.

Butler, T. (2007) 'For gentrification?', *Environment and Planning A*, 39: 162–81.

Clifford, J. (1988) 'Introduction: partial truths', in J. Clifford and G.E. Marcus (eds) *Writing Culture: The Poetics of Ethnography.* Berkeley, CA: University of California Press, pp. 1–26.

Clifford, J. (1994) 'Diasporas', *Cultural Anthropology*, 9: 302–28.

Crenshaw, K.W. (1995) 'Mapping the margins: intersectionality, identity politics, and violence against women of color', in K. Crenshaw, N. Gotanda, G. Peller and K. Thomas (eds) *Critical Race Theory: The Key Writings that Formed the Movement.* New York: New Press, pp. 357–83.

Daniels, S. and Lee, R. (eds) (1996) *Exploring Human Geography.* London: Edward Arnold.

Davis, M. (1990) *City of Quartz: Excavating the Future in Los Angeles.* London: Verso.

Duncan, J.S. and Duncan, N.G. (2004) *Landscapes of Privilege: Aesthetic and Affluence in an American Suburb.* New York: Routledge.

Dwyer, C. (1999) 'Contradictions of community: questions of identity for young British Muslim women', *Environment and Planning A*, 31: 53–68.

Foucault, M. (1977) *Discipline and Punish: The Birth of the Prison.* London: Penguin.

Fyfe, N. and Bannister, J. (1998) 'The eyes upon the street: closed-circuit television surveillance and the city', in N. Fyfe (ed.) *Images of the Street.* London: Routledge, pp. 254–67.

Gilmore, R.W. (2007) *Golden Gulag: Prisons, Surplus, Crisis, and Opposition in Globalizing California.* Berkeley and Los Angeles: University of California Press.

Gilroy, P. (1993) *The Black Atlantic: Modernity and Double Consciousness.* Cambridge, MA: Harvard University Press.

Gough, J. and Eisenschitz, A. with McCulloch, A. (2006) *Spaces of Social Exclusion*. London and New York: Routledge.

Gramsci, A. (1971) *Selections from the Prison Notebooks* (Trans. and edited Quintin Hoare and Geoffrey Nowell Smith). New York: International Publishers.

Hall, S. and Jefferson, T. (eds) (1976) *Resistance through Rituals: Youth Subcultures in Post-war Britain*. London: Unwin Hyman.

Hamnett, C. (2003) 'Gentrification and the middle-class remaking of inner London, 1961–2001', *Urban Studies*, 40: 2401–26.

Haraway, D. (1985) 'A manifesto for Cyborgs: science, technology and socialist feminism in the 1980s', *Socialist Review*, 80: 65–107.

Harvey, D. (1982) *Limits to Capital*. Oxford: Blackwell.

Harvey, D. (1992) 'Postmodern morality plays', *Antipode*, 24: 300–26.

Harvey, D. (1993) 'Class relations, social justice and the politics of difference', in M. Keith and S. Pile (eds) *Place and the Politics of Identity*. London: Routledge, pp. 41–66.

Hubbard, P. (2005) 'Accommodating otherness: anti-asylum centre protest and the maintenance of white privilege', *Transactions, Institute of British Geographers*, 30: 52–65.

Katz, C. (2001) 'On the grounds of globalization: a topography for feminist political engagement', *Signs: Journal of Women in Culture and Society*, 26: 1213–34.

Katz, C. (2004) *Growing Up Global: Economic Restructuring and Children's Everyday Lives*. Minneapolis, MN: University of Minnesota Press.

Keith, M. and Pile, S. (eds) (1993) *Place and the Politics of Identity*. London and New York: Routledge.

Knox, P.L. and Marston, S.A. (2006) *Human Geography: Places and Regions in Global Context*. Upper Saddle River, NJ: Prentice-Hall.

Kwong, P. (1987) *The New Chinatown*. New York: Hill & Wang.

Lefebvre, H. (1991) *The Production of Space*. Oxford: Blackwell.

Marston, S.A. (2000) 'The social construction of scale', *Progress in Human Geography*, 24: 219–42.

Massey, D. (1984) *Spatial Divisions of Labour: Social Structures and the Geography of Production*. London: Macmillan.

Massey, D. (2005) *For Space*. London: Sage.

Mavroudi, E. (2007) 'Diaspora as process: (de)constructing boundaries', *Geography Compass*, 1: 467–79.

McDowell, L. (1999) *Gender, Identity, and Place: Understanding Feminist Geographies*. Minneapolis: University of Minnesota Press.

McKittrick, K. (2006) *Demonic Grounds: Black Women and the Cartographies of Struggle*. Minneapolis: University of Minnesota Press.

Mitchell, D. (1997) 'The annihilation of space by law: the roots and implications of anti-homeless laws in the United States', *Antipode*, 29: 303–35.

Mitchell, D. (2003) *The Right to the City: Social Justice and the Fight for Public Space*. New York: Guilford Press.

Mitchell, K. (1997a) 'Different diasporas and the hype of hybridity', *Environment and Planning D: Society and Space*, 15: 533–53.

Mitchell, K. (1997b) 'Conflicting geographies of democracy and the public sphere in Vancouver, BC', *Transactions, Institute of British Geographers*, 22: 162–79.

Mohanty, C.T. (1988) 'Feminist encounters: locating the politics of experience', *Copyright*, 1: 30–44.

Panelli, R. (2004) *Social Geographies*. London: Sage.

Parr, H. (1997) 'Mental health, public space and the city: questions of individual and collective access', *Environment and Planning D: Society and Space*, 15: 435–54.

Pile, S. and Keith, M. (eds) (1997) *Geographies of Resistance*. London: Routledge.

Pile, S. and Thrift, N. (eds) (1995) *Mapping the Subject: Cultural Geographies of Transformation*. London: Routledge.

Pratt, G. (2004) *Working Feminism*. Philadelphia: Temple University Press.

Pred, A. (2000) *Even in Sweden: Racisms, Racialized Spaces, and the Popular Geographical Imagination*. Berkeley and Los Angeles, CA: University of California Press.

Professional Geographer (2002) 'Focus section on race, racism and geography', *The Professional Geographer*, 54(7).

Rose, G. (1990) 'Imagining Poplar in the 1920s: contested concepts of community', *Journal of Historical Geography*, 16: 45–47.

Said, E. (1978) *Orientalism: Western Conceptions of the Orient*. Harmondsworth: Penguin.

Scott, J.C. (1990) *Domination and the Arts of Resistance: The Hidden Transcript*. New Haven, CT: Yale University Press.

Sharp, J.P., Routledge, P., Philo, C. and Paddison, R. (eds) (2000) *Entanglements of Power: Geographies of Domination and Resistance*. London: Routledge.

Sibley, D. (1995) *Geographies of Exclusion*. London: Routledge.

Smith, N. (1984) *Uneven Development: Nature, Capital and the Production of Space*. (2nd edn, 1990). Oxford: Basil Blackwell.

Smith, N. (1992) 'Geography, difference and the politics of scale', in J. Doherty, E. Graham and M. Malek (eds) *Postmodernism and the Social Sciences*. London: Macmillan, pp. 57–79.

Smith, N. (1996) *New Urban Frontier: Gentrification and the Revanchist City*. New York and London: Routledge.

Smith, S.J. (1987) 'Residential segregation: a geography of English racism?', in P. Jackson (ed.) *Race and Racism: Essays in Social Geography*. London: Allen Unwin.

14

Environmental Systems: Philosophy and Applications in Physical Geography

Stephan Harrison

⒟efinition

A system can be defined as a set of related objects along with their attributes. In simple terms, three main types of system may be identified: (1) closed systems, where the system does not exchange mass or energy with its surroundings (these probably never occur in nature); (2) isolated systems, which exchange energy but not mass with their surroundings (the planets approximate such systems); and (3) open systems, that exchange both mass and energy with their surroundings (the systems that most geographers and earth scientists are accustomed to work with; Chorley and Kennedy, 1971). All systems (human as well as physical) have three fundamental characteristics – structure, functioning and evolution – which holds out the prospect of unified approaches to analysis and modelling. Systems may also be defined in terms of their complexity (the amount of information that is required to describe the system exactly), and their organization (the number and dynamics of the inter-relationships exhibited by the system). As a result, some systems theorists describe systems as tractable when they display low values of complexity or organization (meaning that they are relatively easy to model) and as intractable when the opposite case applies.

INTRODUCTION

General System Theory and its role in physical geography and geomorphology

Many analyses and uses of systems approaches in physical geography and its sub-disciplines derive from the inspiration and framework provided by General

System Theory (GST). GST has proved to be a remarkably powerful conceptual tool in driving scientific thought, practice and teaching, as well as in linking disciplinary approaches and identities. The biologist Ludwig von Bertalanffy is generally seen as the founder of GST, and he developed the ideas in the first part of the twentieth century to counter what he saw as the increasingly reductionist nature of biological research (see von Betalanffy, 1973). Von Betalanffy recognized that the reductionist approach produced detailed understandings and explanations of the workings of individual organisms, but frequently failed to produce the scientific laws which could be used to test hypotheses and generate robust and coherent theories. We will see that such failings still bedevil reductionist analyses in physical geography.

In geomorphology and other parts of physical geography, the initial adherents of systems theory became enormously influential in the discipline, and their influence has been maintained by their acolytes. Two principal aims of GST can be identified: one has been to develop unifying principles throughout the sciences; and another has been to simplify, and therefore make accessible, the complexity we see in nature. This chapter provides an introduction to the history, development and context for application of systems approaches in physical geography, and to some of the methodological and philosophical problems associated with this.

Historical background to systems approaches in geomorphology and physical geography

Over the last two centuries, geomorphology (allied to geology) has played a central role as the main driver of physical geography and it has achieved this by sustained attempts to construct models of landscape evolution. These models are based upon both a physical understandings of geomorphological process, and the importance of time as an agent of landscape change. Other physical systems (which we would now regard as forming important subdisciplines in physical geography), such as soil and biogeographical systems, were relegated to subsidiary positions (see Harrison, 2005). As a result, in the hands of practitioners such as W.M. Davis and G.K. Gilbert, geomorphology gained a pre-eminent position within physical geography, a position it may hold today.

Davis (1850–1934) is often seen as the founder of pre-second World War Anglo-American geomorphology. His work on cycles of erosion stressed the large-scale historical and evolutionary nature of landscape evolution (Davis, 1899) and was heavily influenced by Hutton's theory of gradualism in understanding landscape change and by Darwin's development of evolutionary theories to explain biological change. Davis argued for three important concepts: first, the idea that the drainage basin was the fundamental geomorphic unit and that geomorphological understanding was appropriately focused at the regional scale; second, that landscape change occurred over long (geological) timescales; and third, that change was evolutionary, with landscape change being driven by denudation processes moving towards an end-state through stages of youth, maturity and old age.

Davis's work was widely criticized, falling foul both of the critical rationalists' criterion for falsifiability as a mark of a science and, associated with this, new understandings of the philosophy of scientific practice developed by the Logical Positivists (for a brief introduction to these issues in the context of physical geography, see Haines-Young and Petch, 1986). His work was further criticized by later geomorphologists as being too qualitative, and for its attention to processes. From the mid-twentieth century onwards, computational and statistical advances and positivist philosophies were employed by physical geographers such as Strahler (1950, 1952, 1980) and Chorley (1962) to develop new understandings of physical geography based largely upon systems theory and employing the language of mathematics and physics. Most frequently, the manifestation of systems approaches was the production of conceptual models of environments and landforms, most of which took the form of box-and-arrow diagrams (Figure 14.1). In these, systems were constructed based upon the observed (or assumed) correlation structure between landscape or environmental variables, joined by the flows of energy and material between them. Subsystems could be isolated to examine particular aspects of linkages within the larger system, and rejoined to represent cascades of energy/material between them, which governed the boundary conditions for the next-related subsystem. The diagrams could be used to suggest investigation of particular linkages in the systems; as flow charts to scenario model and predict changes (given appropriate numerical parameterization and calibration of the linkages) in analogy with flow charts and algorithms; and as pedagogic devices to represent and simplify natural environments using numerical or physical analogues (see Chorley and Kennedy, 1971).

The application and championing of systems theory was thus an attempt to place geomorphology into a more scientific, empirical and rigorous framework, as well a conceptual and methodological device to simplify complex natural systems and model them as accumulations of transfers and storages of mass and energy. From this approach came the dominant, and enduring, physical geography paradigm in English and American universities – that of process–form geomorphology. It is an intellectually seductive approach. By using detailed mesoscale process studies (allied to sophisticated laboratory and field experiments), the statistical treatment of the results and the application of computer modelling, the aim has been to understand the complexities of landscape-forming processes, boundary conditions and change at small spatial and temporal scales and to use this knowledge to extrapolate in order to explain change at the landscape scale (Harrison, 2005). This process–form geomorphology can be seen as the antithesis of the qualitative narratives constructed by Davis, and underpinned systems approaches which have been used in many branches of physical geography. Examples include ecosystem dynamics and change (e.g. Whittaker, 1998), fluvial geomorphology (e.g. Richards, 1982) and coastal geomorphology (e.g. Trenhaile, 1987). We can use the example of a glacier as a system to explain some of the important concepts in systems theory.

Glaciers can be seen as the product of the interplay of accumulation and removal of mass and energy. Mass (mainly snow, ice and rock) is added throughout the system, but has a net contribution to the mass balance of a glacier in the accumulation zone. As snow and ice are the dominant factors in glacier mass, the accumulation zone is usually found at the higher altitudes on the glacier where temperatures are lower and precipitation is greater. Downslope of this, in the

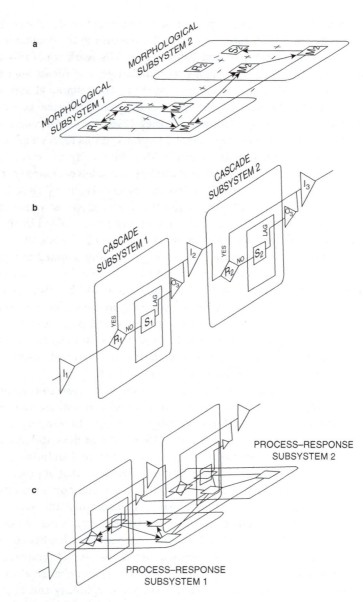

Figure 14.1 Diagrams illustrating the character of morphological (lansdscape systems): (a) is a morphological system containing eight variables linked to form a correlation structure; (b) is a cascading system containing two subsystems connected by inputs and outputs which are regulated into storages from which lagged outputs occur; (c) is a process–response system formed from the linkage of a and b

Source: Reproduced from Chorley and Kennedy (1971: 6, Figure 1.3)

ablation zone, mass is lost from the glacier through melting of the snow and ice, runoff of water from the system and deposition of rock and sediment (Figure 14.2). In addition, in cold, dry environments considerable ice mass can be lost through a phase change called sublimation, and in areas where the glacier terminus ends in water, considerable ice mass can be lost through calving processes. There are

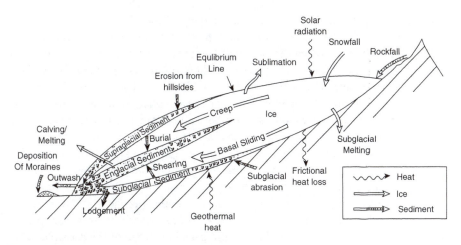

Figure 14.2   A cirque glacier as a system

also complex energy gains and losses through the system from solar radiation, albedo effects and gravity. The junction between the accumulation and ablation zones is called the 'equilibrium line', where the glacier mass is balanced. Overall, if the glacier displays positive mass balance it will tend to thicken and/or advance; conversely, if the glacier has a negative mass balance (as most mountain glaciers have experienced throughout the end of the twentieth and beginning of the twenty-first centuries) the glacier will tend to recede and/or thin. The glacier therefore has inputs into the system and outputs from the system, and the glacier itself (including ice, snow, water and rock material) constitutes the storage of mass. Using this glacier system model, we can see how seductively simple this is as a teaching and, perhaps, research tool.

This simplification of glaciers as systems allows us to assess the controls on glacier behaviour and hence use glacier behaviour as a proxy for changes in climate (e.g. Oerlemans, 2005). However, glaciers may not always respond in linear ways to climate forcing, and there are many places in the simple glacier system model where small changes in (perhaps unknown) variables will play an important role in changing system behaviour. For instance, glacier behaviour is influenced by variations in the rheology of the underlying sediments, by changes in subglacial meltwater fluxes and by changes in subglacial temperatures. All these impact upon subglacial permeability and, in turn, affect each other in ways which are not fully understood. More importantly, calving of ice into water bodies constitutes the major mass-loss element in many glaciers. The behaviour of such glaciers is partly decoupled from the regional climate, and substantially driven by topographic factors such as changes in water depth at the terminus and pinning points along fjord sides (e.g. Warren, 1992). Calving dynamics also vary between tidewater- and lake-calving glaciers. As a result, calving forms a second-order control on glacier behaviour. Added to all this, glacier behaviour also displays natural, and possibly chaotic variation, from which the climate signal has to be extracted (see Phillips, 1996, for a discussion of this).

We can therefore see a systems approach to understanding and explaining geomorphological processes and landscapes as a useful teaching and research tool. Systems provide an epistemological device (identifying attributes, flows and exchanges of mass and energy, and connections between process and form) with potentially ontological implications – they reflect or suggest what we actually believe exists in the landscape and how the landscape actually operates. However, while systems theory and model-building have formed a central methodological approach for many physical geographers and geomorphologists, it is clear that system behaviour is complex and, at times, counter-intuitive (see Chapter 7). Relatively little attention has been focused on the uncertainties and limitations inherent in this approach. Certainly, the formal systems approach is unable to account for many of the behaviours associated with chaotic and non-linear behaviour, especially where chaos and sensitivity to initial conditions are present but ultimately unknowable. As a result, some approaches to systems theory provide a simplistic view of cause-and-effect relationships. It is also in these issues of chaos and uncertainty that lie glimpses into some of the possible future directions and developments of physical geography.

UNCERTAINTY AND LIMITATIONS TO SYSTEMS APPROACHES

One of the main issues raised by GST is how do we assess, analyse and manage uncertainty, and how do we evaluate our predictions? In other words, how do we know if our models are right (see Beck, 2002)? Answers to some of these questions have been developed by atmospheric physicists to understand and quantify limitations in the use of General Circulation Models (GCMs). However, such assessments are also useful for understanding uncertainty issues more generally in system behaviour, and also relate to the frequently encountered phenomenon known as equifinality.

Uncertainty

There are at least five types of uncertainty: forcing uncertainty; microscopic initial condition uncertainty; macroscopic initial condition uncertainty; model inadequacy; and model uncertainty (see Stainforth et al., 2007). These interact with each other at certain times and places and in non-linear ways, and are therefore not independent.

Forcing uncertainty relates to what drives or initiates changes in physical systems. One of the important variables in most environmental systems is changes in climate. This has a major influence on geomorphological systems. As a result, uncertainty in future climate change must form one of the major uncertainties in geomorphic system behaviour. Similarly, changes in land-use or tectonic stability will play a major role in future landform development, and thus, prediction of these may be problematic.

Initial condition uncertainty can be subdivided into those uncertainties associated with macroscopic, and those with microscopic, initial conditions (IC). Macroscopic IC uncertainty dominates when the scale of enquiry is macroscopic and where the boundary conditions of the system are not well known. Some of the uncertainties associated with microscopic IC are irreducible, associated with the practical and logical limits in measuring all the variables contributing to system behaviour and their variations over the timescales required to produce complete information on the system. One way in which Microscopic IC uncertainty has been analysed in climatology is by using ensembles of models, each with slightly different initial states. The outcome from this is a Probability Distribution Function (PDF) which displays all the possible model outcomes and these are regularly used as predictive tools for future climate change (see Stainforth et al., 2005). Similar analyses used by geomorphologists might enable the boundary values within which landform development occurs to be explored. We can see this as the likely *outcome space* giving probabilities of the direction of landform change. As in climate systems, initial condition uncertainty plays a role in determining the predictability of the future evolution of the system, but scale plays an important role here too. Much of the small-scale behaviour of a system might be highly non-linear and sensitive to initial conditions, while the broad-scale system may display predictable trajectories, and this explains many of the problems of system convergence and equifinality which are discussed later.

Given the long timescales over which most landforms and landscapes develop, it is rarely possible to evaluate the accuracy of the model used to describe the system evolution. Modellers have divided such issues into model inadequacy and model uncertainty.

Model inadequacy reflects the degree to which models capture all the physical processes which are relevant to the landscape under study. There are at least three problems associated with this. First, we can have only partial understanding of all the processes that may be relevant for landscape development, especially those that occur over long timescales. Second, our understanding of small-scale processes must involve a form of parameterization, where the scale of model analysis is too coarse to capture all the relevant processes. Our understanding of how such small-scale processes affect the large-scale evolution of the system is, again, partial. Even if our model accurately described past landform evolution, there is no guarantee that future change will similarly be defined (this is a general problem of inductive reasoning). Third, because of the scale of enquiry adopted, there may be elements of the system behaviour that follow from the physical processes driving the model, but which are not represented at all. As a consequence, our model gives us only a partial account of real-world processes and landforms.

Model uncertainty arises because, while most models of complex landform systems can be expected to plausibly mirror landform development at the largest scale (and we can use these to explore likely future macroscopic system trajectories) the range of possible outcomes at the small scale is, however, large. Although ensembles of models could be run and PDFs generated, this would provide us with a sense of uncertainty only within the parameter space. The resultant shape of the distribution of model probabilities may not have any clear relationship with the system probabilities in the real world. As such, a model will be essentially subjective (Stainforth, et al., 2007).

As a result of these uncertainties, an important component of many systems analyses is the 'black box' (see Chorley and Kennedy, 1971: 7–8), usually seen as an epistemological device reflecting uncertainty in our understanding of the operations of the system. In a black-box approach, the whole system is treated without any internal structure, so that the emphasis is on predicting or characterizing output reponses to given inputs. By contrast, grey-box approaches involve a partial view of the system components, with a simplified or limited number of internal subsystems or 'workings' whose own internal operations are not considered. White-box approaches attempt to identify and analyse as many storages, flows and components of the system as possible, so as to understand how and why certain system outputs result from the operation of components and their linkages. The choice between these approaches is not simply methodologically – or practically – governed. A black-box system model may also represent ontological uncertainty – that is, a set of system attributes and operators that cannot be known. In here may lie the workings that produce a range of counter-intuitive outcomes from system behaviour, including equifinality with associated problems of prediction and retrodiction of system trajectory. It is this problem that exemplifies several of the difficult philosophical and practical issues facing physical geography, and it is to these that I now turn.

Equifinality

Equifinality occurs where significantly different initial states evolve to indistinguishable final states through many potential trajectories. This can be seen to oppose the mathematical notion of chaos where the system trajectory is sensitively dependent upon initial conditions. Equifinality is also called 'convergence' and is a characteristic of system behaviour in many complex sciences. However, chaos and equifinality can co-exist; at the small scale, the detailed dynamics of a process may display chaotic behaviour, yet significantly different initial states may evolve to indistinguishable final states. Most physical processes can be modelled at the macroscopic or microscopic level (with the caveats attached to modelling discussed earlier). While at the microscopic level the system displays chaos, at the macroscopic level there exists an emergent property of the system whose evolution fails to display chaos but displays the property of equifinality.

In a number of instances in physical geography, equifinality has been used to explain the outcomes of system evolution (Culling, 1987; Phillips, 1997). For instance, in geomorphology, the classic example of the development of tors is often used to highlight many of the issues associated with equifinality. Here, tor development reflects two distinct models of landform development in the same landscape. Tors associated with two stages of development (produced largely by Tertiary weathering) co-exist with tors associated with one stage of development, constructed during Pleistocene periglacial episodes (Figure 14.3). Similar equifinality is also seen when assessing the problem of classifying rock glaciers, where both periglacial and glacial models co-exist, and in the problem of drumlin formation where different models stress the role of subglacial accretion processes or the role of erosive subglacial water pressures.

Defining an equifinal system appears to be a major practical and philosophical problem in much of physical geography. For example, if we argue that two different sets of processes can produce tors (as in the one-stage and two-stage example above), then we would argue that tors are equifinal landforms. However, this equifinality might just be a consequence of our classification system and

Figure 14.3 A classic case of equifinality? Tors developed by periglacial action or by deep chemical weathering

therefore of how 'coarse-grained' we have become in our attempts at landform discrimination. In other words, if we were to classify tors more carefully, we would find that two distinct forms existed. The equifinality problem would then cease to exist (see Haines-Young and Petch, 1983. Continued, and more fine-graining of our classification system would then allow us to identify even more types of tor. The obvious outcome of this process is that eventually our classifications become so fine-grained that all tors are sufficiently different from each other that we are unable to identify any broader classifications. Such a scheme involves an infinite regress which must be overwhelmingly reductionist and which must destroy all classifications of landforms and processes. At this point, geomorphology would become an exceptionalist and idiographic subject. It therefore seems clear that one of the central problems with equifinality in geomorphology is the problem of clas-sification of processes *and* the landforms that result. The degree of fine-graining which geomorphologists are prepared to employ and the associated issues are also of more general relevance to systems theorists in other parts of physical geography and in other scientific fields.

From this, several questions present themselves. First, how have physical geographers decided to adopt a particular scale of fine-graining? Historically, this has been largely determined by technological considerations, the degree of fine-graining changing through time as a result of increasing computer resources, remote-sensing abilities and analytical techniques. However, the answer to this also depends upon the questions that the researcher is asking. Process–form geomorphology requires a scale of enquiry that is smaller than that needed by landscape-scale investigations. As a result, early models of landscape and landform evolution employed a large-scale approach which did not consider process in detail.

As we have seen, since the middle of the twentieth century there has been a move towards increasing analysis of process and the application of a largely reductionist (and therefore increasingly fine-grained) methodology. The choice of how far to fine-grain the analysis therefore varies according to the predilection of the researcher and the nature of the research problem that is being addressed.

Second, does the degree of fine-graining change depending upon the nature of the landforms under observation? There are certainly treatments of landforms which stress the interconnectedness of landform elements. Perhaps the most successful of these is the land-system approach employed by glacial geo-morphologists to examine the range and disposition of landforms within differ-ing glaciological, topographic and geological settings (e.g. Evans, 2003).

Third, does equifinality also represent a distinct class of uncertainty? Model-building under the assumption of system equifinality may have to recognize that there are no formal ways in which to decide whether a given system model is optimal since a number of different models may operate equally successfully. To move forward, we might be able to reject models based upon the assessment of new data. However, as Beven (2001) cautions, if we are unable to decide between successful models, then the only way to progress might be to reject them all, which means that we would have no predictions! In the final analysis, it may be that we have to be more realistic in our view of what constitutes an acceptable model.

Finally, we can ask whether the level of fine-graining employed by phys-ical geographers will change in the future? We can state with some certainty that our understanding of the trajectory of landform analysis in the past and the nature of the process/form paradigm which is prevalent in much of geomorphology, means that, up to now, fine-graining has been regarded as a recognized route to geomorphological explanation. However, for several decades there have been intriguing movements towards new analyses of physical systems which stress the interrelationships between various system elements, and the emergent structures that result. We can identify such a movement in physics (Anderson, 1972; Deutsch, 1998), complexity theory (Wolfram, 1984) and in much of physical geography (Harrison, 2001; Summerfield, 2005).

The identification of emergence as a characteristic of natural systems places potentially important restrictions on the likely success of GST in explain-ing the development and trajectories of such systems. It shows that explanation is likely to be scale-dependent, that understanding of how systems operate may not be increased by a simple reductionist programme, nor by uncritically extrapolating across the spatial (and temporal?) scales (see Beven, 1995). As a con-sequence, there is now the recognition that systems may possess an evolution-ary trajectory which is not amenable to description or analysis via a simplistic systems theory (Harrison, 1999, 2001).

PHILOSOPHICAL PROBLEMS IN THE APPLICATION OF SYSTEMS APPROACHES

We can see that the overriding (although little discussed) aim of a systems approach is that the system which is developed must be simpler than the dataset or phenomena it is trying to describe and model. This idea is a restatement of the scientific approach described by Liebniz in the late seventeenth century, and provides the basis for much research in complexity and algorithmic information theory. If the phenomenon under analysis is only perfectly describable using an algorithm as complex as the phenomenon, then we can say that the phenomenon is incompressible, and that it has no redundancy. Thus, physical information of a phenomenon is the measure of the system's form and dynamics. Clearly, if GST is to work efficiently as a simplifying device, then the phenomena under analysis must be compressible and display redundancy. However, it is not at all clear that this is the case for many natural systems.

Thus, in the final analysis, equifinality or system convergence (if it exists at all) may form a class of a more general set of ways in which (perhaps all?) natural systems display irreducible uncertainty and incompressibility. GST may not be able to capture adequately the undecidability of system evolution, and will forever provide only limited explanatory power. There are a range of similar problems in mathematics which relate to the limits of our knowledge of natural systems (e.g. algorithmic complexity, computational intractability, and NP-complete problems), although there has been, as yet, little discussion of these in the geographical literature.

We therefore have a major philosophical problem. On the one hand, there is the distinct feeling in science that simplicity and elegance are crucial attributes of successful scientific theories. On the other, there is the recognition that certain attributes of natural systems display irreducible uncertainty. Much of the latter is derived from research in algorithmic information theory by workers such as Kolmogorov and Chaitin. Despite this, we have seen that, within the confines of GST, there is always the search for increased explanatory power. Any resulting theory can have only conditional status as it is always possible that a simpler, more elegant theory will have more explanatory power. So, while systems are developed to simplify complexity, we can never know whether they are the simplest (and therefore most likely to be right) explanations of complex phenomena and interactions.

Further philosophical problems exist which relate to the ontological consequences stemming from quantum mechanics (our most successful and verifiable physical theory). There is compelling theoretical, experimental and mathematical evidence that shows that quantum physics is probabilistic, and that, as a result, some argue that an infinite amount of complexity is created in quantum systems. From this, the world and universe is essentially incomprehensible. As a result, GST would hold formally only if the world had finite complexity. This is highly unlikely and important ontological implications flow from this as it means that systems actually have no explanatory power. The opposite view is taken by Wolfram (2002), who believes that we can describe, recreate and account for much of the complexity we see in natural systems by using simple deterministic algorithms; complexity therefore derives from simplicity. Reconciliation between these views may not be possible and we are then left with the unsettling suspicion that GST may be based upon an insecure ontological footing.

CONCLUSIONS

The preceding discussion has tried to show some of the important practical and philosophical issues facing those who would want to develop further a systems approach to physical geography. The enormous (and perhaps infinite) complexity of natural systems places significant barriers in the way of explanation. However, it is an appropriate human response to attempt to simplify the complexity we see in nature and make sense of it, and the systems approach has been a resilient feature of scientific research and teaching for much of the last century. Indeed, many of the very concepts which now imply that GST is inherently limited as a predictive tool (complexity, emergence, trajectories and equifinality) would not make sense outside a systems conception of the world. It remains to be seen whether the problems outlined here prove to be ultimately insurmountable and restrict the future development of GST, or, conversely, whether they end up illuminating the way to a more sophisticated physical geography. For the latter scenario, the optimist will see the development of Earth Systems Science as a pointer to the ways in which physical geography has recently moved. The integration of many of the subjects previously deemed to have formed the realm of 'physical geography' is a welcome one (see Clifford and Richards, 2005) and shows us the ways in which future scientific collaboration and the creation of communities of knowledge may be used to investigate such 'metaproblems' as climate change.

SUMMARY

- Three main types of system may be identified: closed, isolated and open systems. All systems have three fundamental characteristics: structure, functioning and evolution.

- Systems may also be defined in terms of their complexity and organization. Systems are tractable when they display low values of complexity or organization (meaning that they are relatively easy to model), and as intractable when the opposite case applies.

- Many analyses and uses of systems approaches in physical geography and its subdisciplines derive from the inspiration and framework provided by General System Theory (GST).

- Two principal aims of GST can be identified: one has been to develop unifying principles throughout the sciences; another has been to simplify, conceptualize and build models of the 'real world'.

- The application of GST in physical geography and geomorphology was part of a changing paradigm from dominantly qualitative to dominantly quantitative research.

- Relatively little attention has been focused on the uncertainties and limitations inherent in the systems approach. Uncertainties are of many kinds, and environmental systems also display complexities, non-linearities, evolutionary trajectories and emergent behaviour in common with social and biological counter- parts. These attributes are not amenable to simple and standard systems theory approaches, and raise philosophical as well as practical issues relating to the role and use of systems theory.

Further Reading

Chorley (1962) and Chorley and Kennedy's (1971) *Physical Geography: A Systems Approach* are dated, but are still foundational references in the literature on systems theory and its use in physical geography. The first sets a context for the changing paradigms of landform analysis, and for the application of a new quantitative approach using GST. The second is a detailed exposition of GST methodology and its application to physical geography. The three articles by Harrison (1999, 2001, 2005) provide more detail regarding philosophical and practical issues in landscape study, and should be read in conjunction with Summerfield (2005) which provides an up-to-date perspective on the long-standing, and different, traditions of geomorphology within which GST should be evaluated.

Note: Full details of the above can be found in the references list below.

References

Anderson, P.W. (1972) 'More is different: broken symmetry and the nature of the hierarchical structure of science', *Science*, 177: 393–6.

Beck, B. (2002) 'Model evaluation and performance', in A.H. El-Shaarawi and W.W. Piegorsch (eds) *Encyclopedia of Environmetrics*, 3: 1275–9.

Beven, K. (1995) 'Linking parameters across scales: sub-grid parameterisations and scale-dependent hydrological models', *Hydrological Processes*, 9: 507–26.

Beven, K. (2001) 'How far can we go in distributed hydrological modeling?', *Hydrology and Earth System Science*, 5: 1–12.

Chorley, R.J. (1962) 'Geomorphology and General Systems Theory', *United States Geological Survey, Professional Paper 500B*.

Chorley, R.J. and Kennedy, B.A. (1971) *Physical Geography: A Systems Approach*. London: Prentice Hall.

Clifford, N.J. and Richards, K.S. (2005) 'Earth System science: an oxymoron?', *Earth Surface Processes and Landforms*, 30: 379–83.

Culling, W.E.H. (1987) 'Equifinality: modern approaches to dynamical systems and their potential for geographical thought', *Transactions, Institute of British Geographers*, NS 12: 57–72.

Davis, W.M. (1899) 'The geographical cycle', *Geographical Journal*, 14: 481–504.

Deutsch, D. (1998) *The Fabric of Reality*. London: Penguin.

Evans, D.J.A. (2003) *Glacial Landsystems*. London: Edward Arnold.

Haines-Young, R. and Petch, J. (1983) 'Multiple working hypotheses: equifinality and the study of landforms', *Transactions, Institute of British Geographers*, NS 8: 458–66.

Haines-Young, R. and Petch, J. (1986) *Physical Geography: Its Nature and Methods*. London: Harper & Row.

Harrison, S. (1999) 'The problem with landscape: some philosophical and practical questions', *Geography*, 84(4): 355–63.

Harrison, S. (2001) 'On reductionism and emergence in geomorphology', *Transactions, Institute of British Geographers*, NS 26: 327–39.

Harrison, S. (2005) 'What kind of science is physical geography?', in N. Castree, A. Rogers and D. Sherman (eds) *Questioning Geography: Fundamental Debates*. Oxford: Blackwell, pp. 45–60.

Oerlemans, J. (2005) 'Extracting a climate signal from 169 glacier records', *Science*, 308: 675–7.

Phillips, J.D. (1996) 'Deterministic complexity, explanation, and predictability in geomorphic systems', in B.L. Rhoads and C.E. Thorn (eds) *The Scientific Nature of Geomorphology*. Chichester: John Wiley and Sons, pp. 3: 115–35.

Phillips, J.D. (1997) 'Simplexity and the reinvention of equifinality', *Geographical Analysis*, 29: 1–15.

Richards, K.S. (1982) *Rivers: Forms and Processes of Alluvial Channels*. London: Methuen.

Stainforth, D.A. et al. (2005) 'Uncertainty in predictions of the climate response to rising levels of greenhouse gases', *Nature*, 433: 403–6.

Stainforth, D.A., Allen, M.R., Tredger, E.R. and Smith, L.A. (2007) 'Confidence, uncertainty and decision-support relevance in climate predictions', *Philosophical Transactions of the Royal Society*, 365: 2145–61.

Strahler, A.N. (1950) 'Davis' concept of slope development viewed in the light of recent quantitative investigation', *Annals of the Association of American Geographers*, 40: 209–13.

Strahler, A.N. (1952) 'The dynamic basis of geomorphology', *Geological Society of America Bulletin*, 63: 923–38.

Strahler, A.N. (1980) 'Systems theory in physical geography', *Physical Geography*, 1: 1–27.

Summerfield, M.A. (2005) 'A tale of two scales, or the two geomorphologies', *Transactions, Institute of British Geographers*, NS 30: 402–15.

Trenhaile, A.S. (1987) *The Geomorphology of Rock Coasts*. Oxford: Clarendon Press.

Von Bertalanffy, L. (1973) *General System Theory*. London: Penguin.

Warren, C.R. (1992) 'Iceberg calving and glacioclimatic record', *Progress in Physical Geography*, 16: 253–82.

Whittaker, R.J. (1998) *Island Biogeography: Ecology, Evolution, and Conservation*. Oxford: Oxford University Press.

Wolfram, S. (1984) 'Cellular automata as models of complexity', *Nature*, 311: 419–24.

Wolfram, S. (2002) *A New Kind of Science*. Champaign, IL: Wolfram Media.

15

Landscape: The Physical Layer

Murray Gray

Definition

Landscape can be viewed as comprising three, predominantly superimposed, primary layers – physical, biological and cultural. Of these, the physical layer (the geodiversity of rocks, sediments, land form, soils and physical processes) is often undervalued in its contribution to the character of landscapes. However, it is impossible to have a sensible approach to the understanding of landscape, or to landscape management and planning, that ignores the physical layer. An integrated approach to landscape management, in which the character, processes and materials of the physical landscape plays a central role, is required to meet contemporary environmental challenges as well as to help broaden perspectives of landscape studies.

INTRODUCTION: LANDSCAPE LAYERS

Much has been written about landscape, but it has normally focused on landscape ecology or on cultural landscapes or combinations of these (e.g. Hoskins, 1955; Rackham, 1986; Whitehand, 1992; Cronon, 1995; Wiens and Moss, 2005). Surprisingly little has been written about the direct and indirect contribution of abiotic elements to landscape character. This chapter is therefore mainly concerned with the characterization, conservation and restoration of the physical foundations of landscape.

Landscape can be studied in many ways, but it is useful to think of it as comprising three primary layers, each of which includes a number of secondary

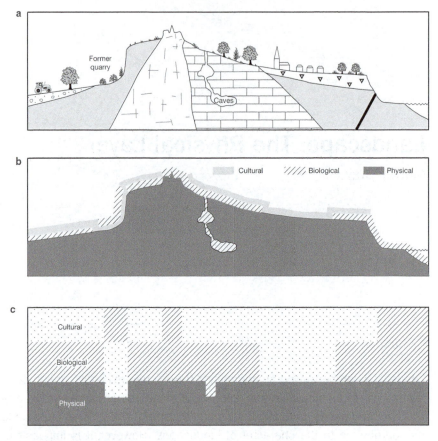

Figure 15.1   The three layers of landscape (a) as a real landscape,
(b) as initial representations of the layer types, and (c) as a further abstraction
into layers

layers (Table 15.1 and Figure 15.1). At the base is a physical layer, comprising the
rocks, sediments and soils, the landscape topography and the physical processes
operating on the landscape. Above this is the biological layer comprising flora
and fauna, wildlife habitats and ecosystems. Finally, there is a cultural layer
involving land-use, buildings and infrastructure, all of which have evolved
through time and therefore include historical land-uses. This layer also includes
the human experience of landscape through the senses of sight, sound, smell and
touch, and through memories and associations of landscape. Any one of the three
primary layers may dominate a landscape. The cultural layer dominates most
obviously in towns and cities, where the urban landscape is predominantly
human-created, or in agricultural rural areas where the natural vegetation has
been cleared and, in the case of the rice terraces of Indonesia or the vine terraces
of the Douro Valley in Portugal, the hillside slopes have been terraced to improve
growing conditions. The biological layer dominates in many non-urban and non-
agricultural areas, particularly where there is no obvious landform, for example
the tropical rain forest of the Amazon or Congo basins. Megafauna may be part
of the biological landscape, as in the case of caribou herds migrating across the

Table 15.1   Primary and secondary landscape layers

Primary landscape layers	Secondary landscape layers
Cultural	Experiences and associations Current land-use Historical land-use
Biological	Megafauna Trees Shrubs Herbs and grasses
Geological	Soils Physical processes Land form Rocks and sediments

Alaskan or Canadian tundra. The physical layer can dominate in natural areas where vegetation is scant or absent. This would include:

- high latitudes, e.g. Antarctica, Greenland
- high altitudes, e.g. Himalayas, Urals
- glaciers, recently deglaciated areas and outwash plains
- arid areas, e.g. Sahara Desert, Kalahari Desert
- semi-arid areas, e.g. Southwest USA
- beaches, rocky coasts and cliffs
- braided river channels
- active or recently active volcanic areas, e.g. central Iceland, Montserrat.

The layers are clearly interrelated and may intermingle. For example, soil is a complex mixture of weathered regolith, organic matter and a living soil ecology. Wildlife may recolonize anthropogenic landscapes, as in the case of Inca, Aztec and Mayan archaeological remains recolonized by the tropical vegetation of Central and South America, or the Norfolk Broads in England, which began as medieval peat diggings but have become flooded and recolonized by wildlife. Archaeological artefacts may be found within river gravels, macrophytes grow in rivers, and towns and cities are predominantly constructed from physical resources (e.g. stone, brick, tile, steel, concrete, tarmac, glass, etc.). But despite this intermingling, the three layers are still clearly distinguishable and they combine to form many modern landscapes (Figures 15.1a and 15.2).

THE PHYSICAL LAYER

If we analyse the physical layer into its components we find that it comprises several elements – rocks, sediments, minerals, fossils, topography, soils and physical processes. In turn, these physical elements vary in character and this means that just as the living world has a biodiversity, so the non-living world has a geodiversity (geological and geomorphological diversity). Geodiversity has been defined as 'the natural range (diversity) of geological (rocks, minerals, fossils), geomorphological

Figure 15.2   The landscape of south Antigua, looking westwards from Shirley Heights across English Harbour and Falmouth Harbour. The physical landscape is represented by the coastal morphology and processes, the rock outcrops and the hill topography. The biological layer is represented by the tree and scrub cover. The cultural layer is represented by the boats and buildings, which include the historical eighteenth-/ nineteenth-century dockyard at English Harbour

Photo: Murray Gray

(land form, processes) and soil features. It includes their assemblages, relation-ships, properties, interpretations and systems' (Gray, 2004: 8). Minerals and fos-sils need not concern us in considering the landscape scale, which is generally taken to extend from hundreds of metres to many kilometres, but the following paragraphs discuss the contributions of the other physical or abiotic elements of geodiversity to landscape.

Rocks and sediments

Rocks (including sediments) are ubiquitous elements of the landscape but not always ubiquitously visible since they can be obscured by soil, vegetation or buildings and human infrastructure. As outlined above, they are therefore most obvious in areas where these elements are sparse due to low temperatures, low rainfall or active physical processes. Even where the rocks are buried below sediments, soils and vegetation, they often have a major influence on topogra-phy through structural controls or variable rock resistance to erosion. For example, geological factors often determine the location and form of uplands, hills, valleys, etc. (e.g. the basin and range landscapes of the US Rockies).

Where rocks and sediments do outcrop, they display a wide variety of characteristics including the following:

- Rock type – igneous, sedimentary, metamorphic and their subdivisions. There are internationally accepted classification schemes for these rock groups, though in practice, in nature there is usually a continuum of rock diversity (e.g. from ultra basic igneous rocks to highly silicic ones).
- Rock colour – rocks vary in colour (e.g. white chalk, red sandstones, grey limestones, multicoloured sediments and conglomerates). Colour includes rock weathering character.
- Rock structure – bedding, jointing, folding, etc.; sediment character – size, sorting, colour, etc.

These characteristics help to give the landscape its distinctive character. Examples include the white chalk cliffs of southern England, the vertically-bedded red sandstone rocks of Uluru, Australia, and the horizontally bedded and varying erosional resistance of the sedimentary rocks of the Grand Canyon.

Land form

Like rocks and sediments, land form is ubiquitous in the landscape and can even be detected, albeit in modified form, within towns and cities. In many places, specific landforms can be named and there are several landform classification schemes (e.g. Gardiner and Dackombe, 1983). However, the term 'land form' rather than 'landforms' is preferred here since most landscapes cannot be described in terms of standard landform categories, yet they have a physical form that remains essentially natural over large parts of the world. The common statement that there are no natural landscapes left in densely-populated countries like England, with a long history of human land-use, applies to the biological landscape layer but often not to the physical layer. It is true that in some areas engineering works, quarrying and urban development have destroyed the natural land form and in other areas farming has smoothed out some irregularities. But in many rural areas the basic form of the land is still natural, the result of millions of years of earth history, with a particular final flourish of landscape evolution during the Quaternary period. In many parts of the world the land form as a whole is natural and it helps to give the landscape its local distinctiveness and provides landmarks that become familiar to, and are valued by, local people (Figure 15.3).

Processes

Physical processes are also ubiquitous and continue to fashion the landscape when allowed to do so. The main processes include volcanic, tectonic, coastal, fluvial, glacial, periglacial, slope and aeolian processes. Volcanic and tectonic processes occur predominantly at plate margins but also at hot-spots at plate interiors, for example Hawaii and Yellowstone in the USA. Glacial and periglacial processes occur where climatic conditions are conducive. Aeolian processes occur where lose sediment is prone to wind action and this is predominantly in arid areas or coastal dune systems. Fluvial and slope processes occur at many locations throughout the terrestrial landscape and coastal processes at its margins.

Figure 15.3   A rolling landscape in Shropshire, England, looking towards the Longmynd, part of the Shropshire Hills Area of Outstanding Natural Beauty. Land form is an important aspect of this landscape but there are few specific landforms to be seen other than general ones such as hills, slopes and valleys. Note that the land form is essentially natural, but the vegetation/land-use is not

Photo: Murray Gray

Within each type of process there are complex internal processes taking place. Of course, there are no processes without agents, and running water, ponds, lakes, the sea and glacial ice are the materials of processes and need to be included as part of landscape character. There are also temporary physical manifestations in the landscape, including snow, sea-ice and volcanic activity, and more remote physical effects, including sunshine, rain, cloud cover and other weather effects.

Soils

Soils also occur everywhere and do have formal classification systems. For example, the USA recognizes a classification hierarchy of soil orders, sub-orders, great groups, subgroups, families and series. There are 19,000 soil series in the USA (Brady and Weil, 2002). In England, there are an estimated 800 main soil types and although many have been modified by human land-use practices over the centuries, they remain at least semi-natural. Soils are generally not visible except where vegetation is sparse or absent, and so make little direct visual contribution to landscape. But they have a major influence on vegetation and therefore have an important indirect impact on landscape. For example, Chalk grassland has a very different landscape character from an acidic heathland. Soils, in turn, are strongly influenced by the parent material (Jenny, 1941) and in some cases this link becomes obvious, particularly through soil colour, as in the red soils developed on red sandstones or dark grey soils developed on coal-bearing rocks. These become visible when land is ploughed.

Until recently soils have tended to be ignored by nature conservation agencies, partly because they fall between the geological and biological worlds and, to some extent, have been regarded as the preserve of agricultural land management. While the latter is still the case, more attention is being given to soil conservation in response to soil losses and degradation resulting from poor land management practices, for example wind and water erosion of bare soils.

Putting it all together

In summary, many elements of geodiversity exist in the wider landscape. Within the visible landscape and at the landscape scale, land form is the most obvious element of the physical landscape. In addition, where soils and vegetation are absent, rock and sediment outcrops contribute significantly to the scene, as do the operation of slope and fluvial processes. In particular locations, volcanic, coastal, glacial, periglacial and aeolian processes give a particular prominent character to the landscape through their operation and influence on landforms, rocks and sediments.

It is common for these three elements of land form, materials and processes to combine to produce distinctive parts of the landscape. Figure 15.4 is an example of a semi-arid landscape in the southwestern USA, dominated by the physical layer as the geomorphological processes and semi-arid environment allow little soil and vegetation to develop. On the coastline, it is the combination of material, process and form that create distinctive coastal landscapes.

LANDSCAPE CHARACTERIZATION

Landscapes vary in character from place to place (landscape diversity) and 'landscape characterization' is the term often used for the processes of mapping and describing the character of the different landscapes of particular parts of the world. Different countries have approached this process in different ways and this section gives some examples, paying particular attention to the contribution of the physical layer.

In the 1990s, English Nature, the government agency charged with nature conservation in England, divided the country into 120 Natural Areas (97 terrestrial and 23 maritime) on the basis of similarities in geology, landscape, landform, vegetation and soils. At about the same time, the Countryside Agency, another government agency charged with promoting the interests of rural areas, was involved in a similar exercise, including the cultural layer and concluding with a finer series of subdivisions. The two approaches were combined into a landscape character map of England with 159 Joint Character Areas combining the physical, biological and cultural layers. The aim of this approach was to recognize that there was much of natural and cultural conservation value outside protected sites and areas and that a 'wider landscape' approach was necessary. The result was, however, rather disappointing as far as geoconservation in

Figure 15.4   The 'Goosenecks' incised meanders on the San Juan River, Utah, USA. The physical layer is dominant in this landscape with materials (sedimentary rocks layers, slope sediments, fluvial sediments), processes (slope evolution, fluvial erosion, transportation, deposition) and landforms (incised meanders, strata terraces, slope forms)

Photo: Murray Gray

general, and geomorphological conservation in particular, was concerned (Gray, 2001). The descriptions of each Natural Area were criticized on several counts but particularly as generally continuing to promote the site-based approach rather than wider landscape conservation. However:

> A few Natural Area Profiles do start to give the essence of what is required. ... In these descriptions we begin to see how the diversity of English landform, surficial deposits and processes – the distinctive geomorphological character of each area – can be recorded. Maps and descriptions of this type would help to crystallise what is valued in the local landscape character, the sensitivity of landscapes to geomorphological change and what needs to be done to protect the landscapes from inappropriate development. (Gray, 2001: 1019)

It remains the case that there are no systematic descriptions of the geology and physical landscape for each of the Natural or Joint Character Areas, nor for England as a whole, that could guide future management of this physical layer by the new integrated agency, Natural England.

An example of good practice in natural area description is the report from Scottish Natural Heritage that describes the geology, palaeontology, geomorphology, soils and physical landscapes of both Scotland as a whole and each of the 21 Natural Heritage Zones (Gordon et al., 2002). It is a good model to follow in implementing a wider landscape approach to geodiversity using natural units as the base.

In the UK, Landscape Character Assessment (LCA) is related to the Natural/Joint Character Areas approach but at a finer scale. Like the Joint Character Areas described above, it takes into account the cultural elements of landscape. In other words, a Joint Character Area, can be further subdivided into smaller landscape areas. This allows for individual landscape types, such as valleys and plateau tops, to be identified or even, valley sides, valley floors, etc. In England, however, LCAs have been carried out for individual local authority areas (counties, districts or unitary authority areas) rather than producing nationally recognized and consistent landscape zones. This creates some boundary problems where local authorities adjoin and this is exacerbated by the absence of a nationally co-ordinated approach. To some extent this is being addressed through national guidance (Swanwick and Land Use Consultants, 2002) and the development of the Landscape Character Network (www.landscapecharacter. org.uk), an informal organization of anyone interested in sharing ideas and approaches to LCA. However, in Scotland, Wales and Northern Ireland there is greater national co-ordination of the individual assessments. The aim of LCA is to raise awareness of the local distinctiveness or diversity of the landscape and encourage appreciation of the differences between places. In turn, this can lead to policies/decisions on protection of key landscape characteristics, restoration strategies or identification of landscape sensitivity.

The approach has great potential to promote the conservation of geodiversity in the wider landscape, though this potential is seemingly only partially being met at present. Thus, in many Landscape Character Assessments in the UK descriptions of geodiversity is cursory or limited to site-based comments, and there is little appreciation of all the values of, or threats to, the physical layer. For example, in the Cumbria LCA (which excludes the Lake District National Park) (Cumbria County Council, 2004), there is some description of the geology and geomorphology of each landscape type or subtype but descriptions of 'key landform and rock evolution interests' are given for only 11 of the 37 landscape subtypes, yet all 37 have key points for ecology and historical/cultural associations. Furthermore, the key points that are given for geology/geomorphology are very brief and refer only to some coastal processes, drumlins, caves/karst and important geological exposures (Table 15.2). The Cumbria LCA is a typical example of landscape character descriptions and it follows from this that LCA in general is also not being applied as it should be to geodiversity in the wider landscape and specifically to policy development, decision-making, restoration strategies or identification of landscape sensitivity. Yet the LCA approach could lead to conservation of the distinctive character of the physical layer, more authentic landform design and remodelling, and the restoration of rivers, coasts, etc. to more naturally dynamic process conditions.

In Wales, the LANDMAP Information System is different from the landscape characterization programmes used elsewhere in the UK in being a multilayered approach in a geographical information system (GIS) that includes at its base a 'Geological Landscapes' layer. Above this are Landscape Habitats, Cultural Landscape, Historic Landscape and Visual and Sensory layers. The layers can be viewed separately or overlain with other layers or other compatible datasets to

Table 15.2 Landscape types and subtypes in Cumbria (excluding the Lake District National Park)

Landscape type	Landscape subtype	Ecology	Historical/cultural	Geology/geomorph	Key interests
1 Estuary and marsh	1A Intertidal flat	x	x	x	Dynamic system; mudflat siltation; shifting river channels; longshore drift
	1B Coastal marsh	x	x	x	Saltmarsh accretion; shifting river channels; front cliff erosion; pans; terraces
2 Coastal margins	2A Dunes and beaches	x	x		
	2B Coastal mosses	x	x		
	2C Coastal plain	x	x		
	2D Coastal urban fringe	x	x		
3 Coastal limestone	3A Open farmland and pavement	x	x		
	3B Wooded hills and pavement	x	x		
	3C Disturbed areas	x	x		
4 Coastal sandstone	4A Coastal sandstone	x	x	x	Permo-Triassic red sandstones
5 Lowland	5A Ridge and valley	x	x		
	5B Low farmland	x	x		
	5C Rolling lowland	x	x		
	5D Urban fringe	x	x		
	5E Drained mosses	x	x		
6 Intermediate land	6A Intermediate land	x	x		
7 Drumlins	7A Low drumlins	x	x	x	Drumlins of sand and gravel or 'boulder clay' as evidence for glaciation and ice direction
	7B Drumlin field	x	x	x	Drumlins from last glaciation; direction NW/SE at Kendal, NNW/SSE in Furness

(Continued)

Table 15.2 *(Continued)*

Landscape type	Landscape Subtype	Ecology	Historical/cultural	Geology/geomorph	Key interests
	7C Sandy knolls and ridges	x	x	x	4 km ridge at Brampton is gravel kame
8 Main valleys	8A Gorges	x	x		
	8B Broad valleys	x	x		
	8C Valley corridors	x	x		
	8D Dales	x	x	x	Fluvial landforms; past mining activity
9 Intermediate moorland and plateau	9A Open moorlands	x	x		
	9B Rolling farmland and heath	x	x	x	Permian rocks, Eden district
	9C Forests	x	x		
	9D Ridges	x	x		
10 Sandstone ridge	10A Sandstone ridge	x	x	x	Permian sandstone
11 Upland fringes	11A Foothills	x	x		
	11B Low fells	x	x		
12 Higher limestone	12A Limestone farmland	x	x	x	Carboniferous limestone; till and fluvioglacial deposits
	12B Rolling fringe	x	x		
	12C Limestone foothills	x	x		
	12D Moorland and commons	x	x		
13 Fells and scarps	13A Scarps	x	x		
	13B Moorland, high plateau	x	x		
	13C Fells	x	x	x	Caves and other karst features

allow integration and interrogation of the data. It is developed as a partnership between the Countryside Council for Wales, the unitary local authorities of Wales and the Welsh National Parks, aimed at providing a spatial information resource for use in sustainable landscape decision-making (Page et al., in press). Geological Landscapes are examined at several scales or levels (after Crofts, 1981):

- Level 1: General landscape character
- Level 2: Large-scale terrain or topography
- Level 3: Medium-scale typifying terrain or topography
- Level 4: Small-scale landform.

The system also allows full descriptions of the geological and topographical character to be included in the database along with other geological information. The methodology also requires the intrinsic value, current condition, assessment of current management regimes, recommendations for future management and tolerance to change to be included in the database (Page et al., in press). As such, the scheme is an excellent example of how physical landscapes can be mapped, described and applied to integrated and sustainable landscape management.

In Italy, a GIS map of landscape physiographic units has been produced from satellite and aerial photographs plus some field checking. The units are defined as homogeneous portions of land characterized by geological structure, morphology, lithology and land-use. Like LCA, both generic landscape types (at the landscape scale) and specific local areas are recognized and the local and unique features are then defined and described. In central Italy, 21 landscape types have been described (Amadio et al., 2002) that combine geological and geomorphological influences, for example 'siliclastic hills', 'volcanic mountains', 'crystalline massif'.

In Malaysia, Komoo and Othman (2002) also used the term 'geological landscape' to describe 'an assemblage of landforms and geomorphic features'. They argued that each landform can be characterized according to six elements:

- general landforms (e.g. mountains, hills, plains, islands and their subdivision)
- geologic terrains (e.g. igneous, metamorphic, sedimentary and their subdivision)
- internal processes (e.g. uplift, compression, wrenching, rifting, volcanism and their subdivision)
- external processes (e.g. weathering, erosion, deposition, mass movement extraterrestrial and their subdivision)
- temporal evolution (e.g. geological age, maturity, types (active or fossil))
- special features.

Geological landscapes are then examined at a range of scales from the regional scale, where only the general landform element is included, down to the local scale where all six elements are included in the landscape characterization. Like the LANDMAP programme in Wales, the scheme also includes an assessment of the scientific value of the geological landscapes as 'outstanding', 'high', 'medium' or 'low'. Komoo and Othman (2002) illustrated this geological landscapes approach in relation to the Kinabalu Mountain area, inscribed by UNESCO as a World Heritage Site, but only on the basis of its biodiversity.

In Australia, several approaches to landscape classification have been used. In the 1950s and 1960s the Commonwealth Scientific and Industrial Research Organization (CSIRO) developed the land-systems approach in which landscapes were classified and mapped according to recurring patterns of geology, landform, soils and vegetation. Land-systems are landscape-scale areas comprising series of landforms, often geologically controlled and which in turn influence soils and vegetation. Thus the key criterion for the recognition of land-systems was land form (Cooke and Doornkamp, 1990). More recently, Australian geomorphologists, particularly in Tasmania, have developed a 'regionalization' approach in which the inventories of landforms or systems in a particular region or on a particular theme are compiled, mapped and described. For example, Jerie et al. (2001) describe a project to classify and map river geodiversity in Tasmania with the aim of furthering river conservation.

Parks Canada has divided the country into 39 Natural Regions based on physical and biological characteristics. The six Natural Regions represented in the province of Alberta have been further subdivided into 20 subregions and three further levels of subdivision to create a natural landscape hierarchy that is being use for nature conservation management, state of environment reporting and integrated natural resource planning.

These brief outlines from very different countries illustrate that classification, mapping and description of whole landscapes or the physical layer/ geological landscapes alone are entirely possible and can be an important input into conservation of geodiversity in the wider landscape as well as more generally into integrated landscape characterization and landscape management. It is to the issue of landscape conservation that we now turn.

LANDSCAPE CONSERVATION

Nature conservation has, until recently, focused on protecting the most important sites or areas. Many of these cover large landscape areas as in the case of some of the National Parks of the western USA such as Yellowstone, Yosemite, Zion, Grand Canyon, Glacier Bay and Denali. Wrangell–St Elias National Park, straddling the Alaska/Canada border, covers 5 million hectares. However, many protected sites are very small (a few hectares), as in the case of most Sites of Special Scientific Interest (SSSIs) in Britain, and would not be classified as protected landscapes.

In the last 20 years, there has been a growing dissatisfaction within nature conservation circles with an approach that relies solely on protected areas to conserve nature. For example, Myers (2002: 54) argued that 'setting aside a park in the overcrowded world of the early twenty-first century is like building a sandcastle on the seashore at a time when the tide is coming in deeper, stronger and faster than ever'. In other words, protected areas were seen as becoming isolated from each other and vulnerable to human impacts. What was needed was a less fragmented approach to nature conservation that values nature wherever it occurs.

What has become known as the 'wider landscape' approach began in relation to protecting fauna, which is dynamic and cannot identify when it is leaving the protection of a designated area. Thus the idea of 'wildlife corridors' or 'greenways' linking protected areas emerged which could allow wildlife to move along these corridors. In this way, the matrix in which protected sites sit is given greater significance in landscape ecology. In some countries, for example the Netherlands, whole ecological networks of protected areas linked by these corridors have been or are being created, at least on paper (Jongman and Pungetti, 2004). Subsequently, the concept of biodiversity has extended nature conservation philosophy to the whole landscape, including urban areas, identifying the need to protect habitats and species wherever they occur. Adoption of the Convention on Biodiversity (CBD) at the Rio Earth Summit in 1992, required the 160 signatory countries to develop plans, strategies or programmes for the conservation and sustainable use of biodiversity.

Many of these arguments apply equally to the physical layer. Geodiversity occurs and has value everywhere, not just in protected areas. To protect the physical layer at the landscape scale there is a need to conserve:

- rock outcrops, where they contribute to landscape character or are scientifically important
- the natural form of the land surface (land form or topography)
- the operation of natural processes in the landscape
- soils.

This approach does not mean that no element of geodiversity can be altered or lost within the wider landscape, but what it does mean is that there should be an understanding of the physical context of areas where development is proposed so that change is carried out in the full knowledge of the implications for geodiversity; that is the threat of ignorance needs to be removed.

There is a particular issue concerning human remodelling of land form that is often overlooked. Kiernan (1997: 9) argued that: 'Landforms are defined by their contours. Hence any unnatural changes to the contours of a landform by definition damages the natural geomorphology. ... The geoconservation significance of the damage is what is important.' These are important words in the context of conserving geodiversity in the wider landscape since they reach to the core of the argument related to conserving geomorphological character and local distinctiveness. The changes to land form include remodelling during golf course construction, landraising for waste disposal sites, lake/reservoir formation, wildlife habitat creation and soil bunding (for an example, see Figure 15.5). Many of these redesigns of the natural land form create features that are out of character with the local topography, and as such are incongruous in the landscape (Gray, 1997, 2002). Linear soil bunds have been described as 'the geomorphological equivalent of the leylandii hedge' (Gray, 2004: 305) in being nonnative landforms. There are many examples of their increased use in the wider landscape and some evidence of a recognition that their use is inappropriate. For example, a planning inspector recently described a 300 m long soil bund in the Kent Downs Area of Outstanding Natural Beauty in England as strongly linear and contrived, forming an incongruous ridge at odds with the natural contours. He therefore dismissed the planning appeal on the grounds of inappropriateness and landscape harm (Anon. 2006).

The need for an authentic approach to land form design was recognized by the Countryside Commission in its *Golf Courses in the Countryside* (1993), which has a chapter on 'Topographical change'. The conclusion was (1993: 25) that 'large-scale remodelling is not essential to the quality of a golf course and can be highly inappropriate. In particular, topographical changes should reflect the local topographical character so that the final landform is indistinguishable from the surrounding landform'. The guidance goes on to note that mounding is often the most alien aspect in the specific landscape setting but much will depend on the local topographical character. 'A flat landscape can accept very little by way of grading and almost certainly no mounding', whereas in other places significant earthworks can be justified. An example of this would be a links-type golf course where natural mounds and hillocks reflect the character of stabilized sand dunes and additional mounding would not be out of character. Gray (2004: 307) argued that:

> what is needed is an intelligent and aware approach to what we are changing and why we are changing it. ... Key characteristics of landform and process character need to be identified and included in character guidelines, geomorphological sensitivity to development should be recognised, and opportunities for restoration or enhancement should be taken.

An important tool with the potential to help in conserving the physical layer is the land-use planning system that exists in most developed countries. Its use in this respect is perhaps most advanced in England where government policy has recently been revised to recognize geoconservation more explicitly. Planning Policy Statement 9 (PPS9) on Biodiversity and Geological Conservation was published in August 2005, with an accompanying Guide to Good Practice published in March 2006 (ODPM, 2006). It contains several encouraging statements of 'The Government's Objectives', which include:

- to promote sustainable development by ensuring that biological and geological diversity are conserved and enhanced as an integral part of social, environmental and economic development, so that policies and decisions about the development and use of land integrate biodiversity and geological diversity with other considerations.

- to conserve, enhance and restore the diversity of England's wildlife and geology by sustaining, and where possible improving, the quality and extent of natural habitat and geological and geomorphological sites; the natural processes on which they depend; and the populations of naturally occurring species which they support.

Figure 15.5 A soil bund surrounding a new agricultural building at Wattisfield, Suffolk, England. In this case the bunding is an alien landform because of its rectilinear form, steep slopes and sharp breaks of slope and simply adds to the landscape impact of the building that it is attempting to partially screen

Photo: Murray Gray

There is, however, a danger that in creating a diversity of wildlife habitats in order to increase biodiversity, geomorphologically artificial landscapes will be produced. There is a long history of cultural land form modification from Capability Brown (1716–83) in England through Frederick Law Olmstead (1822–1903) in the USA and on to the work of modern landscape architects (Cronan, 1995). We may now be about to enter a new phase of land form redesign to suit the needs of biodiversity.

Also relevant to landscape conservation and management is the European Landscape Convention, a Council of Europe initiative that was opened for signature in Florence on 20 October 2000 and entered into force on 1 March 2004. As of August 2008, 35 of the 46 European countries were signatories, the most recent being Serbia on 21 September 2007. Among the countries yet to sign are Germany, Austria and Iceland. Of the 35 signatories, 29 have gone on to ratify the Convention. The aims of the Convention are to promote the protection, management and planning of European landscapes through the adoption of national measures and co-operation between the European nations. These aims are to be promoted through four general and five specific measures. The four general measures are:

- to recognize landscapes in law as an essential component of people's surroundings, an expression of the diversity of their shared cultural and natural heritage, and as a foundation for their identity
- to establish and implement landscape policies aimed at landscape protection, management and planning
- to establish procedures for the participation of the general public, local and regional authorities, and other parties with an interest in the definition and implementation of landscape policies
- to integrate landscape into its regional and town planning policies and in its cultural, environmental, agricultural, social and economic policies, as well as in any other policies with possible direct or indirect impact on landscape.

The five specific measures are:

- awareness-raising among the public, authorities and organizations of the value and role of landscapes and changes to them
- training and education of specialists, professionals, school and university courses
- identification and assessment of a nation's own landscapes, analysing its characteristics, forces and pressures, taking note of changes and assessing the value of the landscapes
- framing landscape quality objectives after public consultation
- implementing measures to protect, manage and/or plan the landscapes.

LANDSCAPE RESTORATION

Restoration of the physical layer is becoming increasingly important as human society seeks environmental improvements and beneficial reinstatement and re-use of sites. It provides the opportunity to assist the aims of geodiversity conservation in the wider landscape by returning landscapes to a more natural state. The phrase 'more natural' is used here since by definition it is often impossible for humans to recreate a natural landform though natural processes can be restored. In fact, wherever possible georestoration should not create static places but should allow process dynamics and natural landform change to occur. As Adams (1996: 169–70) put it: 'what we are doing is facilitating nature and not making it ... We must allow nature space to be itself, to function to build and tear down.'

River restoration is a good example of this approach. It is estimated that 96% of English lowland rivers have been modified in some way, but over the last 15 years there have been growing attempts to reverse the process. Consequently, there is a large body of theory and practice on river restoration, now co-ordinated in the UK by the River Restoration Centre, an independent non-profit partnership whose aim is to encourage good practice in river restoration (www.therrc.co.uk). Examples include the removal of concreted channels and piled river banks, remeandering of straightened rivers, and removal of flood embankments to reconnect rivers with their floodplains.

A similar approach has been taken to coastal management where, rather than continued use of hard sea defences such as sea-walls, revetments and groynes, soft engineering approaches are being used such as beach replenishment and the breaching of flood defence embankments. Hard engineering solutions are expensive to install and maintain, have a limited life, are visually intrusive and often obscure important coastal geological exposures. Therefore a managed realignment approach is being advocated by various bodies responsible for managing the coastline in many countries.

Quarry restoration is another example. Quarried land can be restored to its previous contours by infilling with waste (allowing for settlement), covering with topsoil, and returning the land to agriculture or other beneficial use. An alternative design approach was pioneered by Gunn (1993) and Gagen et al. (1993), who describe the use of restoration blasting techniques in limestone quarries in Derbyshire, England, aimed at replicating the form of the local dry valleys. This, together with seeding, allows abandoned quarries to be transformed within a few years to mature and attractive valley features. Such approaches can also incorporate important geological faces into the final design with good access to them.

Other examples of georestoration include contaminated soil rehabilitation, and the blocking of upland bog drainage channels and ditches to restore the bogs to more natural wetland conditions. But perhaps the most powerful example of the potential and significance of the landscape restoration approach is the Hetch Hetchy Valley in Yosemite National Park in California, USA. This once spectacular valley was dammed in the 1920s to provide water for the San Francisco area. Since it was constructed within a national park, the scheme was very controversial at the time and for many years there has been a campaign for the removal of the dam and reservoir and restoration of the valley to its previous condition:

> The O'Shaughnessy Dam and the Hetch Hetchy Reservoir were a mistake. They were unnecessary and justified on spurious, self-serving reasons. ... San Francisco could have and should have obtained its water elsewhere. The project would never be approved today, and it should not have been approved in 1913. ... It's time to undo the past and make a profound statement about our present priorities and place on this planet. (Simpson, 2005: 318 and 325)

CONCLUSIONS

This chapter has demonstrated the importance of the physical layer in contributing directly to the character and aesthetics of landscape. It has also had an indirect effect on the biological layer in influencing landscape ecology and the cultural layer in creating the stage and physical materials for humans to utilize over the millennia. It follows from this that it is impossible to have a sensible approach to landscape management and planning that ignores the physical layer. What is needed, but too often is missing, is an integrated approach to landscape management in which the character, processes and materials of the physical foundations play a central role. Ironically, in view of much of the discussion in this chapter, integration will involve stopping thinking of the landscape in terms of layers, but instead understanding the interrelated nature of landscape processes and sustainable landscape management.

SUMMARY

- Much has been written about landscape, but it has normally focused on landscape ecology or on cultural landscapes or combinations of these. Surprisingly little has been written about the direct and indirect contribution of abiotic elements to landscape character.

- Many elements of geodiversity exist in the wider landscape: land form is the most obvious element of the physical landscape. In particular locations rock outcrops dominate while volcanic, coastal, glacial, periglacial and aeolian processes give a prominent character to the landscape through their operation and influence on landforms, rocks and sediments.

- Over the last 20 years, landscape has begun to figure in conservation strategies in a wider landscape approach. Two key themes are the characterization of landscape and the management and restoration of landscape.

- The need for conservation and restoration of landscape character is becoming increasingly important as human society seeks environmental improvements and beneficial reinstatement and re-use of sites.

- Landscape Character Assessment (LCA) raises awareness of the local distinctiveness or diversity of the landscape. LCA is important in formulating and applying policies of landscape management, conservation and restoration.

- An integrated approach to landscape management is required. This should include consideration of the character, processes and materials of the physical foundations to landscape, and underpin more common considerations of social and ecological/environmental factors.

Further Reading

Gray's (2004) **Geodiversity** develops the concept of geodiversity, including summaries of the 'natural areas' and 'landscape character assessment approaches'. Gray (1997) was one of the first papers to draw attention to the issue of geomorphologically authentic design and Gagen et al. (1993) is a classic case study of an attempt to restore a limestone quarry in a gemorphologically authentic manner. Gordon et al. (2002) is well worth examining for the quality of the descriptions of the physical landscape of Scotland.

Note: Full details of the above can be found in the references list below.

References

Adams, W.M. (1996) *Future Nature: A Vision for Conservation*. London: Earthscan.

Amadio, V., Amadei, M., Bagnaia, R., Di Bucci, D., Laureti, L., Lissis, A., Lugeri, F.R. and Lugeri, N. (2002) 'The role of geomorphology in landscape ecology: the landscape unit map of Italy', in R.J. Allison (ed.) *Applied Geomorphology: Theory and Practice*. Chichester: John Wiley, pp. 265–82.

Anon. (2006) 'Bund held to accentuate railway landscape harm', *Planning*, 8 September: 33–4.

Brady, N.C. and Weil, R.R. (2002) *The Nature and Properties of Soil* (13th edn). Englewood Cliffs, NJ: Prentice-Hall.

Cooke, R.U. and Doornkamp, J.C. (1990) *Geomorphology in Environmental Management*. Oxford: Oxford University Press.

Countryside Commission (1993) Golf Courses in the Countryside (CP438). Cheltenham: Countryside Commission.

Crofts, R. (1981) 'Mapping techniques in geomorphology', in A. Goudie (ed.) *Geomorphological Techniques*. London: George Allen & Unwin, pp. 66–75.

Cronan, W. (ed.) (1995) *Uncommon Ground: Toward Reinventing Nature*. New York and London: W.W. Norton & Co.

Cumbria County Council (2004) *Landscape Character*. Technical Paper 5. Kendal: Cumbria County Council.

Gagen, P., Gunn, J. and Bailey, D. (1993) 'Landform replication experiments on quarried limestone rock slopes in the English Peak District', *Zeitschrift für Geomorphologie*, Suppl. Bd, 87: 163–70.

Gardiner, V. and Dackombe, R. (1983) *Geomorphological Field Manual*. London: George Allen & Unwin.

Gordon, J.E., Lees, R.G., Leys, K.F., MacFadyen, C.J., Puri, G., Threadgold, R. and Kirkbride, V. (2002) *Natural Heritage Zones: Earth Sciences*. Edinburgh: Scottish Natural Heritage.

Gray, M. (1997) 'Planning and landform: geomorphological authenticity or incongruity in the countryside', *Area*, 29: 312–24.

Gray, M. (2001) 'Geomorphological conservation and public policy in England: a geomorphological critique of English Nature's 'Natural Areas' approach', *Earth Surface Processes and Landforms*, 26: 1009–23.

Gray, M. (2002) 'Landraising of waste in England, 1990–2000: a survey of the geomorphological issues raised by planning applications', *Applied Geography*, 22: 209–34.

Gray, M. (2004) *Geodiversity: Valuing and Conserving Abiotic Nature*. Chichester: John Wiley.

Gunn, J. (1993) 'The geomorphologicl impacts of limestone quarrying', *Catena Supplement*, 25: 187–97.

Hoskins, W.G. (1955) *The Making of the English Landscape*. London: Hodder & Stoughton.

Jenny, H. (1941) *Factors of Soil Formation*. New York: McGraw-Hill.

Jerie, K., Houshold, I. and Peters, D. (2001) 'Stream diversity and conservation in Tasmania: yet another new approach', in I. Rutherford, F. Sheldon, G. Brierley and C. Kenyon (eds) *Third Australian Stream Management Conference*. Monash: Co-operative Research Centre for Catchment Hydrology, pp. 329–35.

Jongman, R.H.G. and Pungetti, G. (eds) (2004) *Ecological Networks and Greenways*. Cambridge: Cambridge University Press.

Kiernan, K. (1997) *The Conservation of Landforms of Coastal Origin*. Hobart, Tasmania: Forest Practices Board.

Komoo, I. and Othman, M. (2002) 'The classification and assessment of geological landscape for nature conservation', in J.L. van Rooy and C.A. Jermy (eds) *Engineering Geology for Developing Countries*. Durban: International Association for Engineering Geology and the Environment.

Myers, N. (2002) 'Biodiversity and biodepletion: a paradigm shift', in T. O'Riordan and S. Stoll-Kleemann (eds) *Biodiversity and Human Communities: Protecting beyond the Protected*. Cambridge: Cambridge University Press, pp. 46–60.

ODPM (2006) *Planning for Biodiversity and Geological Conservation: A Guide to Good Practice*. London: Office of the Deputy Prime Minister.

Page, K.N., Wimbledon, W.A. and Bullen, J. (in press) 'The LANDMAP Information System (System Wybodaeth LANDMAP): Wales, UK – a new multi-disciplinary approach to evaluating and managing 'geological landscapes'.

Rackham, O. (1986) *The History of the Countryside*. Detroit, MI: Phoenix Press.

Simpson, J.W. (2005) *Dam! Water, Power, Politics and Preservation in Hetch Hetchy and Yosemite National Park*. New York: Pantheon.

Swanwick, C. and Land Use Consultants (2002) *Landscape Character Assessment: Guidance for Scotland and England*. Cheltenham: Countryside Agency and Scottish Natural Heritage.

Whitehand, J.W.R. (1992) *The Making of the Urban Landscape*. Oxford: Blackwell.

Wiens, J.A. and Moss, M.R. (eds) (2005) *Issues and Perspectives in Landscape Ecology*. Cambridge: Cambridge University Press.

16

Landscape: Representing and Interpreting the World

Karen M. Morin

D efinition

In Anglophone cultural geography, landscape tends to refer to a physical area visible from a particular location, as well as an ideological or social process that helps (re)produce or challenge existing social practices, lived relationships, and social identities. Textual representations of landscapes, in paintings, film, advertising, and numerous other media, are key to understanding the processes by which social practices and landscape are mutually constituted.

INTRODUCTION

'Landscape' is a basic organizing concept in Anglophone cultural geography, but is equally foundational in fields as diverse as art and architecture, environmentalism, planning, and the earth sciences. This chapter focuses only on a select number of ways in which the term has been used throughout the twentieth century by Anglophone cultural geographers, that is, those in North America, Britain, and in other English-speaking places (see Chapter 15 on landscape and environment in physical geography). Thus what appears here is as much a *history* as a *geography* of landscape studies. This chapter begins with an historiography of the landscape concept in Anglophone cultural geography. Next it demonstrates the usefulness of the landscape concept to studies of how social and cultural conflict (or consensus) arises and prevails. Finally, the chapter examines the key debates that have taken place in geography over landscape and landscape representation, around what is loosely categorized as Marxism and feminism. The chapter demonstrates overall the mutually constitutive role of landscape, landscape representation, and social practice.

While it is important to keep in mind that landscape traditions differ across disciplines and places, even within Anglophone cultural geography the concept of landscape carries much ambiguity and complexity, with probably hundreds of nuances to the term. Landscape has been (confusingly) conflated with numerous other geographical categories, such as region, area, nature, place, scenery (particularly the rural countryside), topography or landform, and environment. Nonetheless, cultural geographers have tended to emphasize the visual aspects of the physical world when they use the term 'landscape', defining it as a portion of the earth visible by an observer from a particular position or location. (Of course, the position or location of the viewer – both physical location and social location – is never unmediated, as will be discussed below.) Thus one persistent connotation of landscape is as a particularly visual form of spatial knowledge that can be taken 'in a single view', a definition that derived from sixteenth-century Dutch landscape painting, with its emphasis on scenery. Today, landscape continues to connote this visuality, although it is no longer confined to the single framed view or aesthetic pleasure, and also invokes a greater concern for viscerality and experience (Cosgrove, 2006: 51).

Thus landscape may be thought of in the first instance as a 'thing' – an area or the appearance of an area, and the particular ways component parts of that area have been arranged to produce that appearance. From this vantage point, we can talk about 'agricultural landscapes', 'urban landscapes', 'landscapes of consumption', 'modern' and 'postmodern' landscapes, 'symbolic landscapes', 'corporate landscapes', 'heritage landscapes', and so forth.

But it should be quickly noted that landscapes have *both* material and ideological aspects. Landscapes have physical, material form or 'morphologies' that are literally produced through labour and other lived relationships (Mitchell, 1996). But landscapes are also represented in various media (film, paintings, advertising), and they themselves are representations of lived relationships. Discussions and debates about landscape do not simply rest on what landscape *is* then; they also focus on what landscape *does* in social life. One fundamental aspect of this, as the title to this chapter implies, is that landscape always carries with it a set of 'representational practices'. These refer to how people see, interpret and represent the world around them *as* landscape, and how that represented landscape reflects and actually helps produce a set of lived relationships taking place on the ground. Over the past 20 years Anglophone cultural geographers have come to recognize how important representational practices are to the production of landscapes, and hence to social relations and social structure (Cosgrove, 1984, 2006; Duncan, 1990; Rose, 1993; Olwig, 2002).

Importantly, then, landscape is not only a 'thing', but is also an ideological or symbolic *process* that has the power to actively (re)produce relationships among people and between people and their material world. In this sense, landscapes carry symbolic or ideological meanings that reflect back and help produce social practices, lived relationships, and social identities, and also become sites of claiming or contesting power and authority over an area. The largest monumental landscapes in the USA, for example, carry laudatory messages about war heroes and military conquest. We can identify a set of social actors who produce such landscapes – historical and civic societies, town planners, veterans groups, and so on. The messages deployed by the monuments – a particular version of the past that celebrates masculinist values (Monk, 1999) – actively reproduces those values in the present and thus can shape social practice, such as by reproducing a culture of war. Of course such values never go unchallenged and can be undermined in numerous ways. The Vietnam War Memorial in Washington, DC became a site for contesting war as it highlights the suffering and loss of war rather than triumphal conquest.

HISTORICAL TRAJECTORIES

Landscape studies were introduced into American geography in the 1920s by Carl Sauer, especially with his 'The Morphology of Landscape' (1963 [1925]). Sauer, influenced by German geographers such as Otto Schluter and the *Landschaft* school, reacted against the environmental determinism of his day by arguing that it was collective human transformation of natural landscapes that produced what he called 'cultural landscapes'. Sauer's phenomenal influence on geographers' study of landscape over five decades (in the USA at least) cannot be understated. While Sauer himself was more concerned with physical and biological processes set in motion by humans that produced, for example, agricultural practices and patterns, his more enduring influence was on a whole generation of cultural geographers associated with the 'Berkeley School', who used his empirical observation method to study the morphological features of landscapes as evidence of cultural difference. His followers tended to study cultural artefacts such as house types and barn types to trace cultural hearth areas and diffusion of culture groups.

Mid-twentieth-century landscape studies in geography were also greatly influenced by the English historian W.G. Hoskins, who argued for detailed studies of landscape history (1955), and the American geographer J.B. Jackson, who studied popular culture through vernacular landscapes such as trailer parks in the American Southwest (1990). Jackson was founder of the popular *Landscape* magazine, published for 17 years beginning in 1951. In 1979, Donald Meinig edited a collection of works, *The Interpretation of Ordinary Landscapes*, written by some of the most quotable landscape geographers working at the time – himself and J.B. Jackson, David Lowenthal, Marwyn Samuels, David Sopher and Yi-Fu Tuan. This collection demonstrated both the continued interest in ordinary, everyday landscapes in Anglophone cultural geography (such as churches and houses), as well as how landscapes reveal social and personal tastes, aspirations and ideologies. To Meinig, landscapes themselves could be read as collective social ideologies and processes: 'symbols of the values, governing ideas, and underlying philosophies of a culture' (1979: 6).

The Interpretation of Ordinary Landscapes also demonstrated one of landscape studies' most enduring and ultimately contentious metaphors, that of 'reading' and interpreting landscapes as 'texts'. Just as a book (text) is made up of words and sentences arranged in a particular order with meanings that we read, so landscape has elements arranged in a particular order that we can translate into language, grasp meaning and read. Interpreting architectural forms and their arrangement, for example, *as* symbolic interactions among humans and their environment (e.g. the height of skyscrapers as symbols of power, modernity, public protection, etc.) would in many ways structure landscape debates in the 1980s and 1990s.

By the last two decades of the twentieth century, the textual metaphor helped usher in a number of new questions related to not just what landscape is, but how landscape mediates social relations. Informed by critical social theory, geographers first challenged the assumption of their predecessors that cultural groups collectively produced landscapes and read them in the same way. Instead, they insisted on acknowledging the patterns and processes of hierarchical social organization responsible for the morphological features observed. Thus landscape studies began to be focused on the unequal power relations – social, political and cultural – involved in producing landscapes and (in turn) social difference, by both historical and contemporary actors. Denis Cosgrove, for example, in his *Social Formation and Symbolic Landscape*, defined landscape as a 'way of seeing' associated with the rise of capitalist property relations (1984: 13). He argued that the landscape concept enabled an erasure or naturalization of class difference via media such as landscape paintings of landowners and their country property.

The works of James Duncan (1990, 1992) have been instrumental in clarifying the extent to which landscapes contain different meanings to different viewers, and how they act as intertextual media through which often competing interpretations, discourses and knowledges intersect. Other geographers, such as Don Mitchell (1996), argue that landscape studies have relied too much on visuality, and advocate, among other things, a focus on that which has been hidden from view, such as the histories of labourers whose work literally produces landscape.

Finally, geographers began rejecting the basic opposition that had persisted for so long in landscape analysis, that between subject and object, the viewer and that which is viewed. Representation in earlier landscape studies had assumed that some unmediated, transparent reality could be detected in empirically observed landscapes. More recent studies have emphasized that there is an inherent inseparability of the represent-*er* and the represent-*ed*. Thus the worlds we represent, whether as geographers, corporate executives or graffiti artists, reflect our own positionalities, values, interests, motivations and backgrounds.

The attacks on the World Trade Center in New York City in September 2001 highlighted the vastly different meanings that that corporate landscape represented for observers at numerous scales and locations, both before and after the attacks: as emblem of technological ingenuity, modernity, progress, the success of global capitalism and democratic government, and certainly a new wave of American patriotism; to more decentred understandings – US political and economic vulnerability, anti-capitalism, a holy war waged against the USA, just desserts or a wake-up call for unjust American foreign policy and hegemony, and mourning and loss of loved ones and livelihoods in the New York area. The fact that these various meanings and interpretations all co-existed simultaneously forced a recognition that not only could the same landscape carry vastly different meanings to different observers, but that the landscape itself was also a reference to a much larger set of social relationships, domestic and global, that required attention and contextualization.

Feminist landscape critics have been instrumental in exposing the problems associated with the former dualistic thinking (i.e. that an unmediated, transparent reality existed between observer and observed). Geographers' own embeddedness in the process of landscape interpretation and analysis became central to late twentieth-century geographical studies, with feminists such as Gillian Rose (1993) challenging the masculinist gaze of much landscape geography. Such late twentieth-century advancements in the study of landscape representation warrant a more detailed analysis, which follows in the next two sections.

THE POLITICS OF LANDSCAPE

Representation of landscapes can take many forms – narrative descriptions, drawings, paintings, maps, planning documents, engravings, photographs and films, among others. Trevor Barnes and James Duncan's edited collection, *Writing Worlds: Discourse, Text and Metaphor in the Representation of Landscape* (1992), examined numerous such forms of landscape representation. These authors asserted that landscape representation and interpretation required contextualization of author and audience, an outline of the rhetorics and tropes (figures of speech) employed to convey meanings, and an analysis of the processes by which readers become convinced that meanings conveyed are the natural order of things in the world.

Anglophone cultural geographers of the 1980s and 1990s emphasized that landscapes are social products, the consequence of how people, particularly dominant groups of people, create, represent and interpret landscapes based on their view of themselves in the world and their relationships with others. While authority lies with those who can 'produce landscapes as property' (in Don Mitchell's words) as well as control their representation, there is always room for contestation of that authority. In this sense, more recent landscape studies include a decidedly political component as they highlight the social and cultural conflicts and relationships, especially unequal power relations based on race/ethnicity, class, gender and sexuality, that are involved in the creation, representation and interpretation of landscapes (Mitchell, 2003).

Landscapes of graffiti, for example, highlight both hegemonic and subversive representations and interpretations of landscape. Dominant or hegemonic readings of graffiti, by a mayor's office or transportation authorities, might interpret graffiti as simply destruction of property, a crime against the city. But graffiti have also been variously understood as a means by those with no other power to mark and stake out territory, or a means to challenge the existing social order by drawing visual attention to the situation of those marginalized in the city. Alternative readings of landscapes always exist, and landscapes can always be read in ways not intended. Tim Cresswell (1996), for example, shows that many graffiti-makers think of themselves as creating art – an intention behind graffiti landscapes rarely acknowledged by more powerful voices.

Methodologically, it is important to recognize an 'intertextual' approach to reading and studying landscapes. Duncan (1990) discusses social structures (and their accompanying ideologies) that produce landscapes as signifying systems that can be read as metaphorical texts; and the discourses and systems of language and written works that are involved in their production, representation and interpretation as actual texts. He refers to the transformation of ideas from one medium to another as the intertextual nature of landscapes, arguing that the context for any text is other texts. This provides a frame for conceptualizing relationships among an array of phenomena – social structure, social practices, especially the exercise of different forms of power, the physical landscape and landscape representation, which all work to produce and reproduce one another in an ongoing fashion.

Critical social theorists have tended to highlight the extent to which multiple layers of meaning can be embedded within landscapes and their representations (e.g. skyscrapers, graffiti). This is important because landscape as a site of struggle for challenging the dominant social order often rests with their interpretation. Thus one must recognize that meanings are not inherent in concrete objects or the physical world, but that they are socially ascribed to objects and that they change over time, and with the particular perspectives and social positioning of the viewer. Thus not only is every landscape capable of multiple readings, but every landscape has been produced by multiple actors for whom no single intention can be inferred; nor can everything with causal power be observed and experienced (Duncan 1990: 12–13). The notion that landscapes contain multiple layers of meaning has been challenged by some Marxists (see below). Suffice to say at this juncture that 'good' interpretations of landscapes connect contextualized understandings of social relations and practices (particularly of prevailing discourses and ideologies), with the physical morphology on the ground.

James Duncan's (1992) study of the late twentieth-century redevelopment of the Shaughnessy Heights neighbourhood in Vancouver, British Columbia, demonstrated how social ideologies and practices worked to create, represent and reproduce landscape as a genteel, picturesque reproduction of the English country house and garden. Duncan discusses a successful attempt in the 1980s by a small group of elite, mansion-owing families to zone the neighbourhood against multi-family houses and slip-ins. These property owners managed to appropriate a nexus of interests to their own advantage – the City Council and planning commission's commitment to the preservation of green space and historic buildings, as well as to neighbourhood self-determination. Vancouver's working-class people, whose best interests would not seemingly be served by the preservation of such an elite landscape, nevertheless supported it as well. To them it represented a beautiful space in which all Vancouverans could take pride, meanwhile promising the possibility of upward mobility. Duncan effectively shows how representation of landscape became a way not just of seeing the world but of experiencing it and, indeed, 'making' it.

KEY DEBATES

One of the most significant developments in Anglophone landscape studies was the movement in the 1980s towards approaches advocated by the new cultural geographers, a shift that began first in Europe under the influence of an emerging cultural studies paradigm. The shape this discussion took was not so much a conversation between advocates of different approaches as much as it was a one-sided rejection of the old school by adherents of the new. Little resistance seemed to follow, although many geographers continue to study landscapes in the earlier tradition(s). A more significant debate has ensued between the Marxists and post-structuralists, and another by feminist geographers dissatisfied with the seemingly intractable masculinism of much landscape studies.

The first conversation has at its foundation a difference of opinion as to whether or in what sense landscape contains some sort of reality beyond its representation. Much landscape work in recent years has highlighted questions about what exactly is the relationship between the concrete, physical, material world – the morphological aspect of landscape – and its representation. Geographers such as Don Mitchell worry that landscape studies that are concerned only with representations (e.g. Barnes and Duncan, 1992) seem to leave the real world of landscape modes of production and reproduction behind as objects of study. In this way of thinking, meanings produced in and through language, texts, discourses, iconography and symbolism neglect the 'brute reality' of landscapes and thus represent a 'dangerous politics' (Mitchell, 1996: 27).

Materialist approaches early on in the debate tended to emphasize that linguistic or representational expressions are important aspects of landscapes, but that landscapes are not fundamentally linguistic entities; that there is a world outside the linguistic that is experienced (if not 'seen') and that performs a different function from representations. In a debate published in the *Professional Geographer*, Judy Walton, Don Mitchell and Richard Peet argue the point, and to paraphrase Mitchell, the 'morphology of landscape, *no matter how it is represented*' has a role in social life (*Professional Geographer,* 1996: 99, emphasis in original). Elsewhere he argues that if landscape 'is indeed a relation of power' there cannot be multiple interpretations of it, since that would defeat its ideological function, that is one that depends on the imposition of a dominant social order (Mitchell, 1996: 27). Mitchell's *The Lie of the Land* (1996) makes an important point about the role of labour within the expanding capitalist economy in California's San Joaquin Valley. Mitchell uses this example to illustrate how relations of production are involved in shaping any landscape; in effect, that we must pay attention to how landscapes 'get made' in addition to how they are then re-presented *as* landscape. In this case of southern California, that representation is an aesthetic, pastoral depiction of thriving agriculture that is the product of (otherwise invisible, exploited) labour (1996: 16).

Part of the problem with the 'representation versus reality' debate is that it is based on unsupportable binaries and dualistic thinking. In the *Professional Geographer* debate, Walton poses the question, 'Where is the pure materiality or physicality of an object (or landscape) beyond our interpretations of it?' (*Professional Geographer,* 1996: 99). In other words, there is always a cultural filtering process that brings reality to us through language. We only know landscapes, therefore, through our readings of them. As Denis Cosgrove (2006: 50) explains, 'landscapes have an unquestionably material presence, yet they come into being only at the moment of their apprehension by an external observer, and thus have a complex poetics and politics'. Rather than set up a false dualism, it seems more productive to focus on the necessarily discursive constitution of the material world. Representations, then, are not reflective or distortive images of some real, pre-interpreted reality, but they themselves materially constitute reality. Thus 'reality' is indistinguishable from its representation, and in this sense the much more important question is how representations of landscapes are produced and contested.

The real versus representational debate has subtly shifted in recent years to focus more on landscapes as fields of action, questioning how people are able to shape landscapes and be shaped by them, for instance through control of legal systems. Epistemological differences remain between Marxists and post-structuralists; though both 'sides' have moved beyond considering the question of the binary itself a departure point for analysis, there is little agreement on how to conceptually dismiss it (Dixon and Grimes, 2004: 274). Many political-economy orientated landscape geographers have become more attuned to the contingent and arbitrary mechanisms through which landscapes are produced – by examining relevant race and gender discourses and social relations along with that of class, as well as by rejecting universalist and reified explanations of the causes and effects of capitalism.

A second key debate in recent studies of landscape has focused on 'gendered landscapes'. Everything from homes to downtowns to suburbs to shopping malls to workplaces to national monuments to natural environments *as* gendered landscapes have drawn the attention of Anglophone feminist geographers since the 1970s, though this work has yet, unfortunately, to fundamentally change the way that the most geographers of landscape pose questions about landscape production and representation. While Mitchell's (1996) work elegantly problematizes the racialization of the California labour force and its politics of landscape production, for example, it does not go far enough in analysing the constructions of gender difference that created unique problems for women labourers trying to negotiate moveable work sites and their domestic and reproductive work in the labour camps.

Much of the feminist landscape scholarship focuses on how landscapes construct, legitimate, reproduce and contest gendered and sexualized identities, or how women's relationships to landscapes (as experience, representation or interpretation) differ from men's. Several analyses of gender differences in the representation and interpretation of western American landscapes have appeared. Janice Monk and Vera Norwood's edited volume *The Desert is No Lady* (1987) provided one of the first attempts at counteracting the masculine landscape tradition in geography. Scholars of nineteenth-century male settlers, industrialists, politicians, military men and railroad boosters had argued that such men viewed western landscapes either as the setting for the 'great male adventure story' or as a platform for large-scale mastery and subduing of the land and accumulation of wealth. Monk and Norwood showed that women did not necessarily share this masculine (and masculinist) vision of the southwestern desert land-scape, and questioned the appropriateness of these images for women. Their col-lection demonstrates how Hispanic, Native American and Anglo women imaged the American Southwest in a way that was both different from men's and also quite unlike each other's. Women writers, photographers and artists envisaged the desert land not in terms of its material resources to be exploited, a land await-ing metaphorical rape, but as a strong woman, unable to be conquered. The women artists' imagery turns out to be sexual (like men's), though not in terms of domination or suppression but in terms of affinity and connection, of uniting with the productive and reproductive energy of the earth.

Feminist landscape studies in the American West more recently have integrated an analysis of gender constructs with numerous others axes of social identity in their assessments of women's landscape representation. Jeanne Kay Guelke (1997), for instance, demonstrates how women involved in re-creating a Mormon Zion on the Utah frontier were deeply embedded, as faithful religious women, in the economic development of the region. To Guelke (1997: 362), reli-gious constructs pre-empted associations of nature with the female body, which had prevailed as the most common of landscape metaphors: land as Great Mother, enticing temptress, and dangerous or uncontrollable hag or fury. Thus many of these Mormon women perceived themselves as willing and active par-ticipants in the subduing and conquering of nature, and transforming wild land-scapes into productive agricultural ones. Karen Morin's (1998) study of British Victorian women's travel writing shows that attention to mode of transportation, type of engagement with the land, domestic and imperial social relations, and Romantic literary conventions all converged to produce largely negative representations of the same western American landscapes.

One feminist critique that has found its way into more mainstream human geography is Gillian Rose's (1993) study of geography's traditions in fieldwork and landscape analysis. Informed by a larger feminist corpus which highlights situated and partial knowledges and the positionality of the researcher (or observer) of objects, people and landscapes, Rose argues that geographical repres-entation through fieldwork and landscape analysis reveals deeply embedded masculinist cultural values and knowledge.

To Rose, geography's traditions involve a masculinist way of seeing land-scape that is not just one of a relation of mastery or domination, but one of (white, bourgeois, heterosexual) pleasure in looking at landscape that has been constructed as feminine. Part of her commentary revolves around the same paint-ing that Cosgrove deconstructed in his (1984) study, Thomas Gainborough's 'Mr and Mrs Robert Andrews' (c. 1748), which codified a particular way of seeing the land that helped naturalize and celebrate capitalist property and the rights of owners (see Figure 16.1). Rose, however, rightly claims that Cosgrove's interpre-tation misses the different relationships that men and women had to the sur-rounding landscape; the painting reminds us that only men were landowners, and women's role was principally reproductive (Rose, 1993: 92–3). In this and other landscape paintings, women appear passive or prostrate, as commodities of the male gaze. Not only do such landscape images themselves associate women with a feminized landscape, but, as Rose points out, geographers reinforce

Figure 16.1   Thomas Gainsborough's: 'Mr and Mrs Robert Andrews' (c. 1748)

sexism and masculinism by their inattentiveness to the impact that an analysis of gender roles and relations plays in landscape representation. These are not innocent, detached representations but they refract and reinforce lived gender roles and relations.

Other feminist geographers have suggested possibilities for other types of homoerotic and female heterosexual gaze on the landscape. For example, Catherine Nash (1996) examines Diane Baylis's photograph 'Abroad' as a representation of the male body as aesthetic nature. Still, Rose's larger observation holds; geographers have not generally problematized themselves as authors of landscape representation or interpretation. For all their success in carefully contextualizing landscape representation, Barnes and Duncan (1992), for example, allow the geographer himself to remain unmarked and disembodied. To Rose, a feminine resistance to such hegemonic ways of seeing is necessary. Such resistance promises to:

> dissolv[e] the illusion of an unmarked, unitary, distanced, masculine spectator, [while] permit[ing] the expression of different ways of seeing among women. ... Strategies of position, scale and fragmentation are all important for challenging the particular structure of the gaze in the discipline of geography. (Rose, 1993: 112)

While much feminist work has demonstrated the mutual constitution of gendered landscapes and women's gendered identities, a more recent turn in landscape studies has directed attention to relationships between men, masculinity and landscape. Rachel Woodward (2000), for example, examines the processes by which military masculinity and the landscape of Britain's rural countryside are mutually constituted. Woodward examines five sources of information – Army recruitment materials, general publicity, basic training information and videos, mass market paperbacks about military adventures, and television documentaries on military life. She shows how essential a particular construction of rurality itself is to the construction of 'warrior hero' – it is dangerous, rough and

Figure 16.2   Military masculinity and landscape

hazardous. The rural countryside in the Army documentation is not that of idyllic community and nature in harmony, but is rather a harsh, threatening landscape against which the new recruit is pitted, and out of which his requisite physical and mental attributes will arise through its conquest (see Figure 16.2). Thus this representation of the rural serves the dual purpose of articulating and legitimating one hegemonic type of military masculinity, as well as constructing the rural itself as a legitimate place to bear arms.

CONCLUSION

'Traditional' landscape geographers' attention to ordinary and everyday landscapes, especially in attention to their morphological aspects, has ultimately and justifiably endured. The materialist–post-structuralist debate in landscape studies that raged in the 1990s seems less worrisome today. Materialists seem better attuned to the importance of landscape representation in the construction of reality, and those who have been most interested in linguistic or discursive analysis of landscape representation also seem more attuned to what is 'on the ground'. A focus on those who produce landscapes has been fruitful (e.g. Mitchell, 1996), as has the entire corpus of work that sees contests over representational practices as key to challenging the existing social order (e.g. Cresswell, 1996; Duncan, 1999). The relative contingencies of capitalism and other social relations in the production of landscape, however, continue to be debated. The situation on the feminist front is encouraging, but much work remains to be done. Attention to gendered landscapes within which various masculinities are produced seems a fruitful direction, as does attention to the myriad ways in which landscape helps produce and mediate national, ethnic and sexual difference.

SUMMARY

- Landscapes have both material and ideological aspects; they have physical, material form, are represented in various media and are themselves representations of lived relationships.

- Landscapes carry symbolic or ideological meanings that reflect back and help produce social practices, lived relationships and social identities, and also become sites of claiming or contesting authority over an area. Social practices and landscapes mutually constitute one other in an ongoing fashion.

- While early twentieth-century Anglophone landscape studies focused more on morphological features and cultural difference read through them, later studies argued that landscapes are not collectively produced by culture groups but rather act as intertextual media through which competing authority, interpretations, discourses and knowledges intersect. Both the observer and that which is observed require greater contextualization.

- Understanding the power of landscape to challenge or subvert the existing social order has been of primary concern to many cultural geographers.

- Two key critiques appear in geography's landscape studies: Marxian and feminist. The first takes a materialist orientation to argue that many landscape geographers focus too heavily on landscape representation at the expense of morphology (an ultimately false dichotomy), and suggest more attention to those inside landscapes, those who produce it. A second challenge focuses on the masculinism of landscape studies.

Further Reading

Sauer's (1963 [1925]) paper, 'The morphology of landscape', is a foundational statement in American cultural geography. It argues for an empirical observation method to study the morphological features of the landscape as evidence of cultural difference. Subsequently, Marxist, post-structuralist and feminist approaches to geography have led to the emergence of a diverse range of ways of understanding landscape. One of the first statements in Anglophone cultural geography that brought a Marxist sensitivity to artistic representations of landscape is Cosgrove's (1984) **Social Formation and Symbolic Landscapes**. In this book he understands landscape as a 'way of seeing' associated with the rise of capitalist property relations. Duncan's

(Continued)

(Continued)

(1990) **The City as Text** helped usher post-structural linguistic theory into geography's landscape studies by analysing the creation of the urban landscape of the precolonial Kandyan kingdom in Sri Lanka. In this book, Duncan addresses layers of landscape signification, rhetorical devices, power relations and intertextuality. Rose's (1993) **Feminism and Geography** challenges the masculinist foundation of geography's history and geographical knowledge, including a critique of the 'masculinist gaze' embedded in landscape studies. Some of the tensions between these different ways of viewing landscape are evident in a special issue of the journal **The Professional Geographer** (1996), in which Judy Walton, Don Mitchell and Richard Peet debate the tensions between materialist and post-structuralist interpretations of landscapes. Mitchell continues to wrestle with these themes in his *Progress in Human Geography* reviews (2002, 2003).

Note: Full details of the above can be found in the references list below.

References

Barnes, T. and Duncan, J. (1992) *Writing Worlds: Discourse, Text and Metaphor in the Representation of Landscape*. London: Routledge.

Cosgrove, D. (1984) *Social Formation and Symbolic Landscapes*. London: Croom Helm.

Cosgrove, D. (2006) 'Modernity, community and the landscape idea', *Journal of Material Culture*, 11: 49–66.

Cresswell, T. (1996) *In Place/Out of Place: Geography, Ideology, and Transgression*. Minneapolis, MN: University of Minnesota Press.

Dixon, D. and Grimes, J. (2004) 'Capitalism, masculinity and whiteness in the dialectical landscape: the case of Tarzan and the Tycoon', *GeoJournal*, 59: 265–75.

Duncan, J. (1990) *The City as Text: The Politics of Landscape Interpretation in the Kandyan Kingdom*. Cambridge: Cambridge University Press.

Duncan, J. (1999) 'Elite landscapes as cultural (re)productions: the case of Shaughnessy Heights', in K. Anderson and F. Gale (eds) *Cultural Geographies*. Melbourne: Addison Wesley Longman, pp. 53–70.

Hoskins, W.G. (1955) *The Making of the English Landscape*. London: Hodder & Stoughton.

Jackson, J.B. (1990) 'The house in the vernacular landscape', in M. Conzen (ed.) *The Making of the American Landscape*. Boston, MA: Unwin Hyman, pp. 355–9.

Kay Guelke, J. (1997) 'Sweet surrender, but what's the gender? Nature and the body in the writings of nineteenth-century Mormon women', in J.P. Jones et al. (eds) *Thresholds in Feminist Geography: Difference, Methodology, and Representation*. Lanham, MD: Rowman & Littlefield, pp. 361–82.

Meinig, D.W. (ed.) (1979) *The Interpretation of Ordinary Landscapes: Geographical Essays*. New York: Oxford University Press.

Mitchell, D. (1996) *The Lie of the Land: Migrant Workers and the California Landscape*. Minneapolis, MN: University of Minnesota Press.

Mitchell, D. (2002) 'Cultural landscapes: the dialectical landscape – recent landscape research in human geography', *Progress in Human Geography*, 26: 382–9.

Mitchell, D. (2003) 'Cultural landscapes: just landscapes or landscapes of justice?', *Progress in Human Geography*, 27: 787–96.

Monk, J. (1999) 'Gender in the landscape: expressions of power and meaning', in K. Anderson and F. Gate (eds) *Cultural Geographies*. Melbourne: Addison Wesley Longman, pp. 153–72.

Monk, J. and Norwood, V. (eds) (1987) *The Desert is No Lady: Southwestern Landscapes in Women's Writing and Art*. New Haven, CT: Yale University Press.

Morin, K.M. (1998) 'Trains through the plains: the Great Plains landscape of Victorian women travelers', *Great Plains Quarterly*, 18: 235–56.

Nash, C. (1996) 'Reclaiming vision: looking at landscape and the body', *Gender, Place and Culture*, 3: 149–69.

Olwig, K. (2002) *Landscape, Nature and the Body Politic: From Britain's Renaissance to America's New World*. Madison, WI: University Wisconsin Press.

Professional Geographer (1996) Special issue on landscape, *The Professional Geographer*, 48: 94–100.

Rose, G. (1993) *Feminism and Geography: The Limits of Geographical Knowledge*. Minneapolis, MN: University of Minnesota Press.

Sauer, C. (1963) 'The morphology of landscape [1925]', in J. Leighly (ed.) *Land and Life: A Selection of the Writings of Carl Ortwin Sauer*. Berkeley, CA: University of California Press, pp. 315–50.

Woodward, R. (2000) 'Warrior heroes and little green men: soldiers, military training, and the construction of rural masculinities', *Rural Sociology*, 65: 640–57.

17

Nature: A Contested Concept

Franklin Ginn and David Demeritt

ⓓefinition

Nature is a contested term that means different things to different people in different places. Generally, this contestation revolves around three main meanings: the 'nature' or essence of a thing; 'nature' as material place external to humanity; and 'nature' as universal law or reality that may or may not include humans.

INTRODUCTION

Natural food is all the rage. Walk down the aisle of your local supermarket and you'll be confronted by entire ranges of products boasting 'all natural' or 'organic' ingredients. Often the packaging is decorated with pictures of verdant fields dotted with grazing dairy cows – or perhaps it's small children frolicking. Bombarded as we are by advertising, we rarely take the time to interrogate the cascade of associations and myths it echoes and extends. Such images of bucolic countryside draw on a long tradition of pastoral art and poetry celebrating nature and the countryside as the true home of humanity. In the context of food packing, they serve to reassure consumers about the quality, freshness, safety and sustainability of particular commodities by locating them rhetorically in an idealized, Edenic environment of healthy, wholesome and leisurely living that is at once youthful and timeless, familiar and far away. There are no factory farms, pesticides, processing plants or migrant farm workers slaving away from dawn until dusk in the imagined geographies of nature depicted in most supermarkets.

We begin with this example to show that 'nature' and the 'natural' are not always what they seem. Behind apparently simple labels like 'natural' and 'organic' stand a whole array of regulations, and the various state, or increasingly non-governmental, inspectors charged with certifying that those standards have been met. In the UK, the Food Standards Agency (2002) published a 20-page set of 'criteria for the use of terms fresh, natural, etc. in food labelling', while a host of non-governmental organizations, such as the Soil Association, have formulated codes of practice and other certification schemes to assure the sustainable, organic, Fair Trade or other credentials of particular products. The meaning and definition of nature are more than simply academic concerns. They have important implications for what you eat and how you live.

Geographers, more than most other academics, have been centrally concerned with nature. There is, as Castree (2005) notes, a very close and contested relationship between the nature of Geography as an academic discipline and the nature that geographers take as their object of study. Along with space/location, the concept and study of nature holds together physical and human geography in a single integrative discipline. For this reason alone, 'nature' deserves a central place in any discussion of *Key Concepts in Geography*. Tracing the different ways geographers have understood and studied nature, both as concept and object, provides one way to understand the history of geography as a discipline. Indeed, as we shall see, one of the most important trends in recent research is to blur this distinction between concepts and the objects to which they refer. This move challenges long-standing dualisms and the positivist ideals of objective science that depend on them, which is one reason why debates about the social construction of nature have become so heated.

HISTORY OF A CONCEPT

The literary critic Raymond Williams (1983: 219) famously observed that 'nature' is perhaps the most complex word in the English language. He identified three broad but complexly interconnected meanings:

1 Intrinsic nature: the essential characteristics of a thing (e.g. the nature of social exclusion).
2 External nature: the external, unmediated material world (e.g. the natural environment).
3 Universal nature: the all-encompassing force controlling things in the world (e.g. 'natural laws' or 'Mother nature').

All three of these meanings figure in debates about the nature (meaning 1) of Geography as an academic discipline. Turner (2002: 63), for instance, sees study of the environment (meaning 2) as central to Geography's claim to be 'an integrated environmental science' well placed to address real-world problems like flooding. Taking that case, physical geographers have elucidated the natural laws (meaning 3) governing the movement of water through landscapes (meaning 2), needed to predict the nature (meaning 1) and impact of flooding. Similarly, behavioural geographers have developed models to predict the factors controlling (meaning 3) public perceptions of such risks, while a host of critical human geographers have sought to 'take the naturalness out of natural disasters' (O'Keefe et al., 1976: 566) and to show how the nature (meaning 1) of disasters is 'not just an act of God' or a function of 'extreme physical events' (meaning 2) but is socially determined by 'socio-economic conditions that can be modified by' people, if we choose. Against Turner's view that nature is a unifying object of geographical study, it is also possible to draw on other senses of the concept to distinguish human geography, concerned with meaningful human affairs, from physical geography, which studies a brute physical nature in the sense of (2) or (3) or both. This ontological difference between nature and society then forms the basis for distinguishing epistemologically between human geography as a hermeneutic social science of interpretation and physical geography as a positivist natural science of law-like prediction and explanation.

Thus the concept of nature is central not only to Geography and the division between human and physical geographers, but also to science as a whole. Since the dawn of modern science during the seventeenth-century Enlightenment, nature has been critical to various philosophical efforts to distinguish scientific knowledge from other forms of belief. First, science has often been distinguished from religious superstitions on the grounds that its knowledge about the nature of things (meaning 1) is objective in the double sense that it is not based on subjective belief but on direct, impersonal and, in that sense, objective observation of an external and independent reality (meaning 2). Second, positivism defined science in terms of its ability to generate valid predictions from hypotheses. To this view, what human and physical geographers share in common is a search for the essentially necessary and therefore scientifically predictable properties of their respective objects of study. Thus human geographers concerned with the nature (meaning 1) of economic growth would seek to identify the laws (meaning 3) governing its behaviour, while physical geographers explain the nature (meaning 1) of hydrological systems (meaning 2) and the natural laws (meaning 3) governing the behaviour of water in different sized catchments. Though human and physical geographers may study different things, positivists insist that their knowledges are equally scientific, so long as they follow that same scientific method of testing hypotheses about the nature of things against independent observations of those same things.

In so far as all three of these broad meanings invoke a vision of nature that is singular, abstract and personified, there is a central ambiguity about whether or not they encompass humans. Is human nature (meaning 1) determined by some inherent, biological force (meaning 3), like our genes or, as many so-called environmental determinists of the late nineteenth and early twentieth centuries believed, by our physical environment (meaning 2)? Or alternatively, isn't what distinguishes humans from other animals that we can use our rationality to rise above our base biological instincts?

A similar ambiguity runs through the Food Standards Agency's (2002: 10) guidance on the use of the term 'natural' in food labelling: '"Natural" means essentially that the product is comprised of natural ingredients, e.g. ingredients produced by nature, not the work of man or interfered with by man.' Here the natural is defined so as to exclude any trace of humans and their artifice. That, however, is an impossible standard in so far as all food is the product of intentional human selection. Literally speaking, it is impossible for food *not* to involve the work of people. The FSA regulations go on to explain that it is permissible to label as natural 'foods, of a traditional nature' that have been processed using 'traditional cooking processes' rather than 'novel' ones, such as 'freezing, concentration, pasteurization, and sterilization'. In this way, defining the natural is also defining the human. By eliding 'traditional' with 'natural', FSA regulations simultaneously locate 'novel' food-processing techniques outside nature in a purely human realm of culture and technology, while at the same time fixing certain traditional practices in a timeless realm close to nature where change and technical development are impossible without alienation from tradition and nature.

This ambiguity as to whether nature encompasses humans is not new, and an historical focus demonstrates that there are powerful cultural politics at play in these distinctions. For 'nature', far from being a neutral term, has a contested colonial heritage. The life of 'uncivilized man' living traditionally in a 'state of nature' has famously been imagined as 'solitary, poore, nasty, brutish, and short' (Hobbes, 1651) or, alternatively, as the free and innocent one of a 'noble savage' (Rousseau, 1762). In the context of European expansion overseas, the opposition between nature and civilization was easily racialized and, in the guise of scientific racism, provided a rationale for European colonial rule over more 'primitive' cultures and peoples who were said to be 'naturally' (meaning 3) less rational, civilized, and developed (see Chapter 1 on the histories of geography). Drawing on late nineteenth-century ideas of evolution, geographers like Sir Henry Harry Johnston, author of *The Backward Peoples and Our Relations with Them* (1920, quoted in Livingstone, 1992), argued that it was the 'white man's burden' to govern less developed people and places until they became civilized enough to do it for themselves.

Europeans projected their views of nature on to the new landscapes they encountered in the Americas, Asia, the Pacific and most powerfully, perhaps, Africa. For example, early settlers in New Zealand wrote of the South Island's plains:

> But this vast tract is unpeopled; millions of acres have never been trodden by human foot since their first upheavement from the sea. It is a country fresh from nature's rudest mint, untouched by hand of man. (Hursthouse, 1857: 225)

This separation of rational man from 'primitive natives' helped legitimize the imposition of scientific management to bring order to and 'improve' the land. Where lands proved unsuited to cultivation and other economic use, they were often set aside as national parks or reserves, where nature was to be preserved in an unspoilt state for future generations to admire. But the 'preservation' of so-called wilderness areas was really a production of wilderness, in so far as it often involved the forcible expulsion of indigenous peoples. In Africa, Maasai were evicted to create the Amboseli National Park and allowed to remain in the Serengeti only because they were viewed as 'part of nature' (Neumann, 1998); in the USA the Blackfeet continue to be accused of 'poaching' on the lands of Glacier National Park that originally belonged to them (Cronon, 1995).

Wilderness, then, is a culturally and historically contingent expression of a certain colonialist way of seeing nature. It is, in short, a social construction:

> Far from being the one place on earth that stands apart from humanity, [wilderness] is quite profoundly a human creation – indeed, the creation of very particular human cultures at very particular moments in human history. It is not a pristine sanctuary where the last remnant of an untouched, endangered, but still transcendent nature can at least for a little while longer be encountered without the contaminating taint of civilization. (Cronon, 1995: 69)

RETHINKING NATURE IN GEOGRAPHY

Much recent work in critical geography has sought to question traditional understandings of nature and the Enlightenment dualisms associated with them. One of the most important moves in this regard is the claim, articulated by Cronon in the quotation above, that nature is somehow socially constructed and contingent rather than being intrinsic, external and universal. As we will see, this claim takes a variety of different forms in different traditions of critical geography (Demeritt, 2002).

Marxism

Karl Marx was one of the first theorists to suggest that nature was socially 'produced' or constructed. Marx meant this in a material sense, in that people work on the raw matter of nature to transform it into a second, social nature. However, Marx's account of nature's production under capitalism is highly abstract (Castree, 2005). In his book *Nature's Metropolis*, the environmental historian William Cronon (1992: 266) has provided an empirically rich description of how the American Midwest was remade through the operation of the market:

> Bisons and pine trees had once been members of ecosystems defined mainly by flows of energy and nutrients and by relations among neighboring organisms. Rearrayed within the second nature of the market, they became commodities: things priced, bought, and sold within a system of human exchange. From that change flowed many others. Sudden new imperatives revalued the organisms that lived upon the land. Some, like the bison, bluestem, and pine tree, were priced so low that people consumed them in the most profligate ways and they disappeared as significant elements of the regional landscape. Others, like wheat, corn, cattle, and pigs, became the new dominant species of their carefully tended ecosystems. Increasingly, the abundance of a species depended on its utility to the human economy: species thrived more by price than by direct ecological adaptation. New systems of value, radically different from their Indian predecessors, determined the fate of entire ecosystems.

In addition to this material transformation, Marxist geographers have also highlighted the way in which capitalism depends on a false ideology of nature as both external and universal that serves to conceal and thereby to legitimate the social relations involved in the capitalist production of nature. In a landmark paper, Harvey (1974) attacked neo-Malthusian arguments about the natural limits to growth both for ignoring the role of economic systems in causing hunger and local resource shortages and for legitimating technical programmes, like the chemical-intensive agriculture promoted as part of the so-called Green Revolution, as the only way to overcome those problems.

Feminism

Feminists have launched some of the most trenchant critiques of the nature/culture dualism and its implications for the subordination of women. Much like Marxist critiques of the ideology of nature, feminists complain that existing and oppressive gender roles are legitimated because they are seen as natural, in the senses both of (1) and (3) we listed above. For instance, in 2005, the then president of Harvard University, Laurence Summers, sparked widespread protests for suggesting that it was biological differences, rather than sexism and discrimination, that explained why so few women succeed in mathematical and scientific careers. In attacking such claims, feminists have enthusiastically embraced constructionist arguments as a 'strong tool for deconstructing the truth claims of hostile science by showing the radical historical specificity and so contestability of *every* layer of the onion of scientific and technological constructions' (Haraway, 1991: 186, original emphasis). Construction talk enables feminists to argue that apparently innate and therefore immutable differences between the sexes are in fact socially constructed *gender* differences that might be changed.

In an influential critique of the masculine bias in geography, Gillian Rose (1993) argued that the discipline's traditions of scientific fieldwork and objective observation were grounded in an eroticized, 'masculine' gaze that at once objectified and feminized the landscape. But Rose's insistence that those scientific ways of knowing are just one of many possible alternatives, begs questions about the status and credibility of feminists' own claims to knowledge. Feminists, as Donna Haraway (1991) notes, have found themselves trying to hold on to two ends of a slippery pole at once. On the one hand, they have sought to dissolve nature/culture and object/subject dualisms so as to insist that all knowledge is essentially social, situated and relative. On the other hand, however, they have also longed for a strong notion of objectivity on which to base their claims about the reality of women's oppression in male-dominated societies. Torn between these conflicting desires, feminists have experienced constructionism as a sort of 'epistemological electro-shock therapy, which ... lays us out ... with self-induced multiple personality disorder' (Haraway, 1991: 186).

Another issue raised by feminist critique is whether and how we distinguish socially constructed *gender* differences from those of a biological nature. De-naturalizing gender roles can leave open the idea that underneath culture, men and women are biologically different. Against that view, a number of scholars have drawn on the work of social theorist Michel Foucault to argue that the sex too is shaped socially and discursively. Foucault (1980) drew on the memoirs of a nineteenth-century hermaphrodite to argue that sex does not have ontological status, and that we are sexualized as woman/man only by medical, social and political discourses. The hermaphrodite troubled sexual boundary-making practices in France, belying the desire to classify a body as either male or female. Extending that argument, Judith Butler (1993), an influential feminist and Queer theorist, has suggested that the (hetero)sexed body is not determined naturally or biologically, but rather is performed. It comes into being through the repetition of everyday performances and routines that are regulated by wider social discourses and norms and come to shape the body and train its behaviour through an effect she likens to sedimentation.

RELATIONAL GEOGRAPHIES

While feminists and Queer theorists like Butler draw on Foucault to insist that sex and the body have no intrinsic and universal nature, but are instead relational achievements whose precise form and content depend on the social context in which they are shaped, other geographers have made similar arguments about the context dependence of things based on very different theoretical starting points (e.g. Harvey, 1996; Whatmore, 2002).

Interest in such relational geographies reflects a wider concern, among geographers, with ontology. Ontology is the branch of philosophy concerned with the nature of existence. Relational approaches to ontology consider how the nature of things, even reality itself, is context-dependent. As Donna Haraway (1992: 297) explains: 'If the world exists for us as "nature", this designates a kind of relationship, an achievement among many actors, not all of them human, not all of them organic, not all of them technological.' This relational approach to ontology challenges several long-standing Enlightenment presumptions about nature and the world. In particular, the role of relations and context are emphasized over the idea that objects have any intrinsic or universal nature, while the Cartesian idea of external reality as an array of objects located absolutely in the two, separate dimensions of space and time gives way to a sense of space–time as manifold and co-constituted along with what it contains (Massey, 2005).

There are several sources of inspiration for such relational thinking. Within the sciences, developments in complexity and chaos theory emphasize the possibility for systems to become self-organizing as complex higher-order behaviour emerges out of lower-order interactions (Manson, 2001). For instance, a school of fish, containing many thousands of individuals, comes to swim as if it were a single entity, through co-ordination of the lower-order endency of the individuals within it to follow the movement of their nearest neighbours. In addition to emergence, complexity theory also highlights the sensitive dependence of some systems upon their initial conditions and changing external factors. For example, it is difficult to forecast future weather conditions beyond more than a week or two both because of the potential for storm systems to 'emerge' suddenly and because of the difficulties of knowing with any certainty all of the factors to which their future evolution might prove sensitive (Phillips, 1999). Likewise at the sub-atomic scale, the development of quantum mechanics and Heisenberg's uncertainty principle both emphasize the limits of predictability and the dependence of our experimental knowledge of the world on the context in which it is generated.

One effect of this new awareness of emergence, contingence and indeterminacy within the environmental sciences has been to challenge the trend towards ever-greater reductionism. Instead of breaking fields of study into smaller and smaller parts, a new integrationist Earth Systems Science seeks to study the earth as a single integrated physical and social system (Pitman, 2005). Within ecology, another effect of complexity and chaos theory has been to undermine the idea of the 'balance of nature' (Perry, 2002), which environmentalists have often used to critique human disturbance of the environment as unnatural. Many environmentalists fear that these new ecological ideas may lead to relativism by depriving any clear scientific grounds for distinguishing an anthropogenic impact from 'natural' change (Demeritt, 1994).

However, in a world of genetic engineering and global warming, geographers are increasingly sceptical of even using 'natural' and 'social' as categories of analysis. One influential source for the idea that nature and culture are inextricably 'mixed up' is the actor-network theory of Bruno Latour. In a series of influential books, Latour has developed a unique vocabulary to describe agency, material effectivity, even existence itself, as emergent properties that are realized through historically and geographically contingent relations among the heterogeneous 'actants' of a more than human world. Latour uses the term 'actant', which he takes from semiotics, to emphasize, first, that humans are not the only actors in these relationships and, second, that agency is something that is dependent on a wider structure of relations through which it is produced. Rejecting traditional Enlightenment distinctions between nature and culture, objects and subjects, people and machines, material and imaginary, actor-network theory insists that all elements of a network be described in the same symmetrical terms.

Latour speaks of actor-networks as networked assemblages that operate by 'enrolling', or incorporating, various hybrid actants (which are themselves also composites of heterogeneous, networked elements) into longer, stronger and more durable networks. Sailing ships, for example, were able to circumnavigate the globe only by 'enrolling' the power of the wind, the seaworthy designs of experienced shipwrights, and navigational aids developed through trial and error. If any one of those elements of the network breaks down – for instance if poor navigation or crashing waves make the ship flounder – the network making the ship a ship ceases to hold and the ship literally breaks apart into its constituent elements – boards, bodies, ropes and rigging (Law, 1986).

Such an understanding of the world has potentially far-reaching theoretical and political implications. By extending agency to non-humans, actor-network theory challenges human exceptionalism and the long-standing divisions based upon it between the social and natural sciences. While some geographers insist that trees can be said to 'act' in the same way as people do (e.g. Jones and Cloke, 2002), critics of actor-network theory often note that, in practice, actor-network theorists tend to violate their principle of explanatory symmetry by centring their accounts of network building around purely human actors (Murdoch, 1997). Nevertheless by rejecting human exceptionalism, actor-network theory raises important questions about 'how the we of ethical communities is to be renegotiated on account of its heterogeneous, intercorporal composition' (Whatmore, 2002: 166). Rising to that challenge, Latour (2004) has recently outlined an expanded sense of 'cosmopolitics'. In Latour's 'parliament of things', questions must be put not just to non-humans as well:

You want to save the elephants in Kenya's parks by having them graze separately from cows? Excellent, but how are you going to get an opinion from the Masai who have been cut off from the cows, and from the cows deprived of elephants who clear the brush for them, and also from the elephants deprived of the Masai and the cows? (Latour, 2004: 170)

Despite these efforts, critics complain that actor-network provides only a descriptive language and fails to address the pressing moral and political questions about what form our relations should take. To the extent that actor-network theory merely describes rather than also critiquing persistent inequalities, critical geographers complain that such relational geographies remain complicit in reproducing relations of inequality (e.g. Castree and MacMillan, 2001; Smith, 2005).

CONCLUSION

The idea that nature is a 'key concept' rather than the empirical domain of geographic study may have initially seemed rather perverse. But we hope you now appreciate that nature is as a much a concept as it is a biophysical reality. Far from being something located 'out there', nature is also something with us 'in here', in the ways that our bodies, our sense of our selves and our world, and our daily routines are informed by various overlapping concepts of nature. Precisely because of their ubiquity, those concepts are both complex and often hotly contested. Nature, to return to Raymond Williams (1980: 67), 'contains an extraordinary amount of human history', but it also has a geography, though Williams did not remark much upon it. As well as changing over time, concepts of nature, like the things and relations to which they refer, also vary from place to place. Within the discipline of geography, conceptions of nature are closely wrapped up with different ideas about the nature of geography as a science and subject of study. For both those reasons nature is perhaps the most important concept in geography.

SUMMARY

- Nature as a contested concept and as biophysical reality has been central to geography as an academic discipline.

- There is an ambiguity in the concept of nature, in who or what is included and excluded from being labelled 'natural': for example, in organic food, the human body, indigenous peoples, postcolonial 'wilderness', and so on.

- Marxist, feminist and postcolonial geographers have been highly critical of the ideology of external nature (meaning 1) as hiding a politics of exploitative capitalist, gender and colonial relations.

- Relational approaches in human geography aim to blur and bypass the nature/culture dualism. This has far-reaching implications for the physical/human divide in geography and for how we conceive the differences between the human and non-human.

Further Reading

Noel Castree's (2005) **Nature** in the Key Ideas in Geography series offers the most up-to-date and accessible survey of how geographers have studied nature, while Braun's (2004) and Demeritt's (2001) essays provide shorter overviews of issues dealt with at greater length by Castree. Soper's (1995) **What is Nature?** remains an excellent overview of the idea of nature, while Habgood's (2002) **The Concept of Nature** offers an interesting defence of essentialism from a theological perspective. Useful collections of essays include Braun and Castree's (1998) **Remaking Reality** and Castree and Braun's (2001) **Social Nature**. Plumwood's (2002) **Environmental Culture** or Merchant's (1996) **Earthcare: Women and the Environment** provide routes into feminist critiques of nature. On animals specifically, edited volumes by Philo and Wilbert's (2000) **Animal Spaces, Beastly Places** and Wolch and Emel's (1998) **Animal Geographies** remain key texts, though Kalof and Fitzgerald's (2007) **The Animals Reader** offer a wider range of essays. Braun's (2002) in treatment of wilderness **The Intemperate Rainforest** refines and extends Cronon's (1995) original arguments, while Wilson (1992) explores the culture of nature in North America more broadly. For accessible applied actor-network theory in geography, see Burgess et al. (2000), Murdoch and Lowett (2003), or Power (2005). The theoretically dense nature of the relational turn in geography presents a challenge to the undergraduate. Murdoch's (2006: Chapters 2–5) **Post-structuralist Geography** offers an accessible introduction; Hinchcliffe's (2007) **Geographies of Nature** draws more directly on relational thinking, see Castree and Macmillan (2001) on lines of disagreement. On complexity theories, see O'Sullivan (2004). Robbins's (2007) **Lawn People** attempts to reconcile relationality with political ecology. On debates about nature as a unifying concern see Harrison et al. (2004).

Note: Full details of the above can be found in the reference list below.

References

Braun, B. (2002). *The Intemperate Rainforest: Nature, Culture, and Power on Canada's West Coast.* Minneapolis, MN: University of Minnesota Press.

Braun, B. (2004) 'Nature and culture: on the career of a false problem', in J. Duncan, N. Johnson and R. Schein (eds) *A Companion to Cultural Geography.* Malden, MA and Oxford: Blackwell.

Braun, B. and Castree, N. (eds) (1998) *Remaking Reality: Nature at the Millenium.* London and New York: Routledge.

Burgess, J., Clark J. and Harrison, C. (2000) 'Knowledges in action: an actor network analysis of a wetland agri-environment scheme', *Ecological Economics*, 35: 119–32.

Butler, J. (1993) *Bodies That Matter: On the Discursive Limits of 'Sex'.* London: Routledge.

Castree, N. (2005) *Nature*. London and New York: Routledge.

Castree, N. and Braun, B. (eds) (2001) *Social Nature: Theory, Practice, and Politics*. Malden, MA: Blackwell.

Castree, N. and MacMillan, T. (2001) 'Dissolving dualisms: actor-networks and the reimagination of nature', in N. Castree and B. Braun (eds) *Social Nature: Theory, Practice, and Politics*. Malden, MA: Blackwell.

Cronon, W. (1992) *Nature's Metropolis: Chicago and the Great West*. New York: W.W. Norton.

Cronon, W. (1995) 'The trouble with wilderness; or, getting back to the wrong nature', in W. Cronon (ed.) *Uncommon Ground: Toward Reinventing Nature*. New York: W.W. Norton, pp. 69-90.

Demeritt, D. (1994) 'Ecology, objectivity, and critique in writings on nature and human societies', *Journal of Historical Geography*, 20: 22-37.

Demeritt, D. (2001) 'Being constructive about nature', in N. Castree and B. Braun (eds) *Social Nature: Theory, Practice, and Politics*. Malden, MA: Blackwell.

Demeritt, D. (2002) 'What is the "social construction of nature"? A typology and sympathetic critique', *Progress in Human Geography*, 26: 767-90.

Food Standards Agency (2002) *Criteria for the Use of the Terms Fresh, Pure, Natural etc.* http://www.food.gov.uk/multimedia/pdfs/fresh.pdf (last accessed 02/03/07).

Foucault, M. (1980) *Herculine Barbin: Being the Recently Discovered Memoirs of a Nineteenth-century French Hermaphrodite*. New York: Pantheon.

Habgood, J. (2002). *The Concept of Nature*. London: Darton, Longman & Todd.

Haraway, D.J. (1991) *Simians, Cyborgs, and Women: The Reinvention of Nature*. New York: Routledge.

Haraway, D.J. (1992) 'The promises of monsters: a regenerative politics for inappropriate/d others', in L. Grossberg, C. Nelson and P. Treichler (eds) *Cultural Studies*. London: Routledge, pp. 295-337.

Harrison, S., Massey, D., Richards, K., Magiligan, F.J., Thrift, N. and Bender, B. (2004) 'Thinking across the divide: perspectives on the conversations between physical and human geography', *Area*, 36: 435-42.

Harvey, D. (1974) 'Population, resources and the ideology of science', *Economic Geography*, 50: 256-77.

Harvey, D. (1996) *Justice, Nature and the Geography of Difference*. Cambridge, MA: Blackwell.

Hinchcliffe, S. (2007) *Geographies of Nature: Societies, Environments, Ecologies*. London: Sage.

Hobbes, T. (1651) *Leviathan, sive de materia, forma, et potestate civitatis ecclesiasticae et civilis*. Amsterdam: Johannes Blaev. Translation (n.d.) *Leviathan: or, The Matter, Form and Power of a Commonwealth, Ecclesiastical and Civil*. London: Routledge.

Hursthouse, C. (1857) *New Zealand, or New Zealandia the Britain of the South, Volume 1*. London: Edward Stanford.

Johnston, H.H. (1920) *The Backward Peoples and Our Relations to Them*. Oxford: Oxford University Press.

Jones, O. and Cloke, P. (2002) *Tree Cultures: The Place of Trees and Trees in their Place*. Oxford: Berg.

Kalof, L. and Fitzgerald, A. (eds) (2007) *The Animals Reader: The Essential Classic and Contemporary Writings*. Oxford: Berg.

Latour, B. (2004) *Politics of Nature: How to Bring the Sciences in Democracy*, trans. C. Porter. Cambridge, MA: Harvard University Press.

Law, J. (1986) 'On the methods of long distance control: vessels, navigation and the Portuguese route to India', in J. Law (ed.) *Power, Action and Belief: A New Sociology of Knowledge?* London: Routledge, pp. 234-63.

Livingstone, D. (1992) *The Geographical Tradition: Episodes in the History of a Contested Discipline*. Oxford: Blackwell.

Manson, S.M. (2001) 'Simplifying complexity: a review of complexity theory', *Geoforum*, 32: 405-14.

Massey, D. (2005) *For Space*. London: Sage.

Merchant, C. (1996) *Earthcare: Women and the Environment*. New York: Routledge.

Murdoch, J. (1997) 'Inhuman/nonhuman/human: actor-network theory and the prospects for a nondualistic and symmetrical perspective on nature and society', *Environment and Planning D: Society and Space*, 15: 731–56.

Murdoch, J. (2006) *Post-structuralist Geography*. London: Sage.

Murdoch, J. and Lowett, P. (2003) 'The preservationist paradox: modernism, environmentalism and the politics of spatial division', *Transactions, Institute of British Geographers*, 28: 318–32.

Neumann, R.P. (1998) *Imposing Wilderness: Struggles over Livelihood and Nature Preservation in Africa*. Berkeley, CA: University of California Press.

O'Keefe, P., Westgate, K. and Wisner, B. (1976) 'Taking the naturalness out of natural disasters', *Nature*, 260: 566–7.

O'Sullivan, D. (2004) 'Complexity science and human geography', *Transactions, Institute of British Geographers*, 29: 282–95.

Perry, G.L.W. (2002) 'Landscapes, space and equilibrium: shifting viewpoints', *Progress in Physical Geography*, 26: 339–59.

Phillips, J.D. (1999) *Earth Surface Systems: Complexity, Order and Scale*. Oxford: Blackwell.

Philo, C. and Wilbert, C. (eds) (2000) *Animal Spaces, Beastly Places: New Geographies of Human–Animal Relations*. London and New York: Routledge.

Pitman, A.J. (2005) 'On the role of geography in earth system science', *Geoforum*, 36: 137–48.

Plumwood, V. (2002) *Environmental Culture: The Ecological Crisis of Reason*. London: Routledge.

Power, E.R. (2005) 'Human–nature relations in suburban gardens', *Australian Geographer*, 6(1): 39–53.

Robbins, P. (2007) *Lawn People: How Grasses, Weeds and Chemicals Make Us Who We Are*. Philadelphia, PA: Temple University Press.

Rose, G. (1993) *Feminism and Geography: The Limits of Geographical Knowledge*. Cambridge: Polity Press.

Rousseau, J.-J. (1762) *The Social Contract*. Trans. and introduction by G.D.H. Cole; revised and augmented by J.H. Brumfitt and John C. Hall (1986). Markham, Ontario: Fitzhenry & Whiteside.

Smith, N. (2005) 'Neo-critical geography, or, the flat pluralist world of business class', *Antipode*, 37: 887–99.

Soper, K. (1995) *What is Nature? Culture, Politics and the non-Human*. London and Cambridge, MA: Blackwell.

Turner, B.L. II (2002) 'Contested identities: Human–environment geography and disciplinary implications in a restructuring academy', *Annals of the Association of American Geographers*, 92: 52–74

Whatmore, S. (2002) *Hybrid Geographies: Natures, Cultures, Spaces*. London: Sage.

Williams, R. (1980) *Problems in Materialism and Culture: Selected Essays*. London: Verso.

Williams, R. (1983) *Keywords: A Vocabulary of Culture and Society*. London: Flamingo.

Wilson, A. (1992) *The Culture of Nature: North American Landscape from Disney to the Exxon Valdez*. Oxford and Cambridge, MA: Blackwell.

Wolch, J. and Emel, J. (eds) (1998) *Animal Geographies: Place, Politics and Identity in the Nature–Culture Borderlands*. London and New York: Verso.

18

Nature: An Environmental Perspective

Roy Haines-Young

D efinition

Despite its apparent innocence, the word 'nature' can have a number of meanings. It can be used to refer to the set of external laws and regularities that seem to govern the universe, or to denote all that is essentially non-human – that part of the universe not created by people. However, others have argued that 'nature' is more a construct of the human mind and that different people and cultures can mean quite different things when they speak of it. They claim that these 'contested natures' need to be understood if we are to understand what motivates people when they interact with their environment.

INTRODUCTION

You are faced with a choice. Suppose you had to choose to either experience real nature or a simulation of it. The advertising that accompanies the offer says that the simulation would be a good one – allowing you to experience as rich and diverse an environment as exposure to the real thing would – but it would be artificial. And you would know it. Which would you choose?

The choice between real and simulated nature is one O'Neill and others posed in their account of environmental values (O'Neill et al., 2008). We too will look at the way people value nature, but we will also use the 'choice experiment' to consider more generally the kinds of relationship we have with this thing called 'nature'. The aim is to explore some of the ways different people have described nature and how people are connected to it. We will end by looking at how nature can be valued, and how these values motivate people when they act to protect, use or change the world around them.

ONE NATURE OR MANY?

One point that the idea of 'simulated nature' highlights is the question of whether it is possible to capture 'nature' in any simple way. On the face of it, the idea of a simulation seems to work only if there is a single thing called 'nature' that we can experience, a thing external to the human mind that is there to be discovered and understood. While not many of us would claim that the world we perceive is wholly imagined, we would probably have to accept that what we experience or perceive is also partly determined by the mental world that we inhabit or have been brought up in. Thus a simulated nature might be difficult to construct without understanding who the observer is and what he/she expects. For example, we only have to compare the way the 'colonial mind' saw Australia with the view that the Aboriginal people had of that same nature to see that such differences in perceptions can be the root of much conflict.

When European colonizers arrived in Australia they, of course, did not find it uninhabited or 'undiscovered'. However, the indigenous peoples made use of the land in ways that Europeans did not understand or acknowledge. The indigenous peoples were hunter-gatherers, and although they had strong notions of territory, they did not live in villages or appear to 'own' land. Eighteenth-century colonization was legitimized by the concept of *terra nullius*. This doctrine held that lands occupied by 'backward' peoples, who lacked European forms of government and who failed to cultivate their territories, were essentially vacant (Clark, 1999). Under English law at the time, in a territory acquired by treaty or military victory the existing institutions were retained, although the system could be changed by executive or legislative action after conquest. In territory deemed as *terra nullius*, however, the inhabitants were not recognized and English notions of justice and its legal system applied. In the case of Australia this meant no recognition was given to the rights of the Aboriginal people (Gibson and Fraser, 2007).

As Harding (1998: 89) points out, over the thousands of years of their occupation the Aboriginal peoples of Australia had developed their own 'systems of understanding and classification of relations with the environment'. These perspectives, she suggests, were 'holistic' and 'spiritual', and differed markedly from the way the European colonists perceived nature and how people were related to it. The indigenous peoples of the Australian Northern Territories considered themselves 'as one' with nature. Nature provided them with all they needed and they had an intimate relationship with their surroundings. 'The land not only belonged to them but they belonged to the land' (Cole, 1982; after Skertchly and Skertchly, 1999–2000). By way of illustration, Harding (1998) cites the way the Aboriginal people defined their calendar in what is now part of the Northern Territories. Unlike the one we are familiar with, in which seasons are defined by predetermined months, the Aboriginals defined the seasons by indicators

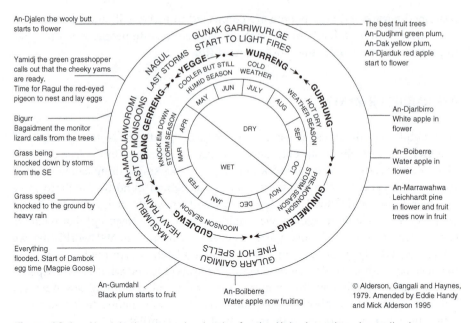

An-Djalen the wooly butt
starts to flower

Yamidj the green grasshopper
calls out that the cheeky yams
are ready.
Time for Ragul the red-eyed
pigeon to nest and lay eggs

Bigurr
Bagaidment the monitor
lizard calls from the trees

Grass being
knocked down by storms
from the SE

Grass speed
knocked to the ground by
heavy rain

Everything
flooded. Start of Dambok
egg time (Magpie Goose)

An-Gumdahl
Black plum starts to fruit

An-Boilberre
Water apple now fruiting

The best fruit trees
An-Dudjhmi green plum,
An-Dak yellow plum,
An-Djarduk red apple
start to flower

An-Djarlbirro
White apple in
flower

An-Boiberre
Water apple in
flower

An-Marrawahwa
Leichhardt pine
in flower and fruit
trees now in fruit

© Alderson, Gangali and Haynes,
1979. Amended by Eddie Handy
and Mick Alderson 1995

Figure 18.1   Aboriginal seasonal calendar for the Kakadu region, Australia, in
Gundjeyhmi (Mayali) language

such as the appearance or behaviour of a particular species or the phenology of
different plants (Figure 18.1). As a result, they could modify their activities to
take account of changing weather patterns over the year, and the different types
of resources that were likely to be available to them. Their lives were thus finely
tuned to the rhythm of the seasons.

The difference in perspectives on nature between Aboriginal people and
colonists, or even between us and our ancestors, suggest that the offer of a simulated nature is possibly not so straightforward as it sounds. If we want to pursue
the idea of constructing such a thing, we have to contend with the fact that there
are probably many natures to be represented, each resulting from the interaction
of people's minds and cultures with whatever is around them. One reason that
we might be suspicious of any offer to experience 'simulated nature' is that it
might not be clear exactly whose representation of nature it actually is, *and what
purposes or interests that representation is designed to serve.*

The contention that there are possibly many 'natures', and that the different ways people construct this idea is at the root of many of environmental
conflicts, is the point that Macnaughton and Urry (1998: 1) make in their
Contested Natures. For them: '...there is no singular "nature" as such, only a diversity
of contested natures ... each is constituted through a variety of socio-cultural
processes from which such natures cannot be plausibly separated'. They attempt
to show how different conceptualizations of nature are 'embedded or produced'
by different forms of social practice. To illustrate their argument they draw on
examples from the latter half of the twentieth century. They describe the way the
'environmentalism' developed partly out of key texts such *Silent Spring* (1962), in

which Rachael Carson highlighted the impact that widespread use of pesticides such as DDT were having both on wildlife and human health, and partly out of the new 'counter-cultures' that were developing at that time. Macnaughton and Urry (1998: 47) argue that the origins of environmentalism cannot be explained only by the insights that science provided, but must be looked at in terms of the way these ideas were taken up as part of a more general rejection by people, and especially young people at the time, of the values generally associated with material and technological progress. The image of nature threatened by progress was, they suggest, one constructed as much by the social processes represented by the protest movements of the 1960s as by contemporary scientific debates.

If we accept the idea of nature as essentially a social construct, then, it could be argued that the conflict between the colonial and Aboriginal minds in Australia was nothing to do with how different groups saw 'nature', but was more to do with the strategies used by one group of people to subjugate or marginalize another. Certainly, Australia's recent attempts to resolve the Aboriginal land rights issue appears to have been framed more in terms of the legality of the appropriation process rather than just an acceptance that different people may see the world in different ways (see, for example, Altman et al., 2005). Nevertheless, in looking at the history of the matter, there is also a sense that we are watching 'modern' Australia recognize and make space for other ways of living and *knowing* nature – and to learn from them. For example, as Skertchly and Skertchly (1999–2000) have argued, the Aboriginal seasonal calendar for the Kakadu region is not merely of interest as some kind of cultural artefact, but, as with other types of indigenous knowledge, provides a means of helping modern societies develop sustainably and cope with natural hazards in Australia's monsoon region. Thus, perhaps these accounts of nature that are strange for some really do capture something outside the human mind.

TAKING NATURE APART

That people can take different views of the same thing is not something that is particular to discussions of nature. However, if, as Williams (1976: 184) has argued, 'nature' is the most difficult word in the language', it is probably possibly an issue that we might need to look at more closely than in other situations.

The difficulties of untangling different perspectives on nature, and what they mean, are compounded by the fact that in the English language at least, the word itself has a number of different meanings. O'Neill et al. (2008) have summarized some of ideas that are packaged up in the word 'nature' by using a series of contrasts or dichotomies. The most important, they suggest, are the juxtapositions that use of the word sets up with ideas about the 'miraculous' and the 'artificial'.

One intention people have in using the word 'nature' has been to suggest that there is an external set of enduring regularities that explain the world around us, and that we do not need to resort to special or exceptional events to account for what we see. Thus they may speak of the 'laws of nature' or 'natural science' to indicate their focus on things other than the 'unnatural' or even 'supernatural'. Many writers (e.g. Pepper, 1996) have suggested that this kind of meaning is embodied in the modern scientific traditions of rationalism and reductionism, and the way it differentiates itself from more spiritual approaches to understanding. Another intention that people have in using the word 'nature' is to bring out a contrast between what seems to exist without people and that which is constructed, 'manufactured' or 'contrived' through the intervention of the human mind and action. Thus we hear people speak of 'natural habitats' as opposed to 'semi-natural' or 'artificial' ones, or landscapes that are 'natural' rather than 'cultural'.

That people use the term in these different ways clearly adds a further layer of complexity to the problem of understanding what they are trying to suggest when they speak of 'nature'. If we agree with those who have argued for 'multiple natures', then both ways of using the term seem to set some potential conflicts. On the one hand, such usage tends to suggest that there really *is* a single 'thing' called nature, represented by a set of enduring processes and laws that constrain and control the behaviour of the non-human world. On the other, it seems to suggest that people are somehow separate from this external world, and that we are essentially observers or explorers of it, and that there is some value or importance to knowing how things work without human intervention.

Although we might agree with Macnaughton and Urry (1998) that different people have different *perceptions* of 'nature', it does not follow that we have to reject the idea that the world around us really does consist of a single set of processes and constraints. It could just be that people describe the 'thing' called 'nature' in different ways. In this context, the offer of a 'simulated' nature is really quite an interesting one because it opens up the question of whether we might expect or require those constructing it to take any account of the behaviour of this external reality – *and to have represented it faithfully*.

If we require only that a simulation takes account of people's perceptions of nature, then such representations would seem pretty arbitrary things. Indeed, if we were to proceed in this way, it is difficult to see how such simulations would differ from any fictional account of a universe, constrained by nothing other than the human imagination. To be worth considering, 'simulated nature' would have to be more grounded on an understanding of the structure and dynamics of this external reality than would a commercial computer game.

Simulations or models of environmental systems and processes, and scenarios describing how environments might change under different combinations of external drivers, are, of course, now part and parcel of the *natural* sciences. Scientists use them not just to explore the consequences of their theories about how nature works, but also increasingly to communicate ideas to non-scientists so that they may be better involved in decision-making. Clearly, in these situations questions about 'authenticity', 'plausibility' and the extent to which these models 'accurately' capture something about the way nature *really works* seem to be critical.

Now it has been argued that the 'reductionist tendency' of modern science to take nature apart in order to try to describe and understand how it works has marginalized or devalued more spiritual or instinctive kinds of knowing. Pepper (1996), for example, traces how the view of nature as a 'machine' developed in Europe from the time of Copernicus, eventually giving rise to the belief that there was a fundamental difference between, or separation of, human society and nature, as expressed most clearly in the writings of Descartes and

Bacon in the early part of the seventeenth century. This 'dualism', Pepper suggests, led not only to the belief in the 'separateness of humans from their objects of study, nature' (Pepper, 1996: 143), but also to the assertion by Bacon, for example, that scientific knowledge represented power over nature. In the modern world these ideas manifest themselves under the banner of *technocentrism* that is the belief that fundamentally science or technology can solve our contemporary environmental problems, and our *utilitarian* approach to the exploitation of natural resources which often seeks to maximize human welfare at the expense of the integrity of environmental systems.

Whether or not it is possible to construct such a thing, the very idea of a 'simulated nature' captures some of the problems that we have to face in coming to terms with what is meant by the word. Could we conceive such a thing only because we have been indoctrinated into this reductionist scientific paradigm that sees nature and society as fundamentally separate? What would the Aboriginal mind make of it? Can the kind of intimate relationship between people and nature that they imagine be represented in such an artificial way?

The question of whether the idea of a simulated nature makes sense only in the context of our 'western' notions of science is an intriguing one, because if it were the case, then it would seem to imply that what we accept as knowledge about nature is to be judged relative only to the beliefs and mores of the social group that we are part of. In fact, the answer turns out not to be as simple as this. To see that it is so we have to distinguish clearly between the *ways* of knowing about 'nature', that is the methods different peoples use to acquire and communicate knowledge, from the 'thing' itself – the 'nature' that we seek *to know*.

Even within the western scientific tradition it is not accepted by everyone that we have to take nature apart to understand it. Indeed, the importance of adopting an holistic perspective is one that many have argued is essential if we are to come to terms with environmental complexity and the problem of living in the world sustainably. Pepper (1996) and others (e.g. O'Riordan, 2000) have, for example, described how these ideas are played out in the contemporary debates surrounding the notion of *ecocentrism* in both the ethical and political arenas. Ecocentrism asserts that the separation between people and nature is a false one and that we are fundamentally part of nature. Thus we should respect nature's limits and attempt to live in balance with it. Moreover, it is claimed that ecological or environmental values have an equal or even superior status alongside human ones. In the context of discussions about appropriate scientific methodologies, Holling (1998), for example, has argued that, particularly for the study of ecological systems, integrated or holistic perspectives are superior to those offered by traditional, reductionist science because nature is more than the sum of its parts. They claim that if we are to understand or recognize the 'emergent' properties of ecosystems, such as resilience and stability, that arise *as a result* of putting the different elements of nature together, then we have to look at things 'as a whole'.

When viewed from a holistic perspective, the idea of a simulated nature might, in fact, be very attractive indeed, for it may be a way of describing just how modern people really are immersed in nature. That society and nature are really one is a key theme in *ecocentrism* and a simulation might *be a useful way of trying to demonstrate what kinds of relationship might exist*. While we think of simulations as modern inventions, they may, in this respect, be no different from the stories our ancestors constructed in an attempt to explain the world around them. Moreover, in the context of Aboriginal cultures, who is to say that the songs, stories and dances used to share and pass on knowledge about how to live in the world are not also 'simulations' of a kind? Their messages were certainly refined by experience – by trial and error – in much the same way that scientific insights are. That they describe nature in different ways compared to ours, how-ever, does not mean that they experience a fundamentally different nature with a different structure and different behaviour.

Thus it seems that we may have to distinguish between the worldviews or paradigms by which people describe nature and nature itself. The importance of making this kind of distinction has also been recognized in the debates about what is meant when we use the tem 'environment' (see, for example, Attfield, 1999). Possibly the problem of multiple perspectives is simply the result of looking at the same complex thing (nature or environment) through different 'cultural lenses' or 'fil-ters' (see Pepper, 1996). These different paradigms or cultures of nature matter because not only do they shape motives and actions, but they also broaden our understanding of nature as a whole. Contemporary science probably ignores local or lay knowledge at its peril if its theories are to be tested rigorously. What is also clear in the context of the offer to experience a simulated nature is that we would prob-ably not be included to accept the different stories people might tell about nature at face value. Any simulation would not do. 'Authenticity', in the sense that the simu-lations *have* to reflect or respect the way nature 'really works' would count a lot too.

VALUING NATURE

When you were first asked to consider the choice between real or simulated nature, the issue of authenticity was possibly the first thing that sprung to mind. Setting all the issues aside, about multiple perceptions of nature and how people relate to it, suppose someone actually accomplished the feat of constructing a perfect, artificial nature: would it really be acceptable? It is not so far-fetched. In the USA, the infamous *Biosphere 2* experiment tried to construct such a thing (Allen et al., 2003). In the UK, the more modest *Eden Project* also represents a simulation of a kind (Bartram and Shobrook, 2000). The question of authenticity opens up the debate about how people value this thing called nature. Would the simulation be acceptable if it provided you with all the *physical* resources that you would need to survive? Or is it that 'real nature' also has some *intrinsic* value that an artificial world could never capture or replace?

The problem of identifying the different values that people apply to nature is probably as difficult as understanding what they mean when they use the word itself. As a way of helping to deal with some of the complexity, O'Neill and his co-authors have argued that we should start by focusing on 'the various ways individuals, processes and places matter, our various modes of relating to them, and the various considerations that enter into our deliberations about action' (O'Neill et al., 2008: 1). To see how this kind of argument works itself out, let us consider a particularly pressing contemporary issue, concerning whether it is possible to assign an *economic* value to nature.

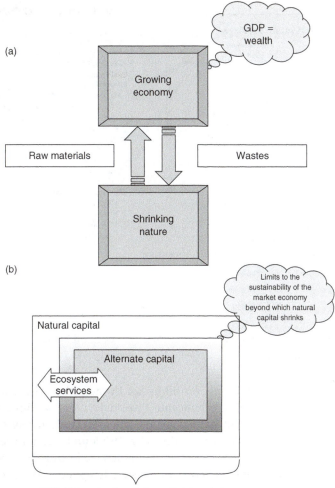

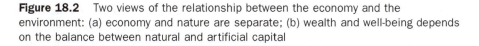

Figure 18.2   Two views of the relationship between the economy and the environment: (a) economy and nature are separate; (b) wealth and well-being depends on the balance between natural and artificial capital

The age-old question of whether 'people are part of nature or somehow outside it' has been resurrected in current debates about the relationship of the human economy and the environment. It is now widely accepted, for example, that modern economic accounting methods do not take the human use of nature and our impact upon the environment fully into account, and that more integrated approaches are urgently required unless we are to fundamentally damage the capacity of the biosphere to support us and the other species that inhabit this planet. Are there ecological limits to the growth of our economy, set by nature? The question is at the heart of the disciplines of environmental or ecological economics (Figure 18.2). The claim that we have to overcome the perceived and mistaken separation between economic systems and nature is also central to understanding what sustainable development is all about.

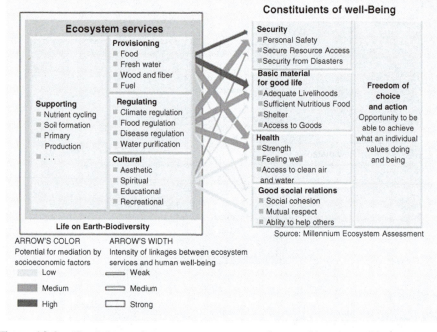

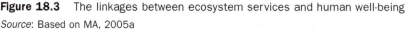

Figure 18.3   The linkages between ecosystem services and human well-being
Source: Based on MA, 2005a

The view that there are no limits set by nature on the growth of our economy is one aspect of the technocentric worldview that we discussed earlier. It is argued that not only will technology and the market ultimately solve all our environmental problems, but also 'artificial' and 'natural' capital are ultimately held to be substitutable in terms of securing human well-being. Natural resources can be consumed and modified and any loss is made up for, or offset by, a corresponding increase in social or economic wealth. By contrast, the ecocentric view not only holds that nature has intrinsic value, that is a value that arises by its very existence, but also that its constraints and limits have to be taken into account. Those who advocate this paradigm argue that resources should be conserved or used in such a way as to minimize our impact upon them. This can be done only if the real costs of the human use of natural resources are understood.

As Turner et al. (1994) and others have pointed out, there is a wide spectrum of views between the extreme, *anti-green*, technocentric position, with its 'very weak' position on the questions of environmental sustainability, and the *deep-green*, ecocentric perspective, which takes a robust or 'strong' position on the matter. Between them others have argued that it is better to try to find ways of accommodating the needs of people and nature, and to look for tools and approaches that can help us pursue what O'Neill and his co-authors have advocated, namely to look for the things that shape our deliberations about the environment.

The Millennium Ecosystem Assessment (MA, 2005a) is an example of a recent, co-ordinated attempt by an international coalition of over 1,300 researchers to document the importance of the contribution that nature makes to human well-being. It did so by arguing that natural systems, or ecosystems, can provide both goods and services to people (Figure 18.3), and that they are significant because they both contribute to the market economy and provide a range of free, 'non-market' services such as the regulation of the quality of air and water, or protection from natural hazards.

There are many case studies illustrating how ecosystems can provide services to people, even though no formal economic value has, in the past at least, been placed upon them. In the uplands of the UK, for example, it is recognized that Blanket Bog provides a significant 'free service' to society by regulating the flow of good-quality water into our reservoirs. The large water-holding capacity of these ecosystems means that the water is released steadily, evening out the peaks and troughs between rainfall events. The ecological processes operating in these ecosystems also ensure that, under 'normal circumstances', the water draining from them is of high chemical and biological quality. The whisky industry of Scotland partly depends on the particular qualities of waters draining from peat-rich soils. However, many of these upland areas have also been used for agricultural purposes and, historically, overgrazing, drainage, reseeding of grasses and fertilizer applications has, in some catchments, led to the conversion of bog to other habitats – with a subsequent loss of their ecological integrity. As a result, there has been a decline in the quantity of water available. Its quality has also declined as a result of oxidized peat entering drainage systems, thereby discolouring the water and making it more difficult to use for human consumption without further treatment. Oxidation occurs when active peat formation is arrested, and the peat dries out. Erosion and oxidation of peat also results in the release of carbon sequestered by the ecosystem, along with other greenhouse gases, such as methane and nitrous oxide.

Arguments about the conservation and restoration of Blanket Bog in the UK have traditionally been made in terms of their importance for biodiversity. Their significance to society clearly goes far beyond this and any assessment must include these other types of service that they provide for us. This can be illustrated by reference to an initiative led by United Utilities, a water supply company in the northwest of England, who own about 57,000 hectares of Blanket Bog, primarily to protect the quantity and quality of water entering reservoirs. They are currently attempting to restore degraded Blanket Bog, by reseeding and blocking the drains previously cut in an attempt to improve their use for grazing. Kettunen and ten Brink (2006) report that these measures to improve water quality are likely to lead to annual savings for the water company of between £1.2 million and £2.6 million, based on the reduced costs of 'end of pipe' water treatment, meaning that the cost–benefit analysis favours ecosystem restoration rather than investment in new plant and infrastructure.

The example of Blanket Bog from the UK is but one illustration of how a more holistic understanding of the ways in which nature contributes to human well-being can help people make better decisions. The lack of such an integrated view, or the failure to understand or take responsibility for the complete relationship that society has with 'natural systems', was probably a major factor leading to the degradation of these ecosystems in the first place. The same kind of story can be told using case studies from both the developed and developing world. The Millennium Ecosystem Assessment looked at the state of 24 key ecosystem services (Table 18.1) at global scales, and concluded that about 60% of them were

Table 18.1   Global status of ecosystem services evaluated by Millennium Ecosystem Assessment

An upwards arrow indicates that the condition of the service globally has been enhanced and a downwards arrow that it has been degraded. Definitions of 'enhanced' and 'degraded' for the three categories of ecosystem services shown in the table are provided in the note below. Supporting services, such as soil formation and photosynthesis, are not included here as they are not used directly by people.

Service	Sub-category	Status	Notes
Provisioning services			
Food	Crops	▲	Substantial production increase
	Livestock	▲	Substantial production increase
	Capture fisheries	▼	Declining production due to over-harvest
	Aquaculture	▲	Substantial production increase
	Wild foods	▼	Declining production
Fiber	Timber	+/–	Forest loss in some regions, growth in others
	Cotton, hemp, silk	+/–	Declining production of some fibers, growth in others
	Wood fuel	▼	Declining production
Genetic resources		▼	Lost through extinction and crop genetic resource loss
Biochemicals, natural medicines, pharmaceuticals		▼	Lost through extinction, over-harvest
Water	Fresh water	▼	Unsustainable use for drinking, industry, and irrigation; amount of hydro-energy unchanged, but dams increase ability to use that energy
Regulating services			
Air quality regulation		▼	Decline in ability of atmosphere to cleanse itself
Climate regulation	Global, regional and local	▲ ▼	Net source of carbon sequestration since mid-century Preponderance of negative impacts
Water regulation		+/–	Varies depending on ecosystem change and location
Erosion regulation		▼	Increased soil degradation
Water purification and waste treatment		▼	Declining water quality
Disease regulation		+/–	Varies depending on ecosystem change
Pest regulation		▼	Natural control degraded through pesticide use
Pollination		▼[a]	Apparent global decline in abundance of pollinators
Natural hazard regulation		▼	Loss of natural buffers (wetlands, mangroves)
Cultural services			
Spiritual and religious values		▼	Rapid decline in sacred groves and species
Aesthetic values		▼	Decline in quantity and quality of natural lands
Recreation and ecotourism		+/–	More areas accessible but many degraded

Table 18.1 *(Continued)*

Note: For provisioning services, we define enhancement to mean increased production of the service through changes in area over which the service is provided (e.g. spread of agriculture or increased production per unit area. We judge the production to be degraded if the current use exceeds sustainable levels. For regulating services, enhancement refers to a change in the service that leads to greater benefits for people (e.g. the service of disease regulation could be improved by eradication of a vector known to transmit a disease to people). Degradation of regulating services means a reduction in the benefits obtained from the service, either through a change in the service (e.g. mangrove loss reducing the storm protection benefits of an ecosystem) or through human pressures on the service exceeding its limits (e.g. excessive pollution exceeding the capability of ecosystems to maintain water quality). For cultural services, degradation refers to a change in the ecosystem features that decreases the cultural (recreational, aesthetic, spiritual, etc.) benefits provided by the ecosystem.
[a] Indicates low to medium certainty. All other trends are medium to high certainty.

Source: Based on MA (2005a)

in decline, largely as a result of human impact upon them (MA, 2005a). As a result, the panel argued that without significant policy intervention, it was unlikely that the world would meet the UN Millennium Development Goals for poverty alleviation and health (MA, 2005b).

How can we better understand and document the ways that the non-market benefits provided by nature matter to people – to show that our economy and nature are not separate but integrated systems? One sort of calculus is provided by the environmental economists, who have developed a schema that classifies how the total economic value (TEV) that resides in environmental systems can be broken down (Figure 18.4).

At the top level in the classification of TEV is the distinction between use and non-use values. The former groups the values or benefits that we gain by directly using or extracting them as consumers, such as when we use timber or fish, together with those kinds of benefit that are gained by a more indirect use of nature through such things as recreation. The non-use values mainly deal with the benefits that people enjoy just by knowing that nature, or some aspect of nature, exists, such as polar bears or tropical rainforests, even though they never see or experience it. A more subtle form of direct use value is the 'option use value', that is the importance of preserving some aspect or component of nature because we might want to use it in the future. The importance that we attach to being able to pass on resources to future generations ('bequest value') captures, according to the schema, aspects of both use and non-use values. Environmental economists go on to suggest different ways of estimating the various components of TEV, and ultimately an aggregate sum.

The term 'total economic value' is sometimes misunderstood because it should not be taken as implying that what we are calculating is some absolute worth for nature. This kind of mistake is at the root of the debate surrounding the estimate of Costanza et al. (1997) that the biosphere is worth $33 trillion per year. Rather, use of the word 'total' just means that the calculation is attempting to be comprehensive in terms of all the different *economic* components that contribute to the overall monetary assessment. Moreover, the individual components

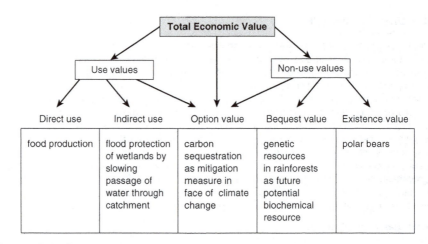

Figure 18.4   The components of total economic value (TEV)

and TEV are used only to understand how the importance that we attach to the nature (or parts of it) might change under different assumptions or influences, that is to understand 'marginal economic values'. This idea of using valuation techniques to calculate 'marginal values' is illustrated by the example of Blanket Bog in the UK, where people have attempted to use monetary estimates of indirect use to understand the *relative* changes in costs and benefits of different management decisions, not to calculate some total worth of the system.

It is likely that estimates of the economic value of different aspects of nature will become more common in public debates as the results of studies such as the Millennium Ecosystem Assessment are digested, and more generally as societies evaluate strategies for dealing with the likely consequences of climate change and the role that ecological systems can play in them. Such trends are interesting because they illustrate the re-emergence of questions about the separation of people and nature, recast in the 'modern language' of economics.

The different ways that those concerned with the economics of the environment have devised to calculate monetary values for nature need not concern us further here. Whether or not we agree with robustness or reliability (the methods currently available), the key question we have to consider is the extent to which attempts to place an economic valuation on nature or its services exhaust the 'various ways' people describe how 'processes and places' matter? As Turner et al. (1994) have pointed out, in the context of making estimates of total economic value, the process is concerned with assessing the preferences held people in a way that is both anthropocentric and instrumental. Thus assessments are human-centred and pragmatic in character. The methods do not consider any value that resides within or that are *intrinsic* to nature and thus probably have clear limitations. So what other dimensions of value need to be considered?

A nature ethic?

The narrowness of economic valuation has long been known. It was described very clearly by Aldo Leopold in his *A Sand County Almanac*, written in 1949, where he discussed the importance of developing a 'land ethic': 'One basic weakness in a conservation system based on wholly economic motives is that most members of the land community have no economic value. Wildflowers and song-birds are examples...' (Leopold, 1966: 247).The 'members of the land community' that he referred to were other species. He notes the 'circumlocutions' of those who have sought to 'invent' an argument for conservation based on economic value, and suggests that such thinking needs to be looked at in the wider context of a 'land ethic' that seeks to change the role of people from 'conqueror' of the land community (i.e. nature) to 'plain member and citizen of it' (Leopold, 1966: 241). He concludes:

> A system of conservation based solely on economic self-interest is hopelessly lopsided. It tends to ignore, and thus eventually to eliminate, many elements in the land community that lack commercial value, but that are (as far as we know) essential to its healthy functioning. It assumes falsely, that the economic parts of the biological clock will function without the uneconomic parts. (Leopold, 1966: 251)

Leopold thus argues that there are issues of value that transcend economics. On the one hand, he echoes the ecocentric position by suggesting that in making arguments about 'what matters' we have to go beyond human values and accept the ethical position that other species are entitled to moral consideration for their own sake. On the other, he highlights the fact that at a more pragmatic level, the very existence of biodiversity (nature) is a *prerequisite* for all other types of value (including human values) and so that it has some instrumental significance – that is, worth because they are a means to some end (O'Neill et al., 2008).

Today, we can see some of the arguments that nature has some instrumental values that are not captured by economic assessments in current discussions of how we should weigh the importance or risks of human action causing ecological collapse, as a result of natural systems being pushed beyond some ecological limit or threshold. Limberg et al. (2002), for example, have argued that estimates of change in marginal economic value make sense only when the coupled 'social-ecological' systems, that we are all part of are near to equilibrium, because there is a high degree of certainty and predictability. In these situations, individuals are well placed to make decisions about the trade-offs and substitutions between different potential outcomes and economics can help us make these decisions. In the proximity of a threshold, however, where systems may be close to or liable to collapse, Limberg et al. (2000) suggest that such considerations no longer really apply – risk is the issue. Questions of risk avoidance may partly be determined by the possibility of financial losses, but they also entail ethical and social dimensions, since questions of rights and environmental justice may arise, if the loss or collapse of a natural system affects the health or welfare of different people or groups, or other species, in different ways (Bührs, 2004).

Leopold (1966: 251) argued for a 'land ethic' to 'supplement and guide the economic relationship to land'. Today, we can see similar arguments embedded in discussions surrounding the virtues of adopting the 'Ecosystem Approach' to decision-making and policy evaluation (IUCN, 2004). The latter emerged as a

Table 18.2   The priniciples of the Ecosystem Approach

1	The objectives of management of land, water and living resources are a matter of societal choice.
2	Management should be decentralized to the lowest appropriate level.
3	Ecosystem managers should consider the effects (actual or potential) of their activities on adjacent and other ecosystems.
4	Recognizing potential gains from management, there is usually a need to understand and manage the ecosystem in an economic context. Any such ecosystem management programme should:

    (a) Reduce those market distortions that adversely affect biological diversity
    (b) Align incentives to promote biodiversity conservation and sustainable use
    (c) Internalize costs and benefits in the given ecosystem to the extent feasible.

5	Conservation of ecosystem structure and functioning, in order to maintain ecosystem services, should be a priority target of the Ecosystem Approach.
6	Ecosystems must be managed within the limits of their functioning.
7	The Ecosystem Approach should be undertaken at the appropriate spatial and temporal scales.
8	Recognizing the varying temporal scales and lag-effects that characterize ecosystem processes, objectives for ecosystem management should be set for the long term.
9	Management must recognize that change is inevitable.
10	The Ecosystem Approach should seek the appropriate balance between, and integration of, conservation and use of biological diversity.

Source: Adopted by The Conference of the Parties to the Convention on Biological Diversity at its Fifth Meeting, Nairobi, 15–26 May 2000. Decision V/6, Annex 1. CBD COP–5 Decision 6 UNEP/CBD/COP/5/23.

topic of discussion in the late 1980s and early 1990s among the research and policy communities concerned with the management of biodiversity and natural resources (see Hartje et al., 2003). Its advocates argued that a new focus was required to achieve robust and sustainable management and policy outcomes. Application of the principles of the Ecosystem Approach (Table 18.2), it was suggested, would deliver more integrated policy and management decisions than existing frameworks, and be more firmly directed towards the needs of both people and biodiversity more generally.

Setting aside the problem of attempting to design a set of ethical principles 'by committee', the Ecosystem Approach could be seen as a counterpart to Leopold's more eloquently formulated land ethic. While the Ecosystem Approach argues that economic assessments of environmental goods and services have their place in decision-making and asserts that they should be taken more seriously, there are also hints that other types of value count, particularly those related to living within limits of 'ecological functioning'. Despite the compromises that must have been involved in agreeing these principles within an international arena, the principles could therefore be seen as a framework to 'shape our deliberations' in the middle ground between the technocentric and ecocentric positions, and therefore to find ways of accommodating the needs of both people *and* nature.

We therefore face an interesting dilemma in relation to the way we value nature and use values to shape our actions. There is little doubt of the importance of initiatives such as the Millennium Ecosystem Assessment and concepts such as the Ecosystem Approach, which use the language of economics to help us see nature as some kind of 'capital'. They help us re-examine the fundamental relationship between people and nature and possibly close the separation between the economy and environment that has been so damaging. By showing how nature can provide 'services' essential for human well-being, these paradigms also provide new terminologies that can help people in the worlds of politics and business take nature into account. However, at the same time, these new ideas should not be used to justify claims that nature's worth resides *only* in its economic value. To represent the value of nature so narrowly, as a set of utilitarian values, tends to overlook its intrinsic importance and the characteristics that make it unique and unsubstitutable.

CONCLUSION: DIFFICULT CHOICES

So would you choose artificial nature over the real thing? As we have seen, the choice partly depends on *who* has constructed the simulation and how successful they have been at capturing the way the world *really* works. The decision also depends on whether we think there is any importance attaching to this idea of authenticity. Can a 'cover version' ever replace the original? Could an imitation rainforest or coral reef ever substitute for the genuine article?

If you are inclined to answer, well 'nature wins every time', then think about what issues shape your decision and what extra you think you gain with the 'real thing'. If you are tempted to go for the simulation because it is safer or more convenient, or too expensive to sustain the real thing, then ask yourself whether the choices we make depend only on human values and wants.

The aim of this chapter has been to suggest that whatever choice you make, neither is straightforward. Lest you think such hypothetical questions are unimportant, consider the fact that increasingly in our daily lives we experience a 'nature' as artificial and constructed as any simulation game might be. 'Real nature' is certainly not some wilderness, unaffected by human action. On our crowded planet even our most remote areas bear the direct or indirect impact of human activities. Most of the landscapes we are familiar with in the developed world have been planned and are, or have been, managed or impacted by people. Natural elements are to be found there, but it is a nature far removed from any pristine state that existed before the intervention of people.

The development of these cultural landscapes or ecosystems, like the creation of any artificial simulation, bears the imprint of the social and cultural values of the peoples who have designed or created them. Thus arguments about 'authenticity' are possibly difficult to make in terms of justifying why you might choose this 'real' nature over an artificial one. In the modern world, it could be argued that the 'real nature' we experience is as artificial as any simulation in which we might immerse ourselves. The only difference is that we cannot walk away from it as we might when our computer worlds go wrong. Can we find ways of creating social-ecosystems that both retain the things in 'nature' that must be valued and at the same time sustain human well-being? The challenge we face today is how far we take this transformation process and how we weigh the importance of the services that nature provides against our other needs and wants.

SUMMARY

This chapter looks at how different people have represented this thing called 'nature' and the sorts of relationship that we have with it. The fundamental questions it poses are:

• Is there one nature or many? And if there are many, what does this mean about the existence of natural laws and people's attempts to discover them?

• If different people have different understandings of nature, how do we deal with these differences in making decisions?

• In what ways can nature be valued? Although increasingly estimates of economic value for nature services are helping to shape decisions, it is important to recognize that social and ethical dimensions count too.

• Is the 'Ecosystem Approach' a modern version of Leopold's land ethic?

The challenge of modern times, it seems, is to understand that the global economy is part of nature and not separate from it. As geographers we need to find ways of measuring nature's worth and using this knowledge to help shape human action.

Further Reading

Castree's (2005) *Nature: Key Ideas in Geography* is an accessible account of the relationships between people and nature, and how the idea of nature has been constructed and used. It is written from a geographical perspective and goes on to examine the way geographers have studied nature and contributed to our understandings of it. Daily's (1997) *Nature's Services* and Daily and Ellison's (2002) *The New Economy of Nature* provide a sound introduction to the idea of natural capital and the importance it has for people. They also provide a range of examples that show how important it is to try to value natures services. Edwards-Jones et al.'s (2000) *Ecological Economics* is a good introduction to environmental economics which sets the ideas and issues in a wide context. Kumar and Reddy's (2007) *Ecology and Human Well-being* provides a source of case studies from the developing world that demonstrates the importance of understanding the links between ecosystem services and well-being.

Note: Full details of the above can be found in the references list below.

REFERENCES

Allen, J.P., Nelson, M. and Alling, A. (2003) 'The legacy of biosphere 2 for the study of bios-pherics and closed ecological systems', *Advances in Space Research*, 31(7): 1629–39.

Altman, J., Linkhorn, C. and Clarke, J. (2005) *Land Rights and Development Reform in Remote Australia*. Melbourne, Vic.: Oxfam.

Attfield, R. (1999) *The Ethics of the Global Environment*. Edinburgh: Edinburgh University Press.

Bartram, R. and Shobrook, S. (2000) 'Endless/end-less natures: environmental futures at the fin de millennium', *Annals of the Association of American Geographers*, 90(2): 370–80.

Bührs, T. (2004) 'Sharing environmental space: the role of law, economics and politics', *Journal of Environmental Planning and Management*, 47(3): 429–47.

Carson, R. (1962) *Silent Spring*. Boston, MA: Houghton Mufflin.

Castree, N. (2005) *Nature: Key Ideas in Geography*. London: Routledge.

Clark, J. (1999) '"Indigenous" people and constitutional law', in P. Hanks and D. Cass (eds) *Australian Constitutional Law: Materials and Commentary* (6th edn). Melbourne: Butterworth, p. 90.

Cole, K. (1982) *A History of Numbulwar*. Bendigo: Keith Cole Publications, pp. 9–10.

Costanza, R., D'Arge, R., DeGroot R., Farber, S., Grasso, M., Hannon, B., Limburg, K., Naeem, S., O'Neill, R., Paruelo, J., Raskin, R., Sutton, P. and van den Belt, M. (1997) 'The value of the world's ecosystem services and natural capital', *Nature*, 387: 253–60.

Daily, G.C. (ed.) (1997) *Nature's Services*. Washington, DC: Island Press.

Daily, G.C. and Ellison, K. (2002) *The New Economy of Nature*. Washington, DC: Island Press.

Edward-Jones, G., Davis, B. and Hussain, S. (2000) *Ecological Economics*. Oxford: Blackwell Science.

Gibson, A. and Fraser, D. (2007) *Business Law* (3rd edn). Frenchs Forest, NSW: Pearson.

Harding, R. (1998) *Environmental Decision Making*. Annandale, NSW: The Federation Press.

Hartje, V., Klaphake, A. and Schliep, R. (2003) *The International Debate on the Ecosystem Approach: Critical Review, International Actors, Obstacles and Challenges*. Bonn: BfN Skripten 80.

Holling, C.S. (1998) 'Two cultures of ecology', *Conservation Ecology*, 2(2): 4 (available online at http://www.consecol.org/vol2/iss2/art4, accessed 01/01/08).

IUCN (2004) *Comparing the Ecosystem Approach with Sustainable Use*. Seventh meeting of the Conference of the Parties to the Convention on Biological Diversity (COP7). Kuala Lumpur, Malaysia, 9–20 February. IUCN Information Paper, February 2004.

Kettunen, M. and ten Brink, P. (2006) *Value of Biodiversity: Documenting EU Examples where Biodiversity Loss has led to the Loss of Ecosystem Services*. Final report for the European Commission. Institute for European Environmental Policy (IEEP), Brussels, Belgium. 131pp.

Kumar, P. and Reddy, S.B. (2007) *Ecology and Human Well-being*. London: Sage.

Leopold, A. (1966) *A Sand County Almanac*. Oxford: Oxford University Press.

Limberg, K.E., O'Neill, R.V., Costanza, R. and Farber, S. (2002) 'Complex systems and valuation', *Ecological Economics*, 41: 409–20.

MA (2005a) *Ecosystems and Human Well Being: Synthesis*. Washington, DC: Island Press. http://www.millenniumassessment.org/en/Synthesis.aspx (accessed 01/01/08).

MA (2005b) *Ecosystems and Human Well Being: Policy Responses* (Vol. 3). Washington, DC: Island Press. http://www.millenniumassessment.org/en/Responses.aspx#download (accessed 01/01/08).

Macnaughton, P. and Urry, J. (1998) *Contested Natures*. London: Sage.

O'Neill, J., Holland, A. and Light, A. (2008) *Environmental Values*. Abingdon: Routledge.

O'Riordan, T. (2000) *Environmental Science for Environmental Managers*. London: Longman.

Pepper, D. (1996) *Modern Environmentalism. An Introduction*. London: Routledge.

Skertchly, A. and Skertchly, K. (1999–2000) 'Traditional Aboriginal knowledge and sustained human survival in the face of severe natural hazards in the Australian monsoon region: some lessons from the past for today and tomorrow', *Australian Journal of Emergency Management*, September: 42–50.

Turner, R.K., Georgiou, S. and Fisher, B. (2008) *Valuing Ecosystem Services: The Case of Multi-Functional Wetlands*. London: Earthscan.

Turner, R.K., Pearce, D. and Bateman, I. (1994) *Environmental Economics*. Harlow: FT/Prentice-Hall.

Williams, R. (1976) *Keywords*. London: Fontana.

19

Globalization: Interconnected Worlds

James R. Faulconbridge and Jonathan V. Beaverstock

Ⓓefinition

Although in its simplistic sense globalization refers to the *widening, deepening* and *speeding up* of *global interconnectedness*, such a definition begs further elaboration. ... Globalization can be located on a continuum with the local, national and regional. At one end of the continuum lie social and economic relations and networks which are organized on a local and/or national basis; at the other end lie social and economic relations and networks which crystallize on the wider scale of regional and global interactions. Globalization can be taken to refer to those *spatio-temporal processes* of change which underpin a transformation in the organization of human affairs by linking together and expanding human activity across regions and continents. Without reference to such expansive *spatial connections*, there can be no clear or coherent formulation of this term. ... A satisfactory definition of globalization must capture each of these elements: extensity (*stretching*), *intensity*, velocity and *impact*. (Held et al., 1999: 14–15, emphasis added).

But, be cautious in your reading of globalization. Defining globalization can bring more questions than answers, as Dicken (2004: 5) comments:

> 'Globalization' is a big problem in every sense of the term. It is, first and foremost, a problem in a *material* sense, insofar as its associated syndrome of processes creates highly uneven geographical and social outcomes. ... It is also a problem in a *rhetorical* or *discursive* sense, in that its meaning and

significance are widely contested. ... 'Globalization' has evolved into a catch-all term, used by many to bundle together all the goods and bads of contemporary society.

INTRODUCTION: GEOGRAPHIES OF GLOBALIZATION

Globalization, its meaning and conceptual value, has long been contested within human geography. As probably *the* most fashionable concept of the 1990s and now the new millennium, the rhetoric surrounding academic and media uses of the term 'globalization', make it easy to lose sight of its multifarious meanings. As Dicken (2004) argues, globalization is inherently geographical. Understanding globalization as a process requires us to consider the way space, place and time are configured and reconfigured as a result of contemporary changes in techno-logical, economic and political practices. For Taylor et al. (2002), this is why 'geography *and* globalization' are so intimately related: all processes of globaliza-tion have geographical dimensions. They expand on this idea in three further ways:

- The 'geography *in* globalization'. As Swyngedouw (1997) argues, globalization is actually a local–global or 'glocalization' process. Instead of focusing solely upon the global as a scale, we also need to recognize the interconnections between different scales (local, regional, national and global) and how these make up the process of globalization.
- The 'geography *of* globalization'. Processes of globalization create new geo-graphical patterns of flows and activity. For example, the New International Division of Labour reconfigures both the geography of manufacturing activi-ties, but also, as a side-effect, geographies of uneven development, poverty and wealth.
- 'Geography *for and against* globalization'. Needless to say the concept of glob-alization has caused great debate within human geography and many other social sciences.

But why do geographers have so much to say about globalization, good, bad or indifferent? Dicken (2003: 1) argues that for geographers studying global-ization, the 'basic aim is to analyse the processes shaping and reshaping the global map'. These processes exist in the form of cultural, economic, political and social practices that have become increasingly transnational in recent times. Key actors driving these processes include, for example, transnational corporations (TNCs) – firms that have expanded outside their home countries and developed command-and-control operations in multiple locations – and governments, both at the national scale where regulations that are friendly to global trade and for-eign direct investment (FDI) are promoted (Department for Trade and Industry, 1998) and also at transnational scales where supranational bodies such as the World Trade Organization promote transnational standards. In addition, global-ization is manifested by technological changes that facilitate transnational flows of media, information, people and goods that enable the emergence of the type of globally interconnected, shrinking world many associate with globalization (Allen and Hamnett, 1995). Together, these different actors mean the interconnection between places has increased in recent times as a 'space of flows' has emerged that many suggest was unparalleled in previous eras (Castells, 2000; Thrift, 2002).

In the following sections we examine two case studies of different forms and impacts of globalization: in economic globalization, the phenomenon of offshoring and call centres; and in cultural globalization, the transnationality of commodity cultures. Throughout these we draw attention to the geography in and of globalization, and the changing processes that have produced new forms of interconnected worlds. In the conclusions, we then reflect critically upon these discussions and consider why such examples have led to debates between those for and against suggestions that we live in a newly globalized world.

ECONOMIC GLOBALIZATION AND INTERCONNECTED WORLDS

In human geography processes of globalization triggered important debates about the spatial organization of economic activities. Here an important starting point were studies of the *Spatial Divisions of Labour* (Massey, 1984) that emerged during the second half of the twentieth century. The concept of spatial divisions originally pointed to the geographical selectivity of labour processes in manufacturing and the way that at both national and international scales it was possible to see firms dividing up the production process into skilled (knowledge-intensive) and unskilled (labour-intensive) works. Selected labour processes were then sited in the developed (core) and developing (periphery) countries of the globe respectively. Often referred to as the New International Division of Labour (Frobel et al., 1980), this resulted in new spatially interconnected geographies of economic activities as different components for any one product (e.g. a mobile phone) began to be produced in different places with core, world locations retaining skilled work processes (e.g. research and development) and the periphery seeing the establishment of new branch plants responsible for labour-intensive assembly and low-skilled functions (e.g. producing the battery for a mobile phone) (see Hudson, 1994; Wright, 2002; Dicken, 2007). As Dicken (2007: 113) describes, locating labour-intensive assembly processes in the periphery in places like the Export Processing Zones can mean production costs are up to 30 times lower, with hourly wages being above US$30 in Denmark and Norway and as less than US$1 per hour in Sri Lanka.

Hornby, the maker of Scaletrix racing sets and model trains is a prime example of how such a strategy can be used to improve the profitability of a firm. The company manufactured all of its products in the UK until 1997 in its factory in Margate, Kent, that employed 750 people. However, falling sales and profit margins led to a New International Division of Labour being created. All manufacturing was moved to China through the establishment of a major manufacturing plant employing 1,500 people. This reduced costs by two-thirds and allowed production levels to be increased at no extra cost to the firm. Since 1997 sales have risen, as have profits, and the value of the company has now increased by one-eighth (*Financial Times*, 2003).

Central to the emergence of these geographies is the TNC. Defined by Dicken (2007: 106) as 'a firm that has the power to coordinate and control operations in more than one country, even if it does not own them'. TNCs emerged rapidly in the second half of the twentieth century, initially from North America and western Europe and more recently from Southeast Asia. Today, some of the best-known consumer brands, such as Ford, Heinz, Sony and Tesco, are leading TNCs. They have tens or hundreds of subsidiaries located throughout the world and serve consumers from Birmingham to Beijing. As Dunning and Norman (1987) argue in their 'eclectic paradigm', competitive advantage and profitability was sought through this globalization as firms exploited three types of advantage: *ownership* of manufacturing expertise that could be used to set up effective plants in foreign markets; *internalization* advantages that meant owning overseas plants and controlling their activities ensured efficiency and profitability; and *location* advantages which meant that being present in overseas locations allowed cheap labour assets and consumer markets to be exploited. As a result, levels of FDI, where firms invest capital overseas to establish new operations, increased rapidly towards the end of the twentieth century, trebling between 1984 and 1987 and reaching US$196 billion by 1989, well in excess of the previous peak of US$57 billion in 1979 (UNCTAD, 1991). The latest data shows FDI reaching US$916 billion in 2005, a 29% increase on 2004, suggesting no sign of a slowdown in the globalization process (UNCTAD, 2006).

Central to these changes were continuous and rapid technological changes that facilitated economic globalization (see Warf, 1989, 1995). These began with the invention of the telegram, but it was the telephone and the fax machine that revolutionized the geography of business activities. More recently, the internet has superseded these devices. All, however, have one thing in common. They allow the co-ordination of economic activities across space with managers at headquarters being able to communicate with and control subsidiaries quickly and efficiently. Alongside these devices, the humble containership and jumbo jet have been similarly influential, allowing components and assembled goods as well as the executives of TNCs to move across space cheaply and efficiently. The effect of this is captured by the idea of the 'spaces of flows' that are said to exist in the contemporary 'network society' where spatial interconnectivity lies at the heart of economic activity (Castells, 2000). Movement and flows of people (Beaverstock, 2004), goods (Dicken, 2007) and knowledge (Faulconbridge, 2006) characterize this new world space economy.

The shift towards services

As the 2004 World Investment Report suggested, the most recent trend in economic globalization has been the emergence of transnational service firms (UNCTAD, 2004). In 2005, 52% of the total share of FDI was in services (with 30% in manufacturing and 18% in primary industries) (UNCTAD, 2006).

This has been especially important for producer services (firms providing services to other businesses) with, for example, accountants, advertisers, management consultants and lawyers following their clients' globalization strategies and themselves opening overseas offices in many cities around the world (see, for example, Beaverstock, 2004; Bryson et al., 2004; Faulconbridge, 2006). In finance, this has resulted in the emergence of various 'off-shore' locations – financial centres – such as, for example, the Cayman Islands, where favourable tax regimes are exploited to minimize the tax incurred by the wealthiest (Roberts, 1994). Together, this has created new 'hot-spots' of service activity, principally in world cities where the command-and-control functions of banking and advanced producer service firms are located and the financial needs of TNCs are met (Taylor, 2003; Sassen, 2006). In a different arena, but equally importantly, the globalization of services has also involved consumer services with, in particular, the process of off-shoring effecting the way day-to-day services in banking, tele-sales and back-office administration are handled (Hudson, 1999) (see Box 19.1).

Significantly, what all of these processes of economic globalization have in common is the continued importance of place (see Chapter 9 on place and human geography). Whether it is because TNCs design their globalization strategies to facilitate access to the resources of a particular location, or because TNCs gain their characteristics from their home country but are also affected by the local cultures of the host countries in which they operate (Dicken, 2007), or because certain locations, such as world cities, remain disproportionately important in global economic activities and in some activities, especially the delivery of services, business simply has to take place in fixed locations to be in close proximity to the consumer.

CULTURAL GLOBALIZATION AND INTERCONNECTED WORLDS

The emergence of the 'spaces of flows' alluded to earlier led to intense debates about the effects of new forms of interconnectivity on, broadly defined, cultural practices. Initially, the emergence of an 'Americanized world', where the McDonald's culture became ubiquitous, dominated discussions of globalization and cultural geographies (Short and Kim, 1999). Flows of people, images and firms were said to be leading to converging consumer practices and tastes with the American model acting as a blueprint for global culture. Indeed, as Short and Kim (1999) describe, the emergence of English as the dominant language in busi-ness and academic work is another example of how one culture can gain world-wide power and legitimacy when economic and political forces, such as the globalization of the English language and US firms, enable the spreading of cul-tures and values. As others have shown, however (Held et al., 1999; Jackson, 2002; Crang et al., 2003), the effects of globalization on cultural practice and commodities are often more complex than stories of the global domination of McDonald's and the English language might suggest. On many occasions intri-cate processes of de- and re-territorialization occur that mean the spread of cul-tures is messier than might be expected.

In order to understand such processes, as with economic geographies, geographers have first identified the types of flow influencing cultural practice. This has often drawn on the work of Appadurai (1996: 33–6), who conceptual-izes cultural globalization by identifying five forms of flow: (1) *ethnoscapes*, the global landscape of persons made up of tourists, workers, refugees and others

Box 19.1　Off-shoring the call centre: success and failure in the globalization process

One controversial form of economic globalization in recent times has been the emergence of off-shoring strategies in relation to back-office activities and, in particular, telephone call centres. While not new as a process (see Hudson, 1999), off-shoring has had significant impacts on the economies of developed and developing countries alike in the past 10 years. The rationale for off-shoring call centres is no different from that of the globalization of manufacturing firms: spatial divisions of labour allow the exploitation of reserves of cheap labour. However, the process of successfully establishing off-shore call centres is more complex than it might first appear.

Advancements in technology means that managing overseas call centres and redirecting calls to overseas locations is a fast, simple and cheap process. Wages in India, where 75% of all overseas call centres are located, are significantly less than in the USA and UK, the countries where these call centres often originate. An average, newly qualified worker in India will often earn around £2,000 a year (Dicken, 2007). Consequently, when the expense of redirecting calls and establishing overseas infrastructures through FDI have been taken into account, this wage differential often means savings of between 55% and 75% can be achieved (Corporation of London, 2005). Interestingly, those employed are not the low-skilled labourers often associated with such globalization strategies. Instead, they are often university graduates fluent in multiple languages. Hiring such graduates would not only cost 10–20 times more in the home countries of call centres, but it would also be difficult as such work is often unattractive to educated workers. In contrast, working in a call centre in India (many of which are in Bangalore, Chennai or Mumbai) is a popular and prestigious form of employment (Bryson and Henry, 2005).

These advantages have led many firms, including American Express, British Airways and the UK's National Rail Enquiry Service, to off-shore their call centres to India. However, this has not been a problem-free process. While workers in India speak English, cultural differences and misunderstandings have plagued operations. New employees were often showed editions of the UK soap opera *EastEnders* or the US soap opera *Friends* to try to help them understand the norms of the countries they were dealing with, while they were also given cards detailing the meaning of slang phrases or local dialects (*Financial Times*, 2004). However, such strategies have been largely unsuccessful, with many customers complaining about misunderstandings and the service received from overseas call centres. As a result, the off-shoring trend has started to slow or even be reversed. Around 1 million people are expected to be employed in the call-centre industry by the end of 2007. Firms that have not off-shored their operations are now more hesitant about embarking on globalization strategies with some, such as the Aviva insurance group, which employs 8,000 call-centre workers in India, feeling they have reached the limit to which they can off-shore work. For example, the remaining call centres (such as breakdown recovery services) in the UK often require detailed 'local' knowledge that off-shore employees do not have. Indeed, some firms, such as the NatWest bank in the UK, actually market themselves as having no off-shore call centres and direct lines to customers' local branches because of the perceived benefits of talking to someone relatively close by. Indeed, those firms in the UK still pursuing off-shore strategies, while sometimes still choosing India, are increasingly turning to locations such as Ireland or parts of Eastern Europe because of their geographical and cultural proximity yet relatively low wages. As suggested, geography really does still matter and processes of economic globalization have to overcome numerous geographical hurdles which, in the case of call centres, seem to be acting as a break on processes of change.

who are mobile; (2) *technoscapes*, the global configuration of technologies that allow movement across space; (3) *financescapes*, the high-speed global circuits of capital that are central to contemporary life; (4) *mediascapes*, the networks capable of producing and distributing information in printed and visual forms; and (5) *ideoscapes*, the political ideologies and values often circulated through mediascsapes. Combined, these flows are said to have the ability to move, reconfigure and reproduce cultural practices, something that was less intense during previous periods in history.

Today, such flows manifest themselves in global media such as CNN or the *Financial Times* (Thrift, 1997), mobile managers and chief executives in firms that are said to be producing a new global executive culture (Sklair, 2001) and the emergence of global architectural gurus, such as Lord Foster and the Foster and Partners architecture agency, that is responsible for numerous landmark buildings worldwide (McNeill, 2005). The most intricate conceptualizations of the effects of such flows highlight the processes and practices that increasingly lead to both the spread of dominant (often Western) cultures, but also the subtle forms of change and reconstitution in cultures as they move across space and time. McDonald's is also a useful example of this. As Watson (2004) suggests, cultural change has been occurring at breath-taking speed in recent decades. However, using the examples of the arrival of McDonald's in Hong Kong, he notes that:

> The people of Hong Kong have embraced American-style fast foods, and by so doing they might appear to be in the vanguard of a worldwide culinary revolution. But they have not been stripped of their cultural traditions, nor have they become 'Americanized' in any but the most superficial ways. (Watson, 2004: 125)

The arrival of McDonald's was greeted in Hong Kong as the arrival of an 'exotic' culture, something different, but not a replacement for Chinese culture. Today, while McDonald's is ingrained in the fabric of consumer life in Hong Kong, it has not replaced but instead complements existing cultures. Indeed, as

Box 19.2 suggests, the meaning and form of cultural commodities often change as they move across space and time. For McDonald's, this has meant the reinvention of menus to respond to local consumer demand, something that means the Big Mac is not always present.

Continuity and change in cultures?

Human geographers have increasingly argued that the result of such complex processes of cultural flow is not only the replication worldwide of dominant cultures, but also the creation of new, reproduced cultural forms that sometimes differ across space and time. Hannerz (1996) suggests using the term 'creolization' to symbolize the process of Western cultures being transformed and reconceptualized outside the Western world. Indeed, Massey (1991) argues that a 'global sense of place' is now needed to understand the cultures that infuse any one community (see Chapter 9 on place and human geography). Drawing on the example of Kilburn High Street in London, she argues that place is infused with cultural values and practices that have multiple geographies. Consequently, a progressive sense of place involves recognizing how 'cultural imports' such as the kebab house and the branch of the Middle Eastern bank are experienced differently depending on the influence of 'race', gender and other forms of positionality. This is not to say that there are not important forms of globally present cultures. As the McDonald's example suggests, the 'golden arches' and fast-food culture that emerged in the USA is now at home throughout the world and accepted by consumers from London to Beijing, Moscow to Manilla. They are accepted hegemonic forms. Nevertheless, as Massey's work reminds us, we should explore both the globalness of such cultures and the experience of them by different people in different places.

This discussion highlights quite clearly, the fact that globalization has had important effects on the geographies of cultures and cultural practices. Undoubtedly there are forms of ubiquitous, homogeneous and hegemonic cultural forms that pervade daily life today. However, there are also important forms of reproduction, interconnection and time–space contingent occurrences of these cultures and associated cultural practices. The role of the geographer is, then, to provide a critical interpretation of the geographies of such 'global' cultures.

CONCLUSIONS: A 'NEW' GLOBALLY INTERCONNECTED WORLD?

While geographers have been keen to study the processes of globalization, they have also been quick to note the contested and imagined nature of globalization (see, for example, Cameron and Palan, 2004). As Dicken (2004) argues, this does not mean engaging in sterile arguments about whether contemporary globalization is something new, or not. Instead, for geographers, the challenge is to provide insight into the qualitative changes occurring in relationships between geographical scales (see Chapter 12 on scale and human geography). In this sense, it is important to negotiate a path between two opposing visions of globalization.

Hyperglobalists (O'Brien, 1992; Ohmae, 1992) have suggested that the current period of globalization has rendered the national and the peculiarity of place

---

Box 19.2   Transnational commodity cultures

One of the most compelling recent studies of the effects of globalization on cultural phenomena has focused upon the way various commodities have moved across space (Dwyer and Crang, 2002; Crang et al., 2003; Dwyer and Jackson, 2003). This has revealed the way that foods, such as the Indian curry, and fashions, such as South Asian 'ethnic' dress and materials, are de- and then re-territorialized by a number of often different economic actors. This process results in varying degrees of change in the cultures, in particular as adaptations are made to tailor them to spatially heterogeneous consumer desires and expectations.

As Crang et al. (2003) reveal, the movement of food such as curry and the popularization of, for example, Patak's sauces in the UK involves multiple actors, including the producers of spices, buyers and retailers who set standards for these goods, and cultural intermediaries such as advertisers and culinary experts who construct both a market but also a standard for such products. In addition, the consumers themselves and their habits and behaviours feed back into the commodities' 'life' through their buying practices. As a result, the recipe, flavour and consumption habits associated with such food often change as products move across space. So while the Indian curry is now a global dish, the experience of it often varies between places. Perhaps the best known example here is the British Tikka Massala sauce, a flavour and recipe designed to 'please' the British palate which led to the mass consumption of curry. Often, the sauces retailed in the West are not direct replications of those traditionally consumed in South Asia. Instead, they are reproductions or hybrids modified by these multiple actors that enable processes of de- and re-territorialization.

Dwyer and Jackson (2003) describe a similar process in the fashion industry. Here, leading designers using South Asian fabrics and designs reproduce both the characteristics of fabrics and designs differently depending on the consumer markets for which they are destined. Interestingly, they point out that in the UK different designers take very different approaches to this. Some, such as the retailer Monsoon, target the mass-consumer market by adapting designs to reduce costs and generate appeal to a wide audience. In effect, they respond to a mass-market, socially constructed view of ethnic fashions. In contrast, specialist retailers, such as Anokhi, target niche markets where traditional hand-made designs that more accurately replicate the fashions of South Asia are desired. Such an approach, however, has much less widespread appeal and attracts a very different audience with a different view of fashion. As this suggests, cultures have multiple meanings not just across space, but also in particular places because of the social processes involved in their reconstitution.

---

insignificant. This is the type of rhetoric often used in the media to portray a borderless world of flows. *Sceptics* offer a very different interpretation of the current period (Held et al., 1999). They argue that contemporary globalization is actually not very different from the forms of interconnectivity that existed in the past, particularly during periods of Empire. The flows and interconnectivity that exist are, according to this group, actually more regional than global (e.g. intra-European) and the nation-state and the national scale continue to play a central role in mediating cross-border flows. Arguably, the most useful views of globalization are those that are able to negotiate a path between these two positions and recognize recent qualitative changes but also the continuities that exist. This often requires more than quantitative analysis of datasets detailing changes in trade flows, but also qualitative descriptions of the intricacies of the way these flows move across space and continue to be affected by cultures, economies, politics and societies that vary within and between places.

Globalization, then, as a process, is an ongoing syndrome. As Held et al. (1999) remind us, we should not accept this as an inevitable and logical process with a clear and identifiable outcome. Instead, globalization is contested whether it be because of the positive and negative impacts on a country and its people or because of continued barriers to a world of global flows, whether these are national borders and regulations, technological gaps (such as in internet coverage in Africa) or socio-cultural complexity (such as the failure for a global consumer culture to emerge). The challenge for geographers is to understand how and why all of this plays out over time and space and to provide examples of the effects on people in their everyday lives.

SUMMARY

- Defining and explaining the contested meanings of globalization is not an easy task. Globalization as a concept or process must be read in a critical fashion and understood as being inherently uneven over time and space.

- Globalization is quintessentially geographical in scope. Globalization is not only about the interplays between local, regional, national and global scales, but also about interconnectedness, flows and uneven development in the world.

- The strategic practices of transnational corporations, aided significantly by technological change, have been central to shaping processes of economic globalization with respect to geographies of production, employment and, ultimately, location.

- Globalization is not just about economy, international trade and employment. Globalization affects all walks of life in many different ways, from the cultural to the political. Contrary to some beliefs, processes of globalization have not produced homogeneity or standardization in everyday life, but have instead brought much difference and diversity to global society.

- Globalization remains a highly emotive and uneven process throughout the world. Geographers have a key role in unpacking and critiquing globalization in order to illustrate to others that globalization is uneven in scope and has negative as well as positive impacts on people in their everyday lives.

- Always contest and think critically about globalization!

Further Reading

There have been significant volumes of words written about globalization in all social science disciplines. An efficient starting point would be Jones's (2006) **Dictionary of Globalization** and the most recent entries in **The Dictionary of Human Geography** (Johnston et al., 2007): globalization, localization and the local–global dialectic. Johnston et al.'s (2002) **Geographies of Global Change** has a very incisive introductory chapter on geography and globalization, and the book includes a number of specialist contributions discussing various facets of economic, social, cultural and political globalization. From an economic perspective, Lee and Wills's (1997) **Geographies of Economies** has several chapters on (re)thinking globalization. At a more generic level, a collection of inspiring essays on many different aspects of globalization are included in **Geographical Worlds** (edited by Allen and Massey, 1995) and **A Shrinking World?** (edited by Allen and Hamnett, 1995). Other key geographical texts on globalization have been written by Cox (1997), Dicken (2007) and Storper (1997). For a Global South view of globalization and its many uneven contradictions, see Gwynne and Kay's (2004) **Latin America Transformed: Globalization and Modernity**. Finally, there exists several seminal texts on approaching and contesting globalization: Castells's (2000) **The Rise of the Network Society**; Held et al.'s (1999) **Global Transformations**; Held and McGrew's (2007) **Globalization Theory**; and Hirst and Thompson's (1996) **Globalization in Question** (see also Burbach et al., 1997; Held and McGrew, 2000; Nederveen Pieterse, 2000; Schuurman, 2001; Sklair, 2002).

Note: Full details of the above can be found in the references list below.

References

Allen, J. and Hamnett, C. (eds) (1995) *A Shrinking World? Global Unevenness and Inequality*. Oxford: Oxford University Press.

Allen, J. and Massey, D. (eds) (1995) *Geographical Worlds*. Oxford: Oxford University Press.

Appadurai, A. (1996) *Modernity at Large*. Minneapolis, MN: University of Minnesota Press.

Beaverstock, J.V. (2004) 'Managing across borders: knowledge management and expatriation in professional legal service firms', *Journal of Economic Geography*, 4: 157–79.

Bryson, J., Daniels, P. and Warf, B. (2004) *Service Worlds*. London: Routledge.

Bryson, J. and Henry, N. (2005) 'The global production system: from Fordism to post-Fordism', in P. Daniels, M. Bradshaw, D. Shaw, and J. Sidaway (eds) *An Introduction to Human Geography: Issues for the 21st Century*. Harlow: Pearson/Prentice Hall, pp. 313–36.

Burbach, R. Nunez, O. and Kagarlitsky, B. (1997) *Globalization and its Discontents*. London: Pluto.

Cameron, A. and Palan, R. (2004) *The Imagined Economics of Globalization*. London: Sage.

Castells, M. (2000) *The Rise of the Network Society*. Malden, MA: Blackwell.

Corporation of London (2005) *Off-shoring and the City of London*. London: Corporation of London.

Cox, K.R. (1997) *Spaces of Globalization*. New York: Guilford Press.

Crang, P., Dwyer, C. and Jackson, P. (2003) 'Transnationalism and the spaces of commodity culture', *Progress in Human Geography*, 27: 438–56.

Department for Trade and Industry (1998) *Our Competitive Future: Building the Knowledge Driven Economy*. White Paper. London: Department for Trade and Industry.

Dicken, P. (2003) *Global Shift* (4th edn). London: Sage.

Dicken, P. (2004) 'Geographers and 'globalization': (yet) another missed boat?', *Transactions, Institute of British Geographers*, NS 29: 5–26.

Dicken, P. (2007) *Global Shift* (5th edn). London: Sage.

Dunning, J. and Norman, G. (1987) 'Theory of multinational enterprise', *Environment and Planning A*, 15: 675–92.

Dwyer, C. and Crang, P. (2002) 'Fashioning ethnicities: the commercial spaces of multiculture', *Ethnicities*, 2: 410–30.

Dwyer, C. and Jackson, P. (2003) 'Commodifying difference: selling EASTern fashion', *Environment and Planning D: Society and Space*, 21: 269–91.

Faulconbridge, J.R. (2006) 'Stretching tacit knowledge beyond a local fix? Global spaces of learning in advertising professional service firms', *Journal of Economic Geography*, 6: 517–40.

Financial Times (2003) 'British tradition made in China', *The Financial Times*, 23 December.

Financial Times (2004) '£100m contract poses threat to jobs, unions say', *The Financial Times*, 5 February.

Frobel, F., Heinrichs, J. and Kreye, O. (1980) *The New International Division of Labour*. Cambridge: Cambridge University Press.

Gwynne, R. and Kay, C. (eds) (2004) *Latin America Transformed: Globalization and Modernity*. London: Edward Arnold.

Hannerz, U. (1996) *Transnational Connections*. London: Routledge.

Held, D. and McGrew, A. (eds) (2007) *Globalization Theory*. Cambridge: Polity Press.

Held, D., McGrew, A., Goldblatt, D. and Perraton, J. (1999) *Global Transformations: Politics, Economics and Culture*. Cambridge: Polity Press.

Hirst, P. and Thompson, G. (1996) *Globalization in Question*. Cambridge: Polity Press.

Hudson, A.C. (1999) 'Off-shores on-shore: new regulatory spaces and real historical places in the landscape of global money', in R. Martin (ed.) *Money and the Space Economy*. Chichester: Wiley, pp. 139–54.

Hudson, R. (1994) 'New production concepts, new production geographies? Reflections on changes in the automobile industry', *Transactions, Institute of British Geographers*, NS 9: 331–45.

Jackson, P. (2002) 'Commercial cultures: transcending the cultural and the economic', *Progress in Human Geography*, 26: 3–18.

Johnston, R.J., Gregory, D., Pratt, G. and Watts, M. (eds) (2007) *The Dictionary of Human Geography*. Oxford: Blackwell.

Johnston, R.J., Taylor, P.J. and Watts, M.J. (eds) (2002) *Geographies of Global Change*. Oxford: Blackwell.

Jones, A. (2006) *Dictionary of Globalization*. Cambridge: Polity Press.

Lee, R. and Wills, J. (eds) (1997) *Geographies of Economies*. London: Edward Arnold.

Massey, D. (1984) *Spatial Divisions of Labour*. Basingstoke: Macmillan.

Massey, D. (1991) 'A global sense of place?', in D. Massey (ed.) *Place and Gender*. Cambridge: Polity Press, pp. 146–56.

McNeill, D. (2005) 'In search of the global architect: the case of Norman Foster (and partners)', *International Journal of Urban and Regional Research*, 29: 501–15.

Nederveen Pieterse, P.J. (2000) *Global Futures: Shaping Globalization*. London: Zed Books.

O'Brien, R.C. (1992) *Global Financial Integration: The End of Geography*. London: Council on Foreign Relations Press.

Ohmae, K. (1992) *The Borderless World*. New York: Harper & Row.

Roberts, S. (1994) 'Fictitious capital, fictitious spaces: the geography of offshore financial flows', in S. Corbridge, N. Thrift and R. Martin (eds) *Money Power and Space*. Oxford: Blackwell, pp. 91–115.

Sassen, S. (2006) *Cities in a World Economy*. London: Sage.

Schuurman, F.J. (2001) *Globalization and Development Studies*. London: Sage.

Short, J. and Kim, Y. (1999) *Globalization and the City*. Harlow: Pearson/Prentice Hall.

Sklair, L. (2001) *The Transnational Capitalist Class*. Oxford: Blackwell.

Sklair, L. (2002) *Globalization: Capitalism and its Alternatives*. Oxford: Oxford University Press.

Storper, M. (2007) *The Regional World*. New York: Guilford Press.

Swyngedouw, E. (1997) 'Neither global nor local: glocalization and the politics of scale', in K. Cox (ed.) *Spaces of Globalization*. New York: Guilford Press, pp. 137-66.

Taylor, P.J. (2003) *World City Network: A Global Urban Analysis*. London: Routledge.

Taylor, P.J., Watts, M.J. and Johnston, R.J. (2002) 'Geography/Globalization', in R.J. Johnston, P.J. Taylor and M.J. Watts (eds) *Geographies of Global Change: Remapping the World*. Oxford: Blackwell, pp. 1-19.

Thrift, N. (1997) 'The rise of soft capitalism', *Cultural Values*, 1: 29-57.

Thrift, N. (2002) 'A hyperactive world', in R.J. Johnston, P.J. Taylor and M.J. Watts (eds) *Geographies of Global Change: Remapping the World*. Oxford: Blackwell, pp. 29-43.

UNCTAD (1991) *World Investment Report 1991: The Triad in Foreign Direct Investment*. New York: United Nations Conference on Trade and Development.

UNCTAD (2004) *World Investment Report 2004: The Growth of Services*. New York: United Nations Conference on Trade and Development.

UNCTAD (2006) *World Investment Report 2006: FDI from Transition and Development Economies: Implications for Development*. New York: United Nations Conference on Trade and Development.

Warf, B. (1989) 'Telecommunications and the globalization of financial services', *The Professional Geographer*, 41: 257-71.

Warf, B. (1995) 'Telecommunications and the changing geographies of knowledge transmission in the late 20th century', *Urban Studies*, 32: 361-78.

Watson, J.L. (2004) 'McDonald's in Hong Kong', in F.J. Lechner and J. Boli (eds) *The Globalization Reader*. Oxford: Blackwell, pp. 125-32.

Wright, R. (2002) 'Transnational corporations and global divisions of labour', in R.J. Johnston, P.J. Taylor and M.J. Watts (eds) *Geographies of Global Change: Remapping the World*. Oxford: Blackwell, pp. 68-78.

20

Globalization: Science, (Physical) Geography and Environment

Nicholas J. Clifford

Definition

Although globalization is a term usually restricted to economics and the social sciences, there are aspects of the phenomenon that are intimately linked to the practice and purpose of the physical and environmental sciences and exemplified through physical geography. At a fundamental level, physical geography has always sought to describe and understand the multiple subsystems of the environment and their connections with human activity: it is global and globalizing at its very roots. Globalization may be seen historically in the global export of Western science, including physical geography, that underpinned colonial resource exploitation and which subsequently laid the foundations for the worldwide conservation movement. It is evident today in the burgeoning productivity and increasing organization of science as well as in the growing accessibility of scientific information. Globalization is also at work in setting contemporary scientific agendas that are focused on larger-scale issues of environment and development and environmental change. These global agendas are not simply shared, but are also co-produced by the public, politicians and commercial interests.

INTRODUCTION

Globalization is a term most discussed by human geographers, economists and social scientists. Globalization involves ways in which all people and societies in the world are progressively connected, whether through trade, labour and migration, flows of capital and/or technology transfers. For some, this marks progressive unification, while for others, the process is marked by increasing divisions and extremes as some are 'included' and others 'excluded' by the phenomenon. As Chapter 19 by Faulconbridge and Beaverstock makes clear, human geographers have drawn attention to the inevitable geography *of* the globalization process as well as to the inevitable geography *within* globalization. As they put it, there is an ongoing syndrome of globalization, marked not just by economic or social change, but also by debates regarding the extent to which globalization affects all aspects of life and the desirability of this.

Couched in these terms, physical geography seems to have little to say (and little need to say much) about globalization which does not at first appear rather artificial or contrived. Geographers have, however, in different ways, drawn attention to the way in which globalization is changed and changing in space and time – they are 'complexifying' the concept – and it may not, therefore, be inappropriate that some consideration is given to the globalization syndrome from the perspective of environmental science and physical geography.

This chapter provides an initial context for associating physical geography within the globalization debate. It is a starting point for discussion and an introduction to some of the potentially relevant literature. It provides some examples of how globalization has affected and is affecting physical geography as an important part of geography as a whole.

GEOGRAPHY AS A GLOBAL AND GLOBALIZING SUBJECT

Geography itself, of which physical geography is a large part (and for a long time, was the major part) is also a globalized and globalizing subject: it is popular around the world and is organized at international, regional and national levels, most notably through the International Geographical Union (IGU) (http://www.igu-net.org/). The IGU exists to promote and disseminate geographical activity and data around the world and was established in 1922 following earlier international congresses from 1871. Globalized scientific activity is not, therefore, a new phenomenon! The IGU is affiliated to the International Council of Scientific Unions (ICSU) and the International Social Science Council (ISSC), as co-ordinating bodies for the international organization of science. Geography, as a formal academic discipline, is thus connected (between its various areas of study, between differing international groupings and between areas of differing subject matter) *via* a global network that has external links to the worldwide scientific community. There are several reasons to claim, therefore, that this network exhibits characteristics of the globalization syndrome and that it also reproduces them. Take, for example, two key aspects of the mission of the IGU:

> to initiate and co-ordinate geographical research requiring international co-operation and to promote its scientific discussion and publication

and

> to promote international standardization or compatibility of methods, nomenclature
> and symbols employed in geography.

The idea that all geographers should share a common structure, methodology and, by implication, a corporate (world?) view is reminiscent of the McDonaldization phenomenon touched upon by Faulconbridge and Beaverstock (Chapter 19). Other characteristics of the globalization syndrome can be seen in the IGU Congresses, which move around the world on a four-yearly cycle. The International Geographical Union has held 30 international Congresses, the latest of which, in 2004, lists among its achievements the following statistics (Gregory and Craig, 2005): 1,966 delegates from 84 countries; 505 parallel subject-specific sessions and 58 media reports. Significantly, this Congress was joined by the Joint International Geomorphology Conference and the International Association of Geomorphologists – further testimony of a will to join up contrasting social and scientific agendas on a global scale. Less positively, but also significant in the context of the globalization syndrome, was the uneven spread of delegates between continents and the developed and developing worlds: 62% of delegates came from the UK and Europe (the Congress was in Glasgow); 11% from North America; 11% from Asia; 4% from South America; 4% from Australasia and 4% from the Middle East; but only 3% attended from Africa and 2% from Central America. Despite globalization, the 'friction of distance' is still apparent and so too are inequalities in access to events such as this. For example, 22% of the IGU travel grants awarded to help with attendance went to the 3% of delegates attending from Africa, while more than one-third of grants went to European delegates. All geographers are not participating in such global projects as the IGU on an even footing – perhaps this is an expected phenomenon in globalization, but it is one which may or may not 'even out' in time.

Other characteristics of globalization may be seen in the fundamental nature of Geography as an academic discipline. Formally, geography (*as a subject*) might be defined as the study of the ways in which space is involved in the operation and outcome of social and biophysical processes. But geography is also *a property* – it describes and defines the spatial properties of these outcomes of the very processes which it studies as a subject (Gregory, in press). Geography is thus, necessarily, a broad and inclusive subject (it needs a range of expertise to focus on the processes involved in spatial transformations of landscape, ecology and human activity) but it also needs ways of describing a broad subject matter which, singly and together, have geographies. As Gregory goes on to point out, while there is no single paradigm of geographical enquiry, geographers have borrowed from across the range of natural, physical and social sciences and the way geography has come into being itself has a geography. Geography's interconnectedness spans subjects, language traditions, technologies (originally cartography, later the quantitative revolution and statistical methods and latterly Geographical Information Sciences), as well as concepts (in particular, scale, place and region). Historically, Geography may in this way be seen as a kind of product of intellectual globalization. Contemporary debates about the coherence and future of the discipline are no exception to these historical trends and may well represent something of an acceleration in them. Much geographical debate is now in response to the apparent need for new ways to tackle environmental and societal problems and issues which seem themselves to manifest greater interconnectedness between the human and the social and which present scales from the global to the local (consider, for example, global warming, the destruction of the rainforests, or uneven development). In the face of this new class of globalized environmental and societal problems, new epistemologies (ways of producing knowledge of the world) are being sought in trans- and multidisciplinary research. These are breaking with old science models of specialism and reductionist methods and drawing on new ideas such as post-normal science and complexity theory. They are also re-inventing old subject areas in new ways, such as Earth System Science (see below).

A good starting point to address some of these issues is the short introduction by Smith (2005: 389), who poses the challenge of ' [d]oing a geography that is everything for a readership embracing everyone'. The emerging response to this challenge is equally intriguing and Smith suggests that the central theme (the globalization syndrome?) is a concern with connectedness:

> ... geography as an enterprise of relatedness whose vitality is secured by forging connections and crossing intellectual horizons; by pulling the world apart, reassembling it and adding to it, in a variety of intriguing, ethically charged, sometimes surprising and frequently controversial ways. ... Geography forms a hub for these networks of relatedness ... positioned awkwardly, but productively, as an interface for the social, natural and biological sciences ... both an interstitial subject and an impulse to interdisciplinarity. It fills the neglected spaces 'in-between' human, physical and medical sciences with potential; it is a creative practical exercise; a mode of inventive intelligence, through which the virtual becomes real as the world unfolds. (Smith, 2005: 389–90)

Some of that inventive intelligence to which Smith alludes has been deployed in recent decades in seminal contributions by physical geographers in the fields of environment-development studies and in calls for new kinds of science and scientific engagement, notably sustainability science and Earth System Science. All of these can be seen as examples of globalization in operation and are discussed later in this chapter. First, however, some aspects of an older kind of globalization and its implications for the development and status of physical geography are outlined.

PHYSICAL GEOGRAPHY AS A FUNDAMENTALLY GLOBAL SUBJECT: LESSONS FROM HISTORY?

Physical geography encompasses the characterization and explanation of geological, hydrological, biological and atmospheric phenomena and their interactions at, or near, the earth's surface. This has often been undertaken in relation to human occupation and activity. Both from the perspective of scholarly and historical completeness and because of concern with the relations between humans and their environment, physical geography should not be separated from considerations of geography as a whole, nor from debates concerning the past, present and future connectedness of environments and human activity. Physical geography is also a subject with clear claims, not just to be global, but globalizing too. This is apparent in the involvement of physical geographers in studies of global warming, conservation and sustainability, and natural disasters and development (see Spencer and Slaymaker, 1998), in their development and application of technologies such as satellite remote sensing, earth observation and geographical information sciences, and through explorations of the nature of physical geography and its historical development.

The nineteenth century and an older globalization of holism

One of the foundational statements of mid-nineteenth-century physical geography set the context for scope, scale and nature as a globalizing science:

> Physical geography ... ought to be, not only the description of our Earth, but the physical science of the globe, or the science of the present life of *the globe in reference to their connection and their mutual dependence*. (Guyot, 1850: 3, emphasis added)

However, as the nineteenth century progressed, more and more became known about the earth, its environment and its peoples. Also, academic subjects were emerging and organizing on a competitive basis within universities and wider society as a whole (see Livingstone, 1992; Unwin, 1992). *Because* physical geography was a global and globalizing subject which tried to cover so much, it was difficult to preserve its unique role as the integrative science and to maintain its identity as a coherent body of knowledge. By the late nineteenth century, what to some remained a worthy and ambitious subject:

> ... a description of the substance, form, arrangement and changes of all the real things of Nature in their relation to each other, giving prominence to comprehensive principles rather than isolated facts. [and having] ... a unique value in mental training, being at once an introduction to all the sciences and summing up of their results ... (Mill, 1913: 3, 14)

to others was now an embarrassment:

> ... too often degraded into a sort of scientific curiosity shop, in which there is a vast collection of isolated facts without the slightest attempt to show how interdependent they are ... (Skertchly, 1878: 2)

Physical geography was, then, caught in an early globalization of the academy, which required specialization and focus (depth, rather than breadth) as the hallmark of academic credibility within the universities. This kind of globalization was thus one of co-ordinating and structuring the academy, rather than one marked by an intellectual project producing new forms of connected knowledge, as most often championed today and which had been sought since the Enlightenment (see Clifford, in press, for further discussion).

The twentieth century and globalization through specialization

One way in which physical geography responded to the dilemma of 'breadth versus depth' was to tie the physical science of the environment more closely to the human occupancy of the earth and hence justify physical geography as a science of the interconnectedness of human–environment relations (see especially Davis's tests of causes and consequences, 1899). This worked well while geography was concerned with larger-scale regional descriptions of the earth, although physical geography remained as a kind of descriptive backdrop to human activity: regional geographies frequently started with physical descriptions of the land surface and often regions were themselves delineated on the basis of landscape. However, this very attribute also functioned to limit physical geography's connectedness with the *other* physical and environmental sciences, which were becoming more focused and reductionist in their methodologies. Thus, physical geography, in effect, bucked the trend (i.e. stood out against a kind of globalization), but not without cost.

The clearest example of this can be seen in the changing role and relations of geomorphology (the scientific study of landscapes and landforms). From the later nineteenth century, this discipline was typified by larger-scale landform and landscape description, which formed an ideal backdrop to other regional geographies. In the absence of dating methods, techniques of palaeo-environmental reconstruction and of detailed geophysical knowledge about endogenic earth processes, what resulted were elaborate chronologies based essentially on stratigraphic correlations and landform shape (morphology), known as denudation chronologies. These had no means of 'independent' scientific testing and involved no other scientific technique which tied them into the contemporary mainstream science. The geomorphological community thus became inward-looking and disconnected from science and was largely focused on a supporting or separate role within geography.

By the mid-twentieth century, this large-scale landform denudation chronology approach had probably reached its limits in the provision of new knowledge. With the quantitative revolution in geography and a move away from the regional geography paradigm, denudation chronology was effectively redundant (and redundancy of old practice is, of course, a consequence of globalization). Geomorphology's response was twofold. On the one hand, particularly far-sighted practitioners, notably R.J. Chorley (Chorley, 1971), realized both the potential and the necessity to redefine and reconnect geomorphology, physical geography and geography as a whole and to connect geography into a new and more enduring scientific framework. In effect, this was an effort to recapture an older form of globalized and globalizing knowledge and methodology. The vehicle for this was to be General System Theory (see Chapter 14), but in some ways, this was ahead of its time and is more appropriate today as a response to global environmental problems (see below on Earth System Science). On the other hand, in the 1960s and 1970s, the second response was to focus on more traditional physical science and its reductionist methodologies. This stressed the investigation of surface landform processes which could be monitored at the smaller scale. At the same time, from the 1960s, enormous advances were made in various forms of environmental dating, palaeo-environmental reconstruction and large-scale (but fine-detail) earth observation. All of these were

well represented in physical geography, but they developed more or less independently *within* the subject. Today, the consequence is that none of them is well
connected to rapidly developing strands of geophysics that have remained *outside*
geography (Summerfield, 2005). Moreover, *none* of them is well placed to exploit
opportunities offered by integrative science or to respond to challenges
demanded by the awareness of global-scale environmental dynamics – that is, by
the new period of scientific globalization (see below).

The current globalization syndrome

The important lesson of this history is that academic subjects change in character
and popularity through time and that a globalizing subject requires particular
circumstances to thrive. The post-Enlightenment period to the mid-nineteenth
century was one such period as 'new' knowledge developed across a very broad
range and required organization and classification: this was an early period of scientific globalization. As this was accomplished, more focused study requiring distinctive methodologies had greater credibility. Judged against this set of new
requirements, physical geography in the nineteenth century lacked both a *focus
for its study* and a *distinct epistemology* for the disclosure of the new knowledge
which it might claim to provide. Its response (denudation chronology) gave it
uniqueness, but isolated it from other environmental and earth sciences. This situation persisted until the mid-twentieth century when the re-tooling of physical
geography, marked by increasing specialization, reliance on technical methodologies and smaller-scale laboratory-like studies, brought it closer to other physical sciences and science methodology. This was despite strong advocacy of
General System Theory (really, an alternative way to reconnect the many strands
of geography and science, using an older, holistic view). It was not until the end
of the twentieth century, however, that the stage was set for the re-emergence of
the subject cast in a General System manner and recovering some of its much
older heritage (Clifford, in press). The key players on this emergent stage were
new classes of environmental issues, the development of ideas in science, such
as complexity theory, debates favouring increasing inter- and transdisciplinary
studies and new technologies capable of global-scale environmental monitoring
with fine-scale local detail. In a globalizing world, globalization has, in some
respects, thus created the conditions for the re-emergence of a more ambitious
physical geography as a force in Earth Systems Science.

PHYSICAL GEOGRAPHY AND THE NEW GLOBALIZATION
OF SCIENCE: SOME EXAMPLES

Beyond historical experiences, there are at least three indicators of the ways in
which science (and thus, physical geography) can be said to be part of the globalization syndrome today. First, is the enormous growth in scientific output and
the rapid (but still differential) electronic 'remote' access to this. Second is the
increasingly global agenda of science. Third is the changing academic foci and

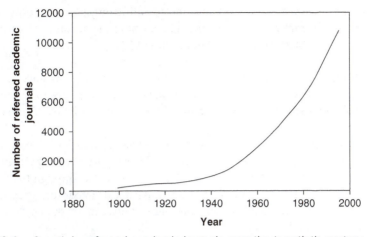

Figure 20.1   Growth in refereed academic journals over the twentieth century

Source: Based upon data from Mabe and Amin (2001)

character of the environmental sciences. Each of these is examined in a little more detail in the sections which follow.

The globalization of scientific output

The enormous growth in scientific output and the ease of access to this is seen in the number and range of scientific journals and in the power of web-based search engines such as Google Scholar and the ISI Web of Knowledge. At the time of writing, for example, Google Scholar (http://scholar.google.co.uk/intl/en/scholar/about.html), which promises to '[s]earch diverse sources from one convenient place; find papers, abstracts and citations; locate the complete paper through your library or on the web', was able to provide 1.37 million references under the keyword 'Geography' and 588,000 under the key word 'physical geography', all in around one-tenth of a second! What this powerful search engine provides is now unremarkable, but it would hardly have been imaginable a generation ago. It is clear testimony to the globalization of knowledge technology – provided, of course, that individuals have access to it. Access (whether to information, wealth or life chances) is a key contested issue in the globalization debate. Some exponents argue for the benefits of global knowledge spread, while others caution against its monopolizing and homogenizing tendencies. People in places also produce knowledge, which is not always fed into the global network. People may thus be empowered or disempowered by the growth of science and technology and by access to it, in common with other aspects of the globlization syndrome.

Underlying the technology of knowledge dissemination is, of course, scientific output – the reporting and publication of research. Figure 20.1 illustrates the exponential growth of academic journals over the course of the twentieth century. This is characterized by an average compound annual growth rate of 3.3% (Mabe and Amin, 2001), but there are also other intriguing aspects (marked in the differing growth rates for refereed scientific journals in particular historical periods) which tie academic and scientific output and activity to deeper definitions and ramifications of the globalization syndrome. Broadly, three periods are suggested (Mabe and Amin, 2001: 157–8):

1   1900–14, where there was little funding for science from government and growth was driven by the collective behaviour of disciplines themselves, with journals almost entirely in the hands of scientific societies. This is called a time of small-scale, 'innocent' science.

2   1944–78, a period of 'Big Science' where advances in science and technology were largely supported by governments. These were driven by concerns for international security following the Second World War through the Cold War, and later by the 'space race'. This was a period of maximum growth in academic output, with publication now mixed between academic societies and commercial publishers.

3   The most recent phase, from the late 1970s, is associated with lower growth rates and a pervasive disappointment or questioning of an overly ambitious science. Issues such as the oil crisis of the 1970s, potential ecological disaster and concerns about nuclear technologies have seen a relative decline in government-funded science and concomitant growth in commercial funding. Growth in the number of journals has slowed (but journal numbers are still increasing) and academic library subscriptions have fallen back more than ever before. This has been termed a period of 'scape-goat' science.

Based upon this analysis it is impossible to separate the growth of academic journals (and science as a whole) from wider issues of funding, from the agendas of government and larger-scale business and from public perceptions of the role and value of science. Discipline-specific behaviours are the source of further debate (in geography, there is discussion of how publication behaviour varies between physical and human geographers; see Ferguson, 2003, for example). What is clear, however, is that scientific publication (and through this, all scientific activity underpinning it) is part of the globalization syndrome. During the twentieth century, science has increasingly left the control of scientists and scientific organizations and been more and more reliant on commercial organizations and the changing economic and political agendas of the time.

The globalization of science agendas

Globalization of science is not just about the scale of scientific output and access to it. Globalization also relates to the way in which science is focused on particular global issues in a context set by changing global views and needs. Physical geography and environmental science offer several examples of this. In the sections below, the historical and more recent involvement of geography with the worldwide conservation movement, with political ecology and, most recently, with Sustainability Science and Earth Systems Science, are presented as illustrations of the globalization syndrome and its effects on these various forms of environmental practice.

WORLDWIDE CONSERVATION

> The connection between people and the environment, built and natural, is the
> crux of both geography and conservation. Geographers study and conservation-
> ists worry about the environment over the same breadth of space and time ...
> Only geographers among the academics who interface with conservation, claim
> to offer the tools from both natural science and social enquiry that conserva-
> tionists are constantly calling for. (Warren, 1987: 322–3)

Warren's statement reflects a common sentiment that geography, environment
and conservation are necessarily and deeply connected, both as intellectual and
as practical projects. Geography and geographers can lay some strong claims to
having provided both paradigmatic or seminal works in the field; they are well-
known chroniclers and popularizers of the conservation/development 'message';
and they are members of international aid and environment agencies (see Butzer,
2002, for a discussion). Crucially, in all of these aspects, physical geographers
have drawn attention to older and newer aspects of the globalization syndrome
and they, themselves, have participated in changing, shaping and enacting the
globalization agenda.

Historically, conservation and geography have been seen as inseparable
from commercial exploration and exploitation of new territories, largely by the
European powers (and the European companies) of tropical islands and continents. In authoritative accounts, Grove (1992, 1995) details how western environmentalism and conservation were recognizable and prominent from the
seventeenth century and developed rapidly over the succeeding century as new
lands were first explored, subjected to scientific evaluation, then colonized and
commercially exploited. The first waves of settlement led to unbridled deforestation and were quickly followed by soil and water resource depletion. In ways
strikingly reminiscent of the critique of globalization today, there was early
recognition that European rule could be environmentally destructive, first, to
local landscapes, lifestyles and people, and, second, to continued resource availability; and subsequently to the local and perhaps global operation and sustainability of climate and the environment.

Early (academic) conservation efforts were championed principally by
naturalists or polymaths, accompanying voyages of discovery financed by private
capital. Their academic background was normally that of medicine and increasingly botany and geology. Significantly, conservation ideas rested on the concepts
of linkage between elements in an environmental system and Grove places the
roots of environmentalism in the work of Alexander von Humboldt, the German
geographer and explorer (see also Chapter 26 in this volume) who was, in turn,
influenced by Hindu philosophers. Eastern and Western traditions did then, to
some extent, interact such that environmental concerns were informed by an
holistic worldview. There was clear recognition that the colonial enterprise meant
the export of a particular style of resource exploitation in which environment
could be threatened, if not destroyed, by unfettered capitalism: a globalization
syndrome of its day.

A more personal account echoes many of Grove's points and brings the historical account to life through the compelling vignette of the African mountain gorilla. *Against Extinction* (Adams, 2004) is a history of conservation and nature which is also a pointer to the need for, and basis of, a changed conservation paradigm. Taking the theme of conservation as essentially a twentieth-century 'western' movement inspired by nineteenth-century ideals, Adams charts the changing scope and fortunes of conservation and of 'nature' since initial encounters with the gorilla by German colonial administrators operating between Rwanda and Uganda (as now). These encounters were initially motivated by 'discovery and capture' – animals hitherto subject to local, indigenous encounters through millennia, were now shot, stuffed and exhibited:

> It is of course, a fiction worthy of those postilions of empire that this was the first sighting people ever had of mountain gorillas. This was the moment when Western science noted them down, locking them into classificatory schemes and giving them an identity in the cabinet of natural history. (Adams, 2004: 2)

Subsequent encounters were for capture of a different kind: the live exhibition in zoos where animals were first exhibited as spectacles in cages and later in enclosures (fragments themselves of a captured nature). Later still, zoos became 'nature's saviours', through captive breeding of rare species in *ex situ* conservation. Zoos became centres not of entertainment, but of conservation, education, science and to some extent, proselytizations of habitat loss and threat of extinction. In a bizarre paradox, it was freedom which thus came to justify captivity, or wilderness to legitimize domestication (Anderson, 1997). Different worlds were actually being appropriated and brought together by such captivation and recreated and symbolically represented to others – another very distinct kind of globalization.

Alongside the zoo, Adams situates the nature and game reserves, sanctuaries and national parks as the big idea of twentieth-century conservation. Again, the mountain gorilla provides a very strong manifestation of this. The first African national park was declared to protect them and the gorilla became an icon for the scientific study of threatened species, through Dian Fossey's (1983) *Gorillas in the Mist* and the subsequent film biopic. While parks were primarily justified on scientific grounds (to protect both landscape and animals), there were important early geopolitical considerations. There were, for example, ideas to link parks transnationally and thus form bridges between competing colonial powers. Now the parks are truly part of a global phenomenon, governed by international conventions and emblematic of a kind of international corporatism. One way or another they signify conservation moving from a minority to a majority concern, increasingly linked with a web of connections around the globe.

The national parks have always wrestled with the question of 'naturalness' and 'community versus conservation'. In both these respects, too, there are explicit links to the globalization syndrome. The question of the degree of acceptable human occupation and use of the park resources by local people (and as well, by outside commercial concerns) has latterly been globalized in at least two ways. First, there is gorilla tourism and a need to habituate gorilla groups to human presence, but second is the need to manage the financial gains from and disturbance caused by growing tourist numbers. For example, some parks are enclosed by walls that protect crops from wild animal encroachment. In Mgahinga Gorilla Park in Uganda, the wall was built by local people, funded as a 'development through conservation programme' as an explicit response to local community problems in part *caused by* conservation efforts (re-location, restriction of activity, tourist incursion, etc.). Thus, the most contemporary phase of conservation is that of a global conservation movement that has shifted emphasis in its instantiation to local communities and local solutions – the globalization syndrome is at work again! For Adams:

> ... at the start of the 20th century, the circle of conservationists was small, a patrician elite network ... who used their influence to try to hold back their more rapacious or short-sighted colleagues, the captains of industry, commerce and colonial governance. By the end of the 20th century, conservation was an integral part of debates of enormous scope, about sustainability, development, poverty and human rights ... although it is now as rooted in economics, canned drinks or human rights in every modern state across the world. ... One of the most important changes ... was the gradual evolution of ideas about conservation, from something that ... reflected the views of colonial white men, to something that could embrace a diversity of ideas of proper relations between humans and other species. By the end of the century, a fairly standard model of conservation had been established worldwide ... something that every government in the world has found space for alongside its other concerns. (Adams, 2004: 12–13, 17)

Finally, Adams asks a basic question: with so much local and global effort in conservation over a century and with so much global emphasis on development, why is there still so little global conservation success and so much local and regional poverty? He suggests that part of the answer to such questions must recognize that: 'Natural science is just one among several ways of understanding nature. ... Conservationists need to recognize that concepts like "biodiversity" shut out other ideas about nature just as effectively as rooms full of Western-salaried conservation scientists mapping hotspots shut out other people' (Adams, 2004: 233–4). Effectively, this is a plea for a pluralist conception of nature, but also of people – what Adams calls nature's neighbours – and a focus on the characterization and restoration of *relationships between the two* in the context of the globalization debate. Writing like this resonates well with the economic and political critiques of globalization, as people and place are brought into a global whole, where lives are affected by globalizing processes, but where people do not necessarily benefit as a consequence and become disempowered to shape their own agendas. Within the environment-development arena, relations between local and global and between empowerment and disempowerment are the central concern of a form of geographical enquiry, championed by those with a strong physical geography training, called political ecology.

POLITICAL ECOLOGY

Political ecology deals with investigations of resource use and abuse, land degradation and marginalization and the environment as over-exploited and undervalued. This line of enquiry began in the 1970s and is most clearly articulated in Blaikie's (1985) *Political Economy of Soil Erosion* and in Blaikie and Brookfield's (1987) *Land Degradation and Society*. These are part of a wider set of geographical contributions to the identification and characterization of uneven development (Smith, 1984, 1996). The key theme is that environmental problems and issues are manifestations of an 'ideology of nature'. This ideology masks social inequality arising from the inequitable distribution and exploitation of resources, while claiming a scientific (i.e. neutral) knowledge of problems arising from human impacts:

> What tends to emerge is a picture of deepening human grip on biophysical processes. ... But what kind of grip? And who is doing the gripping? (Bryant, 2001: 153)

In other words, these writings address some of the more negative consequences of the globalization syndrome in relation to environment and development questions.

Blaikie's central example is the African rangeland landscape and the problem of overstocking and degradation. He reveals that, at different times and to different degrees, the colonial administrations viewed and constructed nature and natives in value-laden terms, leading to often coercive and resisted environmental conservation and management policies, most persuasively seen in soil erosion control measures. 'Western' science, from the 1930s on, adopted a model of erosion based upon field experiments from the USA. These emphasized slope gradient and slope length as primary variables in erosion loss and implicated native tillage practices as an exacerbating factor. The solution was to adopt terracing and to change land-use and landholding practices – at enormous political cost – with a generally paternalistic and modernizing attitude. However, in the ensuing decades, the model of erosion was questioned, leading to an emphasis on rainfall intensity (not controlled by terracing) and to the revaluing of local knowledges as an alternative source of conservation technique. In this reconceptualization of the problem, erosion was controlled by traditional methods of ground-cover, less intensive tillage and intercropping.

This key issue of erosion thus reflected a deeper and wider embeddedness of science within policy-making and bureaucracy, stretching from colonial and postcolonial administrations to world aid organizations. Blaikie (2001) goes on to argue that approaches to nature reflect less about disclosure and adoption of 'truths' and more about understanding political narratives. At best, there are agreements on provisional truths which are locally grounded and negotiated in an accountable manner:

> We can elaborate on the question thus: whose theories and whose reality count?; and why is some knowledge accepted and authoritative while other knowledge is

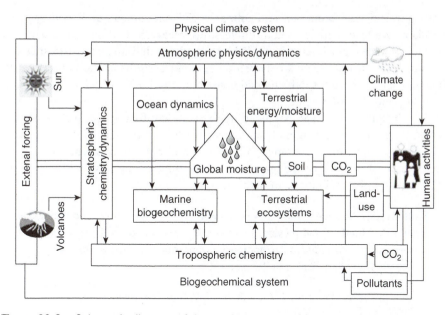

Figure 20.2   Schematic diagram of the earth system and its interactions, encompassing the physical climate system, biogeochemical cycles, external forcing and the effects of human activities

Source: Adapted from Earth System Science: Overview, NASA (http://terra.nasa.gov/Brochure/images/physical_climate.gif)

not? ... These questions encourage us to focus not so much on the production of 'truth', but upon *claims to truth*, which implies a close connection between power, knowledge and theory. (Blaikie, 2001: 36)

From a very simple perspective, for example, the development of narrative accounts help explain how it is that different views of nature are contested. In the sense that such narrative analyses necessitate a globalization perspective, but also a critique of this, then placing 'facts' within a social and political context frequently uncovers new dimensions to environment, conservation and reconstruction problems and to their solutions or responses. It also prompts a political agenda of enabling change through improved representation and explanation (Bryant, 2001) and is something of an antidote to other aspects of globalization.

EARTH SYSTEMS SCIENCE

Earth Systems Science (ESS) is a label which has been variously employed to create structure and coherence to an array of disciplines concerned with aspects of the earth's physical and biological systems (Figure 20.2). ESS was promulgated by NASA in the 1980s for structuring its earth observation activities, but has since broadened to encompass much of the historical territory of physical geography and much of what might be termed 'integrated geography' – that is, consideration of both physical environmental and human-social phenomena as linked systems. ESS is now a term widely used in universities, research funding bodies, environmental agencies and scientific publishing. Its three defining characteristics are: its emphasis on interactions between component systems of the natural world; its emphasis on systems, which encompass and link both natural and social subsystems; and its global scale.

It is clear that ESS has been deliberately configured not just as a 'new' intellectual enterprise (originally and allegedly distinct because of its scope, scale and technological underpinnings) but also as a new unifying theme for the scientific community. Consider, for example, two quotes, one from within the subject of geography and one from outside:

> Earth System Science is the study of the Earth as a single, integrated and social system ... [that] ... will provide solutions to major world problems because it moves away from reductionist approaches and does not recognize disciplinary boundaries. (Pitman, 2005: 6, 10–11)

> The Earth system is, at the highest level of abstraction, represented as

$$E = (N, H)$$

> where N is the ecosphere and H the 'human factor'. Both these main components are subject to further specification, so that $N = (a, b, c ...)$ or an alphabet of intricately linked planetary sub-spheres a (atmosphere), b (biosphere), c (cryosphere) and so on. $H = (A, S)$, where H embraces the 'physical' sub-component A (anthroposphere, as the aggregate of all individual human lives, actions and products) and the 'metaphysical' sub-component, S, reflecting the emergence of a 'global subject'. This subject manifests itself, for instance, by adopting international protocols for climate protection. (Schellnhuber, 1999: C20–C21)

Here is an explicit agenda for the globalization of science: its purpose, practice and products. In some guises, consideration of the opportunities (and, perhaps, necessities?) of improving links between the various earth and environmental sciences is both timely and balanced (Paola et al., 2006). In other guises, ESS represents what might be termed an aggressive project. It has grand and totalizing ambitions: ESS is now prominent, for example, in all US geography programmes, in some cases, replacing the title 'Physical Geography' and exaggerated claims are being made for it (see, for example, Pitman, 2005). Frequently, the appeal to ESS is based on very emotive language concerning climate change, global warming and sea-level rise, and other impending ecological disasters. It is also an example, then, of how the relations between scientists and the public are being reworked. Science agendas are changing so as to shape (but also as a response to) journalistic and public interest in the environment in what has been called the democratization or socialization of knowledge (Gibbons et al., 1994). Science must be 'socially robust' (that is, accepted in society and co-produced) as it is 'scientifically reliable' (Nowotny et al., 2001).

Clifford and Richards (2005) and Richards and Clifford (2008) provide introductory critiques of ESS and draw attention to a more negative aspect of the

ESS which is particularly relevant to the globalization debate. A key consideration is that unlike much physical science and applied physical science where outcomes are predictable, environmental science is often characterized by a lack of certainty and predictability. Add to this that the 'new' science is highly politicized *and* environmental, then all sorts of warning signals regarding power, manipulation of agendas and access to 'solutions' are sounded. Perhaps unwittingly, ESS may represent one more way in which power relations (between scientists, politicians and public) are recast and reinforced in the guise of greater inclusivity, but with a new technocentric method and an *expert* language drawing on General System Theory, GIS and the newer fields collectively known as complexity science (see Chapters 7 and14). The degree to which ESS is used in a 'top-down' manner and is hence a kind of perpetuation of an older order, remains to be seen, but the potential is clearly there. In all of these respects, ESS provides another microcosm of the globalization syndrome. It may be contrasted with the other new and 'big' or global agenda – that of Sustainability Science.

SUSTAINABILITY SCIENCE

Sustainability Science has emerged recently as a formal response to ideas and issues of sustainability developed in the 1980s and 1990s, which examined the connections between environment, development and society at global and regional scales (Clark and Dickson, 2003). These debates emphasized a divide between the North and South not simply in terms of wealth, but in terms of lifestyle, development priorities and with respect to their nature–society–environment relations (see above in the context of political ecology). Sustainability Science is about stating and shaping an agenda that tackles environmental and development issues at a world and world-changing scale and it has been led by the geographer, Robert Kates. It is also the clearest example of how environmental science and physical geography are necessary and fundamental parts of the globalization syndrome:

> A new field of sustainability science is emerging that seeks to understand the fundamental character of interactions between nature and society. Such an understanding must encompass the interaction of global processes with the ecological and social characteristics of particular places and sectors. The regional character of much of what sustainability science is trying to explain means that relevant research will have to integrate the effects of key processes across the full range of scales from local to global. It will also require fundamental advances in our ability to address such issues as the behavior of complex self-organizing systems as well as the responses, some irreversible, of the nature–society system to multiple and interacting stresses. Combining different ways of knowing and learning will permit different social actors to work in concert, even with much uncertainty and limited information. (Kates et al., 2001: 641)

Some of the core questions of this new science are listed as follows and again embody an awareness of both the positive and negative aspects of globalization, whether through its technological or managerial aspects:

- How can the dynamic interactions between nature and society be better incorporated into emerging models and conceptualizations that integrate the earth system, human development and sustainability?
- How are long-term trends in environment and development, including consumption and population, reshaping nature–society interactions in ways relevant to sustainability?
- What determines the vulnerability or resilience of the nature–society system in particular kinds of place and for particular types of ecosystem and human livelihood?
- How can today's operational systems for monitoring and reporting on environmental and social conditions be integrated or extended to provide more useful guidance for efforts to navigate a transition towards sustainability?

This agenda highlights the ethical involvement of geography and geographers in world issues and reflects deeper thinking of the ways in which globalization acts on people and place and is itself formed and changed by and in them, too. Kates et al. (2001) are clear that sustainability demands a new form of science: one that is regionally and locally sensitive and which does not apply science or science policy simply from the 'top down'. In this way, they implicitly distance themselves from what might be termed 'globalization as McDonaldization' and come close to the more complex conceptions of scale and its meanings discussed in Marston et al. (2005). Here, the overarching concept is that of a 'flat ontology' which breaks with hierarchical views of large-scale/small-scale and place/region/globe. To Marston et al., space and place are more like networks and nodes for human–environment interaction, which create a wide range of political opportunities to engage with environment and society. Whereas hierarchical conceptions of space and place (and of globalization within this context) tend to lead to ideas of people and places as passive recipients of globalization, the flat ontology view admits a view of people and place as agents in *its* (i.e. globalization's) change and representation. Globalization is thus produced by people in different cultures and different times, and changes as it progresses not just around the world, but between places. Sustainability Science, then, is one more way in which physical and human geographers, environmental and social scientists are, through new representations of nature–society–development relations, challenging and complicating globalization as a concept, as well as themselves being agents of change in the process of globalization!

CONCLUSION

Globalization, as normally applied in the economic and social sciences, refers to the increasing connectedness or unification of societies through economic, social and cultural transformations. This is often in association with new technologies for the transfer of capital and information. It is not a term which has commonly been applied in the physical and environmental sciences, despite increasing understanding of the connectedness, multiple scales and complexity of many environmental problems and of the essential relatedness of these to human activity, economic development and sustainability.

Geography (and physical geography within this) is a natural subject to examine such complexities and relationships. It is both a vehicle for, and a critical contributor to, the globalization of science, environmental and development policy. Geography and geographers have played key roles in raising awareness of the global scale of human impact on earth surface systems and in providing basic scientific knowledge and developing tools for understanding and managing these impacts. They have also helped to shape and change debates and research agendas associated with globalization in a variety of guises, from esoteric academic critiques, through strategic policies in government and non-governmental organizations, to more popular agenda-setting activities.

Historically, both the structure and content of geography, as well as the practices of geographers, were governed by the export of western capital around the globe and in response to nineteenth-century organizational changes in science and academic institutions, all of which can be seen as early forms of globalization. From a contemporary perspective, geography and physical geography are still part of a wider scientific and academic community which is restructuring and changing in the process of globalization, marked now by the growth in information technologies and by what might be called the democratization of the environment. Neither physical geography nor physical geographers can, or should, be immune to such globalization, since there is as much potential to change and shape the globalization process as there is to simply enact a globalizing agenda.

SUMMARY

This chapter attempts to sketch out several ways in which the concept of globalization may be useful in the physical and environmental sciences with a particular focus on physical geography. Those aspects of globalization discussed in this chapter relate to four key themes:

- the ambition and the nature of geography and physical geography, both historically and today

- the organization of geographical activity and the relationship of physical geography to science and scientific activity more generally

- the politicization of the environment, and the role physical geographers have in this

- the technologies for producing, reproducing and, ultimately, reshaping scientific knowledge.

Following from these themes, there are many possible manifestations of globalization and 'the globalization syndrome', of which the following have been sketched out here:

- the way in which geography has struggled with methods and concepts which integrate an enormous range of different knowledge of the world

- the historical involvement of geography and geographers with colonial resource exploitation and, later, worldwide conservation efforts

- the seminal activities of geographers in drawing attention to the relation between politics, environment and development, most notably through political ecology, Sustainability Science and Earth System Science.

Further Reading

Adams', (2004) **Against Extinction: The Story of Conservation** provides a particularly engaging account of the growth of the worldwide conservation movement and of its links to historic and contemporary aspects of globalization. Blaikie's (1985) **The Political Economy of Soil Erosion in Developing Countries** and Blaikie and Brookfield's (1987) **Land Degradation and Society** are seminal works in political ecology, while Kates et al. (2001) is a short, but no less foundational statement on Sustainability Science. Chorley (1971) is the classic starting point for questioning the role and relations of physical geography to geography as a whole and to General System Theory. Pitman (2005) provides a provocative (if rather uncritical) call for physical geographers to adopt an Earth System Science approach, which may be contrasted with Paola et al. (2006).

Note: Full details of the above can be found in the references list below.

References

Adams, W.M. (2004) *Against Extinction: The Story of Conservation*. London: Earthscan.

Anderson, K. (1997) 'A walk on the wild side: a critical geography of domestication', *Progress in Human Geography*, 21: 463–85.

Blaikie, P. (1985) *The Political Economy of Soil Erosion in Developing Countries*. Harlow: Longman.

Blaikie, P. (2001) 'Social nature and environmental policy in the south: views from the verandah and veld', in N. Castree and B. Braun (eds) *Social Nature: Theory, Practice and Politics*. Oxford: Balckwell, pp. 133–50.

Blaikie, P. and Brookfield, H. (1987) *Land Degradation and Society*. London: Macmillan.

Bryant, R.L. (2001) 'Political ecology: a critical agenda for change?', in N. Castree and B. Braun (eds) *Social Nature: Theory, Practice and Politics*. Oxford: Blackwell, pp. 151–69.

Butzer, K.W. (2002) 'The rising cost of contestation: commentary/response: Turner's "Contested Identities"', *Annals of the Association of American Geographers*, 92: 75–86.

Chorley, R.J. (1971) 'The role and relations of physical geography', *Progress in Geography*, 3: 87–109.

Clark, W. and Dickson, N.M. (2003) 'Sustainability science: the emerging research program', *Proceedings of the National Academy of Sciences*, 100: 8059–61.

Clifford, N.J. (in press) 'Physical geography', in D.J. Gregory et al. (eds) *The Dictionary of Human Geography*. Oxford: Blackwell.

Clifford, N.J. and Richards, K.S. (2005) *'Earth system science: an oxymoron'*, Earth Surface Processes and Landforms, 30: 379–83.

Davis, W.M. (1899) *Physical Geography*. Boston, MA: Ginn and Co.

Ferguson, R. (2003) 'Publication practices in physical and human geography: a comment on Nigel Thrift's *The future of geography*, Geoforum, 34: 9–11.

Fossey, D. (1983) *Gorillas in the Mist*. Boston, MA: Houghton and Mifflin.

Gibbons, M., Limoges, C., Nowotny, H., Schwartzman, S., Trow, M. and Scott, P. (1994) *The New Production of Knowledge: The Dynamics of Science and Research in Contemporary Societies*. London: Sage.

Gregory, D.J. (in press) 'Geography', in D.J. Gregory et al. (eds) *The Dictionary of Human Geography*. Oxford: Blackwell.

Gregory, K.J. and Craig, L. (2005) *Final Report*. International Geographical Congress, Glasgow, 15–20 August 2004. http://www.igu-net.org/uk/documents_download/IGC_2004_report_short_Feb_2006.pdf

Grove, R.H. (1992) 'Origins of western environmentalism', *Scientific American*, 267: 22–7.

Grove, R.H. (1995) *Green Imperialism: Colonial Expansion, Tropical Island Edens and the Origins of Environmentalism, 1600–1860*. Cambridge and New York: Cambridge University Press.

Guyot, A. (1850) *Comparative Physical Geography, or the Earth in Relation to Man*. London: Edward Gover. (Revised edn trans. C.C. Felton.)

Kates, R.W., Clark, W.C., Corell, R., Hall, J.M., Jaeger, C.C., Lowe, I., McCarthy, J.J., Schellnhuber, H.J., Bolin, B., Dickson, N.M., Faucheux, S., Gallopin, G.C., Grubler, A., Huntley, B., Jager, J., Jodha, N.S., Kasperson, R.E., Mabogunje, A., Matson, P., Mooney, H., Moore, B. III, O'Riordan, T. and Svedlin, U. (2001) 'Environment and development: sustainability science' *Science*, 292(5517) : 641–2.

Livingstone, D.N. (1992) *The Geographical Tradition*. Oxford: Blackwell.

Mabe, M. and Amin, M. (2001) 'Growth dynamics of scholarly and scientific journals', *Scientometrics*, 51: 147–62.

Marston, S.A., Jones, J.P. and Woodward, K. (2005) 'Human geography without scale', *Transactions, Institute of British Geographers*, NS 30: 416–32.

Mill, H.R. (1913) *The Realm of Nature*. London: John Murray.

Nowotny, H., Scott, P. and Gibbons, M. (2001) *Re-thinking Science: Knowledge and the Public in an Age of Uncertainty*. London: Polity Press.

Paola, C., Foufoula-Georgiou, E., Dietrich, W.E., Hondzo, M., Mohrig, D., Parker, G., Power, M.E., Rodriguez-Iturbe, I., Voller, V. and Wilcock, P. (2006) 'Toward a unified science of the Earth's surface: opportunities for synthesis among hydrology, geomorphology, geochemistry and ecology', *Water Resources Research*, 42.

Pitman, A.J. (2005) 'On the role of geography in earth system science', *Geoforum*, 36: 137–48. (See also subsequent discussions.)

Richards, K.S. and Clifford, N.J. (2008) 'Science, systems and geomorphologies: Why less may be more', *Earth Surface Processes and Landforms*, 33, 1323–40.

Schellnhuber, H.J. (1999) '"Earth System" analysis and the second Copernican revolution', Nature 402: C19–C23.

Skertchly, S.B.J. (1878) *The Physical System of the Universe: An Outline of Physiography*. London: Daldy, Isbister and Co.

Smith, N. (1984) *Uneven Development: Nature, Capital and the Production of Space*. New York: Blackwell.

Smith, N. (1996) 'The production of nature', in G. Robertson, M. Mash, L. Tickner and T. Putnam, (eds) *Future Natural*. London: Routledge, pp. 56–70.

Smith, S.J. (2005) 'Edtorial: joined-up geographies', *Transactions, Institute of British Geographers*, NS 30: 389–90.

Spencer, T. and Slaymaker, O. (1998) *Physical Geography and Global Environmental Change*. Harlow: Addison Wesley Longman.

Summerfield, M.A. (2005) 'A tale of two scales, or the two geomorphologies', *Transactions, Institute of British Geographers*, NS 30: 402–15.

Unwin, T. (1992) *The Place of Geography*. Harlow: Longman.

Warren, A. (1987) 'Geography and conservation: the application of ideas about people and environment', in M.L. Clark, K.J. Gregory and A.M. Gurnell (eds) *Horizons in Physical Geography*. London: Macmillan Education, pp. 322–36.

21

Development: Critical Approaches in Human Geography

Katie D. Willis

D efinition

Development is used in everyday speech to refer to change. This change is usually viewed in positive terms. However, within geography, development usually has more specific meanings, referring to either national-level processes of economic, political and social change, or the positive change resulting from intentional actions to improve the living conditions of poor or marginal populations. As well as being a process, development can also be defined as a state of being, usually applied to a country or region and implying high levels of urbanization, complex economic activity and standards of living. Such definitions are, however, not neutral as they reflect particular ideologies which vary across time and space. Geographers have been involved in both reinforcing particular concepts of development and revealing the ways in which they are based on the operation of power.

INTRODUCTION

Like many concepts used in geography, 'development' is hard to define. However, a good starting point is to consider the distinction between development as either immanent or intentional (Cowan and Shenton, 1996). Hart (2001) makes a similar distinction between 'little d' development and 'big D' Development. Immanent or 'little d' development refers to the changing nature of capitalism, while intentional or 'big D' development is a particular set of projects and policies aimed at improving the lives of people in the Global South, particularly since the 1940s

Geographers have been concerned both with examining the spatially uneven development of capitalism and its locally-experienced outcomes, as well as the particular spatial patterns and effects of formal development policies. In addition, more recent work has focused on the ways in which such policies are framed by ideologies, which come out of particular contexts and can be spread through the uneven exercise of power, where the idea of 'development' can be presented as natural and therefore immune from challenges.

In this chapter, I will provide an overview of the different ways in which 'development' has been used as a concept within geography. Given the nature of the discipline, it is difficult to place strict borders around it, and many of the authors mentioned in this chapter would not regard themselves as geographers. However, their work has helped inform geographical work on development at a range of scales. As I will argue, different concepts of development have been framed by assumptions regarding the appropriate level of analysis and intervention. Geographers have been key in understanding the ways in which scale has been used within development research and practice.

MODERNIZATION

In the post-Second World War period, many geographers working on and in the Global South, embraced the optimism and hope that the transfer of technology from North to South would be the key to development. Through education and the diffusion of technology, 'underdeveloped' or 'backward' countries, as they were often seen, would be able to progress through agricultural intensification, industrialization and urbanization. These processes, it was argued, would lead to improved living conditions and quality of life. This route from subsistence, rural-based economies, organized around kinship or tribal social structures, to urban industrial societies with formal state institutions, was modelled on the experiences of the Global North and has been termed 'modernization'.

Development as modernization was present in both the policy and academic arenas, with some individuals working in both. The most well-known modernization theorist was the economist W.W. Rostow, whose book *The Stages of Economic Growth: A Non-Communist Manifesto* (1960) outlined five stages through which nations passed on the road to development. The final stage, the 'Age of high mass consumption', represented the situation in 1950s USA and many countries of western Europe. Policy-makers following such a linear approach used aid programmes to promote agricultural intensification (such as the 'green revolution'), large-scale infrastructure projects (such as dams) and industrialization. However, due to their 'one size fits all' approach, these policies often failed or exacerbated existing problems because they did not acknowledge particular environmental, social and cultural contexts.

The spread of development and modernization from North to South, was also seen at a national scale, with theorists such as Hirschman (1958) and Friedmann (1966) discussing the role of geographical inequalities in the development process. Using evidence from Colombia (Hirschman) and Venezuela (Friedmann), they argued that economic development could be achieved by focusing investment in a limited number of core sites that would act as growth poles. As these locations developed through industrialization, the benefits would spread to other parts of the country. Thus, spatial inequalities were a necessary part of the development process, but would eventually be reduced or eliminated. Other theorists, particularly Myrdal (1957), argued that such diffusion would not occur without significant state intervention to reverse the centralizing trends that growth poles encourage.

Detailed studies of diffusion processes became common, particularly in the newly-independent African and Asian nations. For example, Soja (1968) and Riddell (1970) provided an overview of modernization processes in Kenya and Sierra Leone respectively, while Leinbach's work (1972) focused on the spatial dimensions of modernization in Malaya. Such studies tended to focus on quantitative dimensions of development and spatial models encompassing key indicators of modernity, such as industrial production and infrastructure (including road and railway networks).

Theories and policies based on a modernization approach are often criticized for their Eurocentrism and focus on top-down implementation with little or no involvement from the communities affected (see below). However, while such criticisms are often valid, it is important to recognize benefits that have accrued from some modernization-informed policies, including worldwide vaccination programmes against childhood diseases (Corbridge, 1998).

DEPENDENCY AND UNEVEN DEVELOPMENT

In the 1960s and 1970s, new interpretations of development inequalities began to gain popularity. While 'development' was still viewed as including industrialization, urbanization and increasing social complexity, the way in which this was to be achieved was challenged. According to modernization theories, countries would achieve development through following the same path as European countries. Thus a map of the world showing economic inequalities between countries reflected different stages on the same road. In contrast, dependency theories or theories of underdevelopment/uneven development, were based on the belief that 'underdeveloped' countries were unable to advance economically because of constraints within the global capitalist system. According to some arguments, the very state of 'underdevelopment' was caused by capitalist exploitation.

Many of these ideas were based on theorizing from the Global South, particularly from Latin America. Frank (1969) and Furtado (1964) argued that Latin American countries were locked within a situation of dependency within the world economic system. Rather than being able to industrialize and reap the benefits from such economic progress, Latin American countries were exploited by countries in the Global North and would not be able to develop unless these exploitative relationships between the 'metropolis' and 'satellite' countries were ended. Frank outlined a series of chains of exploitation from peasants in rural Latin America, through landowners and middlemen, to the Latin American urban elite and then to Europe. According to Frank, this exploitation dated from the colonial period and had created 'underdevelopment' in Latin America. Rodney (1972) made a similar argument about European colonialism in Africa.

Following this line of argument, development would be achieved only if Southern countries were able to escape from such exploitative relationships. The policy prescriptions relating to this varied from greater use of protectionism to attempts to break from the capitalist system completely. The United Nations Economic Commission for Latin America (ECLA or CEPAL in its Spanish acronym), under Raúl Prebisch, proposed staying within the global capitalist system but increasing tariff barriers for imported goods. This was to try to prevent the already established industries of the Global North undermining infant industries in the South. During the industrial revolution in nineteenth-century Europe, there was no competition from elsewhere, but Prebisch and others noted that the global economic situation was very different for later-industrializing countries, so protectionism was necessary. Such policies, sometimes called 'import-substitution industrialization' (ISI), were adopted in many countries of the Global South during the 1950s–1970s with varying degrees of success.

For more radical theorists, often following a Marxist interpretation of capitalism and inequalities, the solution to global economic inequalities was not an attempt to 'level the playing field', but was to overturn the capitalist system altogether. According to this interpretation, capitalism is based on and reproduces spatial inequalities at all scales (see, for example, Smith, 1990). A number of geographers drew on such arguments in relation to geographies of underdevelopment (see, for example, Slater, 1973, 1977; Cannon, 1975). As Cannon (1975: 215) stated:

> [W]e have to recognize the fact that underdevelopment is not an abstract, indigenous process which each country is automatically capable of under present condition. Instead it is necessary to accept that the development of the advanced countries on the one hand, and the underdevelopment of the Third World on the other, are products of the antagonistic relations between the two groups of countries.

This argument is very similar to the dependency theorists. Cannon's quote also demonstrates the way in which the concept of 'development' is not challenged within this theorizing. Instead, it was the explanation for different levels of development, that is differential incorporation into the global capitalist system, which varied from the modernization school.

'Radical development geography', as this form of thinking has sometimes been classified, has, however, received criticisms, most notably for its often rigid construction of oppositions between 'developed' and 'underdeveloped' parts of the world. Such representations have not helped explain the successful economic and social capitalist development of some previously 'underdeveloped' nations, most notably in East Asia (World Bank, 1993). If capitalism is the cause of underdevelopment, then such processes should be impossible. Spatially, the distinction between 'core' and 'periphery' is a static distinction which does not recognize the diversity of national experiences within each grouping (Corbridge, 1986). The focus on nation-states and class relations has also been critiqued by those adopting a post-structuralist or postcolonial approach (see below).

NEOLIBERALISM

In the 1970s, and particularly the 1980s, prevailing international development policies were again challenged and major changes took place. These changes revolved mainly around the appropriate role for the state. Under modernization approaches and import-substitution industrialization, the state, particularly at a national level, had a key role in service provision, industrial and trade policy. What neoliberal approaches argued was that the involvement of the state in this way was inefficient and inflexible. Theorists such as Balassa (1971) and Lal (1983) argued that economic growth would be much greater and also fairer if market forces were left to operate without state intervention. They drew on the work of economists Adam Smith and David Ricardo, who were working in the eighteenth and nineteenth centuries, to suggest that free trade and global divisions of labour would combat economic stagnation and lead to greater well-being for all.

Global economic recession in the 1970s meant that governments were seeking alternative policies. Margaret Thatcher in the UK and Ronald Reagan in the USA came to power promising economic growth through less state intervention, and such policies were soon being adopted in many parts of the world, particularly the Global South. Adoption of such neoliberal policies became more pressing as the 'debt crisis' hit. During the 1960s and 1970s governments in the Global South had borrowed money from international banks to fund domestic development, but with changes in the external economic climate debts built up until governments were forced to default on their debt.

As commercial banks were unwilling to lend additional funds, Southern governments were left with very little choice but to accept funding from the international financial institutions of the World Bank and the International Monetary Fund (IMF). This funding was conditional on the implementation of Structural Adjustment Policies (SAPs). Thus, particular ideas regarding 'development' were spread throughout most of the globe. Pre-existing power relations made possible the imposition of ideas about development, ideas which were reinforced by these policies (Harvey, 2005; Peet, 2007).

Structural Adjustment Policies were geared around stabilizing and restructuring national economies. Key elements of SAPs included privatization, tax reform and reduced government expenditure on social services to try to reduce government deficits internally. In addition, tariff barriers were lowered or eliminated and national currencies were devalued to promote foreign investment and production for export (Mohan et al., 2000). Such an approach, it was argued, would promote competition so that goods and services would be produced efficiently and would meet the demands of the population. Opening up the economy to foreign investment and promoting international free trade would also help generate wealth for the benefit of the whole of society. Neoliberal approaches were therefore based on ideas of 'development' which were very similar to those which had gone before in that they focused on economic growth, which was assumed to have social benefits, such as improved health and education levels.

While SAPs did help to stabilize many economies and reverse negative economic growth trends in others, they were often associated with increasing social inequalities. Reduced state expenditures meant that the poor were often left without appropriate welfare provision and even the middle classes found themselves in dire economic straits. Such trends were also identified in the ex-communist countries of Eastern Europe, following the collapse of the Soviet Union. Geographers, among others, focused on the impacts of SAPs on different people and places. While neoliberalism was often presented as a homogeneous set of ideas and practices (Barnett, 2005), in reality, national and local conditions resulted in diverse implementations and outcomes.

This realization that SAPs had exacerbated inequalities and levels of poverty in many places meant that the loan packages provided by the World Bank and IMF from the mid-1990s onwards have supposedly been more context-specific. There has also been a shift towards an explicit consideration of 'development' as 'poverty reduction', rather than as economic growth. However, despite many studies demonstrating the multidimensionality of 'poverty' (see, for example, Baulch, 1996; Narayan, 2000), it is usually considered in economic terms (most notably through the concept of an income of 'less than US$1 per day'). The focus on poverty has been codified in the Millennium Development Goals (MDGs) which were agreed by the United Nations in 2000. Despite these shifts in emphasis, neoliberalism remains the main framework within which development policies are understood and implemented (Peet, 2007).

GRASSROOTS DEVELOPMENT

The definitions and approaches to development outlined so far focus particularly on economic aspects and usually assume that economic benefits accrued at a national level will trickle down to poorer and marginalized populations. In reality, this rarely happens and has led to calls to focus on individual communities and how they feel development should be achieved. What has been termed 'grassroots development' or 'bottom-up development' attempts to address development issues at this scale by considering the possible range of development definitions and how such definitions vary both spatially and socially.

This focus has been particularly strong since the 1990s, but it does have its roots in previous policies and theorizing, such as the 'Basic Needs Approach' of the 1960s. Despite these earlier processes, it was only in the late 1980s and early 1990s that bottom-up development really came to the fore. In particular, the role of non-governmental organizations (NGOs) as appropriate agents for development was championed (Drabek, 1987; Edwards and Hulme, 1995). This new focus reflected the fact that grassroots development and NGO activities fitted into both a neoliberal interpretation of development and alternative more radical approaches seeking to empower local communities (Mohan and Stokke, 2000). Theoretical approaches stressing the recognition of diversity of people and perspectives (see below) were also able to embrace much of this move towards greater NGO importance.

Non-governmental organizations are organizations that are neither state-controlled nor profit-driven and are viewed as part of 'civil society' (McIlwaine, 1998). They range in scale from very large international organizations to tiny local groups, and their activities can focus on fields such as human rights, environmental conservation, local economic development and political protest. In relation to development, the 1990s saw a massive increase in the number and range of NGOs. This was mainly in response to neoliberal restructuring. As the state withdrew from service provision, private companies were reluctant to step in due to limited profit-making opportunities. Thus, poor populations were left without services, such as housing, healthcare and education. In many parts of the Global South, NGOs have become key service providers for large sectors of the population who cannot afford either the user fees charged by the government or the costs of private services. In theory, NGOs should be able to provide more effective and appropriate services because they are able to work with local populations and adapt their provision accordingly. However, NGO activities are not necessarily distributed in relation to the need of local communities. Despite calls to 'put the first last' (Chambers, 1997), NGOs are more likely to be found in more physically accessible areas, such as in or near capital cities and not far from main roads (Mercer, 1999). Bebbington (2004) also alerts us to the importance of considering historical economic and political processes, including earlier aid interventions, in explaining the uneven geographies of NGOs.

The ability to work at a local scale is also viewed positively as it provides greater decision-making opportunities for previously marginalized people. It is this dimension which is viewed as particularly positive by those seeking alternative forms of development. Through participation in decision-making and project implementation, it is argued that local people become empowered. This involves both a feeling of greater self-worth and also an awareness of being able to change one's own life. These ideas have been particularly associated with marginalized groups such as women and indigenous peoples (Friedmann, 1992; Rowlands, 1997).

The assumptions about participation have, however, been challenged by researchers who claim that projects are often termed 'participatory' when participation has just been cursory, such as attendance at a meeting where outsiders or elites inform residents of proposed policies. In addition, much participatory work fails to recognize the diversity within communities, such that NGO projects can reinforce rather than challenge existing power relations. 'Participation' has become such a common goal of development policies that even large organizations, such as the World Bank, now claim they promote participatory development. As Cooke and Kothari's edited book *Participation: The New Tyranny?* (2001) has argued, calls for 'participation' have often led to the potential for empowerment and radical change being lost.

The focus on the 'local' scale has also been criticized, particularly from geographers who are rightly wary of the binary division between 'local' and 'global'. As Mohan and Stokke (2000) argue, the privileging of the 'local' is often based on a romantic notion of 'community' and fails to recognize conflict and division. It also does not recognize how this 'local' is constructed and mobilized by different actors to achieve their goals and how places and people are linked into wider social, economic and political processes. Failing to consider these wider processes can limit the possibilities for progressive change. Despite such criticisms, many authors argue that participation and empowerment can be achieved through collaboration and co-operation (Hickey and Mohan, 2004, 2005).

SUSTAINABLE DEVELOPMENT

While 'development' has often been viewed from a human-centric point of view, the ways in which economic and social progress affect the natural environment have become of increasing concern (see also Chapter 22). This relationship is often encapsulated within the concept of 'sustainable development', which, while widely used, is often contested (Elliott, 2006). The most commonly-used definition is that of the World Commission on Environment and Development (WCED), also known as the 'Brundtland Commission'. The WCED (1987: 33) definition of 'sustainable development' is 'development that meets the needs of the present without compromising the ability of future generations to meet their own needs'. This definition has been adopted in international development circles, for example at the United Nations Conference on Environment and Development (the Rio Summit) in 1992 and the Rio+10 Summit in 2002 in Johannesburg.

As many authors, including geographers (see, for example, Adams, 1995; Elliott, 2006), have noted, the ways in which the definition has been interpreted and put into practice vary greatly. In particular, the weight placed on 'development' (viewed as improvements in human living conditions) compared with 'environment' (which is often viewed in very narrow natural environmental terms) differs. For those following 'technocentric' approaches, the solution is to be found in innovation and technical solutions, such as less-polluting industrial technologies or more energy-efficient infrastructure. Such an approach can be followed within an existing capitalist economic development framework so has been widely adopted. Within international development and co-operation, the technology transfer, which was so popular under 'modernization' models, continues with the North sharing 'environmentally-friendly' technology with countries in the Global South. Given the global power relations involved in such exchanges, countries in the Global South have often complained that they are being asked to limit economic growth and development, while the countries of the Global North industrialized and urbanized with no constraints.

In contrast, ecocentric approaches place the natural environment at the centre and would therefore include reduced consumption and the promotion of locally-based livelihoods. Such an approach challenges the focus on growth and capital accumulation which underpins more mainstream ideas. The growing international awareness of the impacts of climate change may be focusing the minds of national governments, but a radical shift in policies to a more 'ecocentric' approach is viewed as politically impossible, not least in those countries where millions of people live in extreme economic poverty.

The grassroots approaches, outlined earlier, have often been claimed to be more favourable to sustainable development because they are based in particular ecosystems and draw on local understandings of the relationship between humans and the natural environment. NGO activity can facilitate both human development and environmental protection, but as with all such practices, there are limitations (see above).

POSTCOLONIALISM, POST-STRUCTURALISM AND POST-DEVELOPMENT

An awareness of how power is implicated in development approaches and practices, as well as the diversity of populations and environments, has led to a significant shift in the ways in which 'development' has been studied by geographers and others in the late twentieth and early twenty-first centuries. This shift has been informed by scholars adopting postcolonial, post-structural and post-development approaches. While the three have similarities, and have been combined by many researchers on 'development' (see Chapter 5 of Lawson, 2007, for an overview), I will consider each in turn.

Postcolonial theory highlights the ways in which prevailing dominant discourses (in this case about development) are based on Eurocentric experiences and norms which are presented as 'natural'. One of the key writings within this field was *Orientalism* (1978), in which Edward Said theorized the ways in which Europeans constructed themselves as 'superior' and 'civilized', through contrasting themselves with the peoples and places of the 'East'. The challenge to Eurocentrism demands that attention is paid to the silenced and marginal (sometimes termed 'the subaltern') rather than the powerful (Blunt and McEwan, 2002). However, it is important that such endeavours do not just replicate pre-existing binaries such as haves/have-nots, North/South and powerful/powerless because this would reinforce prevailing ways of thinking and fail to recognize the diversity of relations between people over time and space (Sharp and Briggs, 2006; Simon, 2006).

Post-structuralism is a similar response to the overarching theories which attempted to use the same explanations to deal with experiences and processes among different communities and in different places. Many earlier ideas about 'development', particularly modernization, but also dependency and neoliberalism, have adopted these metanarratives. Recognizing diversity is key to both understanding people's lives, but also to defining what development can be and how it can be achieved. However, post-structuralist approaches still include an awareness of how power is enacted in development processes along axes such as gender, class and ethnicity (Power, 2003; Radcliffe, 2006), but the ways in which these processes operate are not assumed to be universal; rather they vary across time and space.

Post-development considers issues of power and diversity within the specific context of 'development'. Post-development theorists, as the name suggests, challenge the very idea of 'development', arguing that it is a Northern concept which has been imposed on the Global South with tragic consequences. Rather than being a non-political or neutral term, post-development theorists highlight the ways in which policies have been used to frame people and places as 'development problems' which need to be solved (Sachs, 1992; Ferguson, 1994; Crush, 1995; Escobar, 1995). Such theorists are therefore challenging the ways in which 'development' as a form of discourse and policy process has been framed by those in power and used to justify a range of interventions.

While the dependency and neo-Marxist theorists described earlier stressed the ways in which power differentials within the capitalist system affected how countries could develop, post-development theorists consider power from a different point of view – that of the discursive realm. For some, this focus has been interpreted as 'academic navel gazing', far removed from the material needs and daily fight for survival for millions of people. Another criticism of post-development approaches is the way in which the 'development' that is critiqued is actually an outdated one, based on unreflective modernization policies dating from the 1950s and 1960s. Finally, post-development has been accused of lacking specific alternatives to 'development', other than very general references to 'community-based decision-making', 'grassroots development' and other 'locally-based' solutions which seem to replicate the universalizing of earlier development strategies (Simon, 1998; Sylvester, 1999; Nederveen Pieterse, 2000).

Some post-development work has gone beyond critiques and has sought to examine productive ways of working for progressive change. Gibson-Graham (2005), for example, considers alternative forms of local economic community development in the Philippines. Rather than seeing 'modernization' through engagement in 'global processes', such as overseas labour migration and export agriculture, as the only possible route to improving quality of life and standards of living, Gibson-Graham highlights the ways in which other forms of economic activity are valued within what they term 'the diverse economy'. He argues that: 'One effect of this representation of a diverse economy is that capitalist activity ... is knocked off its perch, so to speak, and with it goes the ordered certainties associated with development dynamics' (Gibson-Graham, 2005: 13).

Gibson-Graham's work certainly provides a productive response to common critiques of the neoliberal development project, but his focus on the 'local' has also left him open to criticism about the exclusion of power relations within the 'community' and the ways in which these 'local' spaces are linked to wider national and global ones (Aguilar, 2005; Kelly, 2005; Lawson, 2005). Such comments resonate with the general critique of community-led development and constructions of place (see 'Grassroots development' section).

CONCLUSION

In this chapter I have provided an overview of the ways in which the concept of 'development' has been used within geography and related disciplines. In particular, I have highlighted how power is implicated in both the definitions adopted and the forms in which such definitions have been mobilized in policy arenas. While the chapter is organized in a roughly chronological sequence, it is important to recognize that this orderly progression is a way of framing the information, rather than a reflection of the actual ways in which development definitions have been used and challenged within geography.

SUMMARY

- Development is not a neutral concept. Its definition and use varies over time and space.

- Development can be used to describe both general societal changes, particularly under capitalism, but has also been used more specifically in relation to policy interventions in the Global South.

- Definitions use different scales of analysis, in particular 'the local' and 'the national'. At each scale the importance of particular actors within development is stressed. These scales are not mutually exclusive.

Further Reading

Two very useful overviews of the ways geographers have used 'development' are Power's (2003) **Rethinking Development Geographies** and Lawson's (2007) **Making Development Geographies.** For particular case studies on how development has been used and abused, see Ferguson's (1994) **The Anti-Politics Machine** and Esobar's (1995) **Encountering Development**. The edited volume by Crush (1995) **Power of Development**, also provides a range of perspectives challenging development as modernization. Two useful collections on geography and postcolonialism are The Geographical Journal (2006) Volume 172, No. 1 and Singapore Journal of Tropical Geography (2003) Volume 24, No. 3. For an overview of development theories see Willis's (2005) **Theories and Practices of Development** and Peet with Hartwick's (1999) **Theories of Development**.

Note: Full details of the above can be found in the references list below.

References

Adams, W.M. (1995) 'Green development theory? Environmentalism and sustainable development', in J. Crush (ed.) *Power of Development*. London: Routledge, pp. 87–99.

Aguilar Jr., F.V. (2005) 'Excess possibilities? Ethics, populism and community economy', *Singapore Journal of Tropical Geography*, 26: 27–31.

Balassa, B. (1971) 'Trade policies in developing countries', *American Economic Review*, 61: 178–87.

Barnett, C. (2005) 'The consolations of "neoliberalism"', *Geoforum*, 36: 7–12.

Baulch, B. (1996) 'The new poverty agenda: a disputed consensus', *IDS Bulletin*, 27: 1–10.

Bebbington, A. (2004) 'NGOs and uneven development: geographies of development intervention', *Progress in Human Geography*, 28: 725–45.

Blunt, A. and McEwan, C. (2002) 'Introduction', in A. Blunt and C. McEwan (eds) *Postcolonial Geographies*. London: Continuum, pp. 1–6.

Cannon, T. (1975) 'Geography and underdevelopment', *Area*, 7: 212–16.

Chambers, R. (1997) *Whose Reality Counts? Putting the First Last*. London: Intermediate Technology Books.

Cooke, B. and Kothari, U. (eds) (2001) *Participation: The New Tyranny?* London: Zed Books.

Corbridge, S. (1986) *Capitalist World Development: A Critique of Radical Development Geography*. Basingstoke: Macmillan.

Corbridge, S. (1998) 'Beneath the pavement only soil: the poverty of post-development', *Journal of Development Studies*, 34: 138–48.

Cowen, M. and Shenton, R. (1996) *Doctrines of Development*. London: Routledge.

Crush, J. (ed.) (1995) *Power of Development*. London: Routledge.

Drabek, A.G. (1987) 'Development alternatives: the challenge of NGOs', *World Development*, 15 (supplement).

Edwards, M. and Hulme, D. (eds) (1995) *Non-Governmental Organisations – Performance and Accountability: Beyond the Magic Bullet*. London: Earthscan.

Elliott, J.A. (2006) *An Introduction to Sustainable Development* (3rd edn). London: Routledge.

Escobar, A. (1995) *Encountering Development: The Making and Unmaking of the Third World*. Princeton, NJ: Princeton University Press.

Ferguson, J. (1994) *The Anti-Politics Machine: 'Development', Depoliticization and Bureaucratic Power in Lesotho*. Minneapolis, MN: University of Minnesota Press.

Frank, A.G. (1969) *Capitalism and Underdevelopment in Latin America: Historical Studies of Chile and Brazil*. New York: Monthly Review Press.

Friedmann, J. (1966) *Regional Development Policy: A Case Study of Venezuela*. Cambridge, MA: MIT Press.

Friedmann, J. (1992) *Empowerment: The Politics of Alternative Development*. Oxford: Blackwell.

Furtado, C. (1964) *Development and Underdevelopment*. Berkeley, CA: University of California Press.

The Geographical Journal (2006) 'Postcolonialism and development: new dialogues', *The Geographical Journal*, 172: 6–77.

Gibson-Graham, J.K. (2005) 'Surplus possibilities: postdevelopment and community economics', *Singapore Journal of Tropical Geography*, 26: 4–26.

Hart, G. (2001) 'Development critiques in the 1990s: *culs de sac* and promising paths', *Progress in Human Geography*, 25: 649–58.

Harvey, D. (2005) *A Brief History of Neoliberalism*. Oxford: Oxford University Press.

Hickey, S. and Mohan, G. (eds) (2004) *Participation: From Tyranny to Transformation?* London: Zed Books.

Hickey, S. and Mohan, G. (2005) 'Relocating participation within a radical politics of development', *Development and Change*, 36: 237–62.

Hirschman, A.O. (1958) *The Strategy of Economic Development*. New Haven, CT: Yale University Press.

Kelly, P. (2005) 'Scale, power and the limits to possibilities', *Singapore Journal of Tropical Geography*, 26: 39–42.

Lal, D. (1983) *The Poverty of Development Economics*. London: Institute of Economic Affairs.

Lawson, V. (2005) 'Hopeful geographies: imagining ethical alternatives', *Singapore Journal of Tropical Geography*, 26: 36–8.

Lawson, V. (2007) *Making Development Geography*. London: Hodder Arnold.

Leinbach, T.R. (1972) 'The spread of modernization in Malaya: 1895–1969', *Tijdschrift voor Economische en Sociale Geografie*, 63: 262–77.

McIlwaine, C. (1998) 'Civil society and development geography', *Progress in Human Geography*, 22: 415–24.

Mercer, C. (1999) 'Reconceptualizing state–society relations in Tanzania', *Area*, 3: 247–58.

Mohan, G. and Stokke, K. (2000) 'Participatory development and empowerment: the dangers of localism', *Third World Quarterly*, 21: 247–68.

Mohan, G., Brown, E., Milward, B. and Zack-Williams, A.B. (2000) *Structural Adjustment: Theory, Practice and Impacts*. London: Routledge.

Myrdal, G. (1957) *Economic Theory and Underdeveloped Areas*. London: Duckworth.

Narayan, D. (2000) *Voices of the Poor: Can Anyone Hear Us?* New York: Oxford University Press for The World Bank.

Nederveen Pieterse, J. (2000) 'After post-development', *Third World Quarterly*, 21: 175–92.

Peet, R. (2007) *Geography of Power: The Making of Global Economic Policy*. London: Zed Books.

Peet, R. with Hartwick, E. (1999) *Theories of Development*. New York: Guilford Press.

Power, M. (2003) *Rethinking Development Geographies*. London: Routledge.

Radcliffe, S. (2006) 'Development and geography: gendered subjects in development processes and interventions', *Progress in Human Geography*, 30: 524–32.

Riddell, J.B. (1970) *The Spatial Dynamics of Modernization in Sierra Leone*. Evanston, IL: Northwestern University Press.

Rodney, W. (1972) *How Europe Underdeveloped Africa*. London: Bogle–L'Ouverture Publications.

Rostow, W.W. (1960) *The Stages of Economic Growth: A Non-Communist Manifesto*. Cambridge: Cambridge University Press.

Rowlands, J. (1997) *Questioning Empowerment: Working with Women in Honduras*. Oxford: Oxfam.

Sachs, W. (ed.) (1992) *The Development Dictionary: A Guide to Knowledge as Power*. London: Zed Books.

Said, E. (1978) *Orientalism*. London: Penguin.

Sharp, J. and Briggs, J. (2006) 'Postcolonialism and development: New dialogues?', *The Geographical Journal*, 172: 6–9.

Simon, D. (1998) 'Rethinking (post)modernism, postcolonialism and posttraditionalism: South–North perspectives', *Environment and Planning D: Society and Space*, 16: 219–46.

Simon, D. (2006) 'Separated by common ground? Bringing (post)development and (post)colonialism together', *The Geographical Journal*, 172: 10–21.

Singapore Journal of Tropical Geography (2003) 'Geography and postcolonialism', *Singapore Journal of Tropical Geography*, 24: 269–389.

Slater, D. (1973) 'Geography and underdevelopment I', *Antipode*, 5: 21

Slater, D. (1977) 'Geography and underdevelopment II', *Antipode*, 9: 1–31.

Smith, N. (1990) *Uneven Development: Nature, Capital and the Production of Space* (2nd edn). Oxford: Basil Blackwell.

Soja, E.W. (1968) *The Geography of Modernization in Kenya: A Spatial Analysis of Social, Economic and Political Change*. Syracuse, NY: Syracuse University Press.

Sylvester, S. (1999) 'Development studies and postcolonial studies: disparate tales of the "Third World"', *Third World Quarterly*, 20: 703–21.

Willis, K. (2005) *Theories and Practices of Development*. London: Routledge.

World Bank (1993) *The East Asian Miracle*. Oxford: Oxford University Press.

World Commission on Environment and Development (WCED) (1987) *Our Common Future*. Oxford: Oxford University Press.

22

Development: Sustainability and Physical Geography

Robert Inkpen

Definition

'Humanity has the ability to make development sustainable – to ensure that it meets the needs of the present without compromising the ability of future generations to meet their own needs' (WCED, 1987: 8, or more commonly known as the Brundtland Report). This quote is the standard starting point for understanding sustainable development. Within this context, the physical environment tends to be viewed as a fragile entity that requires careful management. Concepts such as 'carrying capacity', 'ecological footprint' and 'natural capital' reflect this view of the physical environment as in need of stewardship. Physical geographers have contributed towards sustainable development by establishing baselines from which change can be assessed, by identifying the thresholds and equilibria of the physical environment and by providing an insight into the complexity that locality and scale have on the sustainability of the physical environment.

INTRODUCTION

Sustainable development has emerged as a key issue in policy at national and international levels. Although the concept is often expressed in terms of economics and culture, the concept originated and is firmly based within the physical environment. Views of the physical environment and of how humanity should use or develop it have not been constant and these changing views have greatly affected environmental policies. Early concerns with environmental degradation can be traced back to figures such Evelyn (1661), who wrote about smoke pollution in London and even offered an environmentally sustainable solution of planting trees to ameliorate the problem. Marsh (1864) and Thomas (1956) expressed more contemporary concerns identifying a fragile and interconnected physical environment at risk of destruction through human activities. Such environmental concerns among academics and the intelligentsia should not be taken as indicative of a general view of the nature of the physical environment. In late nineteenth-century North America, for example, the physical environment was seen as an hostile enemy that required taming by human intervention rather than a fragile entity requiring protection.

Concern for the physical environment in the late nineteenth and early twentieth centuries found form in policies such as the development of national parks in the United States and national and international conservation movements, such as the World Wildlife Fund (WWF). Policies were often driven by concerns that seem odd to modern eyes. Establishing Yosemite as the first national park in the USA, for example, was a project that relied heavily upon the determination and connections of John Muir, a champion of the concept of wilderness (Wolfe, 1945; Sargent, 1971), and the representation of the park landscape as a commodity. Uniqueness was seen as being translatable into money and also national pride. Congress heard how vast sums were spent overseas by Americans visiting wilderness areas in Europe such as Switzerland and how American areas could boast greater scenic beauty than their European counterparts (Runte, 1997). Often, policies revolved around the conservation of 'unique' features, such as the Grand Canyon, ably championed by President Roosevelt, or individual species, such as the panda, the symbol of the WWF. A view of the physical environment as an integrated whole may have been gaining currency in academic circles, but it failed to provide an emotive imperative for policy.

Speeding forwards to the late 1960s, a more organized and holistic view of the physical environment started to infiltrate policy-making. *The Limits to Growth* (Meadows et al., 1972), a report from the Club of Rome, presents the environment as a finite resource that requires management to ensure that the physical limits to development that it imposes are not exceeded. The authors modelled the world as a global system in terms of the interaction of a few key parameters. From this modelling approach they were able to show that contemporary activities would soon exhaust the finite resources available. Concepts derived from such analysis, including carrying capacity, became key tools for investigating the ability of this finite physical environment to sustain life. Viewing the physical environment as a finite resource shifted policy emphasis from taming to maintaining. Instead of exploitation, policy has increasingly been couched in terms of balance and stability. The physical environment is increasingly seen as a fragile entity that needs protection from the excesses of humanity – a near role reversal from the nineteenth century. It is within this framework of changing perceptions of the nature of the physical environment that physical geographers have contributed to monitoring and understanding sustainable development.

SUSTAINABLE DEVELOPMENT

Although many discussions of sustainable development begin with the quote from Brundtland (above), there is no consensus definition of what sustainable development is (Adams, 2001; Kates et al., 2001; Kates et al., 2005). Sustainable development has become an ongoing dialogue rather than a fixed entity. There may be a set of guiding principles at the heart of the concept but these too have evolved as the different actors, networks and negotiations have changed. The definition of sustainable development depends on what you believe should be developed (society, people or the economy) and the emphasis you place on the nature of sustainability (as a set of goals, measurements, values or practices). This diverse conceptualization of sustainable development permits both 'light' and 'dark' green activists (see Chapter 18) as well as more conservative economists to buy into, and pledge allegiance to, sustainable development. As Kates et al. (2005: 19) noted:

> One of the successes of sustainable development has been its ability to serve as a grand compromise between those who are principally concerned with nature and environment, those who value economic development, and those who are dedicated to improving the human conditions. At the core the compromise is the inseparability of environment and development.

Official definitions of sustainable development and its basic principles have evolved. From the initial stewardship perspective of the Club of Rome, there has been, through the Brundtland Report, the Rio Declaration (UNCED, 1992; Parson and Hass, 1992) and the Johannesburg Declaration (2002), increasing recognition of the importance of the multitude of links between human institutions and the physical environment. Coupled with this recognition has been the growing prominence of the view of the physical environment as an increasingly frail and fragile resource that requires human institutions to ensure its survival.

From the perspective of physical geography, sustainability has focused on assessing actions as reversible or irreversible within a temporal scale that is limited to a few generations. Within this timeframe, geological processes and patterns, such as species evolution and extinction, are neglected while relatively rapid changes are highlighted as significant. This produces a view of the physical environment that is defined by anthropogenic disturbances rather than biophysical processes for which human intervention is a relatively insignificant factor. When large, geological processes become manifest, as with the tsunami on Boxing Day 2004, humanity is jolted back to a view of the physical environment as hostile and unpredictable.

Within the 'grand compromise' of sustainable development (Kates et al., 2005) physical geography and physical geographers have a large role to play. The nature of this role varies depending on the view of the physical environment employed and the emphasis given to the different aspects of sustainable development.

THE PHYSICAL ENVIRONMENT AND SUSTAINABLE DEVELOPMENT

The Brundtland Report (WCED, 1987: 8) also commented on the importance of the physical environment for sustainable development:

> The concept of sustainable development does imply limits – not absolute limits but limitations imposed by the present state of technology and social organiza- tion on environmental resources and by the ability of the biosphere to absorb the effects of human activities.

The concept of absolute limits is reminiscent of *The Limits to Growth* (Meadows et al., 1972) and the concept of a carrying capacity for an area or volume. The Brundtland Report takes the concept of carrying capacity further by emphasizing the relative nature of these limits. 'Natural' limits to development do not neces- sarily exist but are defined in terms of a balance or emergent relationship between social and technological capabilities and the natural capacity for absorb- ing change. Such a concept retains a distinction between 'natural' and 'human' but does highlight the interactive role of the two in defining limits. Indeed, the Brundtland Report highlights that the two – environment and development – are inseparable. This approach to sustainability implies, however, a relatively, but not infinitely, malleable physical environment. That is, the environment is viewed as at once fragile and in need of human protection but also sufficiently robust to undergo regeneration if adequate guardianship is applied. Scientific analysis, and by extension physical geography, has a clear role in identifying the limits beyond which this regenerative capacity dissipates and in understanding how different scenarios of use interact with the physical processes that set the limits and affect regeneration.

Goudie (2000) notes that human impact can be expressed by the formula:

$$I = PAT$$

where I is the amount of pressure or human impact, P is the population, A is the level of affluence or resource demand made by the population and T is a tech- nological factor. This set of interrelated factors can influence the carrying capac- ity of an area (Figure 22.1). An area or region will have a limit to the population it can maintain but this level need not be constant. For example, changing levels of affluence may mean that demand for resources increases. Similarly, techno- logical advances, such as a switch from coal to nuclear power electricity genera- tion, for example, changes the nature of the resources required. This not only influences the demand for coal, but also affects the nature of pollution in an area and so indirectly influences carrying capacity. Technological changes can also alter the efficiency of resource use, thereby increasing the carrying capacity of an area. The interplay between the three factors means that carrying capacity is a rather fluid property. It is also important to bear in mind that resources and the effects of resource use are not necessarily confined by political boundaries: resources and pollution can be imported and exported.

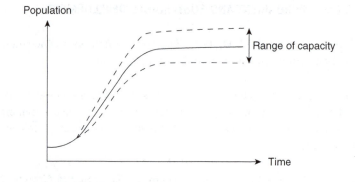

Figure 22.1   Carrying capacity and range of carrying capacity

Ecological footprints and carbon footprints are outcomes of the view that the physical environment is a separate entity under stress. Human activities create a quantifiable impact on a fragile physical environment. These activities can be graded in terms of their impact and so appropriate management strategies employed to ensure that the overall impact of the individual, organization or society is reduced. These concepts are useful tools for extending the understanding of the sensitivity and absorptive capacity of the physical environment and in making abstract concepts comprehensible to individuals and organizations. Likewise, and as important, is the translation of this quantity into a management device for auditing the success or failure of schemes designed to reduce such impacts.

Concepts in sustainable development have also built upon the ideas of environmental economics such as suggested by Pearce et al. (1989) in *Blueprint for a Green Economy* and Pearce (1995) in *Blueprint 4*. Environmental economics uses the concept of maximizing income while maintaining the stock, assets or capital. Environmental economics divides this capital into two types: human-made and natural capital. Natural capital is created by biogeophysical processes and represents the ability of the physical environment to meet human needs. These needs can include the provision of biophysical 'services' that maintain a fit environment for human habitation such as biophysical processes that maintain the capacity of wetlands to absorb pollution. Within environmental economics natural capital has to be converted into a monetary value to allow it to be compared, traded and substituted with other types of capital. Putting a value on natural capital suggests the potential for losing natural capital in some activities while gaining it in others. Maintaining the overall capital of the natural environment implies that human-made capital could replace natural capital in some circumstance where human-made capital provides the same functions (Beckerman, 1995). Although human-made capital may not be able to replace the functions of natural capital in some circumstance, the application of the concept implies that some destruction of the natural environment may be acceptable in economic terms (Barbier et al., 1990; Daly, 1994).

Figure 22.2 illustrates the relationship between natural and human-made capital. Although total capital remains at the same level, the proportion of natural

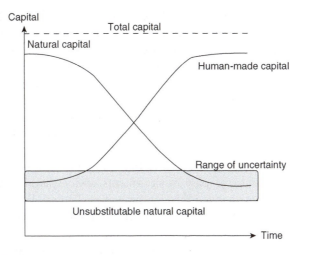

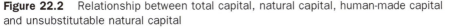

Figure 22.2 Relationship between total capital, natural capital, human-made capital and unsubstitutable natural capital

and human-made capital can alter. There is, however, a lower limit below which natural capital cannot fall. This represents unsubstitutable natural capital, both that part of natural capital for which there is no equivalent human-made capital and a proportion that could be substituted but which if removed would reduce natural capital below a level at which it could reproduce itself. The nature of each curve, its steepness and exact shape, will vary with location and with the factors identified by Goudie (2000) noted above. The range of uncertainty concerning the level and nature of the unsubstitutable portion of natural capital means that a key role for physical geography lies in identifying and understanding the nature of this uncertainty.

Similarly, ecological economics focuses on the links between human and ecological systems and their feedbacks, and tries to merge socio-economic and environmental systems into a holistic entity. It aims to provide practical policies for sustainable development (Berkes and Folke et al., 1994; Berkes and Folke, 1994, 1998). Within this framework, policy is designed to maintain development activities within the 'ecological Plimsoll line', a condition that represents the range of scientific uncertainty about environmental limits, including uncertainty associated with human impacts. According to these theories, development is acceptable for as long as the demands of human activity on the physical environment fall within this zone of uncertainty.

MONITORING, MODELLING AND MANAGEMENT

Pitman (2005) suggests that physical geographers can usefully aid Earth System Science (see below) by improving the scientific understanding of biophysical processes and so provide climatic modellers, for example, with information on the spatial variation of key parameters. Physical geography, however, is much more than a 'little helper' for climate modelling. The discipline has sought to understand the biophysical components of the earth, especially those systems in the near-surface realm (biosphere, geosphere, atmosphere, hydrosphere, cryosphere, etc.), and their process linkages across a large range of spatial and temporal scales.

In turn, physical geographers have made substantial contributions to sustainable development by providing understanding about the nature of the physical environment and how it changes. They have helped to identify the current state of the earth's biophysical system, but also how it changes in space and time and the nature of the limits to that change. Contributions therefore include:

- establishing baselines from which change can be measured
- revealing the nature of past environments and explaining how and why environments have changed through time at particular places
- identifying and explaining differences in temporal and spatial sensitivity to change
- exploring the impact of natural and anthropogenic changes on biophysical systems, often using modelling approaches.

Identifying and mapping the current state of the physical environment is a vital initial step in establishing how close to 'limits' the current environment is. Once a baseline has been identified and represented in an appropriate manner, then changes from the baseline can be monitored. Monitoring change is the second major contribution of physical geography. Physical geographers are trained to use techniques that are useful at very small scales and at very large scales, from remote sensing to *in situ* monitoring using nanoprobes.

The broad range of scales over which environmental processes operate is matched by the ability of physical geographers to integrate techniques from a range of scales to understand a problem. Importantly, physical geographers tend to study the physical environment in all of its complexity, taking into account both general, process laws and the role of factors that are place- and time-specific. That is, geographers are trained to consider both the immanent and configurational elements of scientific explanation, respectively (see Simpson, 1963). Fieldwork and field-based measurements are central to physical geography. This means that physical geographers are well aware of the need to translate often vague and difficult ideas into something that can be identified and measured in the real world (Lane, 2001; Inkpen, 2005). Such an ability is an important skill when dealing with the complexity of the physical environment.

Monitoring of the physical environment needs to be selective both to conserve resources and to enable effective data analysis. Physical geographers understand the spatial and temporal variability and sensitivity of the physical environment. This means that they are able to identify and study areas where the responses to the impact of human activities are likely to be greatest. Pepin and Seidel (2005), for example, found that the ratio of modelled air temperature to recorded surface temperature at high-altitude sites is highly sensitive to the local topography. Similarly, Pepin and Duane (2007) found that the disparity was increased by incised topography. Understanding this local relationship and its geomorphological controls is important for modelling the potential impact of climate change in high-altitude areas.

Physical geographers have also provided a context within which contemporary changes in the physical environment can be assessed. Environmental reconstruction can identify environmental change in the past and the potential triggers for it. Rates of change can be correlated with changes in external stimuli or changes inherent within a system. Rates of change can be used as a guide for the impact of current trends in key parameters, although the past is not necessarily an accurate guide to the future state of a physical system, as Lioubimetseva (2004) notes in relation to arid and semi-arid areas. Goudie (1995) notes that measuring rates of change in the physical environment is difficult because studies are often restricted by the spatial area they can monitor and by the timespans available for monitoring. This often means that areas of different size and characteristics are compared when data are amalgamated. Similarly, he notes that the extrapolation of short-term rates to predict long-term change is fraught with problems, particularly problems concerning relating rates across different spatial scales and comparing change that is episodic, and therefore likely to be rare in monitoring records, with change that is continuous.

Physical geographers have developed concepts that reflect the spatial and temporal complexity of the response of the physical environment. The inherent tendency for physical systems to undergo rapid adjustment as forcing conditions change or thresholds are crossed can be an important system property to identify as this type of change can be mistaken for the result of external stimuli such as climate change. Landscape sensitivity (Brunsden and Thornes, 1979; Brunsden, 2001), for example, aids the researcher in identifying system properties that will resist the impact of external stimuli, such as system structure, coupling efficiency and resilience. Likewise, these concepts help the researcher to understand the complexity of response of the landscape through, for example, divergent pathways of development, connectivity and decoupling.

Slaymaker (2007) illustrates how concepts derived from physical geography can be applied to issues of sustainable development in the context of the MANRECUR project. This project looks at sustainable development in a small Andean watershed in Ecuador, the El Angel River basin. He notes that priority tends to be placed on short-term objectives of sustainability while the geomorphologist might legitimately ask questions concerning sustainability over longer timescales more appropriate for the operation of geomorphological processes.

The role of physical geography in the implementation of sustainable development can be further illustrated using the case study of Hillman and Brierley (2005) and their review of catchment-scale stream rehabilitation. They identify two paradigms to river management: the engineering paradigm and the repair paradigm (see also Downs and Gregory, 2004). The engineering paradigm focuses on a reductionist view of scientific management that understands the river from the principles of fluid dynamics and hydraulics. Solutions within this paradigm are evaluated against the fulfilment of a single objective, usually related to flow management and flood control. Channelization and other 'hard-engineering' methods are favoured responses because these increase the degree of control over hydraulic variables. In this approach to management, the river system's natural variability and complexity is viewed as a major impediment to its management. Within this paradigm expertise and knowledge lies with technical experts, including engineers; local understanding may be considered but there is a tendency to focus on fixing short-term problems using 'cook-book' solutions. If one set of engineering solutions fails, another set is tried, creating a spiral of increasingly technologically 'advanced' solutions. This approach to river management was pervasive in many countries throughout much of the twentieth century but is increasingly being replaced by the repair paradigm. This takes a more holistic approach to river management, partly by changing the focus of study from the river to the whole catchment, thereby altering the spatial and temporal scale of phenomena that should be considered in management (Bravard et al., 1999; Brierley et al., 2002; Everard and Powell, 2002). Any management plan has multiple objectives concerned with the overall ecological 'health' of the catchment. Emphasis is on enhancing the 'natural' dynamics of the system to aid the recovery of the river rather than controlling the river to ensure its mechanical stability. Within this new paradigm, it is vital to identify and understand the pathways of sediment movement and the dynamics of this movement, as well as the role of morphological diversity in supporting system resilience and recovery.

Within the repair paradigm, the human aspect is not neglected. Local and community participation is a key element in defining the objectives of any management project. Given the negotiated nature of sustainable development this is an important consideration. Hillman and Brierley (2005) note, however, that the scale at which geomorphic and ecological processes operate within a catchment might not necessarily be matched by the scale at which local communities and institutions act. This is a general problem with developing sustainable policies for the physical environment. The mismatch between the time and space scales relevant to the operation of biophysical systems and human institutions can produce lags and compromises in policy implementation that negate the original objectives of inclusiveness. Importantly, however, the setting of management objectives requires an understanding of the local and historical context to provide local communities with a feeling that the management goals are of relevance to them. This reduces the overarching acceptance of expert scientific knowledge as the only or final arbitrator of what is 'correct' management. Such a situation is full of difficulties and usually requires a lengthy process of negotiation and compromise to develop an acceptable framework for sustainable development for any specific project.

SUSTAINABILITY SCIENCE AND EARTH SYSTEM SCIENCE

Physical geographers have, as noted above, been concerned with the complexities of the physical environment in space and time, focusing on the local, on the context and thereby drawing out the complexity of place. Recently, there has been a movement towards developing a 'new' science based on the view that addressing problems of intergenerational equity requires a global perspective. Kates et al. (2001) highlight that this 'new' science, Sustainability Science, should allow an understanding of nature and society by focusing on the interaction between global sociological and ecological processes that are characteristic of particular places and sectors. Komiyama and Takeuchi (2006) suggest that this trend towards a global science is partly based in a perceived need for a science of sustainability that recognized the link between science and economy yet was free from the political basis that seemed to dog debates of sustainability at the global scale (Kates et al., 2001; ICSU, 2002; Clark and Dickson, 2003). While it is debatable if politically neutral studies are attainable (Demeritt, 1996, 2001; Cohen et al., 1998; Schneider, 2001), recognition that the global scale is an appropriate level of analysis is significant. Sustainability Science aims to pursue a holistic, transdisciplinary approach to identifying problems at the global scale in these systems and appropriate solutions for sustainability at this scale (Komiyama and Takeuchi, 2006; Swart et al., 2004).

Komiyama and Takeuchi (2006) identified two obstacles to developing such a science: the complexity and interconnectivity of the problem and the fragmentation or specialization of research, an issue echoed by Clifford (see Chapter 20). They suggest that overcoming these twin dilemmas requires knowledge structuring. Knowledge structuring involves the clarification of relationships between problems and then identifying or mapping not only the web of these relations but also the organization and mobilization of the various specialist fields to address the problems. This all-encompassing overview is to be achieved by developing a framework that produces objective and quantifiable criteria and indicators of sustainability that are capable of integration. At the same time, it is expected that the criteria and indicators developed will also be sensitive and flexible enough to recognize the cultural distinctiveness of communities and the need for differentiated solutions.

Clifford and Richards (2005) identify a similar trend towards Earth System Science (ESS) which, citing Pitman (2005: 138), 'is the study of the earth as a single, integrated physical and social system', which will provide solutions to major world problems by eschewing reductionist approaches. Clifford and Richards suggest that despite the avowedly holistic intentions, ESS has substantive reductionist tendencies. They also suggest that the all-inclusive nature of ESS means that there are no boundaries to its scope. The whole globe and presumably all of time is its subject-matter which means it is unclear how the spatial and temporal scales that are relevant to answering any scientific questions can be adequately defined. Clifford and Richards's apprehension concerning the dangers of trying to encompass the plurality and complexity of reality under one scientific enterprise can be equally applied to Sustainability Science. Both are prospects that run counter to the spatially and temporally complex image of the physical environment that physical geographers have been instrumental in identifying and explaining.

CONCLUSION

Definitions of sustainable development reflect the negotiated and evolving nature of the term and the actors involved in that ongoing debate. The physical environment has always played a central role in this debate. Initially, attitudes towards the environment as a hostile entity encouraged an exploitative approach to its use. Slowly, though academic publications and general awareness, the physical environment came to be seen increasingly as a fragile entity in need of care and protection from humanity. This attitude encouraged a managerial view of the physical environment which emphasized limits, the processes that control those limits and the human impact on them. This has influenced what is studied and the concepts used to understand and manage the physical environment.

Physical geography has, and is, making a valuable contribution to the sustainable development debate. Establishing reliable baselines that characterize the current state of the physical environment is an important starting point both for developing sustainable projects and for assessing their impacts. Similarly, monitoring how the physical environment changes and contextualizing this change historically are important aspects of physical geography. When physical geographers have focused on timescales of relevance to managers, their understanding of the local, complex and context-dependent nature of the physical environment have provided an antidote to the more homogenizing tendencies of global modelling. Applying concepts derived from physical geography, such as landscape sensitivity, have helped to instigate more holistic and place-sensitive, as well as sustainable, management practices. This appreciation of the importance of spatial and temporal complexity is increasingly threatened by the modelling- and technocratic-based research of Sustainability Science and Earth System Science.

SUMMARY

- Sustainable development is an evolving concept that is based on a set of negotiations between different groups.

- The physical environment tends to be viewed as a fragile entity requiring management or stewardship to protect it from human activities.

- Physical geography has contributed to this concept by establishing baselines from which to assess change, monitoring change in the physical environment and contextualizing it, by recognizing the appropriate nature and scale of system elements.

- Physical geographers understand that the physical environment and its responses are spatially and temporally complex and historically and contextually bound. This perspective is diluted in global modelling approaches such as Sustainability Science and Earth System Science.

Further Reading

The general literature on sustainable development is large and growing. Good starting points are Adams's (2001) *Green Development,* Elloitt's (2006) *Sustainable Development* and Sayer and Campbell's (2004) *The Science of Sustainable Development*. For sources more explicitly concerned with physical geography, there are a number of project websites that provide a good indication of the role of physical geographers in applying and developing this concept. For river management, useful sites are: Grand River Conservation Authority (www.grandriver.ca/), Plan Loire Grand Nature (www.rrivernet.org/loire/), Mersey Basin Campaign (www.watersnorthwest.org), Murray–Darling Basin Commission (www.mdbc.gov.au/), the River Styles Framework in Australia (http://www.riverstyles.com/outline.php) and the Mekong River Commission (www.mrcmekong.org). For semi-arid environments, useful sites with examples or projects are: MEDALUS, an EU-funded project studying Mediterranean Desertification and Land Use, 1991–99 (www.medalus.demon.co.uk), RECONDES, an EU-funded project studying southeast Spain (www.port.ac.uk/research/recondes/), and Drylands Research, looking at research in Africa (http://www.drylandsresearch.org.uk/), the Department of the Environment, Water, Heritage and the Arts, Australia (www.environment.gov.au/), the United States Department of Agriculture, Agricultural Research Services (www.ars.usda.gov/research/) as well as similar agencies and departments in other countries.

Note: Full details of the above can be found in the references list below.

References

Adams, W.M. (2001) *Green Development: Environment and Sustainability in the Third World*. London: Routledge.

Barbier, E., Markandya, A. and Pearce, D. (1990) 'Environmental sustainability and cost-benefit analysis', *Environment and Planning A*, 22: 1259–66.

Beckerman, W. (1995) 'How would you like your "sustainability" sir? Weak or strong? A reply to my critics', *Environmental Values*, 4; 169–79.

Berkes, F. and Folke, C. (1994) 'Investing in natural capital for the sustainable use of natural capital', in A. Jansson, M. Hammer, C. Folke and R. Costanza (eds) *Investing in Natural Capital: The Ecological Economics Approach to Sustainability*. Washington, DC: Island Press, pp. 128–49.

Berkes, F. and Folke, C. (eds) (1998) *Linking Social and Ecological Systems: Management Practices and Social Mechanisms for Building Resistance*. Cambridge: Cambridge University Press.

Bravard, J.-P., Landon, N., Peiry, J.L. and Piegay, H. (1999) 'Principles of engineering geomorphology for managing channel erosion and bedload transport: examples from French rivers', *Geomorphology*, 31: 219–311.

Brierley, G.J., Fryirs, K., Outhet, D. and Massey, C. (2002) 'Application of the river styles framework as a basis for river management in New South Wales, Australia', *Applied Geography*, 22: 91–122.

Brunsden, D. (2001) 'A critical assessment of the sensitivity concept in geomorphology', *Catena*, 42: 99–123.

Brunsden, D. and Thornes, J.B. (1979) 'Landscape sensitivity and change', *Transactions, Institute of British Geographers*, 4: 463–84.

Clark, W.C. and Dickson, N.M. (2003) 'Sustainability science: the emerging research program', *Proceedings National Academy of Science, USA*, 100: 8059–61.

Clifford, N. and Richards, K. (2005) 'Earth System Science: an oxymoron', *Earth Surface Processes and Landforms*, 30: 379–83.

Cohen, S., Demeritt, D., Robinson, J. and Rothman, D. (1998) 'Climate change and sustainable development: towards a dialogue', *Global Environmental Change*, 8: 341–71.

Daly, H.E. (1994) 'Toward some operational principles of sustainable development', *Ecological Economics*, 2: 1–6.

Demeritt, D. (1996) 'Social theory and the reconstruction of science and geography', *Transactions, Institute of British Geographers*, 21: 484–503.

Demeritt, D. (2001) 'The construction of global warming and the politics of science', *Annals of the Association of American Geographers*, 91: 307–37.

Downs, P.W. and Gregory, K.J. (2004) *River Channel Management: Towards Sustainable Catchment Hydrosystems*. London: Edward Arnold.

Elloitt, J. (2006) *Sustainable Development*. London: Routledge.

Evelyn, J. (1661) *Fumifugium; or the Inconvenience of the Air and Smoke of London Dissipated; Together with Some Remedies Humbly Proposed*. Printed by W. Godbid for Gabriel Bedel and Thomas Collins. London.

Everad, M. and Powell, A. (2002) 'Rivers as living systems', *Aquatic Conservation: Marine and Freshwater Ecosystems*, 12: 329–37.

Goudie, A. (1995) *The Changing Earth: Rates of Geomorphological Processes*. Oxford: Blackwell.

Goudie, A. (2000) *The Human Impact on the Natural Environment* (5th edn). Oxford: Blackwell.

Hillman, M. and Brierley, G. (2005) 'A critical review of catchment-scale stream rehabilitation programmes', *Progress in Physical Geography*, 29: 50–70.

ICSU (International Council for Science) (2002) 'Science and technology for sustainable development', *World Summit on Sustainable Development Report*, 19.

Inkpen, R.J. (2005) *Science, Philosophy and Physical Geography*. London: Routledge.

Johannesburg Declaration on Sustainable Development (2002) www.housing.gov.za/content/legislation_policies/johannesburg.htm. (accessed 12/02/08).

Kates, R.W., Clark, W.C., Corell, R., Hall, J.M., Jaeger, C.C., Lowe, I., McCarthy, J.J., Schellnhuber, H.J., Bolin, B., Dickson, N.M., Faucheux, S., Gallopin, G.C., Grubler, A., Huntley, B., Jager, J., Jodha, N.S., Kasperson, R.E., Mabogunje, A., Matson, P., Mooney, H., Moore, B., O'Riordan, T. and Svedin, U. (2001) 'Environment and development: sustainability science', *Science*, 292: 641–2.

Kates, R.W., Parris, T.M. and Leiserowitz, A.A. (2005) 'What is sustainable development? Goals, indicators, values and practice', *Environment: Science and Policy for Sustainable Development*, 47: 8–21.

Komiyama, H. and Takeuchi, K. (2006) 'Sustainability science: building a new discipline', *Sustainability Science*, 1: 1–6.

Lane, S.N. (2001) 'Constructive comments on D. Massey "Space-time", "science" and the relationship between physical geography and human geography', *Transactions, Institute of British Geographers*, 26: 243–56.

Lioubimetseva, E. (2004) 'Climate change in arid environments: revisiting the past to understand the future', *Progress in Physical Geography*, 28: 502–30.

Marsh, G.P. (1864) *Man and Nature: Or, the Physical Geography as Modified by Human Action*. New York: Scribner.

Meadows, D.H., Meadows, D.L., Randers, J. and Behrens III, W.W. (1972) *The Limits to Growth*. New York: Universe Books.

Parson, E.A. and Hass, P.M. (1992) 'A summary of the major documents signed at the Earth Summit and the Global Forum', *Environment*, October: 12–18.

Pearce, D. (1995) *Blueprint 4: Capturing Global Environmental Value*. London: Earthscan.

Pearce, D., Markandya, A. and Barbier, E. (1989) *Blueprint for a Green Economy*. London: Earthscan.

Pepin, N.C. and Daune, W. (2007) 'A comparison of surface and free-air temperature variability and trends at radiosonde sites and nearby high elevation surface stations', *International Journal of Climatology*, 27: 1519–29.

Pepin, N.C. and Seidel, D.J. (2005) 'A global comparison of surface and free-air temperatures at high elevations', *Journal of Geophysical Research*, 110, D03104, doi:10.1029/2004JD005047.

Pitman, A.J. (2005) 'On the role of geography in Earth System Science', *Geoforum*, 36: 137–48.

Runte, A. (1997) *National Parks: The American Experience.* (3rd edn). Lincoln, NB: University of Nebraska Press.

Sargent, S. (1971) *John Muir in Yosemite.* Yosemite, CA: Flying Spur Press.

Sayer, J. and Campbell, B. (2004) *The Science of Sustainable Development.* Cambridge: Cambridge University Press.

Schneider, S.H. (2001) 'A constructive deconstruction of the deconstructionists', *Annals of the Association of American Geographers*, 91: 338–44.

Simpson, G.C. (1963) 'Historical science', in C.G. Albritton (ed.) *The Fabric of Geology.* London: Addison Wesley, pp. 24–48.

Slaymaker, O. (2007) 'The potential contribution of geomorphology to tropical mountain development: the case of the MANRECUR project', *Geomorphology*, 87: 90–100.

Swart, R.J., Ruskin, P. and Robinson, J. (2004) 'The problem of the future: sustainability science and scenario analysis', *Global Environmental Change*, 14: 137–46.

Thomas, W.L. (ed.) (1956) *Man's Role in Changing the Face of the Earth.* Chicago, IL: University of Chicago Press.

United Nations Conference on Environment and Development (UNCED) (1992) www.un.org/geninfo/bp/enviro.html (accessed 12/02/08).

Wolfe, L.M. (1945) *Son of the Wilderness.* Madison, WI: University of Wisconsin Press.

World Commission on Environment and Development (WCED) (1987) *Our Common Future.* Oxford: Oxford University Press.

23

Risk: Mastering Time and Space

Shaun French

Definition

Risk can be understood as an endeavour to overcome uncertainty, an endemic feature of human existence. Risk represents an endeavour to predict, assess and plan for social, economic or environmental events that are perceived to be dangerous, hazardous or adverse in some sense, either in that they threaten life and limb or that they are likely to incur a financial or some other kind of significant loss. While under conditions of uncertainty the likelihood and outcome of such events is by its very nature unknown, under conditions of risk the probability and consequences of adverse events can be measured and predicted. In turn, the transformation of uncertainty into a quantifiable and objective risk rests on the ability to, and practice of, assigning statistical probabilities to the chances of an event taking place and the range of possible outcomes. Critically, the translation of uncertainty into risk can itself be understood as an endeavour to attempt to tame and master time and space.

INTRODUCTION: GEOGRAPHY, A RISKY SUBJECT

Risk comes in many forms. In addition to economic risks – such as those encountered in the insurance business – there are physical risks and social risks, and innumerable subdivisions of these categories: political risks, sexual risks, medical risks, career risks, artistic risks, military risks, motoring risks, legal risks. ... The list is as long as there are adjectives to apply to behaviour in the face of uncertainty. (Adams, 1995: 21)

When you first picked up this book you might have been forgiven for being a little surprised to find chapters on risk, for it would probably be fair to say that geography does not immediately come to mind when we think of risk. What I mean by this is not that geography is somehow an intrinsically safe and risk-averse subject – as I shall discuss in a moment, our attachment to that most quintessential of geographical activities, the field trip, has in fact meant that geography has increasingly had to grapple with concerns over health and safety – but that while concepts such as space and place represent, to paraphrase Nigel Thrift in an earlier chapter, the 'fundamental stuff of geography', you might well be forgiven for asking where is the geography in risk? Until quite recently risk has, as Adam and Van Loon (2000: 12) stress, been widely considered the 'preserve of insurance experts and a range of specialists in the diverse field of risk assessment' – technical specialists in the emergency services, the fields of health and safety, disaster planning and in financial risk management, for example, but rarely geography. However, geographers have become increasingly interested in, and central to, the study of risk for at least four reasons.

First, is the very fact that it seems that everywhere we look these days we find risk. Or to put it another way, an apparent proliferation and growing obsession with risk – its assessment, management, avoidance, and increasingly its celebration and active embrace – has become one of the defining features of contemporary society. In the UK, a day does not seem to go by without the government, the media, public pressure groups or scientists of one sort or another warning variously of the dangers of bird flu, binge drinking, global warming, mobile phones, nuclear power, smoking, terrorism, flooding, knife crime, globalization, obesity, internet chatrooms, hospital superbugs, GM crops, mobile phone masts, anorexia, gun crime, I could go on and on! Geography and geographers have certainly not been immune in this regard. As I've just suggested, the practice of doing geography, both in terms of teaching and research, has itself become subject to the burgeoning risk industry and its attendant bureaucracy. Most immediately, the demands of health and safety regulations and the attendant burden of responsibility for academics and teachers in universities and schools, to say nothing of the paperwork, has had the very unfortunate effect of making geographers much more cautious when considering arranging field trips (see Cook et al., 2006). Indeed, even as I write, a major report on geography by Ofsted (2008) has specifically cited health and safety concerns and a decline in geography field trip provision as one of the reasons for a downturn in students opting to take the subject in schools. Not only is there a good chance that you may yourself have had fewer opportunities to study geographical processes in the 'real world', but at the very least such regulations would have meant that you will have been obliged on more than one occasion to have completed a detailed risk assessment form before being allowed to venture out of the classroom or the lecture theatre. While such forms might appear to be relatively benign, the annoyance of which is offset by the obvious advantages of minimizing the risk of an accident, the legal obligation to declare any medical condition, such as a simple allergy or asthma, might have in turn subtly altered the ways in which you think about yourself (as healthy or unhealthy, able-bodied or 'disabled', independent or dependent, normal or 'abnormal'), the spaces and places in which you might feel safe and those in which you might feel more fearful, and the ways in which others might also think about and relate to you.

Second, as well as an unintended victim of what the German sociologist Ulrich Beck (1992) has termed the 'risk society', geography and geographers have themselves increasingly come to be interested in directly studying, calibrating, mapping, assessing, interpreting and critically analysing risk in its many forms. As Tobin and Montz illustrate elsewhere in this volume (Chapter 24), a good example of this is the way in which physical geographers have come to play an increasingly important role in helping to understand and plan for natural hazards of a wide variety of kinds. While the nature of engagement, as Adams (1995) above makes clear, is as diverse as there are types of uncertainty, human geographers have in turn been particularly concerned with understanding the ways in which risks, both natural hazards and more social risks such as the danger of getting your car stolen or the likelihood of developing diabetes as a result of your Body Mass Index, are socially constructed. In contrast to technical accounts that understand risk as some 'thing', some danger residing 'out there' in the world – often quite literally a danger lurking in the shadows waiting to be identified, assessed and prevented – the social constructivist approach understands risk quite differently. Rather than something that exists in and of itself, risk is considered to be actively constructed by risk experts themselves, experts who are in turn embedded in particular social, cultural and economic networks and who construct risk in the context of different moral geographies.

But, why, you might ask, has life suddenly become so apparently fraught with danger? Interest in the ways in which hazards are actively constructed and have increasingly become politicized reflects a third reason why human geographers have turned their attention to risk. Put simply, the popular fixation with risk has forced social scientists, human geographers included, to start to take risk more seriously. Some of the most influential current endeavours to understand the nature of contemporary Western societies (and these debates have been almost exclusively concerned with the West) or what has variously been labelled advanced modernity, advanced capitalism, post-Fordist, or post-industrial society have placed the concept of risk, in one form or another, squarely at the heart of their analysis. Three theories, in particular, are important in this regard. First, as previously suggested, Ulrich Beck's (1992) Risk Society thesis. Second, work on the rise in significance of the ideology and practices of neoliberalism. Third, deepening processes of what has come to be termed financialization.

The fourth and most important reason why risk has become so increasingly important to human geographers is, to put it simply, a growing recognition that in spite of the traditional monopoly of risk by all sorts of specialists and technical experts, risk is in fact, and always has been, a geographical concept. In tandem with growing interest in the theories outlined above and that of the study of different types of risk themselves, it has become increasingly apparent that not only can risk only ever be understood in the context of its particular spatialities and temporalities, but more fundamentally risk represents an endeavour to make time and space knowable and controllable. In this sense, the long history of the concept and theory of risk (Bernstein, 1996), and of the practice of risk management, can be understood as part of a much wider human struggle to master time-space.

In what follows I hope also to be able to persuade you of the need for geographers to take risk seriously. In seeking to do so, the rest of the chapter will proceed accordingly. In section two I will begin by first defining exactly what risk is, and in so doing I will trace out some of risk's spatialities and temporalities. I will then go on to look in more detail at the theory of the risk society and how risk has become one of the principal means by which we understand not only our contemporary place in the world, but the world itself. Furthermore, I will illustrate how the proliferation of risk, its increasing individualization and the difficulties experienced in mapping the time-space of many modern risks has served to accentuate our fears and anxieties. The reorientation of societies in the West around the concept of risk must also be understood in the context of the cultural, economic and political transformation, since at least the 1980s, of risk itself. In the fifth and final section I will therefore briefly discuss how the ideology of neoliberalism, with its accent upon entrepreneurialism and markets, has, in tandem with a rise in the significance of financial speculation, helped to transform popular understandings of risk, such that risk has increasingly come to be understood less as threat and danger and more as opportunity and virtue.

PLACING RISK

> The revolutionary idea that defines the boundary between modern times and the past is the mastery of risk: the notion that the future is more than a whim of the gods and that men and women are not passive before nature. (Bernstein, 1996: 1)

Much of what we do in life is clouded in uncertainty. Most fundamentally, and at the risk of being rather maudlin, none of us can be entirely sure how long we are going to live. At a more mundane level we cannot be entirely sure whether the car we drive to work or college will start in the morning, or whether the bus will turn up on time. The future is by its very nature largely unknown. Despite this, endeavours have long been made to try to master the future, religion being a good example, but it is risk that has emerged as one of the most powerful of such social technologies. The roots of the modern science of risk can be traced back to a key moment in the late seventeenth century, the development of the mathematics of probability theory. The birth of probability theory represented a critical turning-point in the history of risk for it made possible the accurate calculation of the likely occurrence of future events, based on data of past experience, and thus of their associated risks. One of the earliest and most important applications of probability theory was in the field of insurance, an industry that has emerged specifically to deal with risk by enabling policy-holders to plan for and insure against the possibility of a range of undesirable, future events (French, 2000, 2002). In discussing the emergence and subsequent evolution of life assurance, Knights and Verdubakis (1993: 731) have stressed that the importance of insurance lies in its ability, through the application of techniques such as probability theory, to 'make possible the conversion of subjective uncertainty into objective risk'. Furthermore, Knights and Verdubakis (1993: 734) go on to argue that 'to calculate risk is to overcome the uncertainty of the future and thus gain some control over it'.

As Knights and Verdubakis make clear, the first important thing we can therefore say about risk is that it represents an endeavour to make the future known and controllable. The calculation of the probability of an adverse event occurring transforms the contingencies of time into known and predictable risks, risks that can then be rationally planned for. In the case of that most adverse of events, death, actuaries (the risk experts of the life assurance industry) are able, for example, to calculate with a high degree of certainty – using variables such as your age, sex, occupation, weight, the area in which you live, whether you smoke or not – the chances of dying prematurely. Moreover, once calculated, such knowledge, like that of any other risk, brings with it new responsibilities. For, critically, once the future is shorn of its uncertainty, then no longer can we consider ourselves hostage to fate, no longer is there an excuse for inaction. In the case of the risk of dying prematurely such knowledge 'imposes a [new] moral obligation on subjects to make provision for their families well into the (uncertain) future' (Knights and Verdubakis, 1993: 734), an obligation, for example, to hedge against such risks through the purchase of life assurance, pensions and other insurance policies (French, 2002). The practices of risk and risk calculation are therefore far from passive, but operate as powerful frameworks for future action.

However, as I have stressed, not only must risk be understood in relation to time, but critically risk must be understood as a technology of both time *and* space, for at least three reasons. First, the responsibilities and obligations that emerge in the wake of the calculation of risk are not universal, but vary considerably across space and from place to place, and are themselves the product of particular moral and social geographies. To take one recent example, the case of current controversies over genetically modified (GM) crops, marked differences exist in the rate of adoption of such crops in Europe compared to the United States, with GM crops experiencing much greater resistance in the former than the latter. However, the explanation for such disparities has much more to do with differing public *perceptions* of risks in terms of human health, biodiversity and ethics, and the power of companies such as Monsanto to shape such perceptions, than it has to do with any real difference in the actual risks themselves. As Knights and Verdubakis (1993: 730) stress, once again in the context of life assurance, 'although there is almost always some agreed state of affairs to which any risk might be said to refer (e.g. old age, premature death, disablement, unemployment), it is the way these are constituted socially as significant events that transforms them into insurable risk'.

Second, while risk experts and the risk industry invariably stress the scientific, objective and rational credentials of risk calculation, social constructivist accounts have drawn attention to the very particular spaces and places in which risk is actually produced and calculated. Some of the most important places of risk production are financial centres – places such as Wall Street, the City of London and Chicago, as well as smaller national centres such as Edinburgh, Leeds and Bristol (Figure 23.1) – and geographers and other social scientists have done much to tease and map out the particular risk cultures that have emerged in such places, and the ways in which such cultures shape the understanding of,

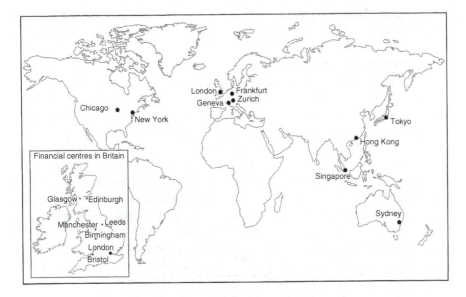

Figure 23.1   Top ten global financial centres

Source: Yeandle et al. (2008)

and calculation of, risk (see, for example, Leyshon and Thrift, 1997; Martin, 1999; French, 2002; Hall, 2008; MacKenzie, 2006).

Third, not only are there important geographies of risk production and interpretation, but risks themselves are inherently geographical. Risk calculation transforms local contingencies into objective geographies of risk and by so doing enables control to be exercised over spaces and places. To take the simple case of travel insurance, for example, when taking out a travel insurance policy the level of premium charged will depend not only on your age, whether you have any previous medical conditions or plan to take part in any dangerous sport while on holiday (the risks of which are in turn calculated using geographically defined datasets), but will also critically depend on the places you plan to travel to and calculations of how risky travel is in different countries. More generally, the importance of geography to the calculation of risk is illustrated by the way in which risks are frequently expressed geographically, and the central role maps and mapping, increasingly by way of the application of GIS technologies, play in the calculation and management of risk (see, for example, Pain et al., 2006). In the case of crime, for example, the city in which I live, Nottingham, has for some time now had to live with the unfortunate label of being the 'capital of gun crime' because of the way in which figures for crime are collected and calculated by the police authority. Another similar example is the manner in which postcode data is used by insurance companies as an important proxy for, and means by which to calculate, the risk of becoming a victim of crime. In the United States the significance of geography to the calculation of insurance risk has been such that entire neighbourhoods have in the past been 'redlined', and those who lived in such areas denied access to insurance by virtue of the fact that such areas were considered too dangerous and risky to insure (Leyshon and Thrift, 1995; Dymski,

2005). Indeed, as a student you may have experienced similar difficulties when trying to obtain home contents insurance because of the perception that student areas are also crime hot-spots.

The mapping of risk is certainly not confined to crime and is a feature of virtually all risk calculation and assessment. To give another example, central to the Department of Health's (2007) recent endeavour to produce a *Health Profile of England* and to formulate new social policies to tackle declining life expectancy has been the mapping of rates of obesity and smoking across the country (Figure 23.2). As in the case of crime, the very act of mapping health risk in turn makes visible and known places that are deemed risky. In particular, the Department of Health report singled out Boston in Lincolnshire as having the highest rate of obesity, and Easington in County Durham the highest percentage of smokers. By highlighting the uneven geography of obesity and smoking, and singling out places such as Boston and Easington, the Department of Health report imposes obligations and responsibilities on the local health authorities of such towns to act to control and reduce such risks and in so doing make Boston and Easington healthier places. Similarly, in the case of crime in Nottingham, the police and local authority have come under considerable public and political pressure to show that they are tackling the local problem of gun crime and thus making Nottingham a safer city.

RISK SOCIETY

While, as we have seen, the modern science of risk can be traced at least as far back as the seventeenth century, Beck's theory of the risk society suggests that the role of risk and its importance as a technology for knowing, controlling and (re)making time-space (for example, the obligation to make safer cities and healthier towns) in Western societies has become greatly enhanced during the latter part of the twentieth century. Writing in the context of his native Germany, Beck (2000: 218) argues that the growing power and application of science and technology has meant that 'risks become the all-embracing background for perceiving the world'. According to Beck, part of the reason why risk has become so important is because 'risks increasingly tend to escape the institutions for [their] monitoring and protection' (Beck, 1995: 5). In risk societies such as the UK, governments, corporations, experts and especially scientists have inadvertently 'become the producers and legitimators of threats they cannot control' (1995: 5). These include environmental hazards, nuclear technology and genetic research. Furthermore, as risks have proliferated so what Beck refers to as the 'dangers of industrial society begin to dominate public, political and private debates and conflicts' (1995: 5). Risk and the 'dichotomy of safe/unsafe' (1995: 11) thus increasingly becomes the central focus of politics.

As well as highlighting the particular importance of risk in the West, Beck's work points towards two other dimensions of contemporary risk societies that are of significance to geographers: the time-space distanciation of risk and individualization. First, one of the most interesting aspects of Beck's theory is the idea that while, as we have seen, risk can be understood as an endeavour to

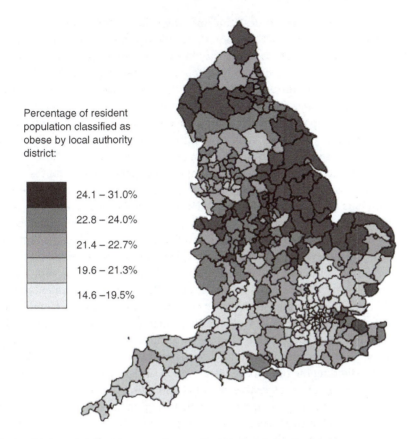

Percentage of resident
population classified as
obese by local authority
district:

24.1 – 31.0%

22.8 – 24.0%

21.4 – 22.7%

19.6 – 21.3%

14.6 –19.5%

Figure 23.2   Mapping the prevalence of adult obesity in England

Source: Department of Health (2007) (www.communityhealthprofiles.info/instant_atlas).

reduce uncertainty and increase control over time-space, the proliferation of risk
has conversely had the opposite effect. Beck explains this contradiction in the
following terms:

> The decisive point, however, is that the horizon dims as risks grow. For risks tell
> us what should not be done but not what should be done. ... Someone who
> depicts the world as risk will ultimately become incapable of action. The salient
> point here is that the expansion and heightening of the intention of control ulti-
> mately ends up producing the opposite. (Beck, 1995: 9)

Through the proliferation of risk, science in particular has created a society that is increasingly characterized by angst, fear and a sense of powerlessness,
and such feelings of angst have found expression in popular culture. Recent
Hollywood movies such as *I am Legend* (2007) and *The Day After Tomorrow* (2004)
have, for example, reflected and expressed contemporary fears about the risks of
genetic engineering and of global warming (Burgess, 2005). Furthermore, such
popular angst has also increasingly become expressed geographically in growing
anxieties over the spaces and places that are perceived to be unsafe and fearful,

especially in connection to crime (Pain and Townshend, 2002; Coleman et al., 2005; Herbert and Brown, 2006), hence the explicitly spatial character of many recent crime initiatives such as anti-social behaviour orders (Belina, 2007), and growing public debate over the safety of children's geographies (Nayak, 2003; Pain, 2006), for example. In turn, Beck argues that our sense of powerlessness to act in the face of such risks comes from the fact that it is increasingly difficult to map their impacts, especially as such risks have increasingly become globalized or at the very least glocalized, with effects frequently distanciated (distanced) from causes across space, and in time (Adam and Van Loon, 2000). Global warming and the continued controversy over attempts to connect human processes with physical phenomenon, such as the retreat of the Arctic ice cap, provides one of the most obvious examples in this regard. However, while Beck's analysis is significant, Beck fails to take proper account of the ways in which the globalization of risk has in turn, just like that of other parallel processes of globalization, also led to the forging of new connections between places of risk calculation and of new networks of experts intent on overcoming the problems of mapping risk. Again, global warming and, in particular, the work of the UN's Intergovernmental Panel on Climate Change offers a good example here, for the network of scientists and experts that make up the IPCC have been instrumental in developing our understanding of the causes and effects of global warming and thus in attempting to build a global political consensus on how to manage and minimize the risks.

Understanding the time-space distanciation of risks and of attempts at controlling and mapping their respective contours is only one of many opportunities and ways in which human geographers are able to engage in, and significantly contribute to, debates about risk society. As well as critically analysing the ways in which the politics of risk society have been played out in the context of children's geographies, human geographers have also investigated other dimensions of risk, including, for example, fears concerning the geographies of environmental risk (Jones, 2004; Burgess, 2005; Carvalho and Burgess, 2005); social risks such as obesity (Evans, 2006; Herrick, 2007), drinking (Jayne et al., 2006; Kneale and French, 2008), and smoking (Brown and Duncan, 2000; Thompson et al., 2007); gendered geographies of fear (Brownlow, 2005; Starkweather, 2007; Whitzman, 2007); financial risks and geographies of speculation (Clark, 1999; Tickell, 2000, 2003; Leyshon and Thrift, 2007; French et al., 2008); and economic risks and geographies of workfare (Peck and Theodore, 2007; Peck and Tickell, 2002). The latter, in particular, brings us on to the second dimension of Beck's work I wish to briefly discuss – individualization. Simply put, Beck argues that another part of the explanation for the fear and insecurity that marks risk societies is the progressive individualization of risk. However, while Beck makes clear, and as borne out by much of the work of the geographers cited, that the expectation is that individuals should increasingly bear the burden of risk, Beck also stresses that individuals are 'expected to master these "risky opportunities", without being able, owing to the complexity of modern society, to make the necessary decisions on a well-founded and responsible basis, that is to say, considering the possible consequences' (Beck, 1995: 8). Thus, for example, while individuals are being increasingly compelled to make their own financial provision against a whole host of risks (including unemployment, sickness and retirement), the complexities of contemporary financial products and markets, coupled with low levels of what has been termed 'financial literacy' (knowledge and understanding of financial matters) have compelled governments to develop policies and pro-grammes specifically designed to improve the financial capabilities of citizens (French et al., 2008). In turn, Beck's emphasis on a shift in the spatial scale of responsibility and risk onto the individual is also mirrored in work on processes of neoliberalism (faith in the power of free markets, individual responsibility and the minimization of government interference), especially in the context of the Anglo-American economies of North America, the UK, Australia and New Zealand. Work on neoliberalism by writers such as Peck and Tickell (2002) has, in particular, emphasized the manner in which welfare risks, most notably unemployment and underemployment, have been progressively shifted from governments and corpo-rations to individuals and markets. Moreover, neoliberalism has succeeded, through the championing of entrepreneurialism, particularly in the guise of Reaganism in the United States and Thatcherism in the UK during the 1980s, but more subtly ever since, to reconfigure popular understandings of risk. It is to such very different meanings of risk that I want now to turn by way of conclusion.

CONCLUSION: RISK TAKING

Much of what I have discussed so far has been based on an implicit understanding of risk as something to be minimized, managed and avoided. Thus the anxiety that Beck ascribes to risk societies follows from our perception as individuals, and of society at large, that time-space increasingly eludes our endeavours at mastery. It seems that the more we endeavour to make our cities safe, through the installation of CCTV, the establishment of drink-free zones and the greater use of anti-social behaviour orders, for example, the more unruly and frightening our cities appear to become. The more sophisticated medical science becomes and the wealthier we become as a society, the more new and unforeseen risks to our health emerge. And the harder science works to find solutions to one set of problems, then the more new risks emerge, risks that are even harder to quantify, map and control.

One important response to such fears has been to radically reconfigure popular understanding of risk, so that rather than threat, risk becomes trans-formed into opportunity. In the UK, for example, successive governments have endeavoured to make us all more entrepreneurial and more willing to take risks. Television programmes such as *Dragons' Den* extol the virtues of personal risk-taking, cities endeavour to throw off the shackles of cautious policy-making and transform themselves, with the help of private capital, into entrepreneurial milieux, and the vitality of UK plc is judged by our willingness to innovate and take risks. Meanwhile, speculation in the financial stock markets is no longer regarded as the preserve of financial risk experts, but has become democratized and transformed into an everyday activity and even into entertainment (Clark et al., 2004). More fundamentally, increasingly sophisticated financial products, such as derivatives, futures, sub-prime loans and mortgage-backed securities, have been progressively championed as means by which risk and time-space can once more be mastered (Tickell, 2000).

In this chapter, I have also sought to persuade you of the opportunities of risk. However, my intention in so doing has been quite different from that of the apostles of neoliberalism and financialization. Far from seeking to extol the virtues of risk-taking, I have sought to persuade you of the need for human geographers to take risk seriously and, in particular, of the importance of further critical, geographical analyses of risk's role in dominating and mastering time-space. Beck (1995) has argued that in the risk society conflicts over 'goods', such as jobs and income, become increasingly displaced by conflicts over the distribution of 'bads', or risks, and whether understood as opportunity or threat human geographers have a central role to play in uncovering and challenging the role risk plays in constituting space, and producing social and economic inequalities.

SUMMARY

- Risk and the practice of risk management can be understood as an endeavour to translate uncertainty into objective and quantifiable probability and by so doing make time and space known and controllable.

- While the study of risk has for a long time remained the preserve of experts and specialists in fields such as disaster planning and health and safety, geographers have increasingly become interested in, and central to, risk studies.

- In particular, human geographers have sought to map the spatial and temporal contours of the risk society and of attendant processes of neoliberalization and financialization.

- As well as processes of time-space distanciation, and of the individualization of risk, human geographers have contributed to an understanding of the ways in which risk has come to be reframed as opportunity to be embraced and virtue to be celebrated.

Further Reading

If you wish to learn more about the theory of the risk society a good place to begin is the edited volume by Adam et al., *The Risk Society and Beyond* (2000). Beck (2006) himself has continued to refine his theory and has recently, for example, sought to distinguish between ecological and financial dangers and risks. In terms of the latter, a large body of literature now exists in the area of geographies of money and finance, and Martin's (1999) *Money and the Space Economy* offers a good introduction. Finally, O'Malley's (2004) *Risk, Uncertainty and Government* represents one of the most sophisticated attempts to theorize and investigate risk and the politics of risk management.

Note: Full details of the above can be found in the references list below.

References

Adam, B. and Van Loon, J. (2000) 'Repositioning risk: the challenge for social theory', in B. Adam, U. Beck and J. Van Loon (eds) *The Risk Society and Beyond*. London: Sage, pp. 1–31.

Adam, B., Beck, U. and Van Loon, J. (eds) (2000) *The Risk Society and Beyond*. London: Sage.

Adams, J. (1995) *Risk*. London: Routledge.

Beck, U. (1992) *The Risk Society: Towards a New Modernity*. London: Sage.

Beck, U. (1995) 'The reinvention of politics: towards a theory of reflexive modernization', in U. Beck, A. Giddens and S. Lash (eds) *Reflexive Modernization: Politics, Tradition and Aesthetics in the Modern Social Order*. Cambridge: Polity Press.

Beck, U. (2000) 'Risk society revisited: theory, politics and research programmes', in B. Adam, U. Beck and J. Van Loon (eds) *The Risk Society and Beyond*. London: Sage, pp. 221–29.

Beck, U. (2006) 'Living in the world risk society', *Economy and Society*, 35(3): 329–45.

Belina, B. (2007) 'From disciplining to dislocation: area bans in recent urban policing in Germany', *European Urban and Regional Studies*, 14(4): 321–36.

Bernstein, P.L. (1996) *Against the Gods: The Remarkable Story of Risk*. Chichester: John Wiley and Sons.

Brown, T. and Duncan, C. (2000) 'London's burning: recovering other geographies of health', *Health and Place*, 6: 363–75.

Brownlow, A. (2005) 'A geography of men's fear', *Geoforum*, 36(5): 581–92.

Burgess, J. (2005) 'Environmental knowledges and environmentalism', in P. Cloke et al. (eds) *Introducing Human Geographies* (2nd edn). Abingdon: Hodder Arnold.

Carvalho, A. and Burgess, J. (2005) 'Cultural circuits of climate change in UK broadsheet newspapers, 1985–2003', *Risk Analysis*, 25: 1457–69.

Clark, G.L. (1999) 'The retreat of the state and the rise of pension fund capitalism', in R. Martin (ed.) *Money and the Space Economy*. Chichester: John Wiley, pp. 241–60.

Clark, G.L., Thrift, N. and Tickell, A. (2004) 'Performing finance: the industry, the media and its image', *Review of International Political Economy*, 11: 289–310.

Coleman, R., Tombs, S. and Whyte, D. (2005) 'Capital, crime control and statecraft in the entrepreneurial city', *Urban Studies*, 42(13): 2511–30.

Cook, V.A., Phillips, D. and Holden, J. (2006) 'Geography fieldwork in a "risk society"', *Area*, 38(4): 413–20.

Department of Health (2007) *Health Profile of England*. London: Department of Health.

Dymski, G. (2005) 'Financial globalization, social exclusion and financial crisis', *International Review of Applied Economics*, 19: 439–57.

Evans, B. (2006) '"Gluttony or sloth": critical geographies of bodies and morality in (anti) obesity policy', *Area*, 38(3): 259–67.

French, S. (2000) 'Re-scaling the economic geography of knowledge and information: constructing life assurance markets', *Geoforum*, 31(1): 101–19.

French, S. (2002) 'Gamekeepers and gamekeeping: assuring Bristol's place within life underwriting', *Environment and Planning A*, 34: 513–41.

French, S., Leyshon, A. and Signoretta, P. (2008) '"All gone now": the material, discursive and political erasure of bank and building society branches in Britain', *Antipode*, 40(1): 79–101.

Hall, S. (2008) 'Geographies of business education: MBA programmes, reflexive business schools and the cultural circuit of capital; *Transactions of the British Institute of British Geographers*, 33(1): 27–41.

Herbert, S. and Brown, E. (2006) 'Conceptions of space and crime in the punitive neoliberal city', *Antipode*, 38(4): 755–77.

Herrick, C. (2007) 'Risky bodies: public health, social marketing and the governance of obesity', *Geoforum*, 38(1): 90–102.

Jayne, M., Holloway, S. and Valentine, G. (2006) 'Drunk and disorderly: alcohol, urban life and public space', *Progress in Human Geography*, 30(4): 451–68.

Jones, V. (2004) 'Communicating environmental knowledges: young people and the risk society', *Social & Cultural Geography*, 5(2): 213–28.

Kneale, J. and French, S. (2008) 'Mapping alcohol: Health, policy and the geographies of problem drinking in Britain', *Drugs: education, prevention and policy*, 15(3): 233–49.

Knights, D. and Vurdubakis, T. (1993) 'Calculations of risk: towards an understanding of insurance as a moral and political technology', *Accounting, Organizations and Society*, 18(78): 729–64.

Leyshon, A. and Thrift, N. (1995) 'Geographies of financial exclusion: financial abandonment in Britain and the United States', *Transactions, Institute of British Geographers*, 20(3): 312–41.

Leyshon, A. and Thrift, N. (1997) *Money/Space*. London: Routledge.

Leyshon, A. and Thrift, N. (2007) 'The capitalization of almost everything: The future of finance and capitalism', *Theory, Culture & Society*, 24(7-8): 97–115.

MacKenzie, D. (2006) *An Engine, Not a Camera: How Financial Models Shape Markets*. Cambridge, MA: MIT Press.

Martin, R. (ed.) (1999) *Money and the Space Economy*. London: John Wiley.

Nayak, A. (2003) '"Through children's eyes": childhood, place and the fear of crime', *Geoforum*, 34(3): 303–15.

O'Malley, P. (2004) *Risk, Uncertainty and Government*. London: Glasshouse Press.

Ofsted (2008) *Geography in Schools: Changing Practice*. Manchester: Ofsted.

Pain, R. (2006) 'Paranoid parenting? Rematerializing risk and fear for children', *Social and Cultural Geography*, 7(2): 221–43.

Pain, R., MacFarlane, R. and Turner, K. (2006) '"When, where, if, and but": qualifying GIS and the effect of streetlighting on crime and fear', *Environment and Planning A*, 38(11): 2055–74.

Pain, R. and Townshend, T. (2002) 'A safer city centre for all? Senses of "community safety" in Newcastle-upon-Tyne', *Geoforum*, 33(1): 105–19.

Peck, J. and Theodore, N. (2007) 'Flexible recession: the temporary staffing industry and mediated work in the United States', *Cambridge Journal of Economics*, 31(2): 171–92.

Peck, J. and Tickell, A. (2002) 'Neoliberalizing space', *Antipode*, 34: 380–404.

Starkweather, S. (2007) 'Gender, perceptions of safety and strategic responses among Ohio university students', *Gender, Place and Culture*, 14(3): 355–70.

Thompson, L., Pearce, J. and Barnett, J.R. (2007) 'Moralising geographies: stigma, smoking islands and responsible subjects', *Area*, 39(4): 508–17.

Tickell, A. (2000) 'Dangerous derivatives? Controlling and creating risk in international finance', *Geoforum*, 31: 87–99.

Tickell, A. (2003) 'Pensions and politics', *Environment and Planning A*, 35: 1381–84.

Whitzman, C. (2007) 'Stuck at the front door: gender, fear of crime and the challenge of creating safer space', *Environment and Planning A*, 39(11): 2715–32.

Yeandle, M., Knapp, A., Mainelli, M. and Harris, I. (2008) *The Global Financial Centres Index*. London: City of London.

24

Risk: Geophysical Processes in Natural Hazards

Graham A. Tobin and Burrell E. Montz

Definition

There is no single, simple and accepted definition of risk; everyone knows what it is, but everyone perceives it in different ways, and therein lies the difficulty. Risk is frequently construed as the potential negative impact of some present or future event on human activities, which can be expressed as the product of the probability of occurrence of such an event and expected loss. With respect to natural hazards, risk has been defined as a function of hazard, vulnerability and the capacity to cope (Wisner et al., 2004), with different ideas on how these variables contribute. These various components of risk are addressed below.

INTRODUCTION

The terms 'risk', 'risk analysis' and 'risk assessment' are pertinent to many fields of study and have generated a sizeable literature (see, for example, the risk series by Flynn et al. 2001; Jaeger et al., 2001; and Linnerooth-Bayer et al. 2001). Risk, of course, is particularly relevant to an understanding of natural hazards where threats of disaster are of great concern because of the effects on human perception and behaviour (Covello and Mumpower, 1985; Mitchell, 1990; Burton et al., 1993; Kasperson et al., 1995; Hewitt, 1997; Wisner et al., 2004). Indeed, the level of disaster risk is an important determinant of whether communities or individuals take action to reduce it. Thus, the concept of risk is complex. It is imperative to understand risk not purely from a technical perspective of the physical processes, but also as it is perceived by those in hazardous areas and as it is managed (or not) through risk mitigation strategies. In order to understand risk, some appreciation is necessary of how risk is determined and what we know about its natural and technological sources.

HAZARD RISK

Frequently, risk of natural hazards is seen as the product of the probability of occurrence of a particular geophysical event and expected losses. The probability of occurrence in this instance is usually assessed through historical trends; from historical records it is possible to determine the approximate size of a 100-year flood and to estimate the probability of certain-sized floods occurring in any given year for a specific place. However, while this information is useful in evaluating technical risk, it does not incorporate the consequences of such an event. Thus, to represent risk fully, details on vulnerability and coping strategies (discussed more fully below) must be incorporated in the analysis. The risk relationship can then be expressed as follows:

Risk = f(probability of occurrence of the event, population exposed, vulnerability, coping strategies)

There is much that remains uncertain in this function because each element varies in time and space and because our knowledge is incomplete. Consequently, uncertainty remains an integral part of risk and must be used in combination with statements of risk.

As is evident from the equation, understanding risk and uncertainty requires examination of physical processes in conjunction with the human use system. This relationship is illustrated in Figure 24.1, which shows the juxtaposition of various forces of the physical environment, on the left, with processes of the human environment, on the right. Of course, in reality the two cannot be so easily separated because they inevitably affect each other. Furthermore, the relationship is complex, with causality reflecting the interaction of many different variables and their changing relationships. Figure 24.1, therefore, does not display the dynamic nature of risk.

In this chapter, we focus specifically on elements of the physical environment in Figure 24.1 – the frequency, magnitude, duration and speed of the onset of events. While these components allow for determination of 'objective' statistical probabilities of technical risk and exposure factors (termed 'statistical risk'), they do not include the human environment, which is fundamental to understanding fully the risk and uncertainty at a location. Indeed, it is not possible to formulate any meaningful response strategy to mitigate risk without examining those socio-economic and political forces that either exacerbate or attenuate hazardousness (Kasperson et al., 1988), specifically population exposure, vulnerability, social norms and coping mechanisms. It is these factors that determine behaviour and decision-making because risk is filtered through various lenses, leading to such concepts as perceived risk, acceptable risk, accepted risk, as well as that risk which is tolerated (Kasperson et al., 1988; Tobin and Montz, 1997; Wisner et al., 2004).

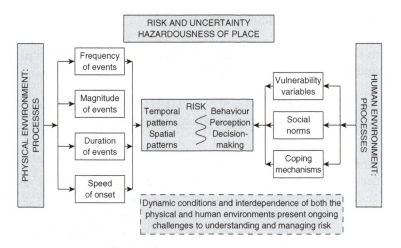

Figure 24.1 Understanding risk and uncertainty in context: the hazardousness of place

The social factors outlined in Figure 24.1 include vulnerability, social norms and coping mechanisms, which together constitute a major component of societal risk. In this context, coping includes those actions that can mitigate exposure to hazardous events, such as flood walls or dams, forecasting and warning systems, and measures like building and planning regulations that help minimize losses and protect citizens. How society deals with threats from the physical environment therefore will influence risk and risk perception. Many examples can be cited whereby societal norms or cultural influences affect risk, from the greater access to healthcare of more privileged groups during natural disasters to the myths associated with HIV/AIDS that permeate sub-Saharan Africa; all condemn people to higher levels of risk. Indeed, any marginalized group, whether through economics, politics or social norms, can be at higher risk than those in the mainstream of society. This, of course, collectively incorporates the issues of vulnerability – hazards do not affect people equally, but rather some suffer more than others irrespective of the magnitude of the event. The poor, elderly, disabled and others are at greater risk, invariably suffer more, and take longer to recover from the ravages of disasters.

In looking at Figure 24.1, therefore, there is often a lack of communication and understanding between the two components, indicated by the wavy line in the middle box. It is not unusual, for example, for physical scientists to express a need for further research to enhance our understanding of, say, meteorological or hydrological processes, or seismic forces, so that early warnings of risky conditions can be issued. However, early warnings do not necessarily translate into reduced risk and effective remedial action. For that we need research by social scientists into risk perception, human behaviour and choice. Physical processes alone do not determine disaster losses and impacts; they help us understand event characteristics. Thus, we also need a consideration of the human processes as shown in Figure 24.1. With that recognition, we explore in this chapter, some of the physical dimensions of risk as they pertain to hazardousness. Aspects of risk in human geography are discussed in Chapter 23 of this volume.

PHYSICAL GEOGRAPHY: PERSPECTIVES ON RISK

Risk from a physical geography perspective has been addressed through parameters of the natural environment, specifically:

- temporal distribution: frequency of occurrence, seasonality, diurnal patterns
- spatial distribution: geographic location, scale of analysis.

Temporal distribution

A traditional approach to estimating risk is to determine frequency and return periods through an analysis of historical records of particular events by calculating the average number of occurrences over a specific period. If there have been 30 blizzards in a 90-year record, the event is said to have a recurrence interval of once every three years on average (90/30 = 3); that is, the blizzard has a 0.33 probability (30/90 = 0.33) or a 33% chance of occurring in any given year. Similarly, 40 floods, earthquakes or tornadoes in a 100-year record would suggest a return period of once every 2.5 years on average, and a 0.4 probability of occurring in any given year. The risk, then, can be stated in simple and yet objective probabilistic terms that reflect the threat to a given place. For example, the probability of a hurricane making landfall in Tampa, Florida, USA, is approximately 0.175 for any given year based on past records, whereas for Miami it is 0.263 (Williams and Sheets, 2001).

This analysis can be further refined by accounting for the size of the potential event. Size is usually determined by the magnitude of the physical process involved: the discharge of a flood; the energy released by earthquake activity (as depicted by the Richter scale); or the wind speed in a tornado or hurricane. We know that large-magnitude events have a lower probability of occurrence than small-magnitude events; there are thousands of low-magnitude earthquakes occurring around the world every day, but only a few that measure seven points or more on the Richter scale in any year. Consequently, there is an inverse relationship between frequency, or probability of occurrence, and magnitude. Two useful sources of information for recent seismic activity are from the United States Geological Survey (USGS Latest Earthquakes, 2008) and the British Geological Survey (BGS Recent Events, 2008).

Flood risk analysis provides another example in which the approach is to calculate recurrence interval based on annual maximum discharge by ranking the data according to flood magnitude. There are various techniques for calculating such probabilities, one of the most simple is that described by Dunne and Leopold (1978). The highest flow for each year is recorded, then these flows are ranked relative to each other (with the largest flood ranked 1). The return period is calculated using the formula:

$$Tr = (n + 1) / m$$

where Tr represents the return period, m represents the rank number of the event, and n is the number of years of record. In a dataset of 125 years, a flood that is ranked ninth can be expected to recur once every 14 years (126/9 = 14), on average. The probability of a flood of this magnitude or greater magnitude occurring in any given year is the reciprocal, that is, 0.07 (9/126 = 0.07). The return period, however, does not give any indication of when an event might occur, it states only the frequency at which we can expect such events to be equalled or exceeded. For example, it is possible that this '14-year flood' could occur twice in one year and not again for 50 years.

For a long record of flow data and for events in excess of a 10-year recurrence interval, this formula provides a useful guide to risk. What it does not capture so well is the frequency of smaller events, for which an alternative dataset is more useful. This is called the partial duration series and it incorporates all floods above a specified level irrespective of the years in which they occurred. Our ninth-ranked flood might then become the thirteenth and hence have a probability of 0.103 of occurring in any given year with a 9.7-year return period. Given this different perspective, the annual risk from this size flood is not fixed but dependent on the method and the data series that is used, and of course the quality of those data. In addition to data quality, there are other limitations that add to uncertainty, including the place-specific nature of risk assessments, dependence on historical records, the temporal dynamism of hazardous events and the assumption of randomness in temporal patterns. These are discussed below.

Data Quality

Data quality, or lack thereof, presents many challenges, as illustrated by tornado records. Grazulis (1993) plotted the number of tornadoes per year against time in the USA. The resulting graph shows a significant increase in tornadoes since the 1950s; pre-1950 averages are below 200 per year compared to post-1960 averages of more than 600 per year. However, as Grazulis points out, the actual number of days per year on which tornadoes occurred, based on thunderstorm data, did not change, suggesting that the apparent increase in incidence reflects other factors, notably improved tornado spotting techniques and better data collection and recording.

Similar difficulties have been found for other hazards, particularly when changing circumstances can modify calculations of statistical risk. For instance, when looking at the risk of death from avalanches, the record shows a remarkable rise in the number of deaths in many countries. The cause has not been an increase in incidence of avalanches, but can be attributed to a growth in alpine recreation. In New Zealand, for example, no avalanche deaths were recorded between 1900 and 1909, but by 1970, 16 people had died (Prowse et al., 1981). Similar trends are found with flash floods (Gruntfest and Handmer, 2001). Different datasets, therefore, can alter the apparent risk involved with different geophysical events. It should be noted, however, that changing climatic conditions may well alter the frequency of such events (see below).

Place Specificity

Simple frequency distributions do not translate easily from place to place. For instance, the 14-year flood discussed above will not (normally) be the same magnitude as a 14-year flood in another drainage basin. The distribution applies only to the drainage basin being studied. There are ways to extrapolate to adjacent drainage basins, but this adds considerably to the uncertainty (Dunne and Leopold, 1978). Events of similar frequencies, then, are not necessarily comparable in terms of magnitude, between places, but importantly the risk, as measured by recurrence interval, is.

Historical Records

Frequency distributions are dependent upon the length of historical record and the events that have occurred during that period. Extrapolating records beyond this record increases uncertainty and magnifies the error. A short record will inevitably contain fewer data points and mean greater uncertainty than a long record. Unusually large events, for instance, can disrupt the time sequence. At some places in the Midwest of the United States during the floods of 1993, some flows were so large that they exceeded all known levels. However, because hydrological records were limited to 80 years or less on some streams, the traditional analysis suggested that the probabilities were 0.012 (an 81-year return period). In reality, extrapolation of the data indicates that some of these floods were consistent with a 200-year return period. Techniques to extend such records out to 200 years, however, increase uncertainty because of this lack of long-term data. Similarly, a simple analysis of tornado frequency provides inconsistent information on risk which depends on the length of the record that is examined. Between 1953 and 1991, the USA experienced 768 tornadoes per year compared with 865 for the years 1970–91. An even shorter timeframe could produce other results. Some years have much higher incidences of tornadoes than others, and it is not unusual to find major outbreaks occurring in other years. For instance, on 3–4 April 1974, 148 tornadoes passed through 13 states (Tierney and Baisden, 1979). Thus, frequency data must be interpreted with caution when estimating risk.

Assigning risk probabilities to such large magnitude events as the Asian tsunami of 2004 or major volcanic eruptions such as Tambora in 1815 or Krakatoa in 1883 is problematic. The lack of sufficiently long records presents difficulties in estimating risk and even long records can present problems if data are missing or unreliable. Physical scientists, therefore, often seek to extend such records to enhance understanding of risk. Dendrochronology (tree ring analysis), lake sediments (varves) and glacial ice cores can reveal evidence of hazardous events in the past (e.g. perturbations of climatic norms) and have thus added to our knowledge. Unfortunately, many smaller events may not be recorded in such ways.

Temporal Dynamism

Many meteorological and hydrological events fit into a statistical distribution of randomness over the long term. Tropical cyclones, tornadoes, windstorms, heat

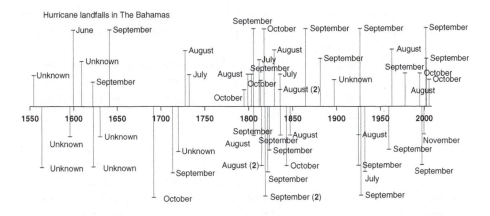

Figure 24.2   Hurricanes making landfall in The Commonwealth of the Bahamas from 1550 through 2007. This timeline depicts known landfalls and probably omits many others that are not in the historic record. Periods of greater intensity can be seen, as can periods of quiescence. Note that the distance from the horizontal line is for legibility purposes only and does not represent anything relating to the hurricanes themselves

Source: Government of The Bahamas (2005), Hughey (2008) and Neely (2006)

waves, cold waves, blizzards, and droughts all occur with a given probability for a specific place for a particular year. However, we might ask whether the same patterns of storms are occurring now as occurred last century, and whether the same probability remains. There is a tendency towards persistence in weather patterns such that particular systems come to dominate for a time. Climatological records show distinct groupings of wet and dry years and of warm and cold years. To some extent, floods and droughts are concentrated in particular periods. For example, the early 1930s and the mid-1950s were warmer than usual for large parts of the United States. Drier and warmer years were recorded throughout much of northern Europe and North America in the late 1980s, and southern Britain experienced droughts in the mid-1970s and early 1990s. This does not affect long-term variability and probability estimates, but it does mean that conditions for specific events may be higher at particular times. There have been periods when Atlantic hurricanes are more active, such as the present period, and when they are less active, as in the 1970s and 1980s (Elsner et al., 1999). The risk for people living along coastal areas in the Caribbean Basin and southeastern USA therefore changes over time, as shown in one county's experience with hurricane landfalls (Figure 24.2). As can be seen, there are periods of relatively high activity, such as between 1813 and 1824, and other periods of quiescence, such as the early part of the twentieth century. Of course, data limitations associated with the latter half of the sixteenth and the seventieth centuries present problems in estimating risk. Thus, despite some persistence in weather patterns, frequency distributions assume that patterns remain the same over the long term; yet variations at different temporal scales can influence the pattern of extreme geophysical events.

A further complication is presented with global climate change, the true consequences of which are not fully known at this time. Global warming will undoubtedly alter global atmospheric circulation and concomitantly the distribution of storms and precipitation around the world, hence raising uncertainty. For example, there is now a debate as to whether warming will lead to a higher incidence of Atlantic hurricanes or an increase in the intensity of storms. Either way, the risk will change. Climate models that predict physical changes are important for improving our understanding (Stern and Easterling, 1999), but careful consideration of how these changes will affect the risks 'on the ground' are also critical. The research in this area has been confounded to some degree by political meddling, especially in the USA, and this has influenced the communication of risk assessment (Mooney, 2005; Shulman, 2006).

Randomness and Periodicity

While it is possible to determine a simple frequency of occurrence for all hazards by dividing the number of years of record by the number of incidents, this would not necessarily give a true picture of the risk patterns for all physical processes because not all geophysical events have this element of randomness. Some geophysical events exhibit periodicity that stems directly from the build-up and release of pressure. For example, many earthquakes are associated with the build-up of strain on fault lines, and the risk of fault failure can increase accordingly; thus, the probability of occurrence may change from year to year. The probability of a major earthquake along the San Andreas fault in San Francisco has increased since the 1906 event, so scientists predict that another major event is imminent. Indeed, the probability of occurrence of a major earthquake in the San Francisco Bay area over a 30-year period was estimated to be 0.5 in 1980 but rose to 0.67 based on 1990 calculations. Some scientists have estimated the probability of the same magnitude earthquake in the San Francisco Bay area at 70% for the period 2000–30 (USGS, 1999), although new data and approaches have recently reduced this probability to 0.62, which further exemplifies the uncertainty associated with such estimates (USGS, 2003).

Explosive volcanoes probably follow a similar pattern, but risk assessment is hampered by a dearth of in-depth studies. The lack of recent major events limits our ability to monitor and hence model such activity, and we are forced to rely on studies of historical events. An historical analysis of Mount St Helens, for example, shows that it is the most active volcano in the Pacific Northwest (Crandell et al., 1979). Four major periods of activity have been identified, each lasting several hundred years before total collapse. It is interesting to note that Crandell et al. (1979) predicted activity on Mount St Helens before the 1980 eruption, including explosive eruptions, lava domes, pyroclastic flows and lahars.

Less explosive volcanoes also have periods of activity and dormancy. Mauna Loa erupted on average every 3.5 years (Hodge et al., 1979) until 1950 when an explosive eruption occurred. Since then, it has erupted sporadically (USGS, 2005). Kilauea was almost constantly active between 1823 and 1924, then it erupted 21 times between 1924 and 1965, before becoming constantly active again between 1983 and 2000 on its east rift zone (USGS 2007a). This illustrates that risk can be associated with chronic dangers; that is, events that continue for some time. For instance, Tobin and Whiteford (2004) show that there have been ongoing risks from ashfalls, pyroclastic activity and lahars for communities located around the volcano, Mount Tungurahua in Ecuador, since 1999. In all these cases, despite the high frequency of events, the risk remains because no distinct pattern can be discerned that would allow meaningful forecasts of impending eruptions. The same can be said of tsunami on a longer timescale. For example, Japan has experienced over 65 destructive tsunami since the year 684 (Pararas-Carayannis 1986); a reported 1,000 people were killed in 869, another 1,700 in 1361, 500 in 1498, and 27,122 in 1896. However, a regular pattern is not discernible. Forecasting these temporal trends may always lie beyond our capabilities, although we may be able to refine our predictive models.

The temporal pattern of these geophysical events is important because of what it tells us about potential magnitude in some instances, and about recurrence intervals and planning needs in others. It is clear that simple frequency of occurrence may not be indicative of any trends, but it is also clear that we need to improve our understanding of temporal patterns if effective forecasting and warning are to occur.

Spatial distribution

A major feature of risk is geographic location. Many extreme geophysical events can be spatially defined, so their mapping enhances our knowledge of risk. Although the occurrence of an extreme geophysical event does not mean that a similar event will necessarily occur again at the same location, the correlation between the two is often strong and this understanding can be used to map regions that are at risk. Such regions often extend beyond the boundaries of the physical event itself and help to identify the locations of populations at risk. It is possible, for instance, to clearly demarcate the potential areas of tsunami run-up in Asia, and provide information to communities with little or no recent experience of such events. By evaluating such spatial distributions, areas can be delineated by risk, with concomitant implications for forecasting, warning and mitigation.

Geographic Distributions

The source areas of natural hazards can be identified relatively easily, providing guidance as to the potential risk in specific locations. For instance, temperate latitudes experience by far the greatest number of snowstorms, cold waves and ice storms, so a simple climate map can indicate areas at risk from such events around the world. Tropical cyclones, on the other hand, are a feature of lower latitudes dependent upon the heat and moisture of warm oceans that help to generate them. Floods are generally confined to floodplains and coastal areas,

Figure 24.3 Flood insurance map for a portion of Broome County, New York. This is typical of a digital flood insurance rate map. AE refers to zones within the 100-year floodplain where base flood elevations are determined. X500 are those areas outside the 100-year floodplain, but within the 100–500 year floodplain, for which flood insurance is not required

Source: Broome County GIS portal (2008)
htp://broomegis.co.broome.ny.us/Website/GISWeb/Portal.htm.last accesed 22 May 2008

mudflows and landslides to steeper slopes, and sinkholes and subsidence to particular geologic conditions, while zones of tectonic and volcanic activity can be identified. We know where such geophysical events are prevalent. However, in looking at geographic distributions, it is important to consider spatial scale because different patterns of risk can emerge, depending on the scale of analysis.

Assuming suitable data are available, mapping hazards at a larger cartographic scale (i.e. smaller spatial areas) allows for relatively well-refined definitions of risk as well as consideration of other important information. Floodplains in the USA are mapped for the more than 20,000 communities with known flood hazards. These maps allow communities to identify locations and regulate land-use based on flood risk. They also enable insurance companies to assign premiums relative to that risk, as defined by location within the 100-year floodplain, or the area that has a 1% chance of being flooded in any given year (Figure 24.3). The 100-year floodplain has been established for these communities based on historical flood records, topographic maps and hydrologic models that allow predictions of the extent of potential future flooding. Given the uncertainty of much of the input data, many of the resulting maps are subject to error, and they certainly do not designate a division between safe and unsafe areas, although they do depict an element of risk. Other communities have further developed these maps and incorporated aerial photographs to delineate the boundaries of potential flood areas more clearly. Figure 24.3 shows a flood insurance map for part of Broome County, New York, which experienced four floods between April 2004 and November 2006.

Similar maps exist for coastal areas, portraying storm surge zones, again based on probabilities determined through models. In the United Kingdom, the Environment Agency (EA) provides an online service that allows owners, developers and buyers to examine the risk of flooding for individual properties (EA Flood Map, 2008). In other places, hazards have been mapped to provide planning tools and to guide development based on risk (USGS, 2001). In Salem, Oregon, for instance, maps were prepared to identify areas that may be susceptible to landslides, with different approval processes in force for these areas (City of Salem, 2007). The scale at which these hazard zones can be identified lends itself to such mapping and planning applications.

At a smaller scale, and thus a larger area, geographic distributions remain important, but less detail is available. For example, tornadoes show a distinct regional distribution, and a map of their incidence helps to define the risk for a region. Identifying the risk at a given location within that region is problematic because, unlike floods and landslides, tornadoes are not restricted to specific geomorphic or geological conditions. Still, knowing areas of high risk facilitates mitigation because resources can be deployed where they are needed most.

The mapping of geological processes presents difficulties associated with spatial scale. We know that earthquakes and volcanoes are most frequent along plate boundaries, but this does not help in planning for such events. The cartographic scale is too small to define the conditions prevailing at a specific place. Because of the many different processes associated with tectonic activity, other, more localized information is required to determine actual risk. Various studies have mapped risk zones for seismic events, such as in Indonesia, Germany, New Zealand and the United States, at various scales, including local and national levels (see, for example, Association of Bay Area Governments 2007; USGS 2007b; Center for Disaster Management and Risk Reduction Technology (CEDIM), 2008).

Identifying locations at risk is important to hazard mitigation and management. However, the maps need to be interpreted and used carefully. It matters what is mapped. For instance, tornado maps of incidences per 25,900 km^2 (10,000 mi^2) across the United States, produce an annual average for each area that does not reflect the magnitude of the events. The resulting pattern shows the expected high occurrences in the Great Plains, but also high incidence in Florida. This is misleading because Florida is characterized by many 'weak' tornadoes that generally present fewer problems than those in the Mid West (Grazulis, 1993).

Spatial and Temporal Interface

Showing risk areas on a map can suggest high levels of both spatial and temporal accuracy that may be misleading. Maps are based on the best available data, but data quality varies significantly from hazard to hazard and from place to place. Frequently, models rather than field measurements are used to estimate risk, as was shown with floodplain maps in the USA. Model predictions also contain different levels of uncertainty that depend on the quality of the model and the data being used to drive it.

In general, greater predictive accuracy is achieved when an understanding of spatial and temporal patterns are combined. Take, for example, tornado risk which changes spatially and temporally. In the USA, high-risk zones shift from the south to the north as storm tracks progress during the spring and summer in the northern hemisphere. As another example, analysis of a number of years of record shows seasonal variation in the spatial pattern of hurricanes in the North Atlantic. The area of formation shifts eastwards gradually as the season progresses, with early-season storms generally confined to the western Caribbean and the Gulf of Mexico. From mid-August to mid-September, the area of formation is at its easternmost boundary near the Cape Verde Islands, after which there is a westward movement of the formative area (Neumann et al., 1987). The origin and movement of hurricanes is also dependent on various factors such as (1) regional air patterns that act as steering currents, (2) internal forces of the storm (Holland, 1983), and (3) outflow jets from the storm (Pielke, 1990).

Although we can identify locations according to risk, we must be careful in interpreting what is portrayed because of the inherent inadequacies of the maps, which result from the lack of accurate data, uncertainties associated with the hazards as well as the dynamic nature of the events. Assessing risk for a given place for a given time presents many challenges but remains an important component of managing that risk.

TECHNOLOGICAL ADVANCES AND COMMUNICATION OF RISK

Technological developments have significantly altered the risk landscape, not only providing new tools to facilitate research and understanding, but also broadening the inventory of potential mitigation strategies. Geographical information systems (GIS), remote sensing and satellite imagery are now fundamental components of hazard and risk research that have created new opportunities and enhanced our scientific approaches (Tobin and Montz, 2004). At the same time, other technological advances, including those in communication systems, have greatly improved the range of detection, warning and mitigation techniques available to hazard managers.

Improved technological capacity has played a major role in understanding and communicating risk. Development of Doppler radar, for instance, has greatly improved weather monitoring and forecasting, while satellite-based remote sensing provides enhanced images of hurricanes and wildfires as they develop and spread. Studies of virtually all natural hazards have benefited from technological advances, from computer databases to remote sensing and GIS (Amdal, 2001). For example, technological advances have increased our knowledge of the potential risks of volcanism, which ultimately may permit more accurate warnings of volcanic activity over time and place. Specifically, Connor et al. (1993, 2001) develop an ash dispersion model to map the deposition of tephra from volcanic eruptions through refined remote sensing techniques, while Hepner and Finco (1995) model gaseous contaminant pathways over complex terrains using GIS. These models and others like them facilitate data collection and significantly enhance mitigation planning through a better understanding of risk.

Forecasting and warning systems have significantly reduced fatalities and injuries from many natural hazards. Warnings depend on the collection of geophysical data, the careful analysis of those data, the broad dissemination of the warning message and, ultimately, effective response and remedial action. The accuracy of such warnings depends, in turn, on the speed and reliability of the scientific models. All elements of the forecasting and warning system must function efficiently and effectively if such mitigation measures are to work. For example, for flash flood warnings both meteorological monitoring and communication systems must be in place and ready at all times. In the United States, real-time meteorological and hydrological data are available that can be combined with soil moisture data and other factors in sophisticated models, to produce information from which accurate forecasts can be made and warnings issued. Unfortunately, as pointed out by Sorenson (2000), this has not been true for all countries, especially the less wealthy nations. The problem remains that the technology required for sophisticated forecasting and warning systems is expensive to develop and maintain. Of course, success of any forecasting and warning system is contingent upon those at risk taking remedial action; they must perceive themselves at risk and have the ability to respond. It is at this point that we cross over to the human realm of risk, as mentioned above. For a more complete discussion of the communication of risk, see the special volume of *Environmental Hazards* edited by Faulkner and Ball (2007).

Specific physical criteria often form the basis for determining risk and therefore appropriate warning and mitigation strategies. In the USA, for example, the 100-year floodplain has become the standard for flood alleviation, such that the National Flood Insurance Program utilizes it with respect to planning regulations. Defining the precise location of the 100-year floodplain depends on accurate land–surface elevation data, which was originally approximated from mapped contour lines. More recently, light detection and ranging (Lidar) technologies are providing more accurate floodplain elevation data (Committee on Floodplain Mapping Technologies, 2007). In Florida, new storm surge zones are being established using Lidar data, which may significantly change the spatial extent of these hazard areas. In this way, technology is leading to improved products with the potential for updates as hydrological conditions change.

Geographical information systems has been particularly important for fostering risk analyses at particular locations. Although it has long been recognized that any given location is subject to multiple hazards (Hewitt and Burton, 1971), analytical tools to evaluate their combined effects are a relatively recent development. Several studies have used GIS to determine risk from multiple hazards, sometimes based on spatial extent of a combination of events, and in other cases based on probabilities of occurrence (Montz, 1994, 2000; Monmonier, 1997; Cutter et al., 1999, 2000; Emrich, 2000).

Along with the benefits associated with additional data availability, improved accuracy and more dynamic platforms for portraying risk and updating risk analyses, advances in technology have also led to greater access to risk information. GIS technology has been applied to interactive mapping on the internet, addressing both natural and technological hazards (Hodgson and Cutter, 2001). Users can supply geographical information, such as a post- or zip-code, and a tailored map is provided. In the United States, for example, such websites are provided by federal government agencies that include the Environmental Protection Agency (EPA Enviromapper, 2008), NASA (NASA Natural Hazards, 2008) and the USGS (USGS Hazards, 2008), as well as public/private collaborations like the one between the Federal Emergency Management Agency (FEMA) and the Environmental Systems Research Institute (ESRI Hazards, 2008). Other websites promote public awareness of risk which, in turn, can help foster an understanding of it. International websites that provide such information include those for the Centre for Research on the Epidemiology of Disasters (CRED Homepage, 2008), the Estrategia Internacional para la Reducción de Desastres (EIRD, Homepage, 2008), and the International Disaster Data Base (EM-Dat database, 2008), to cite a few.

CONCLUSIONS AND CHALLENGES

As discussed at the beginning of this chapter, determination of risk is critical to management of natural hazards. We need to have an understanding of the risks that we face from the geophysical environment if we are going to be motivated to take action. Yet understanding risk is fraught with difficulty and uncertainty because of the complex nature of the processes that combine to create that risk. We understand a great deal about the physical processes that determine the risk at a location, but there remains much that is unknown or at least not yet proven. By definition, risk is a probability with spatial and temporal variations; it is not a certainty. Thus, it is how we interpret the available information that defines how we respond to the risks we face. Physical geographers collect and manipulate data, develop and refine models of the physical processes that generate hazards, and advance ideas that facilitate risk analysis and assessment, whether contributing to prediction, forecasting, warning, mapping, or some other component of hazard and risk management.

Yet, at the end of the day, it is how this information is used for mitigation that is of critical importance, and here we should return to the importance of human factors in risk modelling, as shown in Figure 24.1. Hazard managers must ultimately decide on appropriate courses of action, and in this regard, physical criteria may not be as clearly defined as managers would wish. These decisions come with costs, so decisions such as determining the design levels of mitigation, whether the height of levees for flooding, the building materials for homes in earthquake- or tornado-prone areas, or land-use regulations in landslide-prone areas, all carry political and economic risks. Thus, the relative risks must be weighed alongside the costs and, in general, this requires abundant, accurate information. However, interpretations of that information may vary considerably, adding to the complexity of the problem. As an example, Harwell (2000) demonstrates that identical, remotely-sensed images of forest fires in Indonesia produced significantly different interpretations as to the crux of the problem. Competing analyses of the images actually served to increase uncertainty about the extent of damage to traditional food crops. As a result, a number of questions concerning the hazard remained unanswered. With differing interpretations of data, therefore, come differing responses, which may or may not reflect the true nature of the risk.

Technological advances have certainly increased the amount and accuracy of information available on risk, but there are significant constraints and limitations associated with them (McMaster et al., 1997). GIS offers many advantages, but has been criticized as generating 'great expectations' while not always delivering (Carrara et al., 1999), in large part due to misconceptions about the technology and/or knowledge of hazard theory. With regard to the former, computer-generated maps are considered to be more accurate, credible and objective than hand-drafted ones, even though the conversion of data to digital form may incorporate random or systematic errors. With respect to the latter, handling of geophysical data in a GIS is simplified for users who are not experts and thus are not knowledgeable of the theory behind the data they are mapping. Thus, technology can lend an air of infallibility and scientific validity that goes far beyond the reliability and accuracy of the data, which, in turn, can lead to misdirected mitigation decisions.

As shown in this chapter, the temporal and spatial scales at which risk analyses are undertaken go a long way to determining their outcomes, so caution in both the methods utilized and in the interpretation of results is critical. Improved understandings of physical criteria are important, but that is only part of hazard assessment and mitigation practices. Risk is only a part of hazards, but we must understand risk in order to grasp the complexities of hazards. As illustrated in Figure 24.1, this must be combined with the human dimension in any comprehensive risk assessment.

SUMMARY

- Risk can be explained through greater understanding of the probability of occurrence of particular events, the populations exposed, their various vulnerabilities and the availability and effectiveness of coping strategies to mitigate negative impacts.

- Risk is dynamic, varying over time and space. Some risks are random while others are more predictable.

- Risk can be fully assessed only through an examination of both the physical forces, such as those associated with extreme events, and the human environment, including societal norms and structures that influence the behaviour and perceptual outlooks of individuals who influence decision-making.

- Technology has greatly improved our understanding of statistical risk and our ability to evaluate risk spatially and temporally. Communicating that risk is still a major challenge, yet a critical one, because risk and uncertainty are part of life.

Further Reading

There is a vast literature on risk and risk communication that would add to this chapter. Readers are encouraged to explore some of the hazard texts, such as Smith's (2000) *Environmental Hazards*, Tobin and Montz's (1997) *Natural Hazards: Explanation and Integration*, Wisner's et al.'s (2004) *At Risk: Natural Hazards, People's Vulnerability and Disasters*. Books focused specifically on risk include Flynn et al.'s (2001) *Risk, Media and Stigma*, Jaeger et al.'s (2001) *Risk: Uncertainty and Rational Action*, Kasperson et al.'s (1995) *Regions at Risk*. Paustenbach's (1989) *The Risk Assessment of Environmental and Human Health Hazards*, Slovic's (2000) *The Perception of Risk* and Tulloch and Lupton's (2003) *Risk in Everyday Life*. Bernstein's (1996) *Against the Gods* traces the story of risk. Of course, all the websites listed in the text also serve as important sources of information.

Note: Full details of the above can be found in the references list below:

References

Amdal, G. (2001) *Disaster Response: GIS for Public Safety*. Redlands, CA: ESRI Press.

Association of Bay Area Governments (2007) *Earthquake Maps and Information*. San Francisco: ABAG. http://quake.abag.ca.gov/ (last accessed 01/01/08).

Bernstein, P.L. (1996) *Against the Gods: The Remarkable Story of Risk*. New York: John Wiley and Sons.

BGS Recent Events (2008) http://www.earthquakes.bgs.ac.uk/recent_events/UK_events_map. html (last accessed 01/01/08).

Burton, I., Kates, R.W. and White, G.F. (1993) *The Environment as Hazard*. New York: Guilford Press.

Carrara, A., Guzzetti, F., Cardinal, M. and Reichenbach, P. (1999) 'Use of GIS technology in the prediction and monitoring of landslide hazard', *Natural Hazards*, 20: 117–35.

Center for Disaster Management and Risk Reduction Technology (CEDIM) (2008) University of Karlsruhe, http://www.cedim.de/english/1017.php (last accessed 01/01/08).

City of Salem (2007) *Landslide Hazard Maps*. http://pacweb.cityofsalem.net/permits/landslide_hazard_maps.htm

Committee on Floodplain Mapping Technologies (2007) *Base Map Inputs for Floodplain Mapping*. Washington, DC: National Academy of Sciences. Pre-publication available at http://books.nap.edu/catalog.php?record_id=11829.

Connor, C.B., Hill, B.E., Winfrey, B., Franklin, N.M. and La Femina, P.C. (2001) 'Estimation of volcanic hazards from tephra fallout', *Natural Hazards Review*, February: 33–42.

Connor, C.B., Powell, L., Strauch, W., Navarro, M., Urbina, O. and Rose, W.I. (1993) 'The 1992 eruption of Cerro Negro, Nicaragua: an example of Plinian-style activity at a small basaltic cinder cone', *EOS: Transactions of the American Geophysical Union*, 74: 640.

Covello, V.T. and Mumpower, J. (1985) 'Risk analysis and risk management: an historical perspective', *Risk Analysis*, 5(2): 103–18.

Crandell, D.R., Mullineaux, D.R. and Miller, C.D. (1979) 'Volcanic-hazard studies in the Cascade Range of the Western United States', in P.D. Sheets and D.K. Grayson (eds) *Volcanic Activity and Human Ecology*. New York: Academic Press, pp. 195–219.

CRED Homepage (2008) http://www.cred.be/ (last accessed 01/01/08).

Cutter, S.L., Mitchell, J.T. and Scott, M.S. (2000) 'Revealing the vulnerability of people and places: a case study of Georgetown County, South Carolina', *Annals of the Association of American Geographers*, 90(4): 713–37.

Cutter, S.L., Thomas, D.S.K., Cutler, M.E., Mitchell, J.T. and Scott, M.S. (1999) *South Carolina: Atlas of Environmental Risks and Hazards*. Department of Geography Hazards Research Laboratory, University of South Carolina, Columbia, SC: University of South Carolina Press.

Dunne, T. and Leopold, L.B. (1978) *Water in Environmental Planning*. San Francisco, CA: Freeman.

EA (2008) UK flood map, http://www.environment-agency.gov.uk/maps/info/floodmaps/ (last accessed 01/01/08).

EIRD (2008) Homepage, http://www.eird.org/ (last accessed 01/01/08).

Elsner, J.B., Kara, A.B. and Owens, M.A. (1999) 'Fluctuation in north Atlantic hurricane frequency', *Journal of Climate*, 12: 427–37.

EM-Dat database (2008) http://www.em-dat.net/ (last accessed 01/01/08).

Emrich, C.T. (2000) *Modeling Community Risk and Vulnerability to Multiple Natural Hazards: Hillsborough County, Florida*. Master of Arts thesis, Tampa, FL: University of South Florida.

EPA (2008) Enviromapper, www.epa.gov/enviro/html/em (last accessed 01/01/08).

ESRI (2008) Hazards, www.esri.com/hazards/index.html (last accessed 01/01/08).

Faulkner, H. and Ball, D. (eds) (2007) 'Environmental hazards and risk communication', *Environmental Hazards*, special edition, 7(2): 71–172.

Flynn, J., Slovic, P. and Kunreuther, H. (eds) (2001) *Risk, Media and Stigma: Understanding Public Challenges to Modern Science and Technology*. London: Earthscan.

Government of The Bahamas (2005) *Department of Meteorology Report on the Recent Hurricane History*. Government Document. Nassau: The Commonwealth of the Bahamas.

Grazulis, T.P. (1993) *Significant Tornadoes 1680–1991: A Chronology and Analysis of Events*. St Johnsbury, VT: The Tornado Project of Environmental Films.

Gruntfest, E. and Handmer, J. (2001) 'Dealing with flash floods: contemporary issues and future possibilities', in E. Gruntfest and J. Handmer (eds) *Coping with Flash Floods*. Dordrecht: Kluwer Academic Publishers, pp. 3–10.

Harwell, E.E. (2000) 'Remote sensibilities: discourses of technology and the making of Indonesia's natural disaster', *Development and Change*, 31(1): 55–72.

Hepner, G.F. and Finco, M.V. (1995) 'Modeling dense gaseous contaminant pathways over complex terrain using a geographic information system', *Journal of Hazardous Materials*, 42: 187–99.

Hewitt, K. (1997) *Regions of Risk: A Geographical Introduction to Disasters*. Singapore: Longman.

Hewitt, K. and Burton, I. (1971) *The Hazardousness of Place: A Regional Ecology of Damaging Events*. Toronto: Department of Geography, University of Toronto.

Hodge, D., Sharp, V. and Marts, M. (1979) 'Contemporary responses to volcanism: case studies from the Cascades and Hawaii', in P.D. Sheets and D.K. Grayson (eds) *Volcanic Activity and Human Ecology*. New York: Academic Press.

Hodgson, M.E. and Cutter, S.L. (2001) 'Mapping and the spatial analysis of hazardscapes', in S.L. Cutter (ed.) *American Hazardscapes: The Regionalization of Hazards and Disasters*. Washington, DC: Joseph Henry Press, pp. 37–60.

Holland, G.J. (1983) 'Tropical cyclone motion: environmental interaction plus a beta effect', *Journal of Atmospheric Science*, 40: 328–42.

Hughey, E.P. (2008) 'A Longitudinal Examination of the Effectiveness of the Comprehensive Emergency Management System on Response in The Commonwealth of the Bahamas. Doctoral dissertation', Department of Geography, University of South Florida, Tampa, FL.

Jaeger, C.C., Renn, O., Rosa, E.A. and Webler, T. (2001) *Risk: Uncertainty and Rational Action*. London: Earthscan.

Kasperson, J.X., Kasperson, R.E. and Turner B.L. (eds) (1995) *Regions at Risk: Comparisons of Threatened Environments*. Tokyo: United Nations University Press.

Kasperson, R.E., Renn, O., Slovic, P., Brown, H.S., Emel, J., Goble, R., Kasperson, J.X. and Ratick, S. (1988) 'The social amplification of risk: a conceptual framework', *Risk Analysis*, 8(2): 177–87.

Linnerooth-Bayer, J., Löfstedt, R.E. and Sjöstedt, G. (eds) (2001) *Transboundary Risk Management*. London: Earthscan.

McMaster, R.B., Leitner, H. and Sheppard, E. (1997) 'GIS-based environmental equity and risk assessment: methodological problems and prospects', *Cartography and Geographic Information Systems*, 24(3): 172–89.

Mitchell, J.K. (1990) 'Human dimensions of environmental hazards: complexity, disparity, and the search for guidance', in A. Kirby (ed.) *Nothing to Fear*. Tucson, AZ: University of Arizona Press, pp. 131–75.

Monmonier, M. (1997) *Cartographies of Danger: Mapping Hazards in America*. Chicago, IL: University of Chicago Press.

Montz, B.E. (1994) *Methodologies for Analysis of Multiple Hazard Probabilities: An Application in Rotorua, New Zealand*. Fulbright Research Scholar Report. Prepared for the Centre for Environmental and Resource Studies, University of Waikato, Hamilton, New Zealand.

Montz, B.E. (2000) 'The hazardousness of place: risk from multiple natural hazards', *Papers and Proceedings of Applied Geography Conferences*, 23: 331–9.

Mooney, C. (2005) *The Republican War on Science*. New York: Basic Books.

NASA (2008) Natural Hazards, http://earthobservatory.nasa.gov/NaturalHazards/ (last accessed 01/01/08).

Neely, W. (2006) *The Major Hurricanes to Affect The Bahamas: Personal Recollections of Some of the Greatest Storms to Affect The Bahamas*. Bloomington, IN: Author House.

Neumann, C.J., Jarvinen, B.R. and Pike, A.C. (1987) *Tropical Cyclones of the North Atlantic Ocean, 1871–1986*. Coral Gables, FL: National Climate Data Center and National Hurricane Center.

Pararas-Carayannis, G. (1986) 'The effects of tsunami on society', in R.H. Maybury (ed.) *Violent Forces of Nature*. Mt Airy, MD: Lomond Publications, pp. 157–68.

Paustenbach, D.P. (ed.) (1989) *The Risk Assessment of Environmental and Human Health Hazards: A Textbook of Case Studies*. New York: John Wiley and Sons.

Pielke, R.A. (1990) *The Hurricane*. New York: Routledge.

Prowse, T.D., Owens, I.F. and McGregor, G.R. (1981) 'Adjustment to avalanche hazard in New Zealand', *New Zealand Geographer*, 37(1): 25–31.

Shulman, S. (2006) *Undermining Science: Suppression and Distortion in the Bush Administration*. Berkeley, CA: University of California Press.

Slovic, P. (ed.) (2000) *The Perception of Risk*. London: Earthscan.

Smith, K. (2000) *Environmental Hazards*. London: Routledge.

Sorenson, J.H. (2000) 'Hazard warning systems: a review of 20 years of progress', *Natural Hazards Review*, 1(2): 119–25.

Stern, P. and Easterling, W.E. (eds) (1999) *Making Climate Forecasts Matter, Panel on the Human Dimensions of Seasonal-to-Interannual Climate Variability*. Washington, DC: National Academy Press.

Tierney, K.J. and Baisden, B. (1979) *Crisis Intervention Programs for Disaster Victims: A Source Book and Manual for Smaller Communities*. US Department of Health, Education and Welfare. Rockville, MD: National Institute of Mental Health.

Tobin, G.A. and Montz, B.E. (1997) *Natural Hazards: Explanation and Integration*. New York: Guilford Press.

Tobin, G.A. and Montz, B.E. (2004) 'Natural hazards and technology: vulnerability, risk and community response in hazardous environments', in S.D. Brunn, S.L. Cutter and J.W. Harrington (eds) *Technoearth: A Social History of Geography and Technology*. Dorecht: Kluwer Academic Publishers, pp. 547–70.

Tobin, G.A. and Whiteford, L.M. (2004) 'Chronic hazards: health impacts associated with ongoing ash-falls around Mt Tungurahua in Ecuador', *Papers of the Applied Geography Conferences*, 27: 84–93.

Tulloch, J. and Lupton, D. (2003) *Risk in Everyday Life*. London: Sage.

USGS (2008) Hazards, http://www.usgs.gov/hazards/ (last accessed 01/01/08).

USGS (2008) Latest Earthquakes, http://earthquake.usgs.gov/eqcenter/recenteqsww/(last accessed 01/01/08).

USGS (1999) *Earthquake Probabilities in the San Francisco Bay Area: 2000–2030 – A Summary of Findings*. United States Geological Survey, Working Group on California Earthquake Probabilities. *United States Geological Survey Open-File Report 99–517*. http://geopubs.wr.usgs.gov/open-file/of99-517/ (last accessed 01/01/08).

USGS (2001) *Seismic Landslide Hazard for the Cities of Oakland and Piedmont, California. United States Geological Survey Miscellaneous Field Studies Map MF-2379*. http://pubs.usgs.gov/mf/2001/2379/ (last accessed 01/01/08).

USGS (2003) *Earthquake Probabilities in the San Francisco Bay Region: 2002–2031*. United States Geological Survey, Working Group on California Earthquake Probabilities. *US Geological Survey Open-File Report 03-214*. http://pubs.usgs.gov/of/2003/of03-214/ (last accessed 01/01/08).

USGS (2005) *Summary of Monitoring Data (1970–May, 2005)*. United States Geological Survey. http://hvo.wr.usgs.gov/maunaloa/current/longterm.html (last accessed 01/01/08).

USGS (2007a) *Kilauea Eruption History*. Hawaiian Volcano Observatory. United States Geological Survey. http://hvo.wr.usgs.gov/kilauea/history (last accessed 01/01/08).

USGS (2007b) *Seismic Hazard Maps*. United States Geological Survey. http://earthquake.usgs.gov/research/hazmaps/products_data/index.php (last accessed 01/01/08).

Williams, J. and Sheets, R. (2001) *Hurricane Watch: Forecasting the Deadliest Storms on Earth*. New York: Random House.

Wisner, B., Blaikie, P., Cannon, T. and Davis, I. (2004) *At Risk: Natural Hazards, People's Vulnerability and Disasters* (2nd edn). London: Routledge.

CONCLUSION:
PRACTISING GEOGRAPHY

25

Relevance: Human Geography, Public Policy and Public Geographies

David Bell

Definition

'Relevance', as I am talking about it in this chapter, means the extent to which academic research is seen to matter in and impact on the 'real world'. This can include 'policy relevance' – the main focus of my chapter – which means the extent to which geographical research has any impact on shaping or informing public policy, such as government policies. Relevance is something that academic research is increasingly being asked to evidence, in part to justify its continued funding and support as a worthwhile activity that benefits the wider world.

INTRODUCTION

Being an academic geographer is, for many of us who more or less identify with that name, a curious experience.[1] The professions of which we are a part – being an academic, being a geographer, with all that each entails – are full of ambiguities, quirks and tensions. A lot of academic geographers spend a lot of time publicly and privately pondering this, wondering what we're here for. This isn't always academic navel-gazing, either; we are in a position where we *cannot not* think these things, and in this chapter I want to work through some key ways through which a number of pressing questions have been asked and answered in the process of thinking about what it is that we do. These questions include: What are universities for? What are academics for? And what is geography for? I Will approach these big questions through a series of debates in geography (and indeed elsewhere), debates that have rumbled on for the past 50 years or more, intensifying due to internal and external pressures since the 1970s, and maybe even more so in the present climate. These are debates about 'relevance'; or, to put it the other way round, about how to hold off what Noel Castree (2002) calls the 'spectre of irrelevance'.

This spectre looms large in academic life, as we shall see. Academics, perhaps especially those in the arts, humanities and social sciences, have felt its breath on their neck for a long time now, in part due to the ways in which universities have been funded, the comparative status of different classes of scholarly knowledge and academic disciplines, and the position that higher education and academic research occupies economically, socially and politically. This even bigger set of stories is beyond the reach of this chapter, but we need to keep hold of its main point, which is about a tension between two ways of understanding what universities are: (1) public goods and public servants, or (2) self-financing, marketized, entrepreneurial businesses. In the UK – and I apologize for basing much of this short discussion mainly in the contexts I am more familiar with – universities occupy something of a 'space between' these two poles, still rooted in the public sector, and hence closely bound to the state (though often in a one-way relationship, as we shall see), but increasingly being tasked with acting more entrepreneurially, being more 'market facing', more 'business-like' (Robson and Shove, 2004). As we shall see, this odd in-between location has serious implications for what university academics are in turn tasked with delivering, and one recurring trope of that is the delivery of 'relevance'. Relevance is a virtue expected of all aspects of academic life, but in this chapter I will focus only on the relationship between relevance and research (on different, sometimes competing articulations of relevance in geography, see Staeheli and Mitchell, 2005).

HISTORIES OF RELEVANCE

As already noted, these issues are by no means new; debates about relevance have been visible in academic geography for 40 years or more (Johnston and Sidaway, 2004). Especially since global economic and political restructurings that occurred in the 1970s, which coincided with (indeed, resulted in) changes in government policies about higher education, questions of relevance have become ever more pressing. Of course, those profound structural changes also provided new opportunities for academics to try to intervene in the world 'out there', and geographers began to make a case for their unique potential to diagnose, predict and suggest alternative futures in the face of global problems (Zelinsky, 1970). Geographers confronting this changing world began to debate relevance in the context of this need to relevantly address pressing issues: what questions could (or should) geography ask? What kinds of answers might the application of geographical skills and knowledges provide? In particular, how could geographers position themselves as relevant experts equipped with relevant skills and knowledges, in order to maximize their contribution – and hence their relevance?

Now, this last question needs to be understood in two key ways: first, it asks geographers to get involved in the 'real world', to think about how best to utilize their knowledges and skills to solve 'real world' problems; second, it asks that geography be seen as specially placed, specially equipped, to ask and answer the big questions. This second reading is about intellectual turf wars, therefore: about the position, role and status of different forms of expertise, different disciplines (and this is a very big issue when resources flow to disciplines that are best able to prove their usefulness, as we shall see). The tense interplay between these two 'calls to relevance' remains very important, and is arguably even more acutely felt under the new conditions in which university academics now labour (Castree, 2002).

A number of potential 'domains of relevance' emerged in debates in geography in the 1970s, each making different demands on how academic geographers might redirect their professional, intellectual energies away from the narrow, sealed-off concerns of 'pure' research. The opposite of this 'pure' activity – thinking, theorizing, writing, debating – came to be labelled 'applied research'. Now, this is a very large category, containing many and varied potential forms of application, and also many different routes to applicability. To what, how, and by whom is 'applied research' actually applied? Is application something which comes early or late in the process of formulating, carrying out and disseminating research? How much 'appliedness', of what kinds, is necessary for research to be classed as applied? And who is to do the applying, to what ends?

David Harvey crystallized many of these concerns and questions in his 1974 paper, 'What kind of geography for what kind of public policy?' Harvey's particular reading of the situation didn't hold back in its critique of the discipline for its complicity in upholding rather than challenging the '*status quo*', for its 'moral masturbation' and 'emotional tourism', nor in its rallying cry for 'revolutionary theory' as a key tool for social justice and social change (Harvey, 1974: 144). Harvey thus frames what we might call 'Marxist relevance' as a particular form of applied research aimed at producing a 'people's geography' in the service of class struggle. Later branchings of so-called radical and critical geographies continue this tradition, in some form or another, pondering what Lynn Staeheli and Don Mitchell (2005) rightly tag the 'complex politics of relevance in geography' (on critical geography and relevance, see Castree, 2002).

Staeheli and Mitchell (2005: 358) note that relevance debates in geography in the 1970s were remarkable in that 'charges of irrevelancy were directed at a mode of knowing that was positivist and statistical' – Harvey and his like-minds argued that 'spatial science' was an inappropriate approach to solving the most pressing global problems of the day. A counter-trend in the relevance debates, marked by liberalism rather than Marxism, took applied geography off in a different direction, growing in the 1980s through the application of evolving techniques such as geographical information systems (GIS), and reinserting largely empiricist and positivist approaches through, as Johnston and Sidaway (2004: 335) put it, the 'pragmatic application to technical skills'. Critics detect in this turn a move away from social problems (and explanations) to economic issues (and solutions), and an emphasis on data-gathering and description rather than causal explanation. They also called for greater reflexivity on the part of applied researchers, asking that the mask of objectivity be removed, in part in acknowledgement that relevance itself is political (Clark, 1982). By the 1990s, moreover, a renewed interest in questions of justice – social, environmental, cultural, etc. – brought issues of morality and ethics into contact with issues of relevance, as part of a broader, ongoing project of 'moralizing' geography (see Valentine, 2003).

RELEVANCE NOW

The tension over the meaning of relevance and the shape and uses of applied research are also spectres that haunt geography to this day, especially on the pages of geography journals. Under prevailing conditions that some folk label 'neoliberalism', the possibilities and problems of policy work are reshaped once more (Johnston and Sidaway, 2004). I want to pick it back up in 2001 via two key papers, Doreen Massey's (2001) 'Geography on the agenda', and Ron Martin's (2001) 'Geography and public policy: the case of the missing agenda', where we can find the calls for greater engagement with the policy terrain loud and clear as ever (see Johnston and Sidaway, 2004, for a fuller history). Yet here we can also detect new anxieties, new missed opportunities: geographers, it seems, are missing out on chances to prove their relevance, due to their lack of engagement with one key arena where relevance rules: public policy. So Massey (2001) presents a picture of geography being largely ignored by policy-makers, which leads her to ponder the status of the academic as a knowledge-producer, asking about the kinds of knowledges that geography produces, the sites where those knowledges circulate, and the languages with which we speak: 'I do sometimes wonder at all the masses of stuff we produce, and question what really drives it', she writes (Massey 2001: 11). Noting the heavy pressure on research output and the kinds of metrics used to assess it, notably the Research Assessment Exercise (RAE), she urges academic geographers 'to embed what we are doing in wider debates, rather than only to talk to each other' (2001: 12). Now, Massey is contributing here to a growing critique of the effects of exercises like the RAE on the culture of research in higher education (see also Robson and Shove, 2004). The pressures to produce 'cutting-edge' or 'world-class' research highly ranked in terms of peer esteem is, for critics like Massey, at odds with – and at the expense of – the question of relevance. While the finessing of the audit processes of the RAE has attempted to remedy this, there remains a yawning gap in status (and reward) between top-drawer academic papers and the 'grey' areas of policy relevance (on 'grey geographies', see Peck, 1999). This in turn is part and parcel of the restructuring of higher education and the birth of the 'entrepreneurial university', which has profoundly reshaped the business of being an academic, leading to casualization matched by incentivization, and increasingly competing demands on academic labour (Allen Collinson, 2004). One flash point here concerns the relative merits and rewards of 'RAEable' research outputs against winning applied research contracts – a tension experienced at the level of professional identity and workplace relations for those pulled in different directions by these imperatives (Allen, 2005). It is in this context that Massey sees an understandable reluctance to enter the applied, policy-relevant arena, with its uncertain rewards, and the concomitant risk-averse retreat to focusing on RAE outputs – even if, as she says, those outputs have an impact on only a small band of fellow academic geographers.

Martin's (2001) paper covers much of the same ground as Massey's, though with a more direct focus on policy, and a more thoroughgoing critique of some possible reasons for geography being allergic to policy. Martin takes the relevance of geography as a key point, noting that it should be more relevant, more useful than it currently is. Making geography useful, for Martin, means a number of things:

> It requires us to interrogate and evaluate existing policies and policy-making practices to reveal their limitations, biases and effects. And it means seeking to exert a direct influence on policy-making processes, at all scales, with the aim to produce more appropriate and more effective forms of policy intervention. (Martin, 2001: 190)

This is an important reorientation of the discipline for a number of reasons, including the very pertinent observation that 'policy-making of one kind or another is a prominent and pervasive feature of modern society, affecting the daily lives of us all' (Martin, 2001: 190). Policy-making has become a central arm of government, and especially notable is the emphasis on so-called 'evidence-based policy'. Everybody needs evidence nowadays, in order to prove a case for policy intervention, as well as to prove the success (or otherwise) of that intervention via techniques of evaluation. This turn to evidence means for some writers a great new opportunity for academic geographers to become relevant and useful (Robson and Shove, 2004). Yet the changing policy environment also poses new challenges to those academics trying to get involved; as Robson and Shove (2004) write, becoming policy-relevant isn't just a matter of doing research on a topic with implicit or potential policy relevance, it means knowing how to translate one's work and transform oneself into a 'policy person' (see also Bell, 2006). Performing policy relevance is a skill that many academics lack, and we are arguably much better at critiquing policies than informing or formulating them.

Here we encounter another key problem: Martin (2001: 189) writes that geography has a lot potentially to offer policy-makers – if only it could stop being so cultural, so postmodern, so 'sexy' (2001: 189). Academic geography has been overly infected by 'trendy' theories and 'trendy' methods, he complains, none of which has much by the way of policy purchase. And while he produces a crude cartoon of the 'cultural turn', concluding that 'it is difficult to envisage how the vague abstractions and epistemic and ontological relativism of much human geography research ... can form the basis of critical public policy analysis' (2001: 196), Martin also highlights the snobbishness of the discipline in refusing to take 'applied' research seriously. Ultimately, Martin tries to carve out a role for academics which doesn't involve the selling of souls; this invokes a sleight of hand, however, that moves the question away from engagement with policy-makers and towards policy *analysis*, in what reads as an (understandable) attempt to 'prove' that policy-work is brain-work, too. Yet in this move, he returns to a safer terrain, where academic geographers scrutinize already existing policies, highlight their weaknesses to each other (but not to their originators), and feel very pleased with themselves for their clever deconstructive critique.

This raises a further key dilemma, one that Harvey spotted back in 1974: does policy-work necessarily reproduce the *status quo*, because it is commissioned by and for those already in power? Addressing the potential for complicity with and subservience to dominant agendas for anyone involved in policy-making, Martin suggests that rigorous, empirical and relevant research can be used to shape policy without compromising academic integrity, raising questions beyond the remit of this chapter about what counts as academic integrity anyway (see Castree, 2002).

Massey's and Martin's papers reinvigorated the debate in geography about policy relevance, which remains live to this day (cynics might speculate that writing about the debate has become another way of not engaging head-on with policy). Danny Dorling and Mary Shaw (2002) picked up the thread, again noting the intellectual snobbery that rates policy or 'applied' work as second-class geography and the insularity of too many debates in the discipline, while at the same time tracking a misplaced arrogance: 'many geographers think that what they are doing or saying is of relevance when its impact is in fact minimal' (Dorling and Shaw, 2002: 630). While that may seem like a low blow, it is important to remember that relevance is a different currency in different contexts, and that while we might imagine that the things we write about should have all kinds of applicability, other people might not agree. So becoming policy-relevant, Dorling and Shaw argue, means that researchers must consider '(a) what they say, (b) how and in what form they say it, and (c) to whom they speak – quite simply, the message, its medium and its audience' (2002: 633). This is a vital, if not unproblematic, formulation that has, I think, absolutely crucial bearing on how we 'do' policy, for example in its call to modify what Robson and Shove (2004: 363) call the 'cagey, inward-looking terminology' of the *lingua franca* of academia, in an effort to make it more audible and comprehensible to policy-makers (see also Johnston and Plummer, 2005). And Dorling and Shaw end with a similarly telling set of observations about 'doing' policy, calling for academics and their institutions to 'agree not to penalize those who do take part in and contribute to policy debates' and to acknowledge that 'doing work that might be useful in influencing policy is ... far harder than writing four papers in four years in geography journals [for the RAE]' (Dorling and Shaw, 2002: 638–9). In her response to Dorling and Shaw, Massey (2002) rebuffs some of their criticisms, but more importantly disagrees with their claim that policy engagement is the best way of being useful or relevant, seeing 'usefulness' in many places. And she questions the straightforward equation of applied research with relevance and usefulness: 'Just bringing in loads of contracts is not the best means, necessarily, to advance debates nor to produce original thinking that will change ideas' (Massey, 2002: 646). There are, of course, a lot of assumptions to unpack here, too – not least that what we are or should be about is producing original thinking or changing ideas (who says what's original, what kinds of change?).

The debate on policy relevance has had a lot to say about the role of academic geographers, but it is also important to remember the various 'policy communities' that those pursuing policy relevance are engaging with. Often the 'user' remains an uninterrogated figure, whose characteristics (and intellectual capacity) are assumed. There is, therefore, a need for an 'evidence base' here too, in terms of telling stories about policy engagement, fleshing out user groups and policy communities, reflecting on the detail of the acts of translation, the performances on both sides, and the whole process from tender to report and action. 'Confessional' tales about policy engagement have important lessons to tell us, though there are problems of confidentiality, ownership of contract research, ongoing relations with users, and so on to consider (Bell, 2006). And as with academic publishing generally, of course, those tales may get read only by academics – and only those sufficiently interested in the policy relevance debates – reintroducing all those questions of priorities for publishing, audiences for our work, and so on. Such confessional tales may be one way of translating policy work back into RAEable outputs, of course – a way of parlaying the 'grey geographies' (Peck, 1999) of consultants' reports, feasibility studies and evaluations into something carrying at least some academic capital, though there are risks here, too, in a Google-able world where potential (and past) clients can track down academic outputs, including confessions, at a mouse-click. The levels of impression management required, both in terms of maintaining one's status among one's academic peers, and in winning and delivering applied research contracts, can be very hard to handle (Allen, 2005). Every step brings new quandaries; no wonder so many academic geographers steer clear, playing safe in the relatively protected, not to mention rewarded, waters of RAE ratings.

FROM GREY GEOGRAPHIES TO PUBLIC GEOGRAPHIES

As mentioned earlier, the debates in geography about relevance are of course part of much bigger debates about universities and about academics and intellectuals. Relevance doesn't just mean being relevant to select audiences, such as policy-makers; it means relevance in a broader, societal setting. When you tell people you're studying a geography degree, what questions do they ask (or think) about relevance? Part of the transformation of higher education has been to emphasize the vocational aspects of higher education, so students have also been 'relevantized', especially now they are fee-paying customers who are making a time-and-money investment in their own futures. Studying at university has to be about gaining relevant skills and knowledges, where relevance is a market attribute, a way of trading academic capital. Learning and teaching are hereby 'relevantized'. In fact, teaching is quite often cited as a key way in which academics perform relevance: we may not in ourselves be engaging with 'outside' user communities or debates, but we equip our students with the tools to go forth and do so, having been appropriately trained by us (Staeheli and Mitchell, 2005). In cultural studies, Tony Bennett (1992) made a similar proposition in his discussion of usefulness, talking of the use value of training what he called 'cultural technicians' (on the parallel debates in geography and cultural studies, see Bell, 2006).

Mention of teaching helps us broaden out the relevance debates, then, by reminding us of the different audiences (or users) we interact with. Other disciplines have been asking similar questions about communicating relevance, perhaps most notably the sciences, where the subdiscplines of science communication and public understanding of science have developed diverse theories and methods of analysis in an effort to improve the 'public image' of science, the accessibility of scientific findings, and the 'scientific literacy' of non-scientists (Gregory and Miller, 1998). This agenda has arisen in part as a result of the so-called legitimation crisis in the sciences (Lyotard, 1984), through which, allegedly, the public lost its faith in science as a force for good. A parallel legitimation crisis has arguably affected the rest of academia, too, rising out of the changing political economy of higher education (Robson and Shove, 2004): the need for accountability in the allocation of resources, financial constraints, competing knowledge claims, and so on. Arguments are being made, therefore, for a parallel project of re-enchantment, for analysis of the *public understanding of social science* in an attempt to revivify these disciplines and exorcise the spectre of irrelevance that haunts them.

Geography, it is commonly held, has an image problem as an academic discipline. It is poorly understood, ill-defined, definitively 'unsexy' – and this image problem has contributed significantly, it is argued, to the sidelining of geographical expertise in policy arenas. While geography has arguably fared better than some of its neighbours, such as sociology, it is still struggling to shake off associations of maps and anoraks.

Sociology, just mentioned, had a particularly hard time in the UK during the years of the Thatcher administration and its Conservative successors, whereby its assumed left-wing liberalism became a tabloid shorthand for all that was wrong with 'trendy leftie' politically correct intellectuals. While the Blair administration has seemed less allergic to sociology, with Anthony Giddens a key architect of the Third Way, sociology has had to soul search and work hard to redefine its usefulness, too. Key government agendas, such as social inclusion, have provided such 'relevantizing' to some extent, opening up policy spaces to sociologists. At the same time, in the USA initially but spreading from there, there has been a debate among academic sociologists about public engagement, catalysed by Michael Burawoy's talk about 'public sociologies' (Burawoy 2004).

Kevin Ward (2006) has picked up this thread, asking what it might mean after Burawoy to talk of 'public geographies'. This is, I think, an important (albeit tentative) intervention into the relevance debates in geography; Ward notes, in fact, the 'public' has been a largely missing element of policy debates, despite the strong emphasis on 'public policy'. So, as with the need noted earlier to think more clearly about the policy communities that 'relevantized' academic geographers encounter and (attempt to) engage with, there is also a need to think more about the public of public policy, not purely as some kind of rhetorical figure for classifying policies, but as an actually existing set of individuals and communities involved in the policy process (and not only as its recipients; in these days of evidential policy, public consultation draws the public in as 'warm experts' to help shape policy, too). As Ward (2006: 497) rightly notes, 'there remains more to be done in bringing ... into dialogue' policy debates and broader debates about the redefinition of 'the public' – and it is here that Ward listens in on Burawoy's 'public sociologies' talk. In particular, he picks up on Burawoy's differentiation between public sociologies, policy sociologies and professional (or radicalized) sociologies. Policy sociologies are, for Burawoy, narrowly instrumental; professional sociologies, by contrast, are too

Table 25.1 Comparing policy and public geographies

	Policy geographies	Public geographies
Knowledge Reflexive/communicative	Instrumental/concrete	
Truth	Pragmatic	Consensus
Legitimacy	Perceived effectiveness	Relevance
Accountability	Clients and evaluators	Designated publics
Politics	Interventions in policy discussions	public dialogue
Pathology	Servility	Fads and fashions

invested in their own radicalization, disinterested in external engagement (which can be so easily dismissed as compromise or incorporation; for a discussion of similar debates in geography, see Castree, 2002).

Public sociology, by contrast, means, for Burawoy, bringing 'sociology into conversation with publics, understood as people who are themselves in conversation' (quoted in Ward, 2006: 499) – the relationship is a multi-channel dialogue marked by reciprocity (especially in a subset Burawoy calls 'organic public sociologies', which involve participatory research projects). Translating Burawoy's provocations, Ward offers some initial ideas for similar 'public geographies'. These include removing the Anglo-American bias of our talk about geography, exploring different national formations of public geographies; recognizing ourselves – academic geographers – as one kind of public, and seeing alliances and commonalities with others (e.g. other public sector workers); recognizing our students as another very important public, who 'carry geography into all walks of life' (Ward, 2006: 500); recognizing that academic expertise is only one kind of expertise that contributes to public geographies; working to raise the public profile of geography through numerous channels. Ward retools a table from Burawoy (2005) to summarize the key differentiating features of policy geographies and public geographies (Table 25.1).

While there is much to argue with in Ward's call to move on relevance debates in academic geography – such as its reinstatement of a narrowly instrumental, servile relationship to policy communities – it nonetheless asks the right kinds of question, not unlike those questions, in fact, asked by David Harvey 30-plus years ago, and repeatedly asked since then. The longevity and recurrence of those questions suggests there is much that remains to be thought about and talked about, therefore, in terms of academic geography's relevance.

SUMMARY

- There has been a long and often contentious debate about whether human geography research is or should be 'relevant'.

- Human geographers have argued that our discipline underplays its relevance and has not made a significant impact on public policy.

- The question of relevance means asking 'relevance for whom?' What kinds of relevance should human geography be trying to achieve?

- Recent debates argue for a shift towards broader public engagement with human geography, and call for new 'public geographies'. Parallel debates have taken place in other disciplines, such as sociology and cultural studies.

- These debates have broader resonance, in asking questions about the role of intellectuals and of universities in society.

Further Reading

David Harvey's (1974) early intervention is a good place to start and asks some very important questions about what relevance is or might be. More recently, papers by Martin (2001), Massey (2001, 2002) and Dorling and Shaw (2002) picked up the debate and engaged in a lively argument with each other about the forms, uses and meanings of relevance. Two review articles by Kevin Ward (2005, 2006) cover the background history neatly, and Ward makes the important move towards 'public geographies' here. For a critical commentary on what 'public geographies' mean (and don't mean), which provides some alternative understandings and perspectives, see the conversation between Askins and Fuller (2007). Finally, comparative debates from nearby disciplines can be tracked by reference to Bennett (1992) and Bell (2006) for cultural studies, and Burawoy (2004) for sociology.

Note: Full details of the above can be found in the reference list below

---

NOTE

1 It has to be pointed out from the get-go that my discussion refers only to the sub-discpline called 'human geography'; I have decided not to endlessly reiterate this point in the text by using the term 'academic human geographer', merely because it sounds odd. I use the term 'academic geographer' in acknowledgement that there are many geographies beyond those practised by academic geographers.

References

Allen, C. (2005) 'On the social relations of contract research production: power, positionality and epistemology in housing and urban research', *Housing Studies*, 20: 989–1007.

Allen Collinson, J. (2004) 'Occupational identity on the edge: social science contract researchers in higher eduction', *Sociology*, 38: 313–29.

Askins, K. and Fuller, D. (2007) 'The discomforting rise of "public geographies": a "public" conversation', *Antipode*, 39: 571–601.

Bell, D. (2006) 'Fade to grey: some reflections on policy and mundanity', *Environment and Planning A*, 39: 541–54.

Bennett, T. (1992) 'Putting policy into cultural studies', in L. Grossberg, C. Nelson and P. Treichler (eds) *Cultural Studies*. London: Routledge, pp. 23–37.

Burawoy, M. (2004) 'Public sociologies: contradictions, dilemmas and possibilities', *Social Forces*, 82: 1–16.

Burawoy, M. (2005) 'The critical turn to public sociology', *Critical Sociology*, 31: 313–26.

Castree, N. (2002) 'Border geography', *Area*, 34: 103–12.

Clark, G. (1982) 'Instrumental reason and policy analysis', in D. Herbert and R. Johnston (eds) *Geography and the Urban Environment* (Vol. 5). Chichester: Wiley, pp. 41–62.

Dorling, D. and Shaw, M. (2002) 'Geographies of the agenda: public policy, the discipline and its (re) "turns"', *Progress in Human Geography*, 26: 629–46.

Gregory, J. and Miller, S. (1998) *Science in Public: Communication, Culture, and Credibility*. New York: Basic Books.

Harvey, D. (1974) 'What kind of geography for what kind of public policy?', *Transactions, Institute British Geographers*, 63: 18–24.

Johnston, R. and Plummer, P. (2005) 'What is policy-oriented research?', *Environment and Planning A*, 37: 1521–6.

Johnston, R. and Sidaway, J. (2004) *Geography and Geographers: Anglo-American Human Geography since 1945* (6th edn). London: Hodder Arnold.

Lyotard, J.-F. (1984) *The Postmodern Condition: A Report on Knowledge*. Manchester: Manchester University Press.

Martin, R. (2001) 'Geography and public policy: the case of the missing agenda', *Progress in Human Geography*, 25: 189–210.

Massey, D. (2001) 'Geography on the agenda', *Progress in Human Geography*, 25: 5–17.

Massey, D. (2002) 'Geography, policy and politics: a response to Dorling and Shaw', *Progress in Human Geography*, 26: 645–6.

Peck, J .(1999) 'Grey geography?', *Transactions, Institute of British Geographers*, NS 24: 131–5.

Robson, B. and Shove, E. (2004) 'Geography and public policy', in J. Matthews and D. Herbert (eds) *Unifying Geography: Common Heritage, Shared Future*. London: Routledge, pp. 353–66.

Staeheli, L. and Mitchell, D. (2005) 'The complex politics of relevance in geography', *Annals of the Association of American Geographers*, 95: 357–72.

Valentine G. (2003) 'Geography and ethics: in pursuit of social justice', *Progress in Human Geography*, 27: 375–82.

Ward, K. (2005) 'Geography and public policy: a recent history of "policy relevance"', *Progress in Human Geography*, 29: 310–19.

Ward, K. (2006) 'Geography and public policy: towards public geographies', *Progress in Human Geography*, 30: 495–503.

Zelinsky, W. (1970) 'Beyond the exponentials: the role of geography in the great transition', *Economic Geography*, 50: 499–535.

26

Relevance: The Application of Physical Geographical Knowledge

Michael Church

Definition

Applied physical geography is, most straightforwardly represented, the application of physical geographical knowledge to solve problems of environmental management, including land, water and the atmosphere, and mitigation of environmental hazards, that are of importance to society. In the past, applied geography prominently implied the application of geographical knowledge to statecraft.

INTRODUCTION

There is a widespread notion that applied geography was invented during the latter half of the twentieth century. After about 1970, in particular, growing concern to manage the environment in a rational and ecologically sensitive manner led to applications of physical geographical knowledge in land inventory and planning, in the selection of communication routes, and in atmospheric and water resource management (see Coates, 1971; Cooke and Doornkamp, 1974; Mather, 1974). Increasing concern for the exposure of society to mundane geophysical hazards – landslides, floods, coastal hazards, for example – led to further applications of physical geography in hazard identification and mitigation (e.g. Ives et al., 1976). A similar renaissance was occurring in applied human geography, though with much more introspection about just what 'applied geography' might be (see Chapter 25). This activity intensified with the development of Geographic Information Systems (GIS) as a tool for spatial inventory and analysis.

The first stirrings of modern applied geography developed after the Second World War (see Philipponneau, 2004) and, in 1964, the International Geographical Union (IGU) established a Commission on Applied Geography (later becoming a working group). The first of an annual series of applied geography conferences was held in 1978; professional geographical newsletters made space for applied topics and the journal *Applied Geography* was created in 1981. There was a sense among those involved in these developments that a newly relevant path was being forged for geography.

In fact, all this activity represents something of a return towards geography's original motivation. Geography is arguably the oldest definable intellectual discipline. For the most primitive hunter-gatherers, an intimate knowledge of the landscape around them, however it was codified, was essential for obtaining food, for securing shelter – for survival. The establishment of sedentary societies was closely tied to the establishment of trade in goods, whence a more or less extended knowledge of the sources of materials and routes of exchange was required. Most of this knowledge is geographical knowledge. Ultimately, in the ancient worlds of the Mediterranean, the Middle East and the Far East, geographical knowledge was codified in maps intended to facilitate trade and the development of commerce (Philipponneau, 2004). Thus began an association between geography and the survey of earth's surface that lasted until the twentieth century – an association that largely defined geography through almost the entire history of human civilization and marked it both as a leading intellectual endeavour and as primarily an applied discipline closely associated with statecraft.

In western Europe, this tradition was ambitiously assumed in the fifteenth century with the beginning of European exploration of the wider world. Again, the purpose was chiefly commercial. First the Portuguese (*c.* 1420), then the Spanish, Dutch and British launched a five-century era of exploration and discovery that issued in the modern map of the earth, extensive empires, and various associated attempts to secure military hegemony over the world's trading routes. Exploration, commercial exploitation, military intelligence, and survey of the globe became inextricably mixed together, with geographical enquiry as the principal intellectual tool. In the eighteenth century, increasing interest in the resources of lands that might be colonized from Europe, reinforced by rapidly growing scientific curiosity, led to a broadening of the enterprise to include inventory of natural features. Thus naturalist-scholars began to accompany the voyages of exploration and more comprehensive reports of what amounted to the physical geography of lands overseas began to appear in Europe. A major turning-point was reached in the voyages of James Cook between 1768 and 1780. These voyages were matched by ones from other west European seafaring nations but, for the British, Cook inaugurated a period of scientific voyaging that stretched throughout the following century.

A remarkable feature of Cook's explorations was the attention that he gave to ethnography – to reporting the condition and customs of the people of the lands he explored. This became a characteristic feature of geographical exploration over the following century and a half. Indeed, no substantial distinction was made between human and physical geography until the twentieth century: geography in the earlier period represented effectively all knowledge about lands and their inhabitants so that to speak of human and physical geography as distinctive enterprises makes little sense before the twentieth century. The following account is, accordingly, divided into two unequal parts: applied geography – in fact, most of it 'physical' geography – in the age of European imperialism and applied physical geography in the contemporary world.

APPLIED GEOGRAPHY IN AN IMPERIAL AGE

Geography before 1900 was largely a map-making exercise. The European origins of this tradition coincided with the rise of seaborne exploration from Portugal and Spain in the fifteenth century. Initially, methods for navigation and surveying were adopted from the practices of the medieval Arabs, but steady improvements occurred in instruments for fixing position on earth's surface until, finally, the difficult problem of determining longitude was resolved by Cook's demonstration of the practical precision of John Harrison's chronometer (Sobel, 1995).

Determining positions on a featureless sea by accurately timed reference to celestial bodies is a considerably more straightforward proposition than surveying the complex topography and constructed features of the terrestrial surface. Interest in that project was pressed forward both by military needs and by the increasingly complex needs of civil administration, not least for a system of taxation based on landholdings. From the mid-seventeenth century the French made steady progress in geodetic and topographic mapping. The first national topographical survey was begun in the 1730s under the leadership of César François Cassini de Thury and his son, Jacques Dominique – members of a distinguished family of surveyors – and completed in 1789. This achievement prompted the formation of the British Ordnance Survey in 1791. Mapping methods relied on triangulation (see Figure 26.1) using increasingly sophisticated theodolites, and on plane-table mapping to fill in local details, methods that continued to dominate map-making until the advent of air photographs and photogrammetry early in the twentieth century.

James Cook's voyages (Hough, 1995: see Figure 26.2a) established the European practice of scientific exploration. The first voyage, under the patronage of the Royal Society of London, was undertaken ostensibly to make astronomical observations in the South Pacific. It was an important scientific success. But the voyage also instituted a project directed by the British Admiralty to map the coastlines of the world. The purposes were, as always, to facilitate seaborne commerce, to improve military navigation, and – not least – to extend British dominion over lands newly discovered. Cook was a largely self-taught mathematician and surveyor, and a superb navigator, hence was ideally suited to the task. There were broader scientific purposes as well, partly commercial in intent, partly driven by the spirit of discovery that characterized the eighteenth-century European Enlightenment. Cook deliberately chose learned naturalists and artists to accompany him, including Joseph Banks on his first voyage, a naturalist who

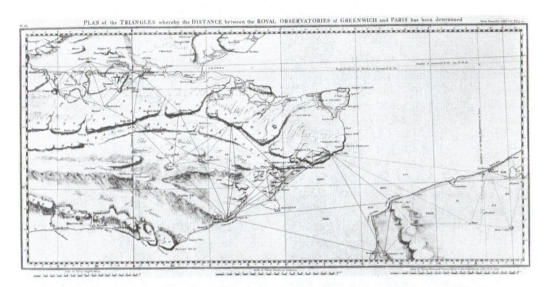

Figure 26.1   Part of the 1790 triangulation survey from Greenwich to Paris, conducted by Major-General William Roy to connect the meridians of Greenwich and Paris. The consequent recognition of how badly England was mapped led to the establishment of the Ordnance Survey

Source: From Roy (1790): plate IX

later became a commanding presence in British science and president of the Royal Society. He returned thousands of plant and animal specimens to Britain. After the second voyage J.R. and J.G.A. Forster – father and son of Prussian origin – wrote, along with Cook himself, extensive ethnographic summaries of the peoples they encountered. Geography as basic science and geography as applied science were one and the same thing.

The nearest equivalent to Cook for terrestrial exploration was Alexander von Humboldt (Figure 26.2b), a Prussian mineralogist who deliberately wished to emulate Cook in the terrestrial sphere. He began his travels in the company of J.G.A. Forster, but it was a five-year expedition to South America (1799–1804) in the company of Aimé Bonpland that established his scientific reputation. In his fieldwork he emphasized measurement – of weather, of topography, of rock and mineral properties – a move towards a precise style of geographical survey that went well beyond topographical mapping. Humboldt was also a grand synthesizer, who dreamed of summarizing all knowledge of earth in a single magisterial work – knowledge of climate, physiography, minerals, geophysical properties such as magnetism, and of plants and animals. The title of the work was *Cosmos*. Some volumes appeared, but the project was not completed in his lifetime – with the torrent of new geographical knowledge arriving in Europe, it probably could not have been completed by one person. Nevertheless, by his activities, Humboldt ushered in an age of greater specialization in physical geography at a time – the early years

Figure 26.2   (a) Captain James Cook (portrait by Nathanial Dance, National Maritime Museum, Greenwich, UK); (b) Baron Alexander von Humboldt

of the nineteenth century – when the whole range of the natural sciences was beginning to assume modern forms, and he also foreshadowed a standard rubric for recording the knowledge of physical geography that emerged fully as the nineteenth century proceeded.

The twin threads of marine and terrestrial exploration continued through the nineteenth century, with decidedly imperialistic overtones. After the Napoleonic era, British dominance on the seas was essentially unchallenged and a stream of surveying expeditions was sent by the Admiralty to chart the world's coastlines. One among those expeditions became a turning-point in the history of science and culture. Between 1826 and 1846 three surveying voyages were conducted in *HMS Beagle*, a modified 10-ton brig, to the coasts of South America, the South Pacific and Australia. Under the command of Robert Fitzroy – reputed to have been the best surveying navigator of his time – the second voyage, from 1832 to 1836, carried, as a 'gentleman guest', Charles Darwin. On that extended trip, Darwin observed the phenomena that eventually suggested to him the radical ideas about the origins and history of life that shook the foundations of western culture. But on *Beagle*, he was concerned with recording physiographic, geological and biological observations – in short, with the project of extending geographical knowledge in the service of British interests. The culmination of this style of work was the *Challenger* expedition of 1872–76, an entirely scientific expedition to chart the physical and biological character of the world's oceans. A number of the members of that expedition and the group that analysed the data, including

H.R. Mill and A.J. Herbertson, became founders of professional and academic geography in Britain.

The history of terrestrial exploration in the nineteenth century is better known. But its connections perhaps are not. In Britain, the reins of official scientific patronage early in the century were firmly held by the Royal Society through the influence of Joseph Banks. After Banks died, however, and in keeping with the increasing specialization of knowledge, a range of scientific societies grew up in London, among them the Royal Geographical Society, founded in 1830 (see Chapter 1 on histories of geography). It quickly attracted a membership with influence and with wide overseas interests, including public officials, traveller-explorers and senior military men. These people pressed exploration forward in the increasingly far-flung British Empire, particularly in the Indian subcontin-ent, in Africa, and in the Middle and Far East. The leadership provided to the Society in mid-century by Sir Roderick Murchison epitomized this fusion of geography – largely applied physical geography – with commercial expansion and military hegemony, and with the 'civilizing mission' associated with missionary activities. Geography was the science that systematically mapped and documented the world's lands for purposes of economic development and European administrative oversight. Many of its observations are chronicled not in books but in diplomatic reports, in military dispatches, in missionary journals, and in the memoirs of colonial administrators. They are, consequently, not highly visible, even today.

It is probably fair to claim that British activities led and exemplified the development of state-directed applied geography in the nineteenth century, but they were by no means exceptional. Other European powers with overseas interests followed the same course, notably the Dutch and French in the East Indies, the French, Portuguese and Belgians in Africa, and the Germans in Africa and the Far East. The use of geography to reinforce imperialist activities became, in the 1870s, the lever to establish geography as a formally recognized discipline in German universities.

A different kind of imperialistic activity was conducted in the guise of geographical exploration in those places where Europeans chose to settle, most notably in North America and Australia. Near the end of the eighteenth century, the newly independent Americans – in fact, transplanted Europeans – were greatly interested in the exploitation of the lands to the west of the Appalachians and were increasingly interested in their coasts and seas, again for purposes of trade and military defence. The US Coast and Geodetic Survey was established in 1807 and the office that became the US Naval Observatory shortly after. Surveys inland initially were left to the individual states, most of which set up some form of geological survey office in the 1820s or 1830s. It is clear from this organization that a significant motivation was the finding of economic minerals, but these early survey organizations conducted topographic surveys, land surveys and general resource surveys – in effect, applied physical geography. The imperialistic caste of this activity derived from the use the information gained to direct European settlers to the west, systematically displacing the native peoples – this was, in truth, a colonizing activity. After the Civil War, the exploration of the far west became a consciously colonizing activity directed by the federal government through its Department of the Interior. Similar developments occurred to the north in Canada, with mandated exploration being directed by the federal government in Ottawa right from the beginning through its Geological Survey. This 'scientific exploration' continued well after the mid-twentieth century as a means to assert sovereignty in the arctic regions.

AN INTERLUDE

Towards the end of the nineteenth century, important changes occurred in the geographical discipline that substantially changed the character of recognizable 'geography'. Most importantly, perhaps, geographical research began to become significantly institutionalized in the emerging modern universities, first in Germany, then in other European countries and in America. Given the promin-ence of the principal chair holders, the subject began to be more clearly led by individuals with no direct practical responsibilities (though, to be sure, many of the early chair holders moved into the universities from practical work) and to be concerned increasingly with more conceptual aspects of the construction of the landscape.

Another significant influence was the emergence of a radical critique of the axis of state–military–commercial interests that had controlled the agenda of 'imperial geography' for nearly a century. Most closely associated with the Russian aristocratic *émigré* Petr Kropotkin and French geographer Elisée Reclus, both prominent political Anarchists, the critique emphasized a search for some dynamical understanding of social conditions through study of the organization of the landscape. It was, therefore, a critique that was profoundly opposed to the mechanistic character of surveying and measuring that characterized 'practical geography' – the description given to the nineteenth-century mainstream by H.R. Mill.

At the same time, a number of other modern scientific disciplines were emerging into institutionalized form, and many of these abstracted portions of the older geographical enterprise away from geography. So, for example, survey and mapping agencies began, in view of the increasingly technical character of modern mapping, to be dominated by engineers and narrowly trained surveyors. Hence a large part – perhaps the cornerstone – of traditional applied geography disappeared from a central place in the discipline, although thematic cartography certainly remained a prominent geographical interest for another half-century. Geology, oceanography and meteorology became recognizably distinct disciplines through the course of the nineteenth century – they employ geographic method extensively, but their increasingly specialized researches and techniques, and – ultimately – academic recognition, defined them as clearly distinct disciplines. The study of plant and animal distributions largely disappeared into the modern biological disciplines. Very significantly, the study of ethnography became a central part of the modern disciplines of anthropology and sociology. Geographical methods were taken up in the emerging social sciences as well, and so the geography of trade and commerce came to be pursued within economics as much as geography.

Within geography, as well, increasing specialization came with the expansion of knowledge. Geomorphology, in particular, with its close links to geology, became a somewhat distinctive subdiscipline. The growing influence of American work, where geomorphology grew out of the early geological surveys, most especially the scientific exploration of the west conducted in the early years of the US Geological Survey, was important in this regard. The discipline was thereby deprived of some of the fluency with which integrated descriptions of landscape and its inhabitants had earlier been given ready practical relevance, in particular for resource appraisal. In this welter of intellectual reorientation, the character of applied geography dramatically changed and the pursuit of an applied physical geography continued on more regional scales within emerging subdisciplines.

Another important change that was looming at the close of the nineteenth century was the exhaustion of new spaces to map, at least in the reconnaissance style that had been adopted for imperial outreach. The habitable world was mapped, to a good first approximation. Expeditionary geography was increasingly floundering in polar snow. For all sorts of reasons, then, geography had to reinvent itself.

All this perhaps gives the impression that the turn of the twentieth century (more specifically, perhaps, the upheaval of the First World War) marked the end of a tradition in applied geography. In considerable measure it did, but applications of physical geography did not entirely disappear: physical geography remained an interest of military establishments. Analytical studies of topography were of particular interest because they formed the basis for studying terrain trafficability – that is, 'where can you drive your tank?' The contrast in scale is instructive. The old tradition of applied geography was pitched on global and continental scales – it was concerned with reconnaissance studies of large regions, suitable for opening up *terra incognita*. The new applied physical geography would be based on topographic and local scales, on contributing to the solution of specific and specifically situated problems of landscape management.

A quite different situation emerged, however, in Eastern Europe. Following the October Revolution of 1917, the establishment of the communist regime in Russia brought to the fore the need for geographical knowledge upon which to base collective planning of land-use and resources. Accordingly, geography remained a discipline with high status in the Soviet Union, producing reports and monographs describing the landscape and its resources, and the spatial organization of the economy. These works contributed to the redefinition of what a modern applied geography might be. Not coincidentally, Soviet geographers contributed significant concepts for the classification and analysis of landscapes, including climate, hydrology, soils and vegetation. They also contributed significant global syntheses on some of these topics (see, for example, work by Budyko (1982) in climatology and by L'vovich (1979) in hydrology).

APPLIED PHYSICAL GEOGRAPHY IN THE CONTEMPORARY WORLD

Contemporary applied physical geography has largely grown out of a convergence of opinion, in the 1960s, that human development is altering earth's envir-onment, globally, regionally and locally, often in undesirable ways. The fact is not new: the entire history of human civilization involves modifying natural envir-onments to better suit the needs and wants of human societies. Humans have made substantial impacts on the environment for thousands of years and virtually nowhere on earth today is entirely untouched by human imprint. But it was increasingly obvious after the mid-twentieth century that these effects were in many places foreclosing future development opportunities, dismantling the fabric of natural environments, and creating human environments less desirable and less healthy than those that were being replaced. Nor was the issue new: as long ago as 1864, the New Englander George Perkins Marsh had produced *Man and Nature, or Physical Geography as Modified by Human Action*. Books with similar themes followed episodically, issuing in the monumental compendium *Man's Role in Changing the Face of the Earth* (edited by W.L. Thomas) in 1956. The event that precipitated the emergence of a consensus on the importance of human disruption of the environment – at least, viewed in retrospect – was the 1962 book *Silent Spring*, written by Rachel Carson, an American naturalist and writer. The book was about the explosive growth in the use of chemicals – fertilizers, pesticides, solvents, cleaning agents and synthetic materials – during the twentieth century and their devastating effects on nature and, prospectively, on humans. It focused attention particularly on the ecological effects, but it awakened concern about the even wider range of changes induced by human activities that have become the focus of much of the activity in contemporary applied physical geography.

There remained, as well, lingering connections with earlier, state-driven applied geographies, possibly best exemplified in Canada, a country still being 'opened up' to the world economy in the mid-twentieth century. At the close of the Second World War, dramatic changes in strategic alliances and rivalries suddenly made the largely empty landscape of Canada – in particular, the Arctic – a potentially contentious region. To assert Canadian sovereignty in the region, a Geographical Branch was established in the Canadian Department of Mines and Technical Surveys with the objective to survey landscape and resources, primarily in the boreal and arctic regions of the country. At the same time, military establishments undertook their own physiographic surveys of this terrain (a series of studies of the Canadian Arctic commissioned by the RAND Corporation, an American paramilitary research organization, is summarized in Bird, 1967). Later, as regional planning issues began to loom, this branch took up the task of regional land-use mapping as well. These activities established a strong tradition of applied geographical studies in Canada that has continued ever since.

City and regional planning, and land-use studies were the foci of interest of geographers who established a formal reference point for applied geography in the International Geographical Union in the early 1960s. Applied physical geography initially developed more informally, in part because much of the work, conducted in North America, was recognized as 'environmental geology'. An important exception, however, was the involvement of geographers in debates about the desirability and methods for integrated water resources development at the scale of entire major drainage basins. Stimulated by the Tennessee Valley developments in the USA for combined flood control, power generation, navigation and water supply, this was a major issue of land resource development from the 1930s on (see Newson, 1997, or Downs and Gregory, 2004, for reviews of river basin management).

The engagement of geomorphologists from the mid-1950s with a mechanistic, process-focused 'dynamical geomorphology' (Strahler, 1952) brought their scale of study into alignment with that of engineers, who typically carry out resource development and hazard mitigation works, making it evident that the work of geomorphologists represents the science that underlies the engineering of the landscape. Accordingly, geomorphology gained practical relevance on a new, more local scale of activity (for early examples, see Coates, 1976). Interestingly, the work of A.N. Strahler and his school was supported by the United States Office of Naval Research. At the same time, climatologists turned their attention from descriptive to physical climatology (e.g. Terjung, 1976), thereby gaining similar leverage in the emerging field of air quality management, and hydrologists made a similar transition to qualify their work as an important basis for water resource management. Except, perhaps, in hydrology, these changes represented developments in the science: they were not initially motivated by practical issues. But they did turn the attention of physical geographers towards a study of the environment on a relatively local scale that admitted ready prac-tical application of the experience gained.

During the last three decades of the twentieth century contemporary applied physical geography has been clearly defined by the nature of the projects that have been undertaken:

- Studies are mostly local in scale, corresponding with the scale of engineering work: this is a dramatic departure from the earlier traditions of applied geography.
- Most work arises from issues of environmental management or hazard mitigation: this means that physical geographers are often dealing with public or semi-public agencies.
- Very often there is no separation between basic research and direct application of the results of research: there is no extended 'research and development' chain in environmental science.
- Applied work that would easily be categorized as 'physical geography' is conducted by an eclectic mixture of professional workers, including engineers, geologists, hydrologists, atmospheric scientists and ecologists: geographers have no monopoly on training and knowledge to prepare one for study of the environment around us.

The last point emphasizes that geography and geographical method are two different concepts, and that many disciplines incorporate geographical method. It also implies that a wide range of expertise must be brought to bear on environmental problems in today's technically specialized world.

Formal public recognition of applied physical geography is growing. The process has perhaps proceeded farther in Canada than anywhere else. In that country, there are autonomously established professional registration bodies for ecologists (including biogeographers) and atmospheric scientists (including climatologists), while geomorphologists and hydrologists are swept up with other earth scientists into the registration process for professional engineers, which is regulated by state legislation. Registration has become a necessary requirement to practise geomorphology and hydrology, an arrangement intended to provide some public assurance of qualification, competence and discipline in the work executed by these practicioners. This development specifically acknowledges the social value and responsibilities of such work. By the particular training that geography provides, the physical geographer may often play the role of linchpin among these various subdisciplines. Similarly, in the United States, geomorphology is increasingly regulated by state boards for 'geoscience' practice.

Because there is little or no gap between research and application in environmental science, the range of work that might be conducted in contem-porary applied physical geography is essentially as broad as the subdisciplines themselves. (This fact makes rather futile any attempt to define applied physical geography in any limiting way.) There have been, however, some notable foci of activity.

Example activities: river restoration

Everywhere in the developed world, and increasingly in the developing world, rivers have been actively managed for human purposes. Channels have been straightened and 'trained' to facilitate navigation and to 'rationalize' property boundaries; banks have been raised and hardened for flood protection and to discourage erosion by the river of adjacent floodplain properties; diversions and dams have been built for water resource development and flood protection, and large dams have been constructed for hydropower development. These last effects more or less radically change the hydrological regime downstream. The changes have dramatic effects upon riverine ecology, both aquatic and riparian (Ligon et al., 1995). With the growth in concern for the human impact on environmental quality that has occurred in developed societies in the last decades of the twentieth century has come a conviction that attempts must be made to restore rivers to something more like their former state. This movement is strong in the United States and in northern Europe, which are relatively wealthy societies where rivers have been dramatically altered (see Wohl et al., 2005, for a statement of river restoration principles).

It is possible to countenance 'restoration' only in limited degree. River floodplains are attractive places on which to settle: fertile soils, good water supply, favourable conditions for the construction of housing and services all promote high population densities. People settled there will not tolerate the thought of being exposed to relatively frequent flooding or loss of property to erosion, nor will they relinquish highly valued land. Furthermore, many changes may be irreversible (Gore and Shields, 1995). The prospects for significant restoration are, therefore, dim. What usually occurs is that opportunities arise for limited rehab-ilitation of riverine ecology by renaturalizing banks and re-establishing a more naturally appearing channel-way along limited reaches of the river where there is some public control over the adjacent land. Geomorphologists and ecologists both become important advisers in the execution of such projects (see Downs and Gregory, 2004). For the geomorphologist, the question is what – for the prevailing hydrological regime – will be a stable channel geometry? The most commonly applied method for channel restoration design employed in the United States (Rosgen, 1997) is based on a morphological classification of channel types keyed to gradient and material type.

In the United Kingdom, recognition of the limited possibility for physical modification of river channels has led to significant emphasis upon channel classifications that relate directly to habitat ecological characterization of river reaches (NRA, 1995; see also Thorne et al., 1997; Padmore et al., 1998). The approach is based on careful field survey and is intended to provide information essential for any management policy or action (Newson et al., 2001). An important function for physical geographers engaged in such practice is to emphasize that the treatment of isolated reaches along a river is a fundamentally flawed approach. Rivers are cascading systems in which conditions at any point along the river represent the integral effect of processes in the entire contributing drainage. What may represent a rational design for habilitation of a given reach of a river will depend on the contributions of water, sediment and nutrients from upstream, and on the movement of biota, both along the river and laterally across the river banks. This question becomes the more important as consideration begins to be given to the removal of old dams from rivers (Doyle et al., 2003). This act modifies hydrological regimes and removes established barriers to travel, hence will require basin-wide anticipation of the effects.

The complexity of issues surrounding river management is well illustrated on lower Fraser River, British Columbia, Canada, where gravel accumulation appears to pose an increasing flood hazard by raising the riverbed level. The pattern of gravel accumulation, however, creates a complex river morphology of bars and islands that represents salmonid habitat of unrivalled productivity (Figure 26.3). Residents along the river and those who would pursue further development of floodplain lands wish to dredge the river to reduce the perceived flood risk, but are opposed by those who value the fishery. Cutting across this conflict is the interest of First Nations people, who claim historic ownership of the river resources in any case. Any reasonable resolution of the issues must begin from a sound base of knowledge: what is the rate of aggradation of the gravels? That is a question for fluvial geomorphologists and, in this instance, a major study was undertaken to determine the sediment budget of the river (Church, 2001; see www.geog.ubc.ca/fraserriver). Surprisingly, the aggradation turned out to be far smaller than had been anti-cipated before the study, opening room to consider a number of management alternatives.

Example activities: forest management

Forests have long been exploited for wood resources (Figure 26.4), so much so that relatively little forest remains in Europe. Today, tropical forests are rapidly being

Figure 26.3 Fraser River, view upstream near Agassiz, British Columbia: a wandering, gravel-bed channel photographed in flood. This river morphology provides an abundance of superior fish habitat through all flow stages

consumed. In the northern hemisphere, extensive forests remain in North America, particularly in Canada and in Siberia (FAO, 2005). The rational management of these forests, both for ecological sustenance and for carbon balance is highly important. Some of the world's grandest forests are found on the Pacific Northwest coast of North America, in the states of Oregon, Washington and southern Alaska, and in the province of British Columbia. Here, trees grow to great age and very large size and carry a very high volume of stemwood, making them highly desirable timber crops. The forests also sustain a rich ecosystem that includes important salmon stocks in the rivers and large carnivores, such as bears and cougars, on land. In recent years, studies have been made to develop sustainable management strategies for these forests. Much of the work represents applied physical geography.

Usual management procedures in forests exploited for timber are based upon maximizing the value of timber yield, assigning lesser importance to non-timber ecological services and often having only inadequate information about such services. Planning is usually based on administrative rather than ecologically meaningful boundaries and the planning process is almost always an expert process with little or no consultation with interested third parties (who may include forest inhabitants). In contrast, a system of forest management has been proposed for coastal forests in British Columbia on the basis of the following principles (synthesized from Clayoquot Sound Scientific Panel, 1995):

Figure 26.4   Aerial view of a clear-cut harvest block in the British Columbia coast forest

- The key to sustainable forest practices lies in maintaining functioning ecosystems.
- [Therefore] planning must focus on those ecosystem elements and processes to be retained rather than on resources to be extracted.
- Long-term, hierarchical planning is required to maintain ecosystem integrity from sub-regional down to site levels, to ensure that the intent of regional planning objectives is reflected in more local planning.
- Drainage basins must be adopted as the fundamental units for planning, and planning is based on assigning ecological functions (including timber harvest) to areal units rather than on a preconceived volume or value of the harvest.
- Cultural values and desires of inhabitants and visitors must be addressed.
- Scientific and traditional [e.g. experiential knowledge of the local peoples] ecological knowledge must be used and encouraged through research, experience and monitoring.
- Both management and regulation must be adaptive, incorporating new experience and information as they develop.

These principles have subsequently been implemented on an experimental basis and extended to incorporate risk-based planning of forest development activities – that is, appraising the desirability of development proposals in comparison with the degree of risk they appear to pose to the continued ecological integrity of the forest.

The basis for these recommendations is quintessentially the stuff of physical geography, entailing as it does an integrated knowledge of how water, soils and biota interact to produce a functioning regional environment. They also represent a geographical system of land management in as much as it is an area-based method of planning land-use. But they go beyond physical geography in their statement that planning must incorporate the concerns of the people who use the landscape – they move towards reintegrating the traditional elements of geography. These principles, and the many operational recommendations that were based on them, were actually developed by a multidisciplinary team of 20 persons, representing the range of forest ecological and management disciplines from soil and water, through forest engineering to worker safety, and included representatives of the local people. Physical geographers contributed physiographic and hydrological expertise fundamental to the exercise, and the critical appreciation of scale effects in planning for the entire landscape.

Example activities: global stewardship

Despite the emphasis of the preceding paragraphs on local problem-solving, there is today a renewed sense of the larger, traditional scale of geography. We recognize that some of our most serious environmental problems are global in scope. Full appreciation of these problems has largely come about because of our ability – won within the last half-century – to view and monitor earth from space, and because of our growing capacity for large-scale numerical computation. By these means, we acquire and analyse the massive amounts of information required to fully comprehend global environmental problems.

The most prominent example of such global analysis is the series of assessments of changing global climate developed by the Intergovernmental Panel on Climate Change (e.g. IPCC, 2007). Led by a small secretariat, this activity is in fact a voluntary global collaboration of atmospheric scientists, climatologists and social scientists who seek to understand the reasons for climatic change and to predict the long-range consequences of it. There is now a clear consensus that the root cause of contemporary climate change is the consequence of human activity modifying the constitution of the atmosphere. But there are other ways, just as important, in which humans are modifying the global environment. For example, global ecosystems have been radically modified by human activity, a fact for the first time brought into quantitative global focus by the Millenium Ecosystem Assessment exercise (2005).

More specialized activities of similar scale include ongoing assessments of the status of world forests (FAO, 2005) and attempts to determine the state of the world's water cycle (Shiklomanov and Rodda, 2003). The status of various fundamental elemental cycles is also under close scrutiny, most importantly those of carbon and nitrogen (e.g. Field and Raupach, 2004). All these activities have dominant geographical elements and are conducted by multipdisciplinary teams that include physical geographers applying the principles of their science.

ISSUES FOR APPLIED PHYSICAL GEOGRAPHY

For all the activities described above, the question 'what is applied geography?' has remained a rather nebulous one. It is a question that has rarely been asked by physical geographers. Sherman (1989) has pursued the issue in relation to geomorphology and concludes that there should be no distinctions drawn between applied and 'pure' geomorphology. Certainly, there is little distance between basic and applied practice in environmental science, no long chain of 'developmental research' between the elaboration of a concept and its potential application. Rather, he sees the important distinction arising in the rationale for, or the sponsorship of, research. Applied research is almost always 'mandated research', meaning research commissioned by another party who has defined the basic question to be answered (which distinguishes the activity from funded academic research).

This situation raises important problems because the sponsor of the research is often seeking a particular answer. The research question, or the scope of the research, may be constrained to predispose a particular answer. The intellectual independence of the investigator is compromised. The ethical problem that this poses is not easily resolved. A related problem is that the researcher, in proposing a solution to some problem, in effect becomes an advocate. In many instances, advocacy of a particular solution represents a partisan or political statement in the larger context of the problem. Whether or not scientists should be (or can avoid being) advocates is a deeply contested issue that is extensively debated in human geography but appears scarcely to have impinged upon physical geography.

These considerations lead directly to the question 'what are the responsibilities of the applied geographer?'

Until 15 or 20 years ago, almost any applied scientist (or engineer) would identify a preferred solution to a particular problem and would expect it to be accepted with little public discussion. In public domain work, public safety would be the dominating issue and widespread community support for this approach would be assumed and, in many cases, assured. Today, a public more sceptical of expert knowledge is apt to question almost any solution to any environmental problem. This situation has developed as an increasingly articulate public has fragmented into a number of communities with different interests: a solution for one interest group may present uncompensated costs for another. In environmental problem-solving, such an impasse characteristically arises in the context of decisions in which public safety or economic development confronts concerns for ecological integrity. The case of lower Fraser River, introduced above, is an excellent example.

In this circumstance, the objectives of technical work may change. In working for particular interests, it may still be possible to identify a solution that is 'best' for the client, but it is also incumbent on the technical investigator to foresee potential objections to the implementation of that solution from other sectors of the community. In public domain issues, the role of the technical investigator becomes that of identifying the comparative advantages and disadvantages of a range of feasible solutions, each of which may be preferred by a different sector of the community: the final decision what to do remains a public–political one. In this sense, technical investigators become, in major degree, educators as much as problem-solvers. It becomes vital for technical investigators to recognize the political process that underlies the work and to realize the potential for bias, in relation to that process, that may arise from their own training and prior experience. It is also vital for the technical advice to be sound and comprehensive, and to be honestly presented, without prejudice to one side or another in the debate. Only in this way will technical advice retain any semblance of authority. On Fraser River, the role of the scientific team that determined the sediment budget is to ensure that all parties gain a clear understanding of it, and that its implications are properly recognized and incorporated into whatever resolution of the problem is eventually adopted.

The character of applied physical geography continues to evolve and expand, in part driven by outside trends that entrain more and more of all scientific activity in applied pursuits. In geography, this more and more blurs the distinction between basic or 'pure' and applied work. In physical geography, applied work increasingly influences the evolution of the entire discipline.

SUMMARY

- Through most of history, geography has been an important subject and mostly an applied subject – the intellectual expression of exploration, trade and statecraft.

- Modern geography developed out of European exploration and colonization, with physical and human geography becoming clearly differentiated only in the twentieth century.

- Contemporary applied physical geography has been defined by the focus on a physically based construction of the discipline and a growing concern for environmental degradation, both trends developing after the mid-twentieth century.

- These characteristics define applied physical geography as a discipline largely concerned with local to regional environmental problems and their resolution, but recent emphasis on planetary environmental change has reinserted global issues into the subject.

- The subject is defined more by what practitioners do than by introspection and it is shared with a wide range of other environmental scientists and engineers whose work entails the application of geographical methods for landscape management.

- Physical geographers, by virtue of their training, are often the synthesizers of various lines of evidence that impinge on a particular problem and, frequently, their key role is that of educator rather than final problem-solver because the actual solution of the problem entrains social and/or economic issues that extend well beyond the more narrowly defined environmental problem.

Further Reading

Much of the early history of applied geography can be inferred from the conventional histories of European exploration. For those interested in the character of British exploration in the early nineteenth century, Keith Thomson's (1995) *HMS Beagle* presents a highly readable chronicle of one of the greatest of all such voyages. David Livingstone's (1992) *The Geographical Tradition* gives a scholarly history that emphasizes the applied motivation of the subject. *The Earth as Transformed by Human Action* (Turner et al. 1990) – the title a deliberate paraphrase of G.P. Marsh's pioneer work – is the latest compendium volume of human impacts on Earth's environment. There is remarkably little reflective discussion of contemporary applied physical geography, but there are relevant essays by Newson, Penning-Rowsell and Warren in Clark, Gregory and Gurnell's (1987) *Horizons in Physical Geography*, and by Lier, Marotz and Sherman in Kenzer's (1989) *Applied Geography: Issues, Questions and Concerns*. Slaymaker (1997) presents an almost isolated philosophical comment, while Richards (2003) examines further the ethical issues raised by applied geography and the importance of an integrated geographical approach in problem-solving. The topics of hazard and risk are reviewed by Chester (2002) in Allison's (2002) *Applied Geomorphology: Theory and Practice*, but there is little reflective discussion in that book of substantive essays, and the same can be said of earlier compilations of papers. Kirk et al. (1999) give a useful view of the institutional context of applied physical geography in New Zealand, an example of experience in one country. A productive exercise is to review the recent contents, in comparison with the contents of 30 years or so ago, of the major physical geography journals and to classify the articles, to the extent possible, as 'basic' research or 'applied' research. This will illuminate both the difficulty of separating the categories and the changing balance of motivation for work in physical geography. The journals *Applied Geography, Physical Geography, Geomorphology and Environmental Geology* contain a substantial volume of applied physical geography but, today, almost any of the specialist geography and environmental science journals include applied papers.

Note: Full details of the above can be found in the references list below.

References

Allison, R.J. (2002) *Applied Geomorphology: Theory and Practice.* Chichester: John Wiley and Sons.

Bird, J.B. (1967) *The Physiography of Arctic Canada.* Baltimore, MD: Johns Hopkins University Press.

Budyko, M.I. (1982) *The Earth's Climate: Past and Future.* New York: Academic Press.

Carson, R. (1962) *Silent Spring.* Boston, MA: Houghton Mifflin.

Chester, D.K. (2002) 'Overview: hazard and risk', in R.J. Allison (ed.) *Applied Geomorphology: Theory and Practice*. Chichester: John Wiley and Sons, pp. 251–63.

Church, M. (2001) 'River science and Fraser River: who controls the river?', in M.P. Mosley (ed.) *Gravel-bed Rivers V*. Proceedings of the Fifth International Gravel-bed Rivers Workshop. Wellington: New Zealand Hydrological Society, pp. 607–31.

Clark, M.J., Gregory, K.J. and Gurnell, A.M. (eds) (1987) *Horizons in Physical Geography*. London: Macmillan Education.

Clayoquot Sound Scientific Panel (The Scientific Panel for Sustainable Forest Practices in Clayoquot Sound) (1995) *A Vision and Its Context*. Report 4: 40pp. *Sustainable Ecosystem Management in Clayoquot Sound*. Report 5: 296pp. Both available at www.cortex.org/dow-cla.html

Coates, D.R. (ed.) (1971) *Environmental Geomorphology*. Proceedings of the First Geomorphology Symposium, Binghamton, NY. Binghamton, NY: State University of New York, Publications in Geomorphology.

Coates, D.R. (ed.) (1976) *Geomorphology and Engineering*. Proceedings of the Seventh Binghamton Geomorphology Symposium. Stroudsburg, PA: Dowden, Hutchinson and Ross.

Cooke, R.U. and Doornkamp, J.C. (1974) *Geomorphology in Environmental Management*. Oxford: Clarendon Press.

Downs, P.W. and Gregory, K.J. (2004) *River Channel Management: Towards Sustainable Catchment Hydrosystems*. London: Edward Arnold.

Doyle, M.W., Harbor, J.M. and Stanley, E.H. (2003) 'Toward policies and decision-making for dam removal', *Environmental Management*, 31: 453–65.

FAO (Food and Agriculture Organization) (2005) 'Global forest resources assessment', *FAO Forestry Paper 147*. Rome: FAO. Available online at: www.fao.org/docrep/008/a0400e/a0400 e00.htm

Field, C.B. and Raupach, M.R. (eds) (2004) 'The global carbon cycle: integrating humans, climate and the natural world', *SCOPE Report 62*. Washington, DC: Island Press.

Gore, J.A. and Shields, F.D. Jr (1995) 'Can large rivers be restored?', *BioScience*, 45: 142–52.

Hough, R.A. (1995) *Captain James Cook*. New York: W.W. Norton.

IPCC (Intergovernmental Panel on Climate Change) (2007) *Fourth Assessment Report*. Available online at: www.ipcc.ch/

Ives, J.D., Mears, A.I., Carrara, P.E. and Bovis, M.J. (1976) 'Natural hazards in mountain Colorado', *Association of American Geographers Annals*, 66: 129–44.

Kenzer, M.S. (ed.) (1989) *Applied Geography: Issues, Questions, and Concerns*. Geojournal Library, Vol. 15. Dordrecht: Kluwer Academic Publishers.

Kirk, R.M., Morgan, R.K., Single, B.M. and Fahey, B. (1999) 'Applied physical geography in New Zealand', *Progress in Physical Geography*, 23: 525–40.

Ligon, F.K., Dietrich, W.E. and Trush, W.J. (1995) 'Downstream ecological effects of dams', *BioScience*, 45: 183–92.

Livingstone, D.N. (1992) *The Geographical Tradition*. Oxford: Blackwell.

L'vovich, M.I. (1979) World Water Resources and their Future. Trans. R.L. Nace. Washington, DC: American Geophysical Union.

Marsh, G.P. (1864) *Man and Nature, or Physical Geography as Modified by Human Action*. (Republished in 1876 as *The Earth as Modified by Human Action*.) New York: Scribner.

Mather, J.R. (1974) *Climatology: Fundamentals and Applications*. New York. McGraw-Hill.

Millenium Ecosystem Assessment (2005) *Ecosystems and Human Well-being: Synthesis Report*. Washington, DC: Island Press.

Newson, M. (1997) *Land, Water and Development*. London: Routledge.

Newson, M., Thorne, C. and Brookes, A. (2001) 'The management of gravel-bed rivers in England and Wales: from geomorphological research to strategy and operations', in M.P. Mosley (ed.) *Gravel-bed rivers V*. Proceedings of the Fifth International Gravel-bed Rivers Workshop. Wellington: New Zealand Hydrological Society, pp. 581–606.

NRA (National Rivers Authority) (1995) *River Channel Typology for Catchment and River Management*. Final Report, Research and Development Project 539/NDB/T. 57pp.

Padmore, C.L., Newson, M.D. and Charlton, M.E. (1998) 'Instream habitat in gravel-bed rivers: identification and characterization of biotopes', in P.C. Klingeman, R.L. Beschta, P.D. Komar and J.B. Bradley (eds) *Gravel-bed Rivers in the Environment*. Proceedings of the Fourth International Gravel-bed Rivers Workshop. Highland Ranch, CO: Water Resources Publications, pp. 345–64.

Philipponneau, M. (2004) 'Historical foundations of applied geography', in A. Bailly and L.J. Gibson (eds) *Applied Geography: A World Perspective*. Geojournal Library Vol. 77. Dordrecht: Kluwer Academic Publishers, pp. 23–46.

Richards, K.S. (2003) 'Ethical grounds for an integrated geography', in S. Trudgill, and A.G. Roy (eds) *Contemporary Meanings in Physical Geography*. London: Edward Arnold, pp. 233–58.

Rosgen, D.L. (1997) *Applied River Morphology*. Pagosa Springs, CO: Wildland Hydrology Inc.

Roy, W. (1790) 'An account of the trigonometric operations, whereby the distance between the meridians of the Royal Observatories of Greenwich and Paris has been determined', *Philosophical Transactions of the Royal Society London*, 80: 111–270, with 11 plates.

Sherman, D.J. (1989) 'Geomorphology: praxis and theory', in M.S. Kenzer, (ed.) *Applied Geography: Issues, Questions, and Concerns*. Geojournal Library Vol. 15. Dordrecht: Kluwer Academic Publishers, pp. 115–31.

Shiklomanov, I.A. and Rodda, J.C. (2003) *World Water Resources at the Beginning of the 21st Century*. Cambridge: Cambridge University Press.

Slaymaker, O. (1997) 'A pluralist, problem-focused geomorphology', in D.R. Stoddart (ed.) *Process and Form in Geomorphology*. London: Routledge, pp. 328–39.

Sobel, D. (1995) *Longitude*. New York: Walker.

Strahler, A.N. (1952) 'Dynamic basis of geomorphology', *Geological Society of America Bulletin*, 63: 923–38.

Terjung, W.H. (1976) 'Climatology for geographers', *Association of American Geographers Annals*, 66: 199–222.

Thomas, W.L. (ed.) (1956) *Man's Role in Changing the Face of the Earth*. Chicago, IL: University of Chicago Press.

Thomson, K.S. (1995) *HMS Beagle*. New York: W.W. Norton.

Thorne, C.R., Hey, R.D. and Newson, M.D. (eds) (1997) *Applied Fluvial Geomorphology for River Engineering and Management*. Chichester: John Wiley and Sons.

Turner, B.L. III, Clark, W.C., Kates, R.W., Richards, J.F., Mathews, J.T. and Meyer, W.B. (1990) *The Earth as Transformed by Human Action*. Cambridge: Cambridge University Press.

Wohl, E., Angermeier, P.L., Bledsoe, B., Kondolf, G.M., MacDonnell, L., Merritts, D.M., Palmer, M.A., Poff, N.L. and Tarboton, D. (2005) 'River-restoration', *Water Resources Research*, 41, W10301, doi: 10.1029/2005WR003985: 12pp.

ACKNOWLEDGEMENT

Kathleen MacDonald provided significant help in the literature analysis that underlies this chapter and Steve Rice provided generous support during its development, for which I am grateful.

Index